BUILT TO THE COMMON CORE

Diya Sehgal

AUTHORS
Carter • Cuevas • Day • Malloy
Kersaint • Reynosa • Silbey • Vielhaber

Bothell, WA • Chicago, IL • Columbus, OH • New York, NY

connectED.mcgraw-hill.com

STEM McGraw-Hill is committed to providing instructional materials in Science, Technology, Engineering, and Mathematics (STEM) that give all students a solid foundation, one that prepares them for college and careers in the 21st century.

Send all inquiries to:
McGraw-Hill Education
8787 Orion Place
Columbus, OH 43240

ISBN: 978-0-02-130152-2 (*Volume 2*)
MHID: 0-02-130152-2

Printed in the United States of America.

7 8 9 10 11 LMN 21 20 19 18 17

CONTENTS IN BRIEF

Units organized by CCSS domain

Glencoe Math is organized into units based on groups of related standards called domains. The Standards for MP Mathematical Practices are embedded throughout the course.

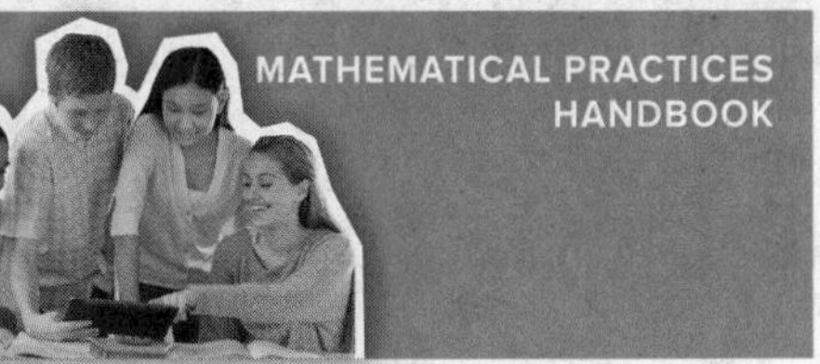

Tetra Images/Getty Images

MP Mathematical Practices

Richard Drury/Photodisc/Getty Images

Ratios and Proportional Relationships

Gerald Nowak/Westend61/Photolibrary

The Number System

UNIT 3
Domain 7.EE

Jill Braaten/McGraw-Hill Education

Expressions and Equations

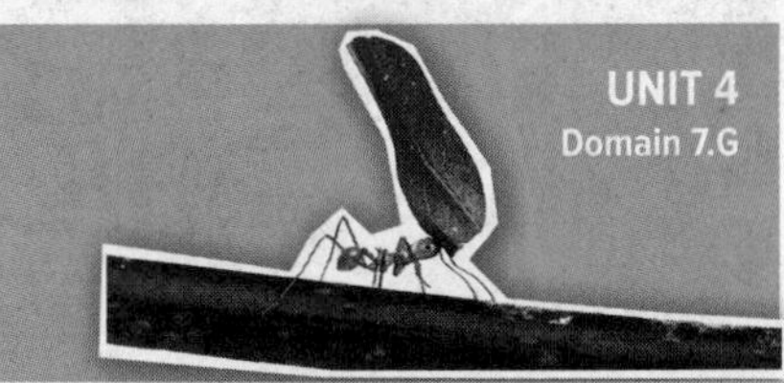

©Photo by Scott Bauer/USDA

Geometry

Back in the Pack dog portraits/Flickr/Getty Images

Statistics and Probability

Everything you need,

anytime, anywhere.

With ConnectED, you have instant access to all of your study materials—anytime, anywhere. From homework materials to study guides—it's all in one place and just a click away. ConnectED even allows you to collaborate with your classmates and use mobile apps to make studying easy.

Resources built for you—available 24/7:

- Your eBook available wherever you are
- Personal Tutors and Self-Check Quizzes to help your learning
- An Online Calendar with all of your due dates
- eFlashcard App to make studying easy
- A message center to stay in touch

Go Mobile!

Visit mheonline.com/apps to get entertainment, instruction, and education on the go with ConnectED Mobile and our other apps available for your device.

Go Online!

connectED.mcgraw-hill.com

your Username

your Password

Vocab

Learn about new vocabulary words.

Watch

Watch animations and videos.

Tutor

See and hear a teacher explain how to solve problems.

Tools

Explore concepts with virtual manipulatives.

Sketchpad

Discover concepts using The Geometer's Sketchpad®.

Check

Check your progress.

eHelp

Get targeted homework help.

Worksheets

Access practice worksheets.

UNIT 1 Ratios and Proportional Relationships

UNIT PROJECT PREVIEW page 2

Chapter 1 Ratios and Proportional Reasoning

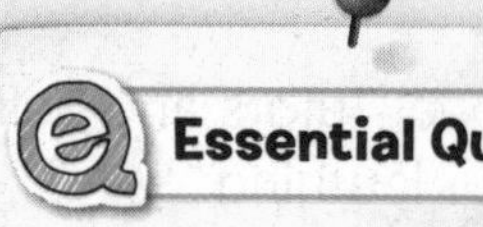

HOW can you show that two objects are proportional?

p. 45

Chapter 2
Percents

Essential Question

HOW can percent help you understand situations involving money?

UNIT PROJECT PREVIEW page 185

Chapter 3 Integers

Essential Question

WHAT happens when you add, subtract, multiply, and divide integers?

Chapter 4 Rational Numbers

Real World p. 319

Essential Question

WHAT happens when you add, subtract, multiply, and divide fractions?

CCSS **UNIT 3** Expressions and Equations

UNIT PROJECT PREVIEW page 344

Chapter 5
Expressions

Essential Question

HOW can you use numbers and symbols to represent mathematical ideas?

Real World p. 387

Chapter 6 Equations and Inequalities

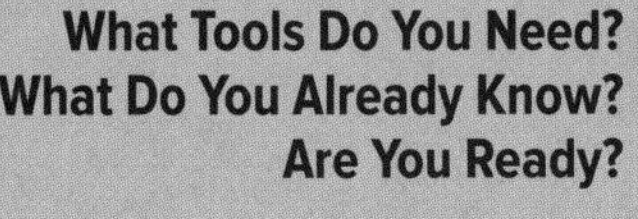

Essential Question

WHAT does it mean to say two quantities are equal?

Stand Up and Be Counted!

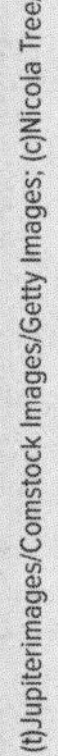

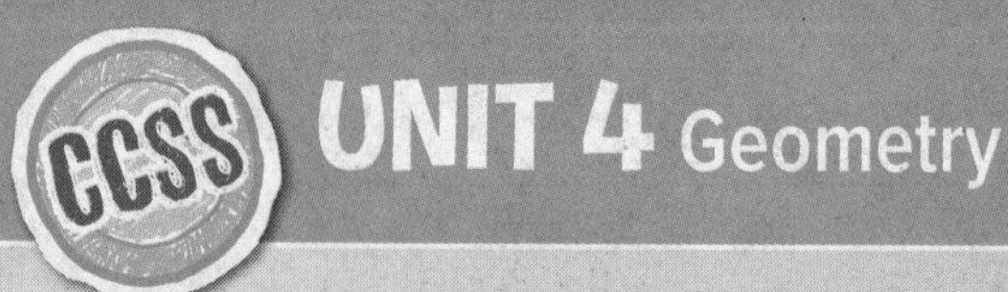

UNIT 4 Geometry

UNIT PROJECT PREVIEW page 530

Chapter 7 Geometric Figures

Essential Question

HOW does geometry help us describe real-world objects?

Real World p. 535

Chapter 8
Measure Figures

Essential Question

HOW do measurements help you describe real-world objects?

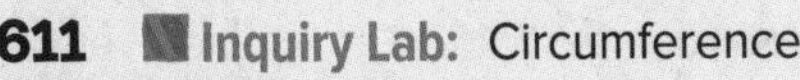

Real World

p. 623

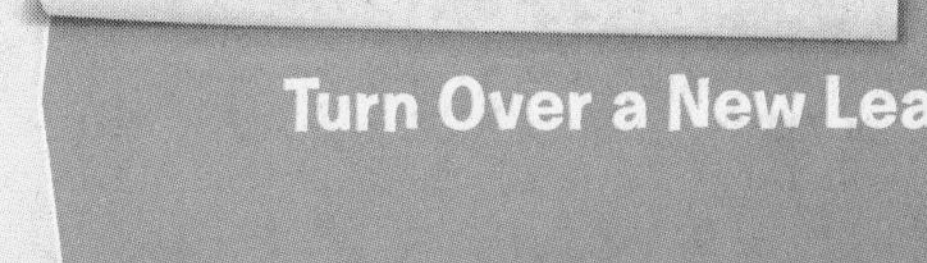

UNIT 5 Statistics and Probability

UNIT PROJECT PREVIEW
page 706

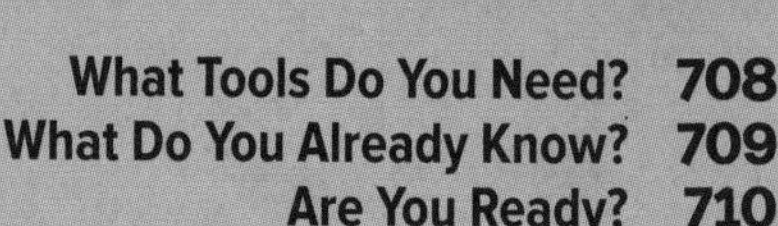

Chapter 9 Probability

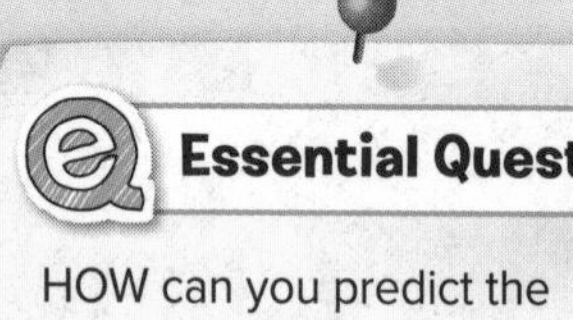

Essential Question

HOW can you predict the outcome of future events?

p. 733

Chapter 10 Statistics

Essential Question

HOW do you know which type of graph to use when displaying data?

Common Core State Standards for MATHEMATICS, Grade 7

Glencoe Math, Course 2, focuses on four critical areas: (1) developing understanding of and applying proportional relationships; (2) operations with rational numbers and working with expressions and linear equations; (3) solving problems involving scale drawings, geometric constructions, and surface area, and volume; and (4) drawing inferences about populations.

Content Standards

Domain 7.RP **Ratios and Proportional Relationships**

- Analyze proportional relationships and use them to solve real-world and mathematical problems.

Domain 7.NS **The Number System**

- Apply and extend previous understandings of operations with fractions to add, subtract, multiply, and divide rational numbers.

Domain 7.EE **Expressions and Equations**

- Use properties of operations to generate equivalent expressions.
- Solve real-life and mathematical problems using numerical and algebraic expressions and equations.

Domain 7.G **Geometry**

- Draw, construct and describe geometrical figures and describe the relationships between them.
- Solve real-life and mathematical problems involving angle measure, area, surface area, and volume.

Domain 7.SP **Statistics and Probability**

- Use random sampling to draw inferences about a population.
- Draw informal comparative inferences about two populations.
- Investigate chance processes and develop, use, and evaluate probability models.

MP Mathematical Practices

1. Make sense of problems and persevere in solving them.
2. Reason abstractly and quantitatively.
3. Construct viable arguments and critique the reasoning of others.
4. Model with mathematics.
5. Use appropriate tools strategically.
6. Attend to precision.
7. Look for and make use of structure.
8. Look for and express regularity in repeated reasoning.

Track Your Common Core Progress

These pages list the key ideas that you should be able to understand by the end of the year. You will rate how much you know about each one. Don't worry if you have no clue **before** you learn about them. Watch how your knowledge grows as the year progresses!

 I have no clue. I've heard of it. 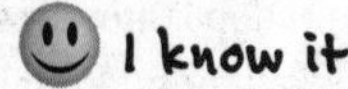

	Before			After		
7.RP Ratios and Proportional Relationships	I have no clue.	I've heard of it.	I know it!	I have no clue.	I've heard of it.	I know it!
Analyze proportional relationships and use them to solve real-world and mathematical problems.						
7.RP.1 Compute unit rates associated with ratios of fractions, including ratios of lengths, areas and other quantities measured in like or different units.						
7.RP.2 Recognize and represent proportional relationships between quantities. **a.** Decide whether two quantities are in a proportional relationship, e.g., by testing for equivalent ratios in a table or graphing on a coordinate plane and observing whether the graph is a straight line through the origin. **b.** Identify the constant of proportionality (unit rate) in tables, graphs, equations, diagrams, and verbal descriptions of proportional relationships. **c.** Represent proportional relationships by equations. **d.** Explain what a point (x, y) on the graph of a proportional relationship means in terms of the situation, with special attention to the points (0, 0) and $(1, r)$ where r is the unit rate.						
7.RP.3 Use proportional relationships to solve multistep ratio and percent problems.						

	Before			After		
7.NS The Number System	I have no clue.	I've heard of it.	I know it!	I have no clue.	I've heard of it.	I know it!
Apply and extend previous understandings of operations with fractions to add, subtract, multiply, and divide rational numbers.						
7.NS.1 Apply and extend previous understandings of addition and subtraction to add and subtract rational numbers; represent addition and subtraction on a horizontal or vertical number line diagram. **a.** Describe situations in which opposite quantities combine to make 0. **b.** Understand $p + q$ as the number located a distance $\lvert q \rvert$ from p, in the positive or negative direction depending on whether q is positive or negative. Show that a number and its opposite have a sum of 0 (are additive inverses). Interpret sums of rational numbers by describing real-world contexts. **c.** Understand subtraction of rational numbers as adding the additive inverse, $p - q = p + (-q)$. Show that the distance between two rational numbers on the number line is the absolute value of their difference, and apply this principle in real-world contexts. **d.** Apply properties of operations as strategies to add and subtract rational numbers.						

7.NS The Number System *continued*	Before ☹	Before 😐	Before ☺	After ☹	After 😐	After ☺
7.NS.2 Apply and extend previous understandings of multiplication and division and of fractions to multiply and divide rational numbers. **a.** Understand that multiplication is extended from fractions to rational numbers by requiring that operations continue to satisfy the properties of operations, particularly the distributive property, leading to products such as $(-1)(-1) = 1$ and the rules for multiplying signed numbers. Interpret products of rational numbers by describing real-world contexts. **b.** Understand that integers can be divided, provided that the divisor is not zero, and every quotient of integers (with non-zero divisor) is a rational number. If p and q are integers, then $-(p/q) = (-p)/q = p/(-q)$. Interpret quotients of rational numbers by describing real-world contexts. **c.** Apply properties of operations as strategies to multiply and divide rational numbers. **d.** Convert a rational number to a decimal using long division; know that the decimal form of a rational number terminates in 0s or eventually repeats.						
7.NS.3 Solve real-world and mathematical problems involving the four operations with rational numbers.						

7.EE Expressions and Equations	☹	😐	☺	☹	😐	☺
Use properties of operations to generate equivalent expressions.						
7.EE.1 Apply properties of operations as strategies to add, subtract, factor, and expand linear expressions with rational coefficients.						
7.EE.2 Understand that rewriting an expression in different forms in a problem context can shed light on the problem and how the quantities in it are related.						
Solve real-life and mathematical problems using numerical and algebraic expressions and equations.						
7.EE.3 Solve multi-step real-life and mathematical problems posed with positive and negative rational numbers in any form (whole numbers, fractions, and decimals), using tools strategically. Apply properties of operations to calculate with numbers in any form; convert between forms as appropriate; and assess the reasonableness of answers using mental computation and estimation strategies.						
7.EE.4 Use variables to represent quantities in a real-world or mathematical problem, and construct simple equations and inequalities to solve problems by reasoning about the quantities. **a.** Solve word problems leading to equations of the form $px + q = r$ and $p(x + q) = r$, where p, q, and r are specific rational numbers. Solve equations of these forms fluently. Compare an algebraic solution to an arithmetic solution, identifying the sequence of the operations used in each approach. **b.** Solve word problems leading to inequalities of the form $px + q > r$ or $px + q < r$, where p, q, and r are specific rational numbers. Graph the solution set of the inequality and interpret it in the context of the problem.						

7.G Geometry	Before :(	Before :|	Before :)	After :(	After :|	After :)
Draw, construct, and describe geometrical figures and describe the relationships between them.						
7.G.1 Solve problems involving scale drawings of geometric figures, including computing actual lengths and areas from a scale drawing and reproducing a scale drawing at a different scale.						
7.G.2 Draw (freehand, with ruler and protractor, and with technology) geometric shapes with given conditions. Focus on constructing triangles from three measures of angles or sides, noticing when the conditions determine a unique triangle, more than one triangle, or no triangle.						
7.G.3 Describe the two-dimensional figures that result from slicing three-dimensional figures, as in plane sections of right rectangular prisms and right rectangular pyramids.						
Solve real-life and mathematical problems involving angle measure, area, surface area, and volume.						
7.G.4 Know the formulas for the area and circumference of a circle and use them to solve problems; give an informal derivation of the relationship between the circumference and area of a circle.						
7.G.5 Use facts about supplementary, complementary, vertical, and adjacent angles in a multi-step problem to write and solve simple equations for an unknown angle in a figure.						
7.G.6 Solve real-world and mathematical problems involving area, volume and surface area of two- and three-dimensional objects composed of triangles, quadrilaterals, polygons, cubes, and right prisms.						

7.SP Statistics and Probability	Before :(	Before :|	Before :)	After :(	After :|	After :)
Use random sampling to draw inferences about a population.						
7.SP.1 Understand that statistics can be used to gain information about a population by examining a sample of the population; generalizations about a population from a sample are valid only if the sample is representative of that population. Understand that random sampling tends to produce representative samples and support valid inferences.						
7.SP.2 Use data from a random sample to draw inferences about a population with an unknown characteristic of interest. Generate multiple samples (or simulated samples) of the same size to gauge the variation in estimates or predictions.						
Draw informal comparative inferences about two populations.						
7.SP.3 Informally assess the degree of visual overlap of two numerical data distributions with similar variabilities, measuring the difference between the centers by expressing it as a multiple of a measure of variability.						

7.SP Statistics and Probability *continued*	Before ☹	Before 😐	Before ☺	After ☹	After 😐	After ☺
7.SP.4 Use measures of center and measures of variability for numerical data from random samples to draw informal comparative inferences about two populations.						
Investigate chance processes and develop, use, and evaluate probability models.						
7.SP.5 Understand that the probability of a chance event is a number between 0 and 1 that expresses the likelihood of the event occurring. Larger numbers indicate greater likelihood. A probability near 0 indicates an unlikely event, a probability around 1/2 indicates an event that is neither unlikely nor likely, and a probability near 1 indicates a likely event.						
7.SP.6 Approximate the probability of a chance event by collecting data on the chance process that produces it and observing its long-run relative frequency, and predict the approximate relative frequency given the probability.						
7.SP.7 Develop a probability model and use it to find probabilities of events. Compare probabilities from a model to observed frequencies; if the agreement is not good, explain possible sources of the discrepancy. **a.** Develop a uniform probability model by assigning equal probability to all outcomes, and use the model to determine probabilities of events. **b.** Develop a probability model (which may not be uniform) by observing frequencies in data generated from a chance process.						
7.SP.8 Find probabilities of compound events using organized lists, tables, tree diagrams, and simulation. **a.** Understand that, just as with simple events, the probability of a compound event is the fraction of outcomes in the sample space for which the compound event occurs. **b.** Represent sample spaces for compound events using methods such as organized lists, tables and tree diagrams. For an event described in everyday language (e.g., "rolling double sixes"), identify the outcomes in the sample space which compose the event. **c.** Design and use a simulation to generate frequencies for compound events.						

UNIT 3

CCSS Expressions and Equations

Essential Question

HOW can you communicate mathematical ideas effectively?

Chapter 5
Expressions

Algebraic expressions can be used to represent real-world situations. In this chapter, you will apply the properties of operations to simplify and evaluate algebraic expressions.

Chapter 6
Equations and Inequalities

An equation is a mathematical sentence stating that two expressions are equal. In this chapter, you will use the properties of equality to solve equations algebraically. Then you will apply what you learn to solve inequalities.

Unit Project Preview

Stand Up and Be Counted! The U.S. Census is a survey of the American people that is taken every 10 years. The census is used to collect data about the United States population and to determine the number of House of Representatives for each state.

Conduct your own mini-census. Survey twenty students about the typical way they come to school: by school bus, by car, on foot, or by some other means. Then make a circle graph of the data.

At the end of Chapter 6, you'll complete a project to find how the U.S. population affects the House of Representatives. This adventure will appeal to your "census."

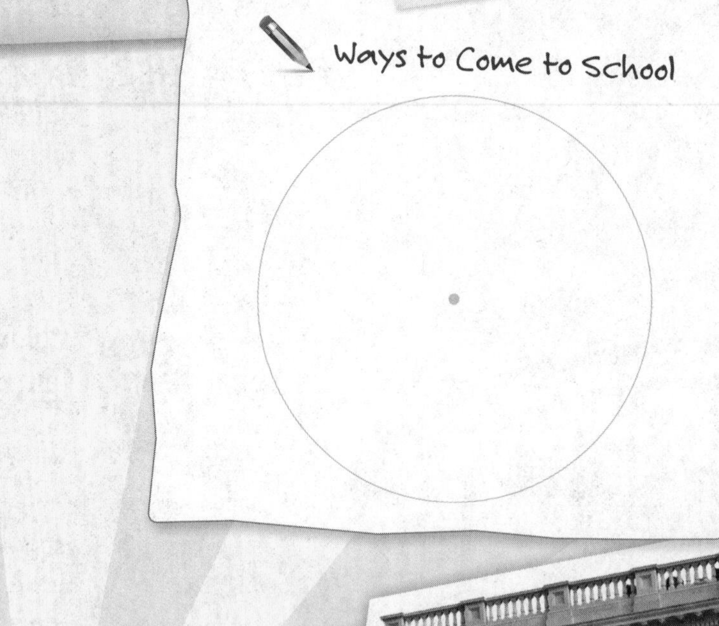

Chapter 5
Expressions

Essential Question

HOW can you use numbers and symbols to represent mathematical ideas?

Common Core State Standards

Content Standards
7.EE.1, 7.EE.2, 7.NS.3

MP Mathematical Practices
1, 2, 3, 4, 5, 6, 7

Math in the Real World

Meerkats live in burrows. Because meerkats have sharp claws, they are able to dig at a rate of 1 foot per second.

Suppose a meerkat digs for 3 seconds. Cross out the expression that does not represent the underground distance dug by the meerkat.

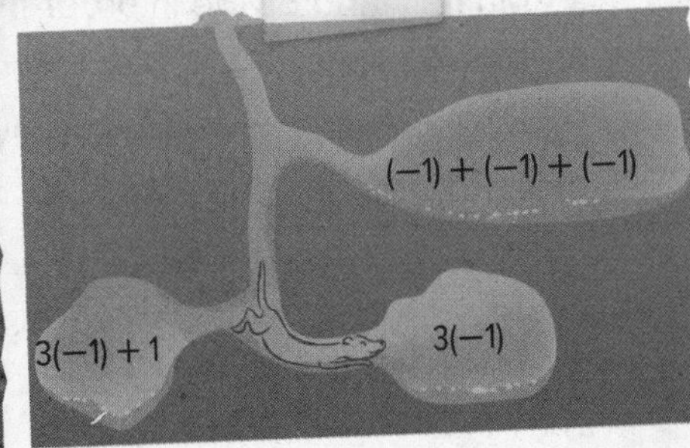

FOLDABLES® Study Organizer

1 Cut out the Foldable on page FL3 of this book.

2 Place your Foldable on page 426.

3 Use the Foldable throughout this chapter to help you learn about expressions.

What Tools Do You Need?

Vocabulary

Additive Identity Property
algebra
algebraic expression
arithmetic sequence
Associative Property
coefficient
Commutative Property
constant
counterexample
define a variable
Distributive Property
equivalent expressions
factor
factored form
like terms
linear expression
monomial
Multiplicative Identity Property
Multiplicative Property of Zero
property
sequence
simplest form
term
variable

Review Vocabulary

Order of Operations The order of operations is a four-step process used to evaluate numerical expressions.

1. Evaluate the expressions inside grouping symbols.
2. Evaluate all powers.
3. Multiply and divide in order from left to right.
4. Add and subtract in order form left to right.

Use the order of operations to evaluate $3 + 5^2(4 + 4)$. Write each step in the organizer below.

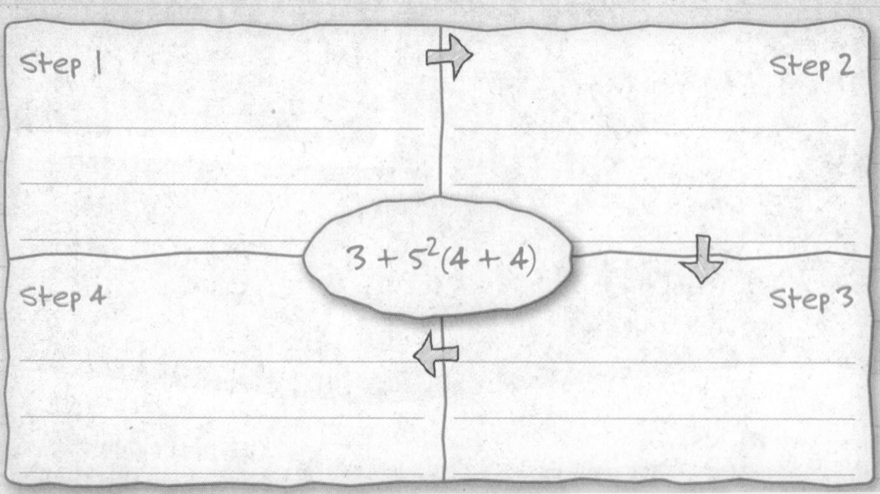

What Do You Already Know?

Read each statement. Decide whether you agree (A) or disagree (D). Place a checkmark in the appropriate column and then justify your reasoning.

Expressions			
Statement	A	D	Why?
Like terms are terms that contain different variables.			
When addition or subtraction signs separate an algebraic expression into parts, each part is called a term.			
An algebraic expression is in simplest form if it has no like terms and no parentheses.			
A property is an example that shows that a conjecture is false.			
When you use the Distributive Property to combine like terms, you are simplifying the expression.			
Equivalent expressions have the same value.			

When Will You Use This?

Here are a few examples of how expressions are used in the real world.

Activity 1 Do you or your parents have a texting plan? If so, how much does it cost per text or per month? Ask your parents to help you research different texting plans. Then compare and contrast each plan.

____

____

Activity 2 Go online at **connectED.mcgraw-hill.com** to read the graphic novel ***Too Many Texts***. How many text messages are included in Dario's texting plan? ______________

Try the Quick Check below.
Or, take the Online Readiness Quiz.

Quick Review

Common Core Review 6.EE.2c, 7.NS.2c

Example 1

Evaluate 2^5.

$2^5 = 2 \cdot 2 \cdot 2 \cdot 2 \cdot 2$

$= 32$

Example 2

Write $3 \cdot 3 \cdot 3 \cdot 3 \cdot 3 \cdot 3 \cdot 3$ in exponential form.

3 is the base. It is used as a factor 7 times. So, the exponent is 7.

$3 \cdot 3 \cdot 3 \cdot 3 \cdot 3 \cdot 3 \cdot 3 = 3^7$

Example 3

Find 4(−2).

4(−2) = −8 The integers have different signs. The product is negative.

Example 4

Find −5(−8).

−5(−8) = 40 The integers have the same signs. The product is positive.

Quick Check

Exponents **Evaluate each expression.**

1. 2^4 = ______
2. 3^3 = ______
3. 4^2 = ______

4. Write $4 \cdot 4 \cdot 4 \cdot 4$ in exponential form. ______

Integer Operations **Multiply.**

5. 5(−10) = ______
6. −9(−4) = ______
7. -5^2 = ______

Which problems did you answer correctly in the Quick Check? Shade those exercise numbers below.

① ② ③ ④ ⑤ ⑥ ⑦

Lesson 1

Algebraic Expressions

Vocabulary Start-Up

A **variable** is a symbol that represents an unknown quantity. An **algebraic expression**, such as $n + 2$, is an expression that contains variables, numbers, and at least one operation.

Write each of the following phrases in the correct section of the Venn diagram: *contains an operation, has variables and numbers, has only numbers*.

Characteristics of Expressions

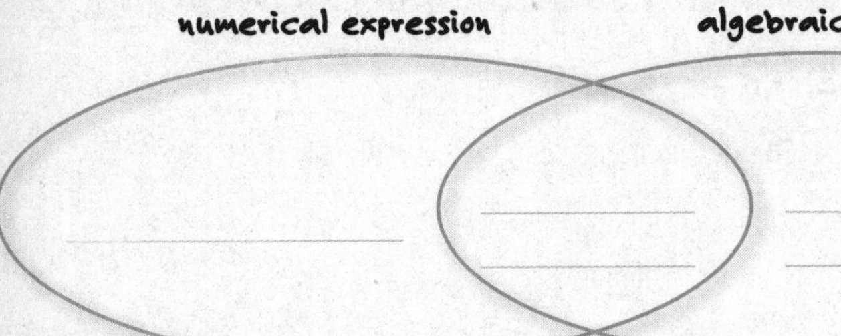

Essential Question

HOW can you use numbers and symbols to represent mathematical ideas?

Vocabulary

variable
algebraic expression
algebra
coefficient
define a variable

Common Core State Standards

Content Standards
Preparation for 7.EE.1 and 7.EE.2

MP Mathematical Practices
1, 2, 3, 4

Real-World Link

The expression $(F - 32) \times \frac{5}{9}$ can be used to convert a temperature from Fahrenheit to Celsius. In this algebraic expression, the variable _______ represents the temperature in degrees Fahrenheit.

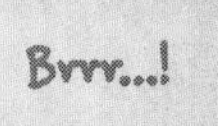

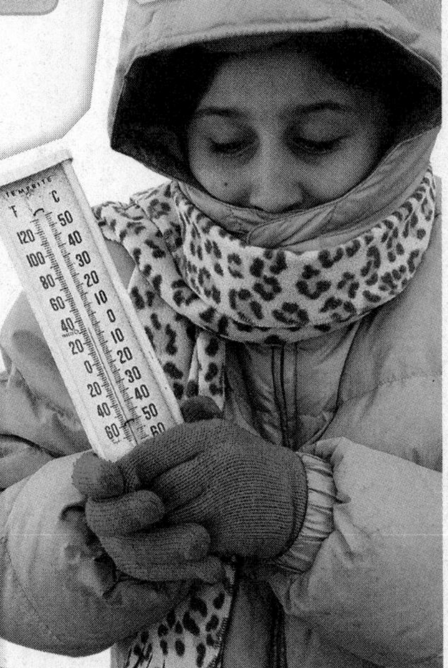

Which MP Mathematical Practices did you use? Shade the circle(s) that applies.

① Persevere with Problems
② Reason Abstractly
③ Construct an Argument
④ Model with Mathematics
⑤ Use Math Tools
⑥ Attend to Precision
⑦ Make Use of Structure
⑧ Use Repeated Reasoning

Work Zone

Evaluate an Algebraic Expression

The branch of mathematics that involves expressions with variables is called **algebra**. In algebra, the multiplication sign is often omitted.

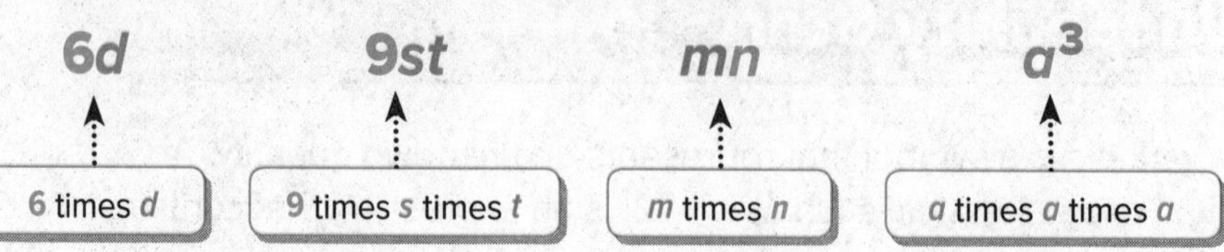

The numerical factor of a multiplication expression that contains a variable is called a **coefficient**. So, 6 is the coefficient of $6d$.

Expressions like $\frac{y}{2}$ can be written as $y \div 2$ or $y \times \frac{1}{2}$.

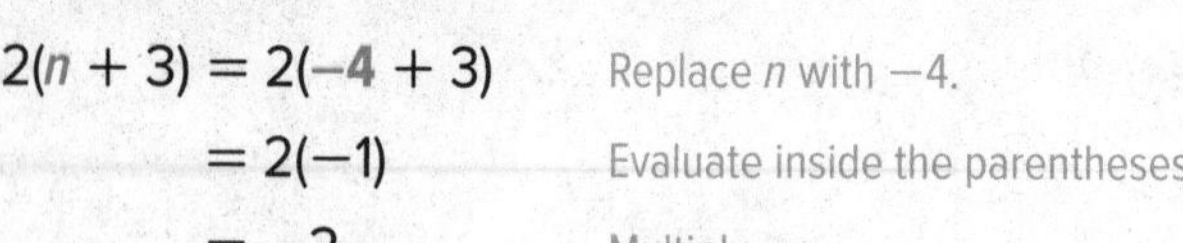

Order of Operations

1. Evaluate the expressions inside grouping symbols.
2. Evaluate all powers.
3. Multiply and divide in order from left to right.
4. Add and subtract in order from left to right.

Examples

Watch Tutor

1. Evaluate $2(n + 3)$ if $n = -4$.

$2(n + 3) = 2(-4 + 3)$ Replace n with -4.

$= 2(-1)$ Evaluate inside the parentheses.

$= -2$ Multiply.

2. Evaluate $8w - 2v$ if $w = 5$ and $v = 3$.

$8w - 2v = 8(5) - 2(3)$ Replace w with 5 and v with 3.

$= 40 - 6$ Do all of the multiplication first.

$= 34$ Subtract 6 from 40.

3. Evaluate $4y^3 + 2$ if $y = 3$.

$4y^3 + 2 = 4(3)^3 + 2$ Replace y with 3.

$= 4(27) + 2$ Evaluate the power.

$= 110$ Multiply, then add.

Got It? Do these problems to find out.

Evaluate each expression if $c = 8$ and $d = -5$.

a. $c - 3$ **b.** $15 - c$ **c.** $3(c + d)$

d. $2c - 4d$ **e.** $d - c^2$ **f.** $2d^2 + 5d$

a. ________

b. ________

c. ________

d. ________

e. ________

f. ________

Example

4. Athletic trainers use the formula $\frac{3(220 - a)}{5}$, where a is a person's age, to find their minimum training heart rate. Find Latrina's minimum training heart rate if she is 15 years old.

$\frac{3(220 - a)}{5} = \frac{3(220 - 15)}{5}$	Replace a with 15.
$= \frac{3(205)}{5}$	Subtract 15 from 220.
$= \frac{615}{5}$	Multiply 3 and 205.
$= 123$	Divide 615 by 5.

Latrina's minimum training heart rate is 123 beats per minute.

Fractions

The fraction bar is a grouping symbol. Evaluate the expressions in the numerator and denominator separately before dividing.

Got It? Do this problem to find out.

g. To find the area of a triangle, use the formula $\frac{bh}{2}$, where b is the base and h is the height. What is the area in square inches of a triangle with a height of 6 inches and base of 8 inches?

g. ____________

Write Expressions

To translate a verbal phrase into an algebraic expression, the first step is to define a variable. When you **define a variable**, you choose a variable to represent an unknown quantity.

Examples

5. Marisa wants to buy a DVD player that costs $150. She already saved $25 and plans to save an additional $10 each week. Write an expression that represents the total amount of money Marisa has saved after any number of weeks.

Words	savings of $25 plus ten dollars each week
Variable	Let w represent the number of weeks.
Expression	$25 \quad + \quad 10 \quad \cdot \quad w$

$25 + 10w$ represents the total saved after any number of weeks.

6. Refer to Example 5. Will Marisa have saved enough money to buy the $150 DVD player in 11 weeks? Use the expression 25 + 10*w*.

$25 + 10w = 25 + 10(\mathbf{11})$	Replace *w* with 11.
$= 25 + 110$	Multiply.
$= 135$	Add.

Marisa will have saved $135 after 11 weeks. Since $135 < $150, Marisa will not have enough money to buy the DVD player.

Got It? Do this problem to find out.

Show your work.

h. An MP3 player costs $70 and song downloads cost $0.85 each. Write an expression that represents the cost of the MP3 player and *x* number of downloaded songs. Then find the total cost if 20 songs are downloaded.

h. ________

Guided Practice

Evaluate each expression if $m = 2$, $n = 6$, and $p = -4$. (Examples 1–4)

1. $3m + 4p$ ________

2. $n^2 + 5$ ________

3. $6p^3$ ________

4. A Web site charges $0.99 to download a game and a $12.49 membership fee. Write an expression that gives the total cost in dollars to download *g* games. Then find the cost of downloading 6 games. (Examples 5 and 6)

____

5. **Building on the Essential Question** Tell whether the statement below is *sometimes*, *always*, or *never* true. Justify your reasoning.

The expressions x − 3 and y − 3 represent the same value.

____

Rate Yourself!

How well do you understand algebraic expressions? Circle the image that applies.

Clear

Somewhat Clear

Not So Clear

For more help, go online to access a Personal Tutor.

Independent Practice

Go online for Step-by-Step Solutions

Evaluate each expression if $d = 8$, $e = 3$, $f = 4$, and $g = -1$. (Examples 1 – 3)

1. $2(d + 9)$ ______

2. $\frac{d}{4}$ ______

3. $\frac{ef}{4}$ ______

4. $4f + d$ ______

5. $\frac{5d - 25}{5}$ ______

6. $d^2 + 7$ ______

7. $\frac{d - 4}{2}$ ______

8. $10(e + 7)$ ______

9. $\frac{2g}{2}$ ______

10. The expression $5n + 2$ can be used to find the total cost in dollars of bowling where n is the number of games bowled and 2 represents the cost of shoe rental. How much will it cost Vincent to bowl 3 games? (Example 4)

______

11. MP **Reason Abstractly** A car rental company's fees are shown. Suppose you rent a car using Option 2. Write an expression that gives the total cost in dollars for driving m miles. Then find the cost for driving 150 miles. (Examples 5 and 6)

______

Car Rental Prices	
Option 1	Option 2
\$19.99 per day	\$50 fee
\$0.17 per mi	\$0.17 per mi

12. Refer to Exercise 11. Suppose you rent a car using Option 1. Write an expression that gives the total cost in dollars to rent a car for d days and m miles. Then find the cost for renting a car for 2 days and driving 70 miles. (Examples 5 and 6)

______

Evaluate each expression if $x = 3.2$, $y = 6.1$, and $z = 0.2$.

13. $x + y - z$ ______

14. $14.6 - (x + y + z)$ ______

15. $xz + y^2$ ______

H.O.T. Problems Higher Order Thinking

16. MP **Reason Abstractly** Write an algebraic expression with the variable x that has a value of 3 when evaluated.

______

17. MP **Model with Mathematics** Write a real-world problem that can be represented by the expression $5x + 10$.

______

18. MP **Persevere with Problems** To find the total number of diagonals for any given polygon, you can use the expression $\frac{n(n-3)}{2}$, where n is the number of sides of the polygon.

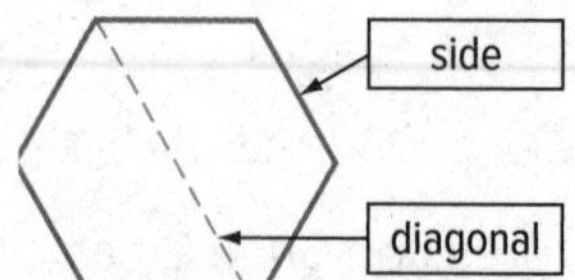

a. Determine the minimum value that n could be. ______

b. Make a table of four possible values of n. Then complete the table by evaluating the expression for each value of n.

c. Check by drawing the diagonals of a pentagon and counting the diagonals.

n	value

19. MP **Persevere with Problems** Franco constructed the objects below using toothpicks.

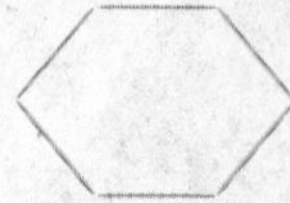
Figure 1

Figure 2

Figure 3

Write two different rules that relate the figure number to the number of toothpicks in each figure.

______

Extra Practice

Evaluate each expression if $d = 8$, $e = 3$, $f = 4$, and $g = -1$.

20. $10 - e$ 7

$10 - e$

$10 - 3 = 7$

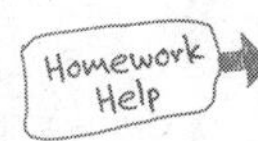

21. $\frac{16}{f}$ 4

$\frac{16}{f}$

$\frac{16}{4} = 4$

22. $4e^2$ ______

23. $8g - f$ ______

24. $\frac{(5 + g)^2}{2}$ ______

25. $e^2 - 4$ ______

26. The expression $\frac{w}{30}$, where w is a person's weight in pounds, is used to find the approximate number of quarts of blood in the person's body. How many quarts of blood does a 120-pound person have?

______

27. MP **Model with Mathematics** Refer to the graphic novel frame below. Let n represent the number of text messages. Evaluate the expression $0.15(n - 250) + 5$ to find the cost of 275 text messages. ______

Price Guide:

Number of Text Messages Sent	Cost
250	\$5.00
252	\$5.30
254	\$5.60
256	\$5.90

Caitlin is helping Dario figure out what his text messaging bill will be this month.

CCSS Power Up! Common Core Test Practice

28. Tonya has x quarters, y dimes, and z nickels in her pocket. Select the appropriate operations to complete the expression that represents the total amount of change Tonya has in her pocket.

(\$0.25 ☐ x) ☐ (\$0.1 ☐ y) ☐ (\$0.05 ☐ z)

Evaluate the expression for $x = 3$, $y = 5$, and $z = 2$. What does this value represent?

☐

29. The prices of magazines and books at the school book fair are shown in the table. Determine if each statement is true or false.

School Book Fair Prices	
Item	**Cost**
Magazines	\$4.95
Paperback books	\$7.95

a. The expression $7.95b + 4.95m$ represents the cost of buying b books and m magazines. ☐ True ☐ False

b. The expression $12.90(b + m)$ represents the cost of buying b books and m magazines. ☐ True ☐ False

c. The total cost of buying 3 books and 4 magazines is \$43.65. ☐ True ☐ False

CCSS Common Core Spiral Review

Define a variable and write each phrase as an algebraic expression. 6.EE.2

30. 8 feet less than the height ______

31. Sarah worked 8 more hours than Paida. ______

32. Kumar has twice the number of goals as Jacob. ______

33. Addison is 3 years younger than Nathan. ______

34. The table shows the costs of different camping activities. Over the summer, Maura canoed 4 times and fished 3 times. Write and evaluate an expression that represents the total cost Maura spent canoeing and fishing. 5.OA.1

Camping Activity Costs	
Activity	**Cost**
Canoeing	\$8
Fishing	\$5

______

Lesson 2

Sequences

Vocabulary Start-Up

A **sequence** is an ordered list of numbers. Each number in a sequence is called a **term**. In an **arithmetic sequence**, each term is found by adding the same number to the previous term.

Complete the graphic organizer below.

Numbers	Words
Continue each sequence.	Describe each sequence.
1, 3, 5, 7, ☐, ...	Add ☐ to the previous term.
1, 1.5, 2, ☐, ☐, ☐, ...	

Essential Question

HOW can you use numbers and symbols to represent mathematical ideas?

Vocabulary

sequence
term
arithmetic sequence

Common Core State Standards

Content Standards
Preparation for 7.EE.1 and 7.EE.2

MP Mathematical Practices
1, 2, 3, 4

Real-World Link

Horseback Riding The number of students who went on each horseback riding trip is shown. Do the numbers represent the terms of an arithmetic sequence? Explain.

Trip	1	2	3	4	5
Number of Students	15	16	18	21	25

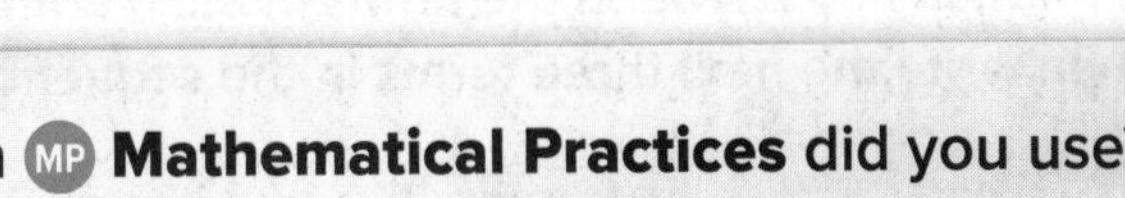

Which MP Mathematical Practices did you use? Shade the circle(s) that applies.

① Persevere with Problems
② Reason Abstractly
③ Construct an Argument
④ Model with Mathematics
⑤ Use Math Tools
⑥ Attend to Precision
⑦ Make Use of Structure
⑧ Use Repeated Reasoning

Work Zone

Describe and Extend Sequences

In an arithmetic sequence, the terms can be whole numbers, fractions, or decimals.

Examples

1. Describe the relationship between the terms in the arithmetic sequence 8, 13, 18, 23, ... Then write the next three terms in the sequence.

8, 13, 18, 23, ...
+5 +5 +5

Each term is found by adding 5 to the previous term.

Continue the pattern to find the next three terms.

$23 + 5 = 28$ $28 + 5 = 33$ $33 + 5 = 38$

The next three terms are 28, 33, and 38.

2. Describe the relationship between the terms in the arithmetic sequence 0.4, 0.6, 0.8, 1.0, ... Then write the next three terms in the sequence.

0.4, 0.6, 0.8, 1.0, ...
+0.2 +0.2 +0.2

Each term is found by adding 0.2 to the previous term.

Continue the pattern to find the next three terms.

$1.0 + 0.2 = 1.2$ $1.2 + 0.2 = 1.4$ $1.4 + 0.2 = 1.6$

The next three terms are 1.2, 1.4, and 1.6.

Show your work.

Got It? **Do these problems to find out.**

Describe the relationship between the terms in each arithmetic sequence. Then write the next three terms in the sequence.

a. 0, 13, 26, 39, ... **b.** 4, 7, 10, 13, ...

c. 1.0, 1.3, 1.6, 1.9, ... **d.** 2.5, 3.0, 3.5, 4.0, ...

a. ______

b. ______

c. ______

d. ______

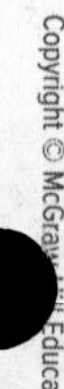

Write an Algebraic Expression

In a sequence, each term has a specific position within the sequence. Consider the sequence 2, 4, 6, 8,...

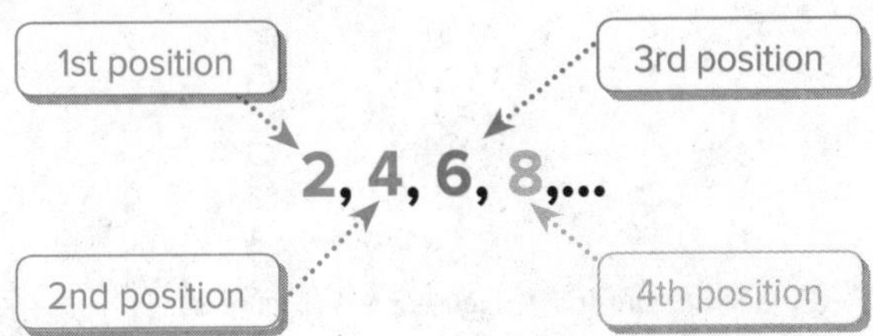

Notice that as the position number increases by 1, the value of the term increases by 2.

Position	Operation	Value of Term
1	$1 \cdot 2 = 2$	2
2	$2 \cdot 2 = 4$	4
3	$3 \cdot 2 = 6$	6
4	$4 \cdot 2 = 8$	8

+1 +1 +1 / +2 +2 +2

Arithmetic Sequences
When looking for a pattern between the position number and each term in the sequence, it is often helpful to make a table.

You can also write an algebraic expression to represent the relationship between any term in a sequence and its position in the sequence. In this case, if n represents the position in the sequence, the value of the term is $2n$.

Example

Tutor

3. The greeting cards that Meredith makes are sold in boxes at a gift store. The first week, the store sold 5 boxes. Each week, the store sells five more boxes. The pattern continues. What algebraic expression can be used to find the total number of boxes sold at the end of the 100th week? What is the total?

Position	Operation	Value of Term
1	$1 \cdot 5$	5
2	$2 \cdot 5$	10
3	$3 \cdot 5$	15
n	$n \cdot 5$	$5n$

Each term is 5 times its position. So, the expression is $5n$.

$5n$ Write the expression.

$5(100) = 500$ Replace n with 100.

At the end of 100 weeks, 500 boxes will have been sold.

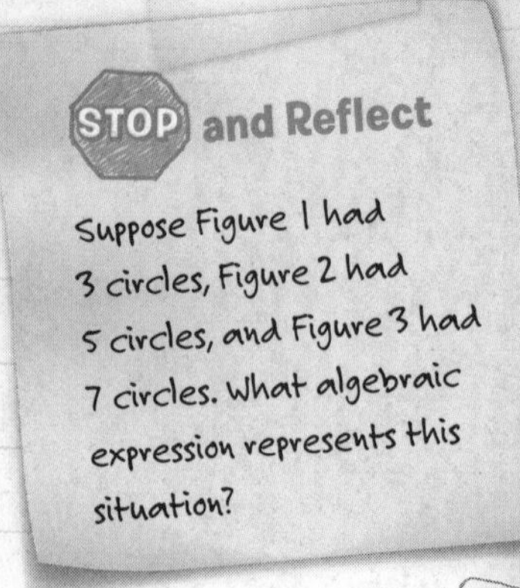

Show your work.

e. ____________

Got It? **Do this problem to find out.**

e. If the pattern continues, what algebraic expression can be used to find the number of circles used in any figure? How many circles will be in the 50th figure?

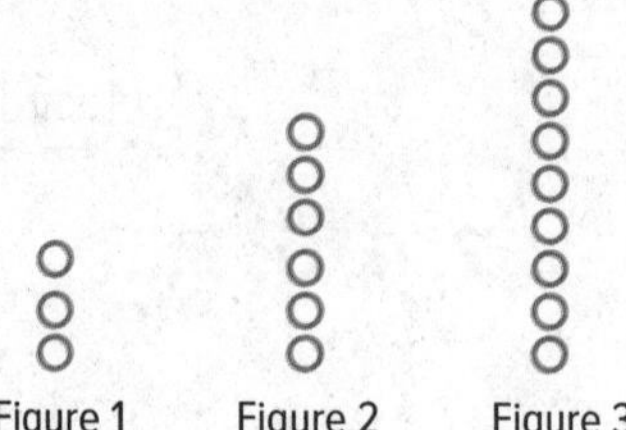

Guided Practice

Describe the relationship between the terms in each arithmetic sequence. Then write the next three terms in each sequence. (Examples 1 and 2)

1. 0, 9, 18, 27, ...

2. 4, 9, 14, 19, ...

3. 1, 1.1, 1.2, 1.3, ...

Show your work.

4. Hannah has a doll collection. The table shows the total number of dolls in her collection for three years. Suppose this pattern continues. Write an algebraic expression to find the number of dolls in her collection after *n* years. How many dolls will Hannah have after 25 years? (Example 3)

Year	Number of Dolls
1	6
2	12
3	18

5. **Building on the Essential Question** Explain why the following sequence is considered an arithmetic sequence.

5, 9, 13, 17, 21,...

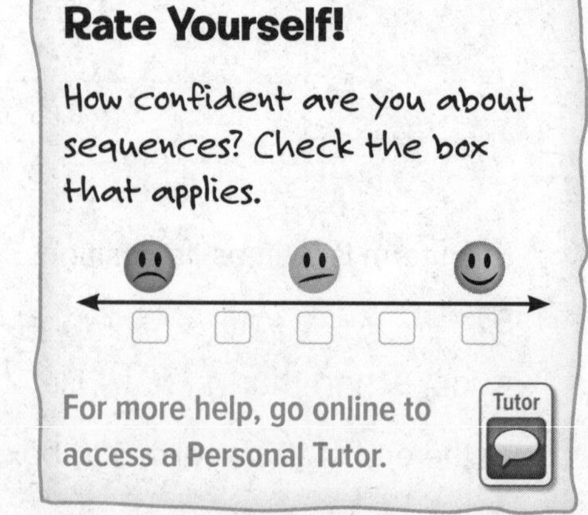

Independent Practice

Go online for Step-by-Step Solutions

Describe the relationship between the terms in each arithmetic sequence. Then write the next three terms in each sequence. (Examples 1 and 2)

1. 0, 7, 14, 21, ...

2. 1, 7, 13, 19, ...

3. 26, 34, 42, 50, ...

4. 0.1, 0.4, 0.7, 1.0, ...

5. 2.4, 3.2, 4.0, 4.8, ...

6. 2.0, 3.1, 4.2, 5.3, ...

7. Refer to the table shown. If the pattern continues, what algebraic expression can be used to find the plant's height for any month? What will be the plant's height at 12 months? (Example 3)

Month	Height (in.)
1	3
2	6
3	9

8. MP **Model with Mathematics** Explain how the number of text messages Dario sent and the cost form an arithmetic sequence. Then write an expression to find Dario's text messaging bill if he sends n text messages over 250.

9. **Multiple Representations** Kendra is stacking boxes of tissues for a store display. She stacks 3 boxes in the first minute, 6 boxes by the end of the second minute, and 9 boxes by the end of the third minute. Suppose the pattern continues for parts **a–d**.

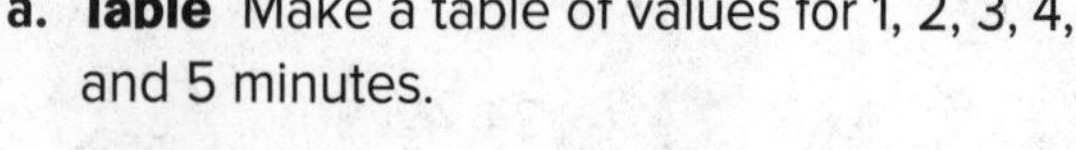
a. **Table** Make a table of values for 1, 2, 3, 4, and 5 minutes.

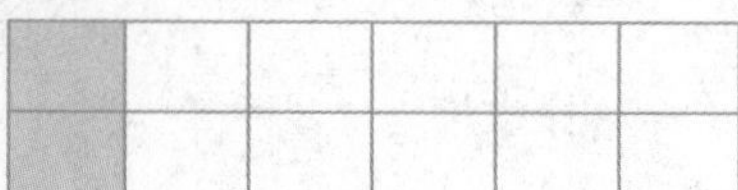

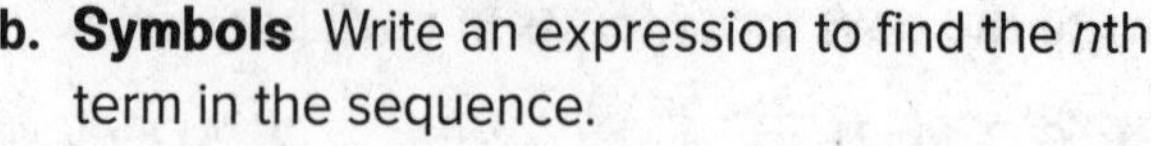
b. **Symbols** Write an expression to find the *n*th term in the sequence.

c. **Graph** Graph the table of values from part **a** on the coordinate plane. Let x represent the number of minutes and y represent the number of boxes. Then describe the graph.

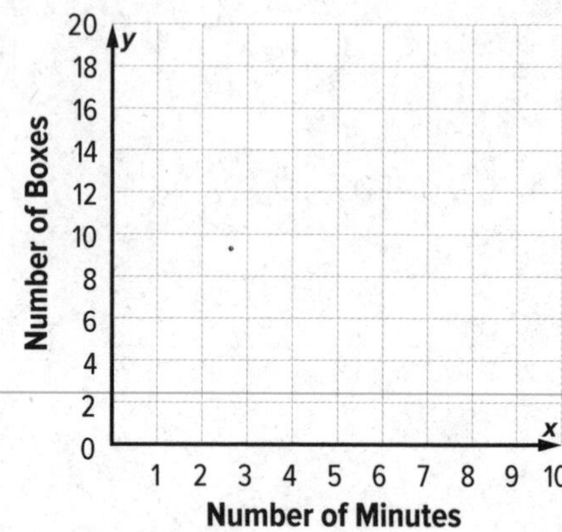

d. **Numbers** How many boxes will be displayed after 45 minutes?

H.O.T. Problems Higher Order Thinking

10. **Justify Conclusions** Write five terms of an arithmetic sequence and describe the rule for finding the terms.

Persevere with Problems **Not all sequences are arithmetic. But, there is still a pattern. Describe the relationship between the terms in each sequence. Then write the next three terms in the sequence.**

11. 1, 2, 4, 7, 11, ...

12. 0, 2, 6, 12, 20, ...

13. **Persevere with Problems** Use an arithmetic sequence to find the number of multiples of 6 between 41 and 523. Justify your reasoning.

Name ______ My Homework ______

Extra Practice

Describe the relationship between the terms in each arithmetic sequence. Then write the next three terms in each sequence.

14. 19, 31, 43, 55, ...

12 is added to the previous term; 67, 79, 91

15. 6, 16, 26, 36, ...

10 is added to the previous term; 46, 56, 66

16. 33, 38, 43, 48, ...

17. 4.5, 6.0, 7.5, 9.0, ...

18. 1.2, 3.2, 5.2, 7.2, ...

19. 4.6, 8.6, 12.6, 16.6, ...

20. 18, 33, 48, 63, ...

21. 20, 45, 70, 95, ...

22. 38, 61, 84, 107, ...

23. MP **Reason Abstractly** Refer to the figures for parts **a** and **b**.

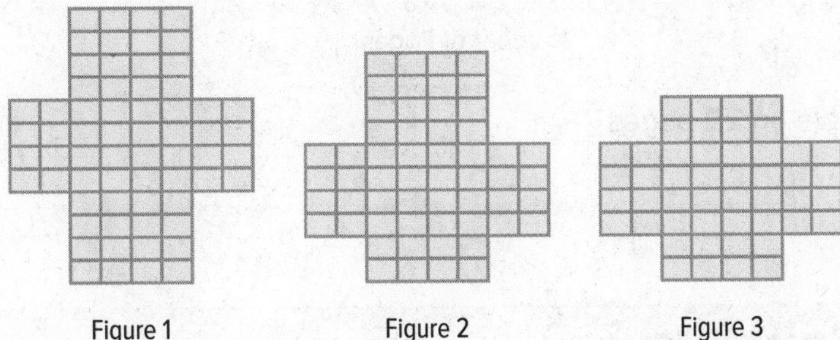

a. Describe the relationship between the figures and the number of rectangles shown. ______

b. If the pattern continues, how many rectangles will be in the next 2 figures? ______

The terms of an arithmetic sequence can be related by subtraction. Write the next three terms of each sequence.

24. 32, 27, 22, 17, ...

25. 45, 42, 39, 36, ...

26. 10.5, 10, 9.5, 9, ...

CCSS Power Up! Common Core Test Practice

27. The table shows the first 5 terms of a sequence. Determine if each statement is true or false.

Position	1	2	3	4	5	n
Value of Term	2	5	10	17	26	■

a. The expression $n^2 + 1$ can be used to find the nth term of the sequence. ☐ True ☐ False

b. The 8th term of the sequence is 65. ☐ True ☐ False

c. The table represents an arithmetic sequence. ☐ True ☐ False

28. Katie is putting photos in an album. She puts five pictures on the first page. Each page after that contains five pictures. Suppose the pattern continues. Complete the table of values for 1, 2, 3, 4, and 5 pages. Then graph the table of values on the coordinate plane. Let x represent the number of pages and y represent the total number of photos.

Number of Pages	Total Photos
1	
2	
3	
4	
5	

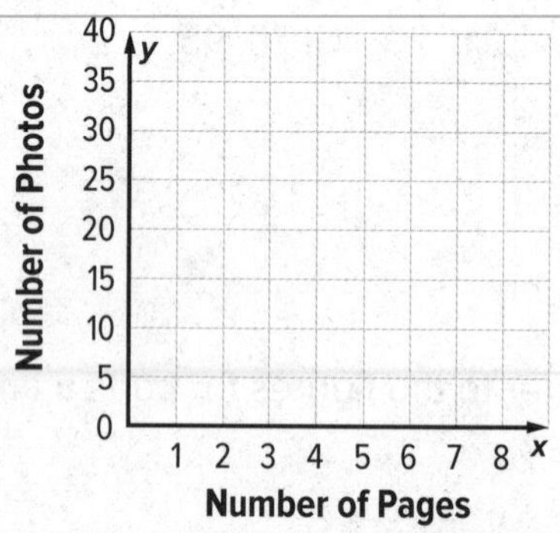

How many photos will Katie have on 20 pages? ☐

CCSS Common Core Spiral Review

Evaluate. **6.EE.1**

29. $1^4 =$ ______

30. $3^3 =$ ______

31. $8^2 =$ ______

32. $10^4 =$ ______

33. $5^1 =$ ______

34. $7^5 =$ ______

35. Jayden goes to the batting cage. He purchases three tokens and rents a helmet. If he spends a total of $6.50, how much is each token? **6.EE.6**

______

Batting Cage Prices	
Tokens	■
Helmet Rental	$2

Inquiry Lab

Sequences

HOW can geometric figures be used to model numerical patterns?

Content Standards
Preparation for 7.EE.2

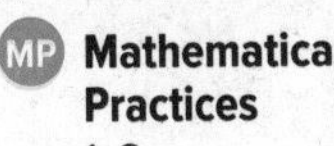

Mathematical Practices
1, 3

A fencing company uses 4 planks of wood for one section of fencing, 7 planks for two sections, and 10 planks of wood for three sections. The fence sections are represented using the toothpicks shown. Determine how many planks would be used to create 5 sections of fencing.

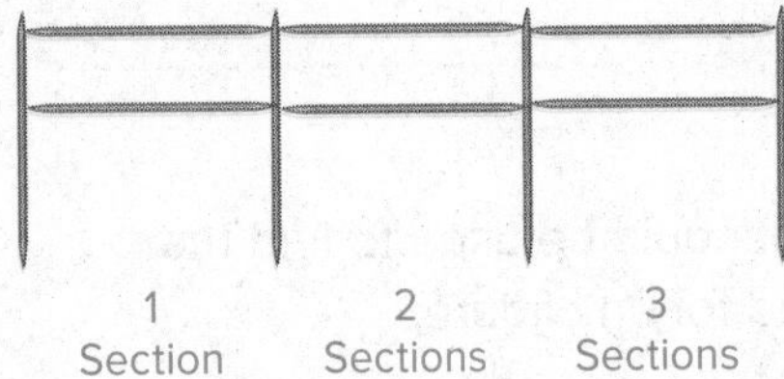

Hands-On Activity

Step 1 Find a pattern in the table. Then fill in the number of planks that would be in 4 and 5 sections of fencing.

Number of Sections	Number of Planks
1	4
2	7
3	10
4	
5	

Step 2 Check your work by using toothpicks to show 5 fence sections. Draw the result in the space below.

So, there will be ☐ planks in 5 sections of fencing.

Investigate

Collaborate

Work with a partner. Complete the table. You can use toothpicks to continue each pattern if needed.

1.

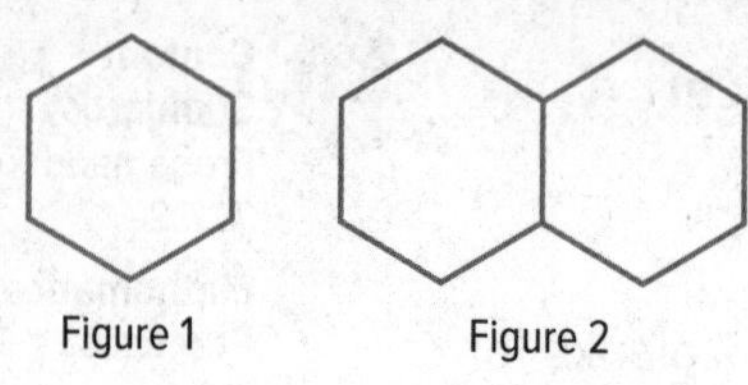

Figure 1 Figure 2

Figure 3

Figure Number	Number of Toothpicks
1	6
2	11
3	16
4	
5	

2. Refer to Exercise 1. Write an expression that could be used to find the number of toothpicks that would be needed for any figure.

3. Use your expression from Exercise 2 to find the number of toothpicks that would be needed to create Figure 10. Explain.

Create

On Your Own

4. MP **Reason Abstractly** Refer to the activity. Write an expression that could be used to find the number of planks in any number of sections.

5. MP **Justify Conclusions** Use the expression in Exercise 4 to find the number of planks that would be needed to create 10 sections of fencing. Explain.

6. Inquiry HOW can geometric figures be used to model numerical patterns?

Lesson 3

Properties of Operations

 Real-World Link

Driving Miss Ricardo drives up and down her street to complete different errands. Some of the places on her street are shown below. The number of blocks between the places are also shown.

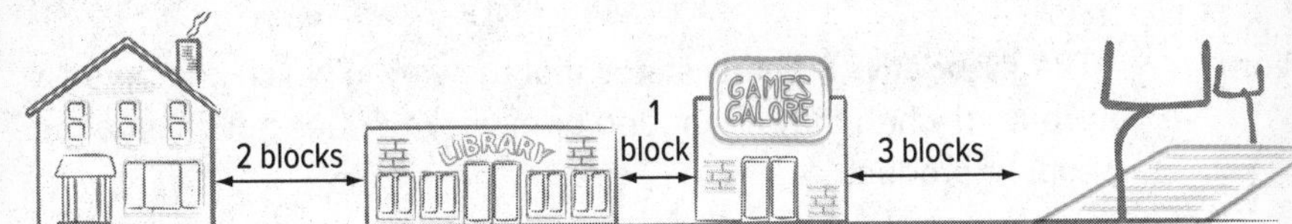

1. Suppose Miss Ricardo drives from home to the game store and back. Write an expression for each distance.

   from home to the game store: ______ from the game store to home: ______

2. Circle the property that is illustrated in Exercise 1.

   Commutative Associative

3. On Monday, Miss Ricardo drives from home, stops at the library, and then drives to the football field. On Tuesday, she drives from home, stops at the game store, and then drives to the football field. Write an expression for each distance.

   Monday: ______ Tuesday: ______

4. Circle the property that is illustrated in Exercise 3.

   Commutative Associative

Which MP Mathematical Practices did you use? Shade the circle(s) that applies.

① Persevere with Problems
② Reason Abstractly
③ Construct an Argument
④ Model with Mathematics
⑤ Use Math Tools
⑥ Attend to Precision
⑦ Make Use of Structure
⑧ Use Repeated Reasoning

 **Essential Question**

HOW can you use numbers and symbols to represent mathematical ideas?

 **Vocabulary**

Commutative Property
Associative Property
property
Additive Identity Property
Multiplicative Identity Property
Multiplicative Property of Zero
counterexample

Common Core State Standards

Content Standards
7.EE.1, 7.EE.2

MP Mathematical Practices
1, 3, 4, 5, 7

Key Concept

Properties of Operations

Watch

Work Zone

	Addition	Multiplication
Words	The **Commutative Property** states that the order in which numbers are added or multiplied does not change the sum or product.	
Symbols	$a + b = b + a$	$a \cdot b = b \cdot a$
Examples	$6 + 1 = 1 + 6$	$7 \cdot 3 = 3 \cdot 7$

	Addition	Multiplication
Words	The **Associative Property** states that the way in which numbers are grouped when they are added or multiplied does not change the sum or product.	
Symbols	$a + (b + c) = (a + b) + c$	$a \cdot (b \cdot c) = (a \cdot b) \cdot c$
Examples	$2 + (3 + 8) = (2 + 3) + 8$	$3 \cdot (4 \cdot 5) = (3 \cdot 4) \cdot 5$

A **property** is a statement that is true for any number. The following properties are also true for any numbers.

Property	Words	Symbols	Examples
Additive Identity	When 0 is added to any number, the sum is the number.	$a + 0 = a$ $0 + a = a$	$9 + 0 = 9$ $0 + 9 = 9$
Multiplicative Identity	When any number is multiplied by 1, the product is the number.	$a \cdot 1 = a$ $1 \cdot a = a$	$5 \cdot 1 = 5$ $1 \cdot 5 = 5$
Multiplicative Property of Zero	When any number is multiplied by 0, the product is 0.	$a \cdot 0 = 0$ $0 \cdot a = 0$	$8 \cdot 0 = 0$ $0 \cdot 8 = 0$

Example

Tutor

1. Name the property shown by the statement $2 \cdot (5 \cdot n) = (2 \cdot 5) \cdot n$.

The order of the numbers and variable did not change, but their grouping did. This is the Associative Property of Multiplication.

Got It? **Do these problems to find out.**

Show your work.

a. $42 + x + y = 42 + y + x$ **b.** $3x + 0 = 3x$

a. ____________

b. ____________

You may wonder if any of the properties apply to subtraction or division. If you can find a **counterexample**, an example that shows that a conjecture is false, the property does not apply.

Example

2. State whether the following conjecture is *true* or *false*. If *false*, provide a counterexample.

Division of whole numbers is commutative.

Write two division expressions using the Commutative Property.

$15 \div 3 \stackrel{?}{=} 3 \div 15$ State the conjecture.

$5 \neq \frac{1}{5}$ Divide.

The conjecture is false. We found a counterexample. That is, $15 \div 3 \neq 3 \div 15$. So, division is *not* commutative.

Got It? Do this problem to find out.

c. The difference of two different whole numbers is always less than both of the two numbers.

c. ______________

Example

3. Alana wants to buy a sweater that costs $38, sunglasses that costs $14, a pair of jeans that costs $22, and a T-shirt that costs $16. Use mental math to find the total cost before tax.

Write an expression for the total cost. You can rearrange the numbers using the properties of math. Look for sums that are multiples of ten.

$38 + 14 + 22 + 16$

$= 38 + 22 + 14 + 16$ Commutative Property of Addition

$= (38 + 22) + (14 + 16)$ Associative Property of Addition

$= 60 + 30$ Add.

$= 90$ Simplify.

The total cost of the items is $90.

Got It? Do this problem to find out.

d. Lance made four phone calls from his cell phone today. The calls lasted 4.7, 9.4, 2.3, and 10.6 minutes. Use mental math to find the total amount of time he spent on the phone.

d. ______________

Simplify Algebraic Expressions

To simplify an expression is to perform all possible operations.

Examples

Simplify each expression. Justify each step.

4. $(7 + g) + 5$

$(7 + g) + 5 = (g + 7) + 5$	Commutative Property of Addition
$= g + (7 + 5)$	Associative Property of Addition
$= g + 12$	Simplify.

5. $(m \cdot 11) \cdot m$

$(m \cdot 11) \cdot m = (11 \cdot m) \cdot m$	Commutative Property of Multiplication
$= 11 \cdot (m \cdot m)$	Associative Property of Multiplication
$= 11m^2$	Simplify.

Got It? **Do this problem to find out.**

e. $4 \cdot (3c \cdot 2)$

e. ______________

Guided Practice

Name the property shown by each statement. (Example 1)

1. $3m \cdot 0 \cdot 5m = 0$ ______________

2. $7c + 0 = 7c$ ______________

3. State whether the following conjecture is *true* or *false*. If *false*, provide a counterexample. (Example 2)

Subtraction of whole numbers is associative.

____

4. Simplify $9c + (8 + 3c)$. Justify each step. (Examples 3–5)

____

5. **Building on the Essential Question** Explain the difference between the Commutative and Associative Properties. ______________

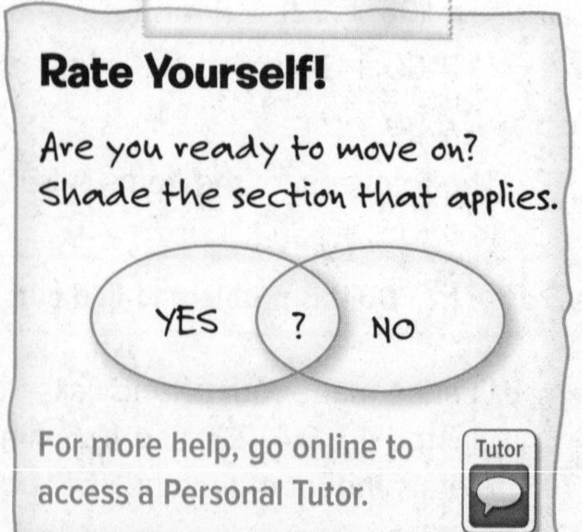

Name ______________________ My Homework ______________________

Independent Practice

Go online for Step-by-Step Solutions 

Name the property shown by each statement. (Example 1)

1. $a + (b + 12) = (b + 12) + a$

2. $(5 + x) + 0 = 5 + x$

3 $16 + (c + 17) = (16 + c) + 17$

4. $d \cdot e \cdot 0 = 0$

5. MP **Use a Counterexample** State whether the conjecture is *true* or *false*. If *false*, provide a counterexample. (Example 2)

Division of whole numbers is associative.

6. Darien ordered a soda for $2.75, a sandwich for $8.50, and a dessert for $3.85. Sales tax was $1.15. Use mental math to find the total amount of the bill. Explain. (Example 3)

Simplify each expression. Justify each step. (Examples 4 and 5)

7. $15 + (12 + 8a)$

8. $(5n \cdot 9) \cdot 2n$

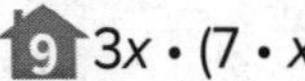

9 $3x \cdot (7 \cdot x)$

10. $(4m \cdot 2) \cdot 5m$

11. Simplify the expression $(7 + 47 + 3)[5 \cdot (2 \cdot 3)]$. Use properties to justify each step.

H.O.T. Problems Higher Order Thinking

12. **MP Model with Mathematics** Write about something you do every day that is commutative. Then write about another situation that is not commutative.

13. **MP Find the Error** Blake is simplifying $4 \cdot (5 \cdot m)$. Find his mistake and correct it.

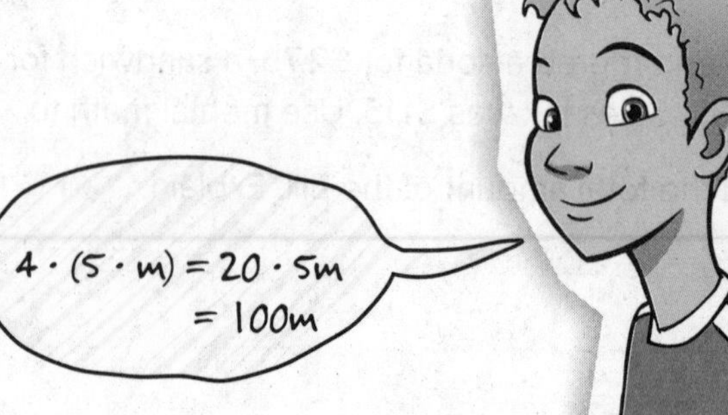

14. **MP Identify Structure** Does the Associative Property *always*, *sometimes*, or *never* hold for subtraction? Explain your reasoning using examples and counterexamples.

15. **MP Persevere with Problems** If you take any two whole numbers and add them together, the sum is always a whole number. This is the Closure Property for Addition. The set of whole numbers is *closed* under addition.

    a. Is the set of whole numbers closed under subtraction? If not, give a counterexample.

    b. Suppose you had a very small set of numbers that contained only 0 and 1. Would this set be closed under addition? If not, give a counterexample.

Name ______________________ My Homework ______________________

Extra Practice

Name the property shown by each statement.

16. $9(ab) = (9a)b$

Associative (×)

17. $y \cdot 7 = 7y$

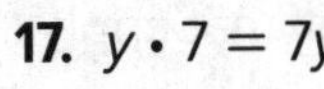

18. $1 \times c = c$

19. $5 + (a + 8) = (5 + a) + 8$

20. State whether the conjecture is *true* or *false*. If *false*, provide a counterexample.

Subtraction of whole numbers is commutative.

21. MP **Use Math Tools** The times for each leg of a relay for four runners are shown. Use mental math to find the total time for the relay team. Explain.

Runner	Time (s)
Jamal	12.4
Kenneth	11.8
Bryce	11.2
Jorge	12.6

Simplify each expression. Justify each step.

22. $(22 + 19b) + 7$

23. $18 + (5 + 6m)$

24. $11s(4)$

25. $10y(7)$

26. $(9 + 31 + 5)[(7 \cdot 5) \cdot 4]$

Power Up! Common Core Test Practice

27. The table shows the cost of different items at a bakery. Yolanda buys 2 doughnuts, a muffin, and 2 cookies. Which of the following expressions represents the total cost? Select all that apply.

Item	Cost ($)
Cookie	2.21
Doughnut	2.29
Muffin	2.50
Roll	1.15

- ☐ 2(2.29) + 2(2.21) + 2.50
- ☐ 2(2.29) + 2.50 + 2(2.21)
- ☐ 2(2.29 + 2.21 + 2.50)
- ☐ 2.50 + 2(2.21 + 2.29)

28. Determine if the two expressions in each pair are equivalent. If they are equivalent, select the property that is illustrated.

- Commutative Property
- Identity Property
- Associative Property
- Multiplicative Property of Zero

	Equivalent?	Property
$9 \cdot 4 \div 20 = 9 \cdot 20 \div 4$		
$3b \cdot 0 \cdot c = 0$		
$35 + 2m + n = 35 + n + 2m$		
$12t \cdot 3v + 0 = 12t \cdot 3v$		

Common Core Spiral Review

Evaluate each expression if $a = 6$, $b = 15$, and $c = 9$. 6.EE.2

29. $a + 2b$ ______

30. $c^2 - 5$ ______

31. $10 + a^3$ ______

32. $8c - 9 + 25$ ______

33. $14 + 8b \div 2$ ______

34. $3^3 \div (3a)$ ______

35. A package of pencils costs $1.25. A new eraser costs $0.45. Write an expression to find the total cost of 3 packages of pencils and 2 erasers. Then find the total cost. 6.EE.2

______

Lesson 4

The Distributive Property

Real-World Link

School Supplies Jordan buys three notebooks that cost $5 each. He also buys three packages of pens for $6 each.

1. Write an expression that shows the cost of three notebooks added to the cost of three packages of pens.

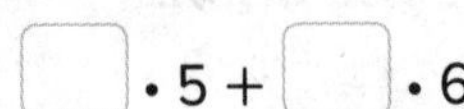

☐ • 5 + ☐ • 6

2. Write an expression that shows three times the cost of one notebook and one package of pens.

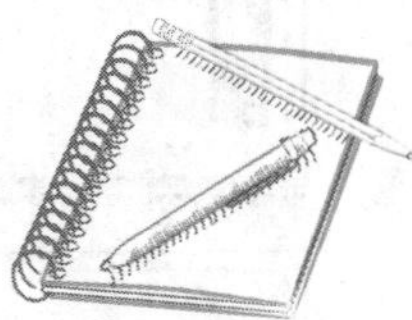

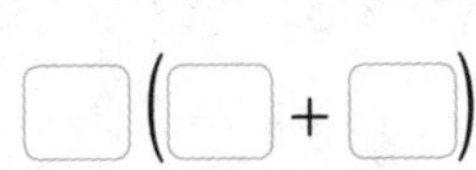

☐(☐ + ☐)

3. Evaluate both expressions. What do you notice?

4. Suppose Jordan buys five notebooks that cost $3 each and five packages of pens that cost $1 each. Circle the expressions that represent Jordan's purchases.

5 • 3 + 5 • 1 　　 5 • 3 • 5 • 1 　　 5(3 + 1)

5. Suppose Jordan buys two rulers that cost $1 each and two folders that cost $1.50 each. Circle the expressions that represent Jordan's purchases.

2 + 1 + 2 + 1.50 　　 2(1 + 1.50) 　　 2 • 1 + 2 • 1.50

Essential Question

HOW can you use numbers and symbols to represent mathematical ideas?

Vocabulary

Distributive Property
equivalent expressions

Common Core State Standards

Content Standards
7.EE.1, 7.EE.2

MP Mathematical Practices
1, 3, 4, 5, 7

Which MP Mathematical Practices did you use? Shade the circle(s) that applies.

① Persevere with Problems
② Reason Abstractly
③ Construct an Argument
④ Model with Mathematics
⑤ Use Math Tools
⑥ Attend to Precision
⑦ Make Use of Structure
⑧ Use Repeated Reasoning

Key Concept

Use the Distributive Property

Work Zone

Words The **Distributive Property** states that to multiply a sum or difference by a number, multiply each term inside the parentheses by the number outside the parentheses.

Symbols $a(b + c) = ab + ac$ $a(b - c) = ab - ac$

Examples $4(6 + 2) = 4 \cdot 6 + 4 \cdot 2$ $3(7 - 5) = 3 \cdot 7 - 3 \cdot 5$

You can model the Distributive Property with algebraic expressions using algebra tiles. The expression $2(x + 2)$ is modeled below.

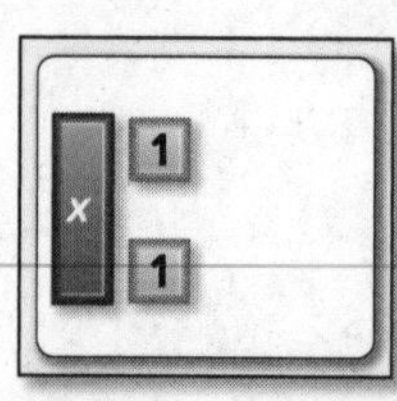

Model $x + 2$ using algebra tiles.

Double the amount of tiles to represent $2(x + 2)$.

Rearrange the tiles by grouping together the ones with the same shapes.

$2(x + 2) = 2(x) + 2(2)$ Distributive Property

$= 2x + 4$ Multiply.

The expressions $2(x + 2)$ and $2x + 4$ are **equivalent expressions**. No matter what x is, these expressions have the same value.

Example

Show your work.

1. Use the Distributive Property to evaluate 8(−9 + 4).

$8(-9 + 4) = 8(-9) + 8(4)$ Expand using the Distributive Property.

$= -72 + 32$ or -40 Multiply. Then add.

Got It? **Do these problems to find out.**

a. $5(-9 + 11)$ **b.** $7(10 - 5)$ **c.** $(12 - 8)9$

a. __________

b. __________

c. __________

Examples

Use the Distributive Property to rewrite each expression.

2. $4(x + 7)$

$4(x + 7) = 4(x) + 4(7)$ Expand using the Distributive Property.

$= 4x + 28$ Simplify.

3. $6(p - 5)$

$6(p - 5) = 6[p + (-5)]$ Rewrite $p - 5$ as $p + (-5)$.

$= 6(p) + 6(-5)$ Expand using the Distributive Property.

$= 6p + (-30)$ Simplify.

$= 6p - 30$ Definition of subtraction

4. $-2(x - 8)$

$-2(x - 8) = -2[x + (-8)]$ Rewrite $x - 8$ as $x + (-8)$.

$= -2(x) + -2(-8)$ Expand using the Distributive Property.

$= -2x + 16$ Simplify.

5. $5(-3x + 7y)$

$5(-3x + 7y) = 5(-3x) + 5(7y)$ Expand using the Distributive Property.

$= -15x + 35y$ Simplify.

6. $\frac{1}{3}(x - 6)$

$\frac{1}{3}(x - 6) = \frac{1}{3}[x + (-6)]$ Rewrite $x - 6$ as $x + (-6)$.

$= \frac{1}{3}(x) + \left(\frac{1}{3}(-6)\right)$ Expand using the Distributive Property.

$= \frac{1}{3}x + (-2)$ Simplify.

$= \frac{1}{3}x - 2$ Definition of subtraction

Got It? Do these problems to find out.

d. $6(a + 4)$

e. $(m + 3n)8$

f. $-3(y - 10)$

g. $\frac{1}{2}(w - 4)$

d. ____________

e. ____________

f. ____________

g. ____________

Example

7. Mr. Ito needs to buy batting helmets for the baseball team. The helmets he plans to buy are $19.95 each. Find the total cost if Mr. Ito needs to buy 9 batting helmets for the team.

Rename $19.95 as $20.00 − $0.05. Then use the Distributive Property to find the total cost mentally.

9($20.00 − $0.05) = 9($20.00) − 9($0.05)	Distributive Property
= $180 − $0.45	Multiply.
= $179.55	Subtract.

The total cost of the helmets is $179.55.

Show your work.

Got It? Do this problem to find out.

h. A sports club rents dirt bikes for $37.50 each. Find the total cost for the club to rent 20 bikes. Justify your answer by using the Distributive Property.

h. ______

Guided Practice

Use the Distributive Property to evaluate or rewrite each expression. (Examples 1–6)

1. $(8 + 11)(-3) =$ ______

2. $-5(2x + 4y) =$ ______

3. $\frac{1}{5}(g - 10) =$ ______

4. A housefly can fly about 6.4 feet per second. At this rate, how far can it fly in 25 seconds? Justify your answer by using the Distributive Property. (Example 7)

______

5. **Building on the Essential Question** Describe how the formula to find the perimeter of a rectangle is an application of the Distributive Property. ______

Rate Yourself!

How confident are you about the Distributive Property? Check the box that applies.

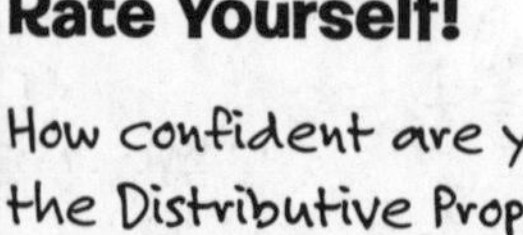

For more help, go online to access a Personal Tutor.

Name ______________________ My Homework ______________________

Independent Practice

Go online for Step-by-Step Solutions

eHelp 

Use the Distributive Property to evaluate each expression. (Example 1)

1. $3(5 + 6) =$ ______

2. $(6 + 4)(-12) =$ ______

3. $-6(9 - 4) =$ ______

4. $5(-6 + 4) =$ ______

5. $4(8 - 7) =$ ______

6. $(5 - 7)(-3) =$ ______

MP Identify Structure **Use the Distributive Property to rewrite each expression.** (Examples 2–6)

7. $3(-4x + 8) =$ ______

8. $4(x - 6y) =$ ______

9. $6(5 - q) =$ ______

10. $\frac{1}{2}(c - 8) =$ ______

11. $-3(5 - b) =$ ______

12. $(d + 2)(-7) =$ ______

13. Amelia bought roast beef for \$6.85 per pound. Find the total cost if Amelia bought 4 pounds of roast beef. Justify your answer by using the Distributive Property. (Example 7)

____

14. The table shows the different prices of items at a movie theater.

Movie Theater Prices	
Item	**Cost (\$)**
box of candy	2.25
drink	3.25
popcorn	4.50
ticket	7.50

a. Suppose Mina and two of her friends go to the movies. Write an expression that could be used to find the total cost for them to go to the movies and buy one of each item.

____

b. What is the total cost for all three people?

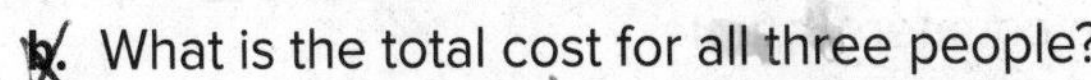

MP Use Math Tools Find each product mentally. Justify your answer.

15. $9 \cdot 35 =$ ______

16. $8 \cdot 28 =$ ______

17. $112 \cdot 6 =$ ______

18. $85 \cdot 8 =$ ______

19. $4 \cdot 122 =$ ______

20. $12 \cdot 64 =$ ______

H.O.T. Problems Higher Order Thinking

21. **MP Reason Abstractly** Write an expression that when using the Distributive Property can be simplified to $12a + 18b - 6c$.

______

22. **MP Identify Structure** Use the Distributive Property to rewrite the expression $7bx + 7by$ as an equivalent expression.

______

23. **MP Persevere with Problems** Use the Distributive Property to write an equivalent expression for the expression $(a + b)(2 + y)$.

______

24. **MP Find the Error** Julia is using the Distributive Property to simplify $3(x + 2)$. Find her mistake and correct it.

correct answer: 3x+6

$3(x + 2) = 3x + 2$

3(x+2)
3·x+3·2
3x+6

Julia was suppose to do 3·x & 3·2. However instead she only did the 3·x and carried the +2.

25. **MP Persevere with Problems** Is $3 + (x \cdot y) = (3 + x) \cdot (3 + y)$ a true statement? If so, explain your reasoning. If not, give a counterexample.

Yes, because when you do distribute property it's (3+x)·(3+y). Also when you check it gives you the same answer.

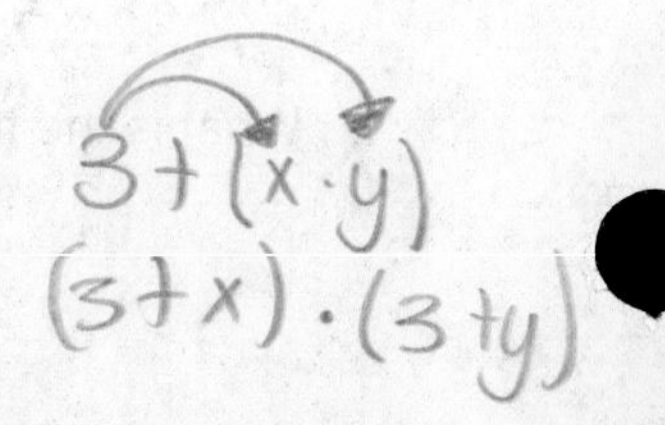

NO!

Name ______________________ My Homework ______________________

Extra Practice

Use the Distributive Property to evaluate each expression.

26. $(3 + 6)(-8) =$ -72

$3 \cdot (-8) + 6(-8) =$

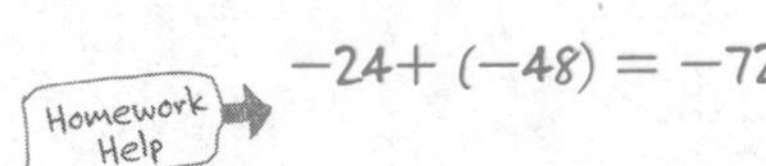

$-24 + (-48) = -72$

Homework Help

27. $4(11 - 5) =$ ________

28. $(12 - 4)(-5)$ ________

Use the Distributive Property to rewrite each expression.

29. $-8(a + b) =$ ________

30. $(2b + 8)5 =$ ________

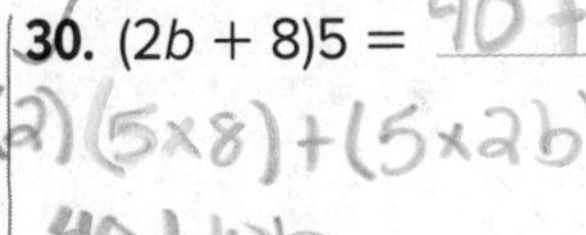

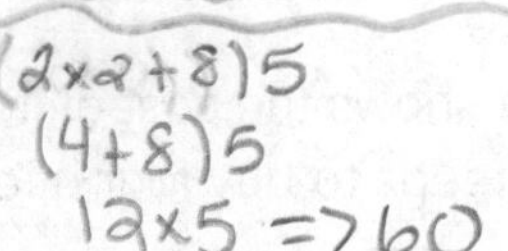

31. $(p + 7)(-2) =$ ________

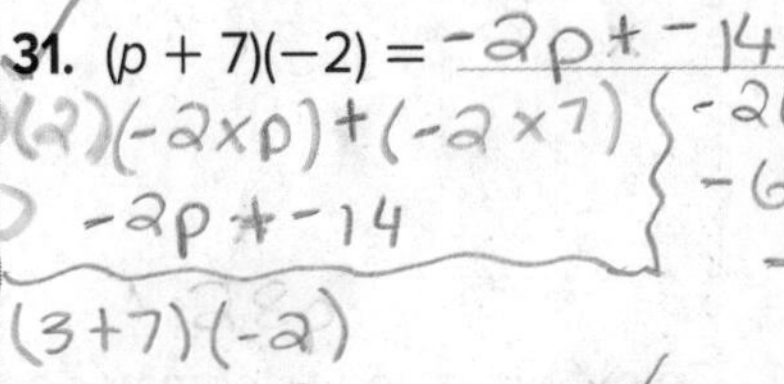

32. MP **Justify Conclusions** Theresa is planning on making a fleece blanket for her nephew. She learns that the fabric she wants to use is $7.99 per yard. Find the total cost of 4 yards of fabric. Justify your answer by using the Distributive Property.

33. You are ordering T-shirts with your school's mascot printed on them. Each T-shirt costs $4.75. The printer charges a setup fee of $30 and $2.50 to print each shirt. Write two expressions to represent the total cost of printing *n* T-shirts.

Use the Distributive Property to rewrite each expression.

34. $0.5x(y - z)$

= ________

35. $-6a(2b + 5c)$

= ________

36. $-4m(3n - 6p)$

= ________

37. $3(2y + 4z)$

= ________

38. $-2(3a - 2b)$

= ________

39. $-6(12p - 8n)$

= ________

40. Write two equivalent expressions for the area of the figure.

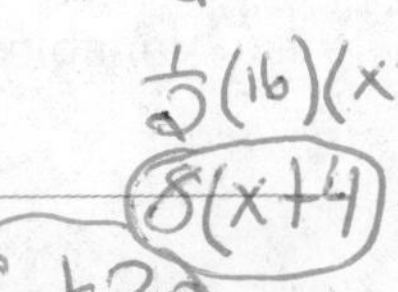

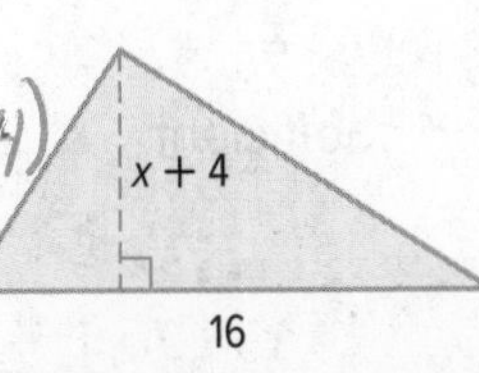

CCSS Power Up! Common Core Test Practice

41. A group of 3 seniors, 3 adults, and 3 children bought tickets to the aquarium.

Type of Ticket	Cost ($)
Adult	18.95
Senior	14.95
Child	9.95

Fill in the boxes to model the total amount spent with an expression.

[3] × ([18.95] + [14.95] + [9.95])

How much did the group spend on tickets altogether? How does applying the Distributive Property make it easier to find this amount?

(3×18.95)+(3×14.95)+(3×9.95)
56.85+44.85+29.85 = $131.55 ← spent altogether
Applying the distributive property makes it easier because I don't have to go through all the math. I only have to do the multiplacation once.

42. Celeste is going to summer camp. The table shows the cost of items she will need to purchase with the camp logo. She needs to buy four of each item.

Item	Cost ($)
T-shirt	8.00
Shorts	4.50
Socks	2.25

Which of the following expressions represents the total cost of the items? Select all that apply.

- [x] 4(14.75)
- [ ] 4(8.00) + 4.50 + 2.25
- [x] 4(8) + 4(4.50) + 4(2.25)
- [x] 4(8.00 + 4.50 + 2.25)

CCSS Common Core Spiral Review

Evaluate each expression if $x = 9$ and $y = 3$. 6.EE.2c

43. $x + y - 58$ ______

44. $y^3 + x^3$ ______

45. $y^4 - 128 =$ ______

46. In the expression below, identify the coefficient and the variable. 6.EE.2

$4x + 450$

coefficient: ______ variable: ______

Problem-Solving Investigation
Make a Table

Content Standards
7.NS.3

Mathematical Practices
1, 3, 4

Case #1 Mountain Biking

Hoshi wants to purchase a membership to a bike park. The cost depends on the number of people on the membership. It costs $55 for 5 people, $65 for 6 people, and $75 for 7 people.

Find the cost of a membership that includes 8 people.

Understand *What are the facts?*

The cost of a membership depends on the number of people included on the membership.

Plan *What is your strategy to solve this problem?*

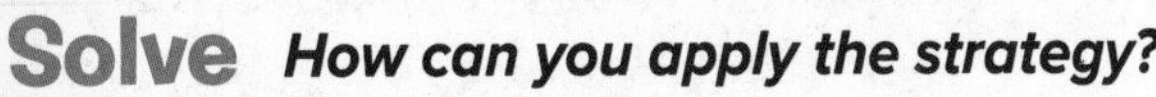

Make a table that shows the number of people and the cost.

Solve *How can you apply the strategy?*

Make a table. Find the cost for 8 people.

Number of People (p)	Cost
5	$55
6	$65
7	$75
8	

+10, +10, +10

So, the cost for 8 people is ______.

Check *Does the answer make sense?*

The expression $10p + 5$ can be used to represent the situation.

Since $\$10(8) + \$5 = \$85$, the solution is reasonable.

Analyze the Strategy

Justify Conclusions Hoshi wants to purchase a membership for four people. Explain how the table would change and then solve.

Case #2 Financial Literacy

Latoya is saving money to buy a saxophone. After 1 month, she has $75. After 2 months, she has $120. After 3 months, she has $165. She plans to keep saving at the same rate.

How long will it take Latoya to save enough money to buy a saxophone that costs $300?

Understand

Read the problem. What are you being asked to find?

I need to find ______________________________

______________________________.

Underline key words and values. What information do you know?

After 1 month, Latoya has ☐. After 2 months, she has ☐.

After 3 months, she has ☐. She continues to save at the same rate.

Is there any information that you do *not* need to know?

I do not need to know ______________________________.

2 Plan

Choose a problem-solving strategy.

I will use the ______________________________ strategy.

Solve

Use your problem-solving strategy to solve the problem.

Months	1	2	3	4	5	6
Amount Saved ($)	75	120	165			

+45 +45 +45 +45 +45

Latoya will have $300 saved in ______________.

Check

Use information from the problem to check your answer.

____

____

Collaborate

Work with a small group to solve the following cases. Show your work on a separate piece of paper.

Case #3 Carnivals

For a carnival game, containers are arranged in a triangular display. The top row has 1 container. The second row has 2 containers. The third row has 3 containers. The pattern continues until the bottom row, which has 10 containers.

A contestant knocks down 29 containers on the first throw. How many containers remain?

Case #4 Budget

Tamara earns \$2,050 each month. She spends 65% of the amount she earns. The rest of the money is equally divided and deposited into two separate accounts.

How many months until Tamara has deposited more than \$2,500 in one of her accounts?

Case #5 Toothpicks

Write an expression that can be used to find the number of toothpicks needed to make any figure. Then find the number of toothpicks needed to make the eighth figure.

Figure 1 Figure 2 Figure 3

Case #6 Diving

A diver descends to −15 feet after 1 minute, −30 feet after 2 minutes, and −45 feet after 3 minutes.

If the diver keeps descending at this rate, what is their position after 12 minutes?

Mid-Chapter Check

Vocabulary Check

1. Fill in the blank in the sentence below with the correct term. (Lesson 1)

   A ____________________ is a symbol that represents an unknown quantity.

2. Define *arithmetic sequence*. Then provide an example. (Lesson 2)

Skills Check and Problem Solving

Describe the relationship between the terms in each arithmetic sequence. Then write the next three terms in each sequence. (Lesson 2)

3. 5, 8, 11, 14, ...

4. 4, 11, 18, 25, ...

5. 5.8, 10.8, 15.8, 20.8, ...

Use the Distributive Property to rewrite each expression. (Lesson 4)

6. $4(x + 9) =$ ______

7. $2(x + 5) =$ ______

8. $3(-2x + 4) =$ ______

9. **MP Identify Structure** What property is shown by the statement $8x + 0 = 8x$? (Lesson 3)

10. **MP Persevere with Problems** A coach bought some baseball bats and five baseball gloves. Let b represent the number of bats. Write an expression that can be used to find the total cost of the bats and gloves. Then find the total cost if he bought three bats. (Lesson 1)

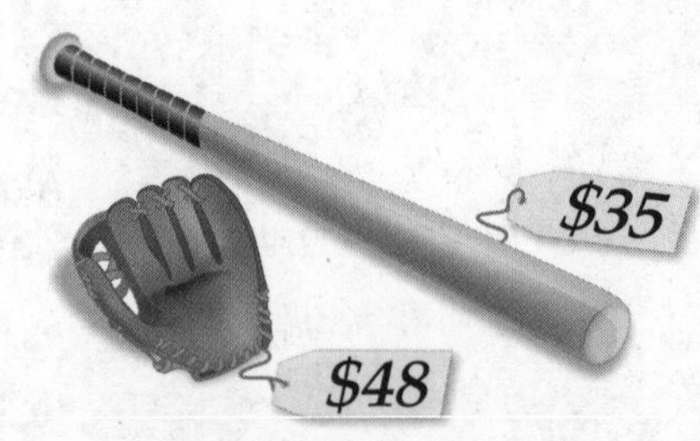

Lesson 5

Simplify Algebraic Expressions

Real-World Link

Music Store Patricia, Hugo, and Sun work at a music store. Each week, Patricia works three more than twice the number of hours that Hugo works. Sun works 2 less hours than Hugo.

1. Let x represent the number of hours that Hugo works each week. The number of hours that Hugo, Patricia, and Sun work can be modeled as shown below. Write an expression that represents each person's number of hours.

Hugo's hours

Patricia's hours

Sun's hours

Expression: ________ Expression: ________ Expression: ________

2. Model the total number of hours that Patricia and Sun work. Draw the result below. Then write an expression for the drawing.

Expression:

____

3. Like tiles are tiles that have the same shape. Group like tiles together and remove the zero pairs. Draw the result below. Then write an expression for your drawing.

Expression:

____

Which MP Mathematical Practices did you use?
Shade the circle(s) that applies.

① Persevere with Problems
② Reason Abstractly
③ Construct an Argument
④ Model with Mathematics
⑤ Use Math Tools
⑥ Attend to Precision
⑦ Make Use of Structure
⑧ Use Repeated Reasoning

Essential Question

HOW can you use numbers and symbols to represent mathematical ideas?

Vocabulary

term
like terms
constant
simplest form

Common Core State Standards

Content Standards
7.EE.1, 7.EE.2

MP Mathematical Practices
1, 2, 3, 4, 6

Work Zone

Identify Parts of an Expression

When addition or subtraction signs separate an algebraic expression into parts, each part is called a **term**. Recall that the numerical factor of a term that contains a variable is called the coefficient of the variable.

Like terms contain the same variables to the same powers. For example, $3x^2$ and $-7x^2$ are like terms. So are $8xy^2$ and $12xy^2$. But $10x^2z$ and $22xz^2$ are *not* like terms. A term without a variable is called a **constant**. Constant terms are also like terms.

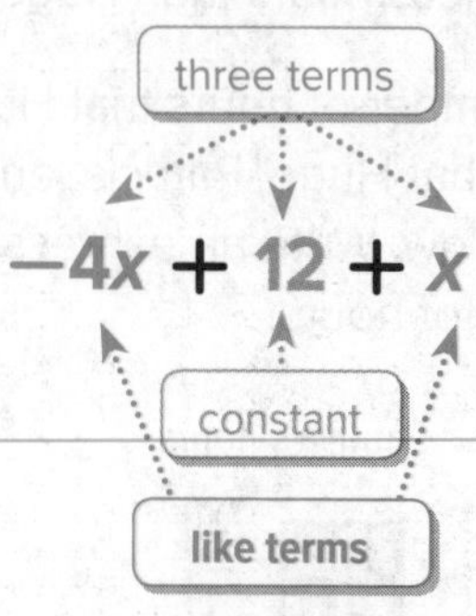

STOP and Reflect

Circle the term below that is a like term with $-4x^3$.

$-4x^2$ x^3 -4

Example

1. Identify the terms, like terms, coefficients, and constants in the expression $6n - 7n - 4 + n$.

$6n - 7n - 4 + n = 6n + (-7n) + (-4) + 1n$ Rewrite the expression.

- Terms: $6n, -7n, -4, n$
- Like terms: $6n, -7n, n$ All of these terms have the same variable.
- Coefficients: $6, -7, 1$
- Constants: -4 This is the only term without a variable.

Got It? Do these problems to find out.

Identify the terms, like terms, coefficients, and constants in each expression.

a. $9y - 4 - 11y + 7$ **b.** $3x + 2 - 10 - 3x$

Show your work.

a. ____________

b. ____________

Simplify Algebraic Expressions

An algebraic expression is in **simplest form** if it has no like terms and no parentheses. Use the Distributive Property to combine like terms.

Examples

2. Write $4y + y$ in simplest form.

$4y$ and y are like terms.

$4y + y = 4y + 1y$ — Identity Property; $y = 1y$

$= (4 + 1)y$ or $5y$ — Distributive Property; Simplify.

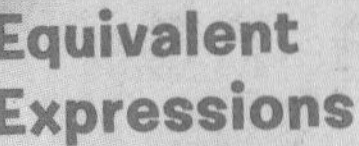

Equivalent Expressions

To check whether $4y + y$ and $5y$ are equivalent expressions, substitute any value for y and see whether the expressions have the same value.

3. Write $7x - 2 - 7x + 6$ in simplest form.

$7x$ and $-7x$ are like terms. -2 and 6 are also like terms.

$7x - 2 - 7x + 6 = 7x + (-2) + (-7x) + 6$ — Definition of subtraction

$= 7x + (-7x) + (-2) + 6$ — Commutative Property

$= [7 + (-7)]x + (-2) + 6$ — Distributive Property

$= 0x + 4$ — Simplify.

$= 0 + 4$ or 4 — Multiplicative Property of zero and Additive Identity Property of zero.

Got It? **Do these problems to find out.**

c. $4z - z$ d. $6 - 3n + 3n$ e. $2g - 3 + 11 - 8g$

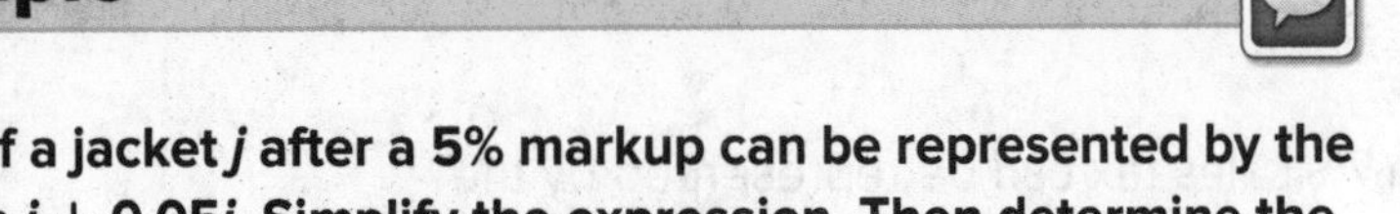

c. ________

d. ________

e. ________

Real World Example

4. The cost of a jacket j after a 5% markup can be represented by the expression $j + 0.05j$. Simplify the expression. Then determine the total cost of the jacket after the markup, if the original price is \$35.

$j + 0.05j = 1j + 0.05j$ — Identity Property; $j = 1j$

$= (1 + 0.05)j$ — Distributive Property

$= 1.05j$ — Simplify.

$1.05j = 1.05(35)$ — Replace j with 35 to find the total cost.

$= 36.75$ — Multiply.

So, the cost of the jacket after a 5% markup is \$36.75.

Got It? **Do this problem to find out.**

f. Write an expression in simplest form for the cost of the jacket in Example 4 if the markup is 8%. Then determine the total cost after the markup.

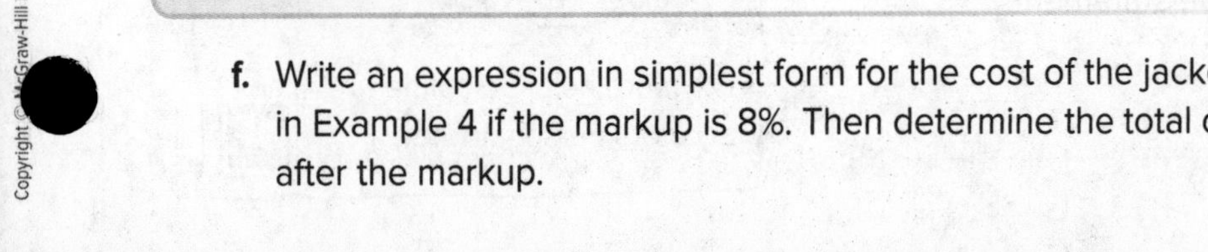

f. ________

Example

5. At a concert, you buy some T-shirts for $12.00 each and the same number of CDs for $7.50 each. Write an expression in simplest form that represents the total amount spent.

Let x represent the number of T-shirts and CDs.

$12x + 7.50x$ — Write the expression.

$12x + 7.50x = (12 + 7.50)x$ — Distributive Property

$= 19.50x$ — Simplify.

The expression $19.50x represents the total amount spent.

Got It? Do this problem to find out.

g. You have some money. Your friend has $50 less than you. Write an expression in simplest form that represents the total amount of money you and your friend have.

g. ______

Guided Practice

1. Identify the terms, like terms, coefficients, and constants in $5n - 2n - 3 + n$. (Example 1)

2. Write $4p - 7 + 6p + 10$ in simplest form. (Examples 2 and 3)

3. The cost of a game g with 7% sales tax can be represented by the expression $g + 0.07g$. Simplify the expression. Then determine the total cost of the game after sales tax if the original price is $52. (Example 4)

4. You go to a basketball game and buy 3 waters that cost x dollars each. Your brother buys a bottle of water and a bag of peanuts that costs $4.50. Write an expression in simplest form that represents the total amount of money spent altogether. (Example 5)

5. **Building on the Essential Question** Explain why $2(x - 1) + 3(x - 1) = 5(x - 1)$ is a true statement.

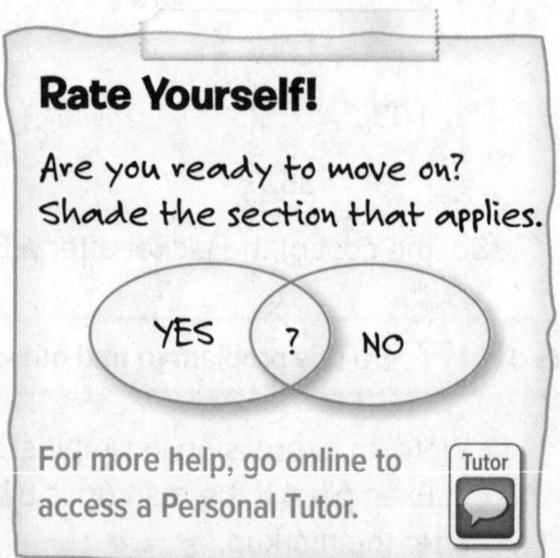

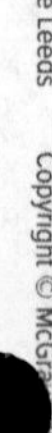

Name ________________________ My Homework ________________

Independent Practice

Go online for Step-by-Step Solutions

Identify the terms, like terms, coefficients, and constants in each expression. (Example 1)

1. $2 + 3a + 9a$

2. $7 - 5x + 1$

3. $9 - z + 3 - 2z$

Write each expression in simplest form. (Examples 2 and 3)

4. $n + 5n =$ ______

5. $12c - c =$ ______

6. $-4j - 1 - 4j + 6 =$ ______

7. The cost of a ticket t to a concert with a 3% sales tax can be represented by the expression $t + 0.03t$. Simplify the expression. Then determine the total cost after the sales tax if the original price is $72. (Example 4)

____

Write an expression in simplest form that represents the total amount in each situation. (Example 5)

8. You rent x pairs of shoes for $2 each. You buy the same number of drinks for $1.50 each. You also pay $9 for a bowling lane.

____

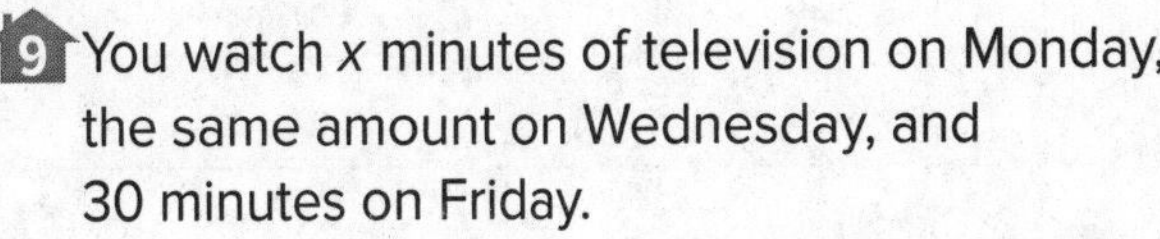

9 You watch x minutes of television on Monday, the same amount on Wednesday, and 30 minutes on Friday.

____

10. In a State Legislature, there were 119 more members in the House of Representatives than in the Senate. If there were m members in the Senate, write an expression to represent the total members in the State Legislature. ______________________

11 Elian and his friends paid a total of $7 for tickets to the school football game. While at the game, they bought 5 hot dogs at x dollars each, 4 boxes of popcorn at y dollars each, and 2 pretzels at z dollars each.

a. Write an expression to show the total cost of admission and the snacks.

____

b. Hot dogs cost $4, popcorn cost $3, and pretzels cost $2. What was the total cost for admission and snacks? ______________________

MP Reason Abstractly **Write an expression in simplest form for the perimeter of each figure.**

12.

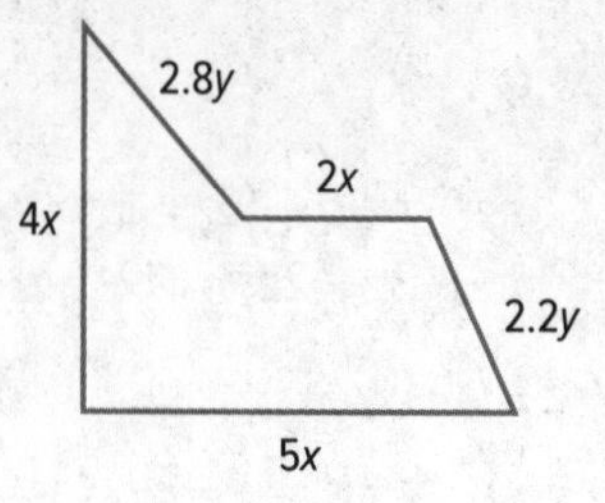

13.

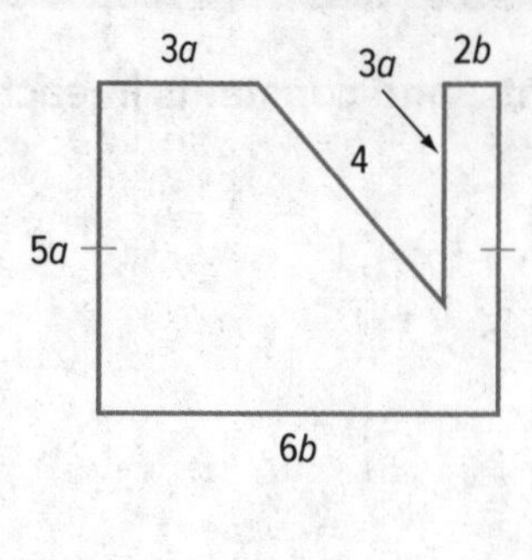

14.

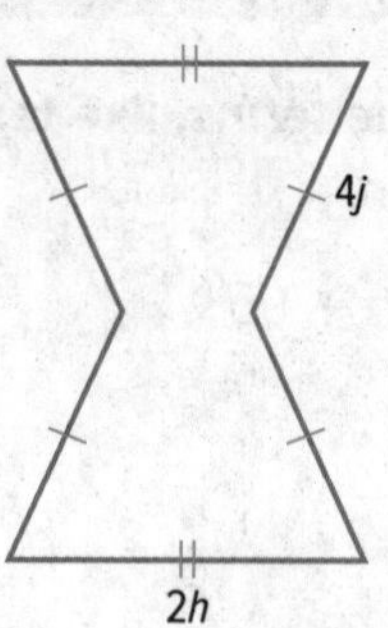

H.O.T. Problems Higher Order Thinking

15. **MP Be Precise** Write an expression that has three terms and simplifies to $4x - 7$. Identify the coefficient(s) and constant(s) in your expression.

16. **MP Which One Doesn't Belong?** Identify the expression that is not equivalent to the other three. Explain your reasoning.

$x - 2 + 3x$	$4(x - 2)$	$-2 + 7x - 3x$	$4x - 2$

17. **MP Persevere with Problems** Simplify the expression $8x - 2x + 12x - 3$. Show that your answer is true for $x = 2$.

18. **MP Justify Conclusions** Determine whether the following statement is *always*, *sometimes*, or *never* true. Explain your reasoning.

When using the Distributive Property, if the term outside the parentheses is negative, then the sign of each term inside the parentheses will change.

Name ______ My Homework ______

Extra Practice

Identify the terms, like terms, coefficients, and constants in each expression.

19. $4 + 5y - 6y + y$

terms: 4, 5y, −6y, y

Homework Help

like terms: 5y, −6y, y

coefficients: 5, −6, 1

constant: 4

20. $n + 4n - 7n - 1$

n+4n+⁻7n+⁻1

⁻2n+⁻1

21. $-3d + 8 - d - 2$

-3d+⁻d +8+⁻2

-4d+6

Write each expression in simplest form.

22. $5x + 4 + 9x$

= 14x+4

5x+9x+4

14x+4

23. $2 + 3d + d$

= 4d+2

3d+d+2

4d+2

24. $-3r + 7 - 3r - 12$

= ______

Write an expression in simplest form that represents the total amount in each situation.

25. You subscribe to *m* different magazines. Your friend subscribes to 2 fewer than you.

______

26. Today is your friend's birthday. She is *y* years old. Her brother is 5 years younger.

______

27. You spent *m* minutes studying on Monday. On Tuesday, you studied 15 more minutes than you did on Monday. Wednesday, you studied 30 minutes less than you did on Tuesday. You studied twice as long on Thursday as you did on Monday. On Friday, you studied 20 minutes less than you did on Thursday. Write an expression in simplest form to represent the number of minutes you studied in all.

______

28. MP **Reason Abstractly** Write a real-world situation for $7.50y + 9$.

______

Simplify each expression.

29. $3(4x - 5) + 4(2x + 6)$

= ______

30. $-8(2a - 3b) - 5(6b - 4a)$

= ______

31. $10(5g + 2h - 3) - 4(3g - 4h + 2)$

= ______

Power Up! Common Core Test Practice

32. Vince, Neal, and Patrick collect baseball cards. Neal has 3 fewer cards than twice the number of cards Vince has. Patrick has 5 more baseball cards than Vince. Let x represent the number of baseball cards that Vince has. Use the algebra tiles to represent the number of cards each person has.

	Vince	Neal	Patrick
Model			
Expression			

Write an expression, in simplified form, for the number of baseball cards the three friends have altogether.

33. The table shows the number of tickets needed and the number of times Talia participated in different activities at a carnival. Write an expression, in simplified form, for the total number of tickets that Talia used.

Activity	Tickets	Times Completed
Balloon Pop	3	a
Dunk Tank	4	b
Ring Toss	2	a
Trampoline	5	b

Common Core Spiral Review

34. Mica spends \$5 for her lunch and \$2 for breakfast each day Monday through Friday. Use the Associative Property to find how much money she spends on lunch and breakfast for 4 weeks. **7.EE.1**

Define a variable. Then write each phrase as an algebraic expression. **6.EE.2**

35. Anna has volunteered 9 more hours than Tricia

36. the cost of a pair of jeans is 4 times the cost of a book

Evaluate each expression if $x = 2$, $y = 10$, and $z = 4$. **6.EE.2c**

37. $5z - 10$

38. $y \div 2 + x$

39. $x^3 + (y \div x)$

Lesson 6
Add Linear Expressions

Real-World Link

Tools

Homework Luke has 20 math problems and 11 science questions for homework. Cameron has 23 math problems and 10 science questions for homework.

1. The expression below represents the types of exercises that Luke has for homework.

   20 math problems + 11 science questions

   Complete the expression that represents the types of exercises that Cameron has for homework.

   ☐ math problems + ☐ science questions

2. Write an expression for the total number of math problems and science questions for both boys.

   ☐ math problems + ☐ science questions

3. Suppose Luke has x math problems and 5 science questions for homework and Cameron has x math problems and 6 science questions. The algebra tiles below represent the total number of math problems and science questions for both boys. Write an expression in simplest form that represents the algebra tiles.

Expression: ______________________

Which MP Mathematical Practices did you use? Shade the circle(s) that applies.

① Persevere with Problems
② Reason Abstractly
③ Construct an Argument
④ Model with Mathematics
⑤ Use Math Tools
⑥ Attend to Precision
⑦ Make Use of Structure
⑧ Use Repeated Reasoning

Essential Question

HOW can you use numbers and symbols to represent mathematical ideas?

Vocabulary

linear expression

Common Core State Standards

Content Standards
7.EE.1, 7.EE.2

MP Mathematical Practices
1, 2, 3, 4

Work Zone

Add Linear Expressions

A **linear expression** is an algebraic expression in which the variable is raised to the first power and variable are not multiplied or divided. The table below gives some examples of expressions that are linear and some examples of expressions that are not linear.

Linear Expressions	Nonlinear Expressions
$5x$	$5mn$
$3x + 2$	$3x^3 + 2$
$x - 7$	$x^4 - 7$

You can add linear expression with or without models. Sometimes you will need to use zero pairs.

Examples

Add.

1. $(2x + 3) + (x + 4)$

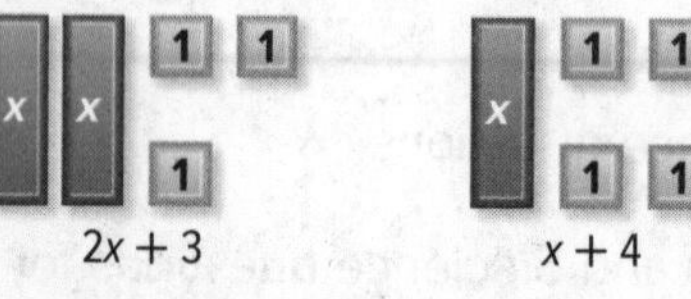

Model each linear expression.

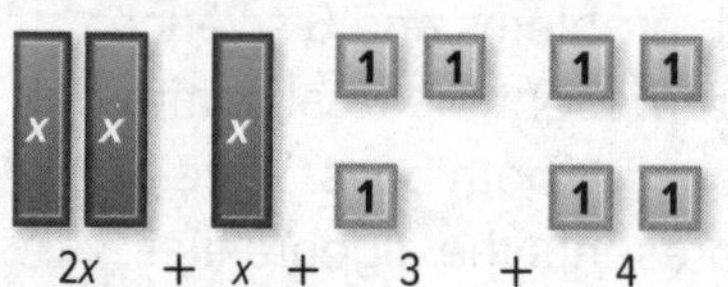

Combine like tiles and write a linear expression for the combined tiles.

So, $(2x + 3) + (x + 4) = 3x + 7$.

2. $(2x - 1) + (x - 5)$

$(2x - 1) + (x - 5) = [2x + (-1)] + [x + (-5)]$ Definition of subtraction

$$\begin{array}{r} 2x + (-1) \\ \underline{+\ x + (-5)} \\ 3x + (-6) \end{array}$$

Arrange like terms in columns.

Add.

So, $(2x - 1) + (x - 5) = 3x + (-6)$ or $3x - 6$.

Show your work.

a. ______

b. ______

Got It? **Do these problems to find out.**

a. $(3x + 5) + (2x + 3)$

b. $(2x - 4) + (3x - 7)$

Examples

3. **Find $(2x - 3) + (-x + 4)$. Use models if needed.**

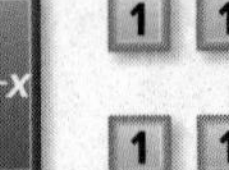

Model each linear expression.

Combine like tiles. Then remove all zero pairs and write a linear expression for the remaining tiles.

So, $(2x - 3) + (-x + 4) = x + 1$.

4. **Find $2(x + 3) + (3x + 1)$.**

$2(x + 3) + (3x + 1) = (2 \cdot x + 2 \cdot 3) + (3x + 1)$ Use the Distributive Property.

$= (2x + 6) + (3x + 1)$ Simplify.

$$\begin{array}{r} 2x + 6 \\ +\ 3x + 1 \\ \hline 5x + 7 \end{array}$$

Arrange like terms in columns.

Add.

So, $2(x + 3) + (3x + 1) = 5x + 7$.

5. **Find $5(x - 4) + (2x - 7)$.**

$5(x - 4) + (2x - 7) = (5 \cdot x - 5 \cdot 4) + (2x - 7)$ Use the Distributive Property.

$= (5x - 20) + (2x - 7)$ Simplify.

$$\begin{array}{r} 5x - 20 \\ +\ 2x - 7 \\ \hline 7x - 27 \end{array}$$

Arrange like terms in columns.

Add.

So, $5(x - 4) + (2x - 7) = 7x - 27$.

Got It? **Do these problems to find out.**

Add. Use models if needed.

c. $(x - 1) + (2x + 3)$

d. $(x - 4) + (-2x + 1)$

e. $6(x + 7) + (x + 3)$

f. $(12x + 19) + 2(x - 10)$

c. ________

d. ________

e. ________

f. ________

Properties

The Commutative Property allows the terms of the expression to be reordered.

Example

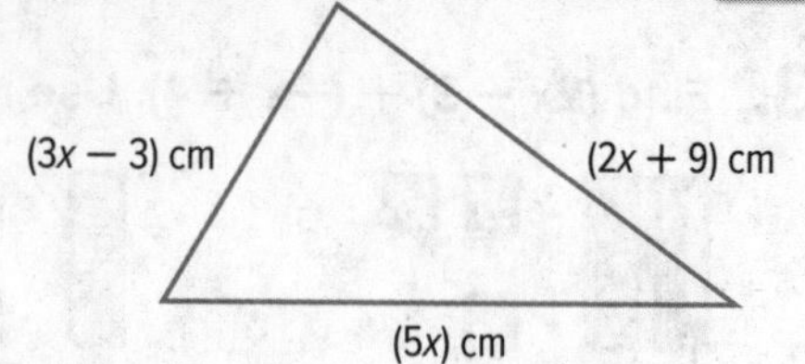

6. Write a linear expression in simplest form to represent the perimeter of the triangle. Find the perimeter if the value of x is 5 centimeters.

Write a linear expression for the perimeter of the triangle.

$(3x - 3) + (2x + 9) + (5x)$	Write each expression.
$(3x + 2x + 5x) + (-3 + 9)$	Rearrange to combine like terms.
$10x + 6$	Add.

Find the perimeter.

$10x + 6 = 10(5) + 6$ or 56 — Replace x with 5. Simplify.

So, the perimeter of the triangle is 56 centimeters.

Show your work.

Got It? Do this problem to find out.

g. A rectangle has side lengths $(x + 4)$ feet and $(2x - 2)$ feet. Write a linear expression in simplest form to represent the perimeter. Find the perimeter if the value of x is 7 feet.

g. ________

Guided Practice

Add. Use models if needed. (Examples 1–5)

Show your work.

1. $(2x + 3) + (x + 1) =$ ________

2. $10(x - 2) + (6x - 6) =$ ________

3. Write a linear expression in simplest form to represent the perimeter of the pentagon. Then find the perimeter if the value of x is 3 yards. (Example 6)

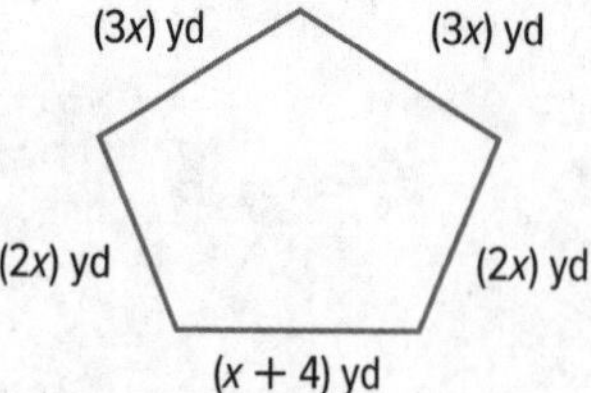

____

4. Building on the Essential Question Explain how adding linear expressions is similar to simplifying expressions.

____

____

Rate Yourself!

How confident are you about adding linear expressions? Check the box that applies.

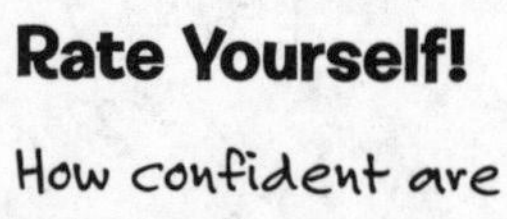

For more help, go online to access a Personal Tutor.

FOLDABLES Time to update your Foldable!

Name ______________________ My Homework ______________________

Independent Practice

Go online for Step-by-Step Solutions

eHelp 

Add. Use models if needed. (Examples 1–5)

1. $(4x + 8) + (7x + 3) =$ ______

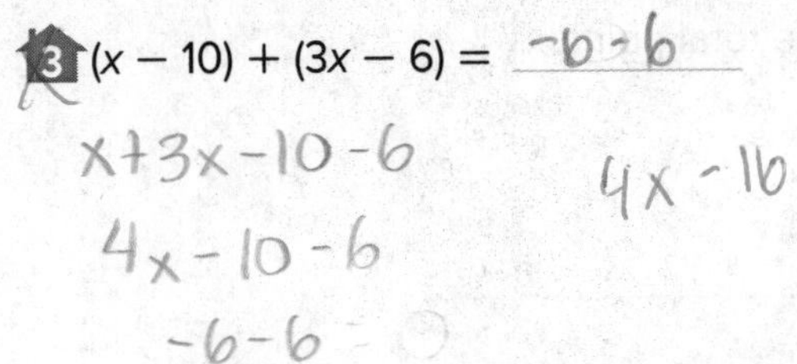

2. $(-3x + 7) + (-6x + 9) =$ ______

3. $(x - 10) + (3x - 6) =$ ______

4. $(-3x - 7) + (4x + 7) =$ ______

5. $2(x + 14) + (2x - 14) =$ ______

6. $(11x - 8) + 7(x - 1) =$ ______

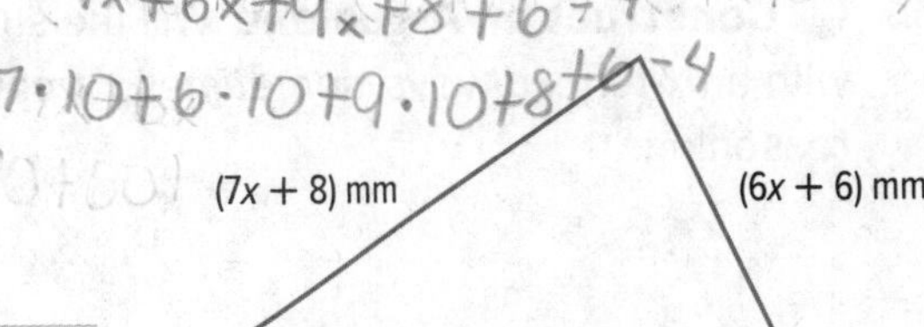

7. Write a linear expression in simplest form to represent the perimeter of the triangle at the right. Then find the perimeter if the value of x is 10 millimeters. (Example 6)

______

$(7x + 8)$ mm

$(6x + 6)$ mm

$(9x - 4)$ mm

8. A rectangle has side lengths $(2x - 5)$ meters and $(2x + 6)$ meters. Write a linear expression in simplest form to represent the perimeter. Find the perimeter if the value of x is 12 meters. (Example 6)

______

9. Find the sum of $(x + 5)$, $(-4x - 2)$, and $(2x - 1)$.

______

Add.

10. $(-3.5x + 1.7) + (9.1x - 0.3) =$ ______

11. $(0.5x + 15) + (8.2x - 16.6) =$ ______

12. **Reason Abstractly** The table shows the breakdown of the points scored in last week's basketball game.

	1st Quarter Field Goal Points	2nd Quarter Field Goal Points	3rd Quarter Field Goal Points	4th Quarter Field Goal Points	Total Free Throw Points
Panthers	$2x - 6$	$x + 2$	$2x$	$x - 6$	9

a. Write a linear expression in simplest form to represent the total field goal points scored in the first two quarters.

b. Write a linear expression in simplest form to represent the total points scored in the game.

H.O.T. Problems Higher Order Thinking

13. **Reason Inductively** Write two linear expressions with a sum of $-5x + 4$.

14. **Construct an Argument** Will the sum of two linear expressions, each with an x-term, *always*, *sometimes*, or *never* have an x-term? Explain your reasoning.

15. **Persevere with Problems** An integer can be represented by x. The next integer can then be represented as $(x + 1)$. Write a linear expression that represents the sum of any two consecutive integers. Show that the sum of any two consecutive integers is always odd.

16. **Reason Inductively** Explain how algebra tiles represent like terms and zero pairs.

Name ______________________ My Homework ______________________

Extra Practice

Add. Use models if needed.

17. $(-x + 10) + (-3x + 6) =$ $-4x + 16$

Homework Help

$-x + 10$

$(+)\ -3x + 6$

$-4x + 16$

18. $(-4x + 3) + (-2x + 8) =$ ______

19. $(-6x + 5) + (4x - 7) =$ ______

20. $(-4x + 5) + (15x - 3) =$ ______

21. $(-5x + 4) + -1(x - 1) =$ ______

22. $17(2x - 5) + (-x + 4) =$ ______

23. Write a linear expression in simplest form to represent the perimeter of the trapezoid at the right. Then find the perimeter if the value of x is 7 yards. ______

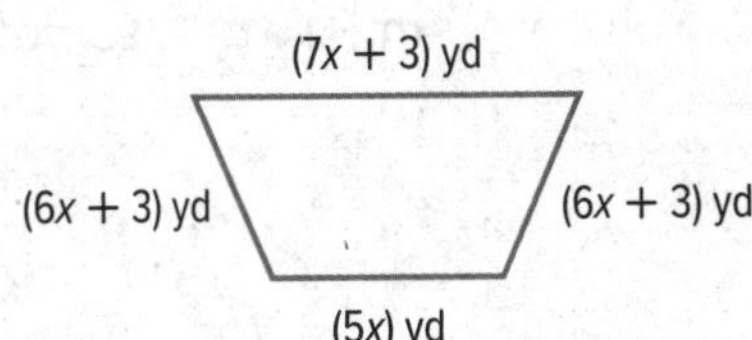

24. MP **Reason Abstractly** The table shows the points earned by a contestant in four rounds on a game show.

Round 1	Round 2	Round 3	Round 4
$2x + 40$	$5x + 12$	100	$6x - 10$

a. Write a linear expression in simplest form to represent the total points earned by the contestant in rounds 1 and 2.

______

b. Write a linear expression in simplest form to represent the total points earned in all four rounds.

______

c. If the value of x is 8, what is the total points earned in all four rounds?

______

Power Up! Common Core Test Practice

25. Karina makes x dollars per hour working at the grocery store. She makes y dollars per hour working at the library. One week she worked 9 hours at the grocery store and 12 hours at the library. Determine if each statement is true or false.

a. The expression $21x$ represents Karina's earnings from the library. ☐ True ☐ False

b. The expression $9y$ represents Karina's earnings from the grocery store. ☐ True ☐ False

c. The expression $9x + 12y$ represents Karina's total earnings for the week. ☐ True ☐ False

26. A triangle has the side lengths represented by the expressions shown in the figure. Select the appropriate numbers and expressions to complete the model representing the perimeter of the triangle.

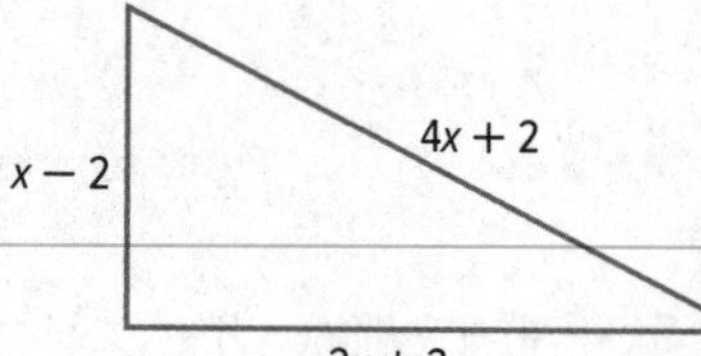

☐ + ☐

☐ + ☐

\+ ☐ + ☐

\+ ☐ + ☐

x	-1
$2x$	1
$4x$	2
$5x$	3
$7x$	7
-2	

Common Core Spiral Review

Use the Distributive Property to evaluate each expression. 6.EE.3

27. $7(9 - 4) =$ ______

28. $(9 + 2)6 =$ ______

29. $5(9 + 8) =$ ______

30. The number of students in each of the seventh grade homerooms that volunteer in the office are shown in the table. Use mental math to find the total number of students who volunteered. Explain. 6.EE.3

______

______

Office Volunteers	
Homeroom	Number of Students
A	6
B	5
C	4
D	8

Lesson 7

Subtract Linear Expressions

Real-World Link

Watch

Dog Sledding The Iditarod is a dog sledding race over 1,150 miles across Alaska. The table shows two winning times.

Iditarod				
	Days	Hours	Minutes	Seconds
Race 1	9	11	46	48
Race 2	9	5	8	41

1. What is the difference in hours, minutes, and seconds between the two races?

☐ h ☐ min ☐ s

2. Explain how you could find the difference in times between any two races, given the days, hours, minutes, and seconds.

3. Describe another situation in which finding the difference involves subtracting like units.

Essential Question

HOW can you use numbers and symbols to represent mathematical ideas?

Common Core State Standards

Content Standards
7.EE.1, 7.EE.2

MP Mathematical Practices
1, 2, 3, 4

Which MP Mathematical Practices did you use? Shade the circle(s) that applies.

① Persevere with Problems
② Reason Abstractly
③ Construct an Argument
④ Model with Mathematics
⑤ Use Math Tools
⑥ Attend to Precision
⑦ Make Use of Structure
⑧ Use Repeated Reasoning

Work Zone

Subtract Linear Expressions

When subtracting linear expressions, subtract like terms. Use zero pairs if needed.

Examples

Subtract. Use models if needed.

1. $(6x + 3) - (2x + 2)$

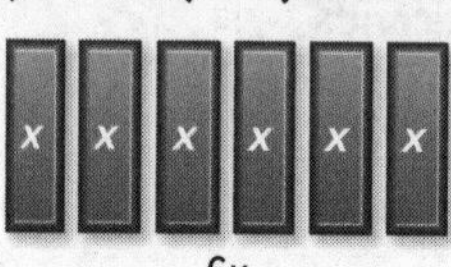

Model the linear expression $6x + 3$.

To subtract $2x + 2$, remove two x-tiles and two 1-tiles. Then write the linear expression for the remaining tiles.

There are four x-tiles and one 1-tile remaining.

So, $(6x + 3) - (2x + 2) = 4x + 1$.

2. $(2x - 3) - (x - 2)$

2x + (–3)

Model the linear expression $2x - 3$.

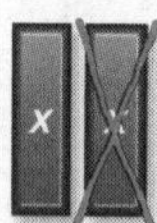

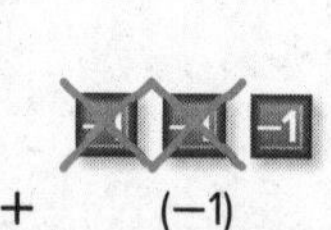

x + (–1)

To subtract $x - 2$, remove one x-tile and two –1-tiles. Then write the linear expression for the remaining tiles.

There is one x-tile and one –1-tile remaining.

So, $(2x - 3) - (x - 2) = x - 1$.

a. ____________

b. ____________

Got It? **Do these problems to find out.**

a. $(5x - 9) - (2x - 7)$

b. $(6x - 10) - (2x - 8)$

Example

3. Find $(-2x - 4) - (2x)$. Use models if needed.

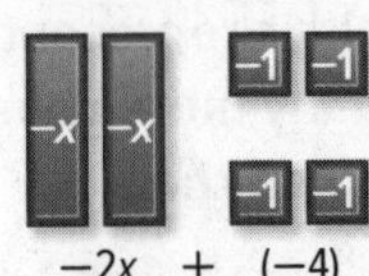

$-2x \quad + \quad (-4)$

Model the linear expression $-2x - 4$.

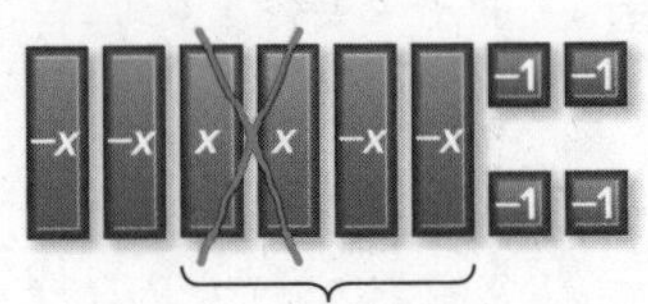

Since there are no positive x-tiles to remove, add two zero pairs of x-tiles. Remove two positive x-tiles.

So, $(-2x - 4) - (2x) = -4x - 4$.

Got It? Do these problems to find out.

c. $(3x - 2) - (5x - 4)$

d. $(4x - 4) - (-2x + 2)$

c. ______

d. ______

Use the Additive Inverse to Subtract

When subtracting integers, you add the opposite, or the additive inverse. The same process is used when subtracting linear expressions.

Examples

4. Find $(6x + 5) - (3x + 1)$.

$$\begin{array}{r} 6x + 5 \\ (+)\ -3x - 1 \\ \hline 3x + 4 \end{array}$$

Arrange like terms in columns.
The additive inverse of $3x + 1$ is $(-3x - 1)$.

5. Find $(-4x - 7) - (-5x - 2)$.

$$\begin{array}{r} -4x - 7 \\ (+)\ 5x + 2 \\ \hline x - 5 \end{array}$$

Arrange like terms in columns.
The additive inverse of $(-5x - 2)$ is $(5x + 2)$.

Additive Inverse
The additive inverse is found by multiplying the linear expression by −1.

Got It? Do these problems to find out.

e. $(4x - 3) - (2x + 7)$

f. $(5x - 4) - (2x + 3)$

e. ______

f. ______

Example

6. **A hat store tracks the sale of college and professional team hats for *m* months. The number of college hats sold is represented by $(6m + 3)$. The number of professional hats sold is represented by $(5m - 2)$. Write an expression to show how many more college hats were sold than professional hats. Then evaluate the expression if *m* equals 10.**

Find $(6m + 3) - (5m - 2)$.

$6m + 3$	Arrange like terms in columns.
$(+)-5m + 2$	The additive inverse of $5m - 2$ is $(-5m + 2)$.
$m + 5$	

Evaluate the expression if $m = 10$.

$m + 5 = \mathbf{10} + 5$	Substitute 10 for *m*.
$= 15$	Simplify.

So, 15 more college team hats were sold.

Guided Practice

Subtract. Use models if needed. (Examples 1–5)

1. $(2x + 4) - (-x + 5) =$ ______

2. $(6x + 9) - (7x - 1) =$ ______

3. The number of runs scored by the home team at a baseball game is represented by $(x + 7)$. The number of runs scored by the visiting team is represented by $(3x - 7)$. Write an expression to find how many more runs the home team scored than the visiting team. Then evaluate the expression if the value of *x* is 6. (Example 6)

______

4. **Building on the Essential Question** How can you use the additive inverse to help you subtract linear expressions?

______

______

______

______

Rate Yourself!

How well do you understand subtracting linear expressions? Circle the image that applies.

Clear

Somewhat Clear

Not So Clear

For more help, go online to access a Personal Tutor.

FOLDABLES Time to update your Foldable!

Name ______________________ My Homework ______________________

Independent Practice

Go online for Step-by-Step Solutions

Subtract. Use models if needed. (Examples 1–5)

1. $(9x + 5) - (4x + 3) =$ ______

2. $(-x + 3) - (x - 5) =$ ______

3. $(3x + 4) - (x + 2) =$ ______

4. $(7x + 5) - (3x + 2) =$ ______

5. $(9x - 8) - (x + 4) =$ ______

6. $(9x - 12) - (5x - 7) =$ ______

7. **MP Reason Abstractly** The number of customers in a store on the first day is represented by $(6x - 3)$. The number of customers on the second day is represented by $(x - 1)$. Write an expression to find how many more customers visited the store on the first day. Then evaluate the expression if x is equal to 50. (Example 6)

8. The perimeter of the garden shown is $(6x + 2)$ units. Find the length of the missing side.

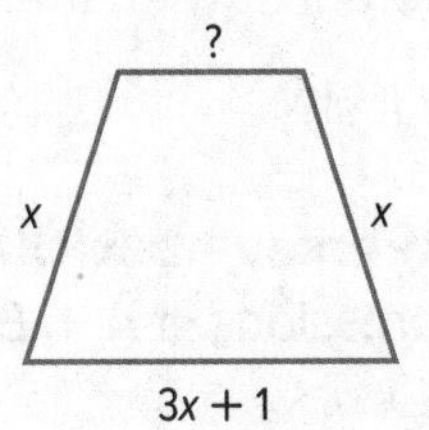

9. The cost for shipping a package that weighs x pounds from Boise to Los Angeles is shown at the right. How much more does Shipping Central charge than Globe Delivery?

Company	Cost ($)
Shipping Central	$3x + 3.50$
Globe Delivery	$2x + 2.99$

10. Find the difference in the given lengths of the polygons. __________

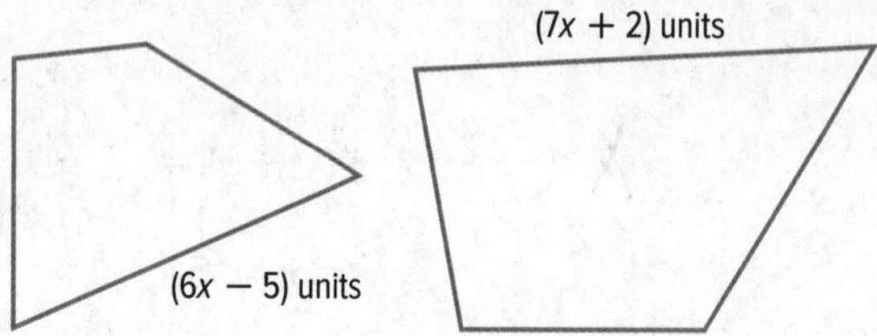

H.O.T. Problems Higher Order Thinking

11. **MP Find the Error** Theresa is finding $(5x + 3) - (2x + 1)$. Find her mistake and correct it.

Theresa's mistake was she did not flip 2x to -2x and 1 to -1. (5x+3)+(-2x+-1)
K C F
(5x+3)-(2x+1)

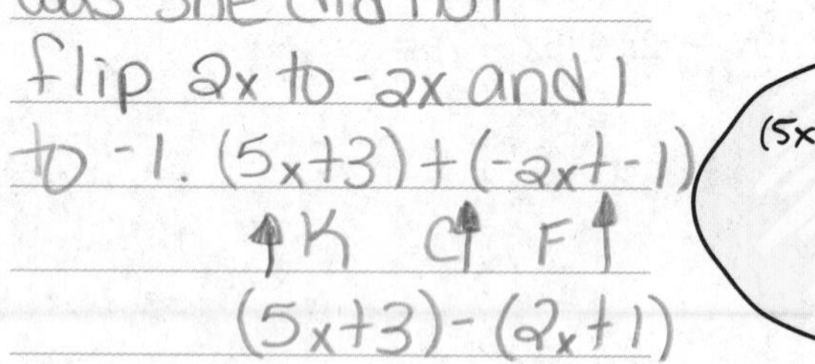

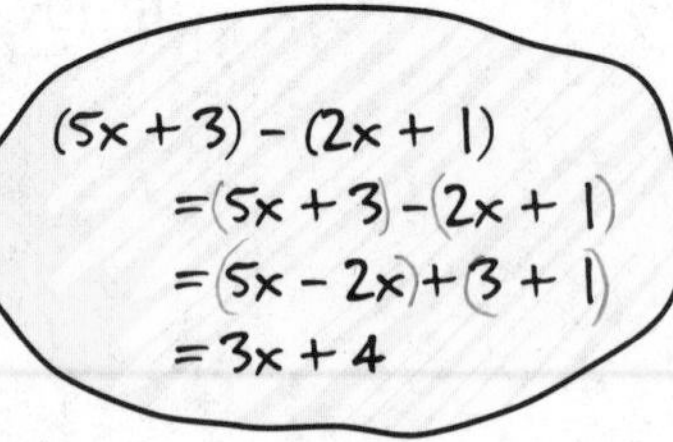

12. **MP Reason Inductively** Name two linear expressions whose difference is $5x - 4$.

(3x+2)-(2+2)=5x-4 (3x+2)-(-2x+6)=5x+-4
(4x+1)-(3+1)=5x-4 (7x+3)-(2x+7)=5x+-4

13. **MP Persevere with Problems** One linear expression is subtracted from a second linear expression and the difference is $x - 5$. What is the difference when the second linear expression is subtracted from the first? __________

14. **MP Persevere with Problems** Suppose A and B represent linear expressions. If $A + B = 2x - 2$ and $A - B = 4x - 8$, find A and B.

15. **MP Reason Inductively** Explain how you can apply the rule for subtracting integers to linear expressions.

Name ______________________ My Homework ______________________

Extra Practice

Subtract. Use models if needed.

16. $(-3x - 2) - (7x + 9) =$ $-10x - 11$

$-3x - 2$
$(+)\ -7x - 9$
$-10x - 11$

17. $(-2x - 1) - (x - 7) =$ $-3x + -8$

K C F
$(-2x - 1) + (-x - {}^{-}7)$
$(-2x + x) + (-1 + 7)$
$-3x + -8$

18. $(9x + 5) - (6x - 8) =$ ______

19. $(-8x + 1) - (8x - 1) =$ $-16x + 0$

K C F
$(-8x + 1) + (-8x - {}^{-}1)$
$(-8x - 8x) + (1 - 1)$
$-16x + 0$

20. $(4x + 10) - (-3x + 5) =$ ______

21. $(-6x - 11) - (-2x - 4) =$ ______

K C F
$(-6x - 11) + ($

22. MP **Reason Abstractly** The number of questions on a math test is represented $(3x + 1)$. The number of questions on a spelling test is represented by $(x + 12)$. Write an expression to find how many more questions were on the math test. Then evaluate the expression if the value of x is 8.

____

Subtract.

23. $(5.7x - 0.8) - (4.9x - 1.4) =$ ______

24. $\left(-\frac{5}{6}x + 5\frac{1}{2}\right) - \left(\frac{2}{3}x + 4\right) =$ ______

25. $2(x + 1) - 3x =$ ______

26. $5(x - 3) - x =$ ______

Power Up! Common Core Test Practice

27. The costs for a large pizza and each topping for two pizzerias are shown in the table.

Pizzeria	Cost per Pizza ($)	Cost per Topping ($)
Mario's Pizza	10	1.25
Pizza Palace	12	1.50

Select the appropriate values to complete the model to show how much more a pizza with t toppings costs at Pizza Palace than at Mario's Pizza.

0.25	2.75	1.25	1.50
2	10	12	22

☐ + ☐ t − (☐ + ☐ t) = ☐ + ☐ t

28. Mei wants to frame a picture. The picture is $(12x + 4)$ units long, and the frame is $(7x + 1)$ units long. Determine if each statement is true or false.

a. The picture is longer than the frame. ☐ True ☐ False

b. The frame is longer than the picture. ☐ True ☐ False

c. Mei will have to trim $(5x + 3)$ units from the picture to fit it in the frame. ☐ True ☐ False

Common Core Spiral Review

29. Camilla wants to attach a string of lights to the edges of her patio for a party. She does not want the string to go across the edge with the steps. Write a linear expression that represents the length of string in feet she will need. Then find the length if $x = 3$. **7.EE.1**

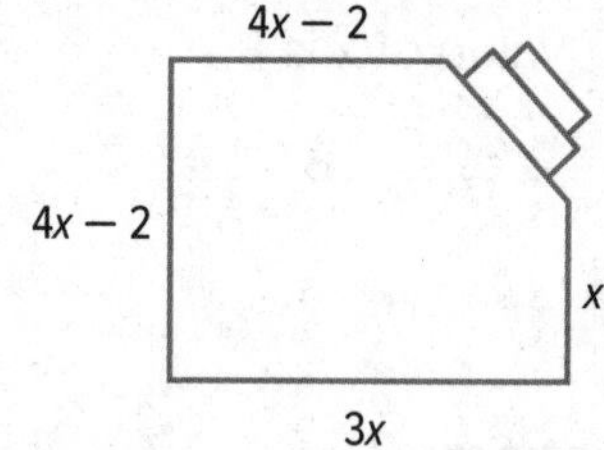

____

Evaluate each expression if $x = \frac{1}{2}$ and $y = \frac{3}{4}$. 6.EE.3

30. xy ______

31. $x - y$ ______

32. $x + y$ ______

33. x^3 ______

34. $3y + 2x$ ______

35. $x \div y$ ______

Inquiry Lab

Factor Linear Expressions

HOW do models help you factor linear expressions?

Content Standards 7.EE.1

Mathematical Practices 1, 3

Max has enough 1 inch square glass tiles to create a rectangular piece of mosaic art that has an area of 24 square inches. Some of the possible dimensions of the rectangle are listed in the table. Write the two missing possible dimensions.

Length (in.)	Width (in.)
24	1
3	8

Each of the dimensions listed are factors of 24. Sometimes, you know the product and are asked to find the factors. This process is called *factoring*.

Hands-On Activity 1

Tools

Use algebra tiles to factor $2x + 6$.

Step 1 Model the expression $2x + 6$.

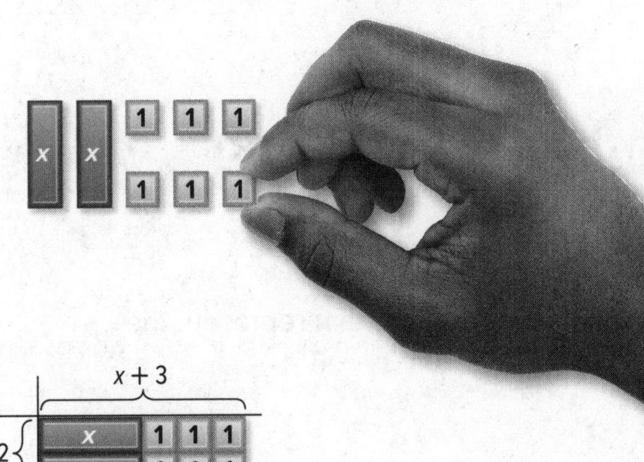

Step 2 Arrange the tiles into a rectangle with equal rows and columns. The total area of the tiles represents the product. Its length and width represent the factors.

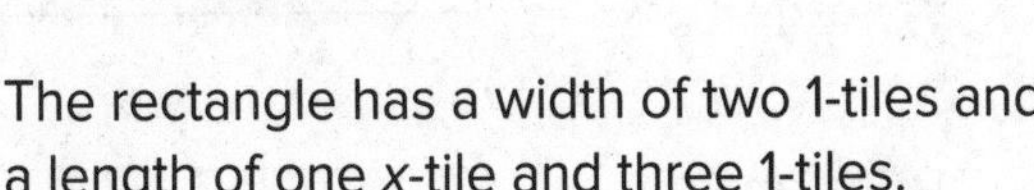

The rectangle has a width of two 1-tiles and a length of one x-tile and three 1-tiles.

So, $2x + 6 = 2(x + \square)$.

Hands-On Activity 2

Use algebra tiles to factor $2x - 8$.

Step 1 Model the expression $2x - 8$.

Step 2 Arrange the tiles into a rectangle with equal rows and columns.

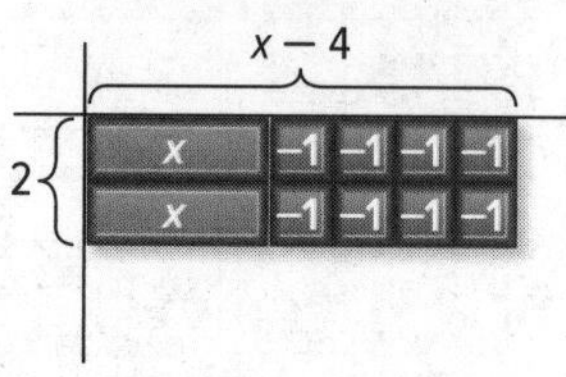

The rectangle has a width of two 1-tiles and a length of one x-tile and four −1-tiles.

So, $2x - 8 =$ ________.

Hands-On Activity 3

Use algebra tiles to factor $3x - 6$.

Step 1 Draw the tiles that represent the expression $3x - 6$.

Step 2 Redraw the tiles into a rectangle with equal rows and columns.

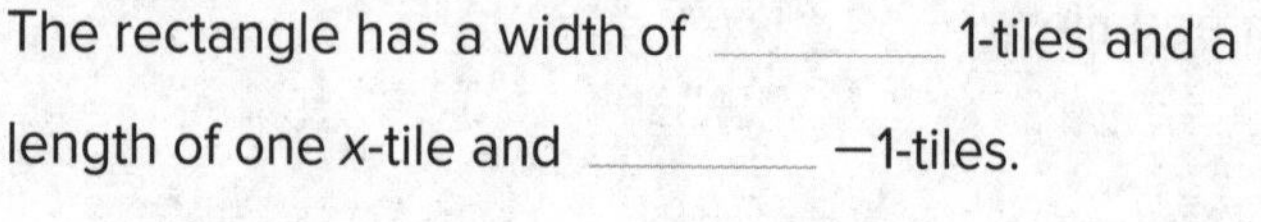

The rectangle has a width of ________ 1-tiles and a length of one x-tile and ________ −1-tiles.

So, $3x - 6 =$ ________.

Investigate

Work with a partner. Factor each expression by arranging the appropriate algebra tiles into equal rows and columns. Draw the finished product.

Show your work.

1. $4x + 6 =$ ______

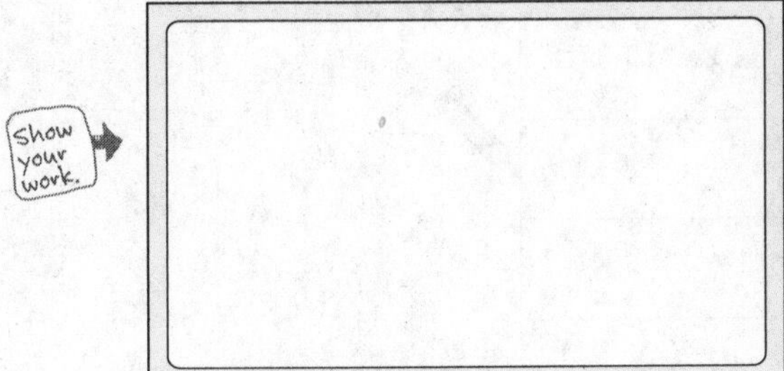

2. $5x + 10 =$ ______

3. $3x + 12 =$ ______

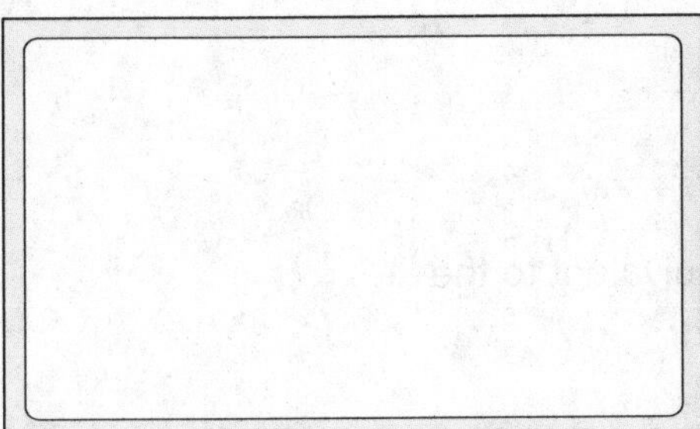

4. $4x - 10 =$ ______

5. $3x - 9 =$ ______

6. $2x - 4 =$ ______

7. $4x + 2 =$ ______

8. $5x - 5 =$ ______

Analyze and Reflect

Work with a partner to complete the table. Use algebra tiles if needed.

	Original Expression	Factored Expression	Distributive Property
	$2x + 8$	$2(x + 4)$	$2(x) + 2(4) = 2x + 8$
9.	$4x - 8$	$4(x - \square)$	$4(x) - 4(2) = 4x - 8$
10.	$6x + 2$	$2(\square x + 1)$	$2(3x) + 2(1) =$
11.	$2x - 10$		$2(x) - 2(5) =$
12.	$8x + 6$		

13. **MP Reason Inductively** How is factoring related to using the Distributive Property?

14. **MP Construct an Argument** Is the expression $2x - 2$ equivalent to the expression $2(x - 2)$? Explain.

Create

15. **MP Justify Conclusions** Explain how you could use algebra tiles to factor $5x + 15$.

16. **Inquiry** HOW do models help you factor linear expressions?

Lesson 8

Factor Linear Expressions

Real-World Link

Yard Sale A rectangular yard is being separated into four equal-size sections for different items at a yard sale. The area of the yard is $(8x + 12)$ square meters.

1. How can you find the area of each section of the yard sale?

2. What is the area of each section? Explain your answer.

3. The algebra tiles represent the area of the entire yard sale. Fill in the length and width. Write an expression that represents the area in terms of the length and width of the model.

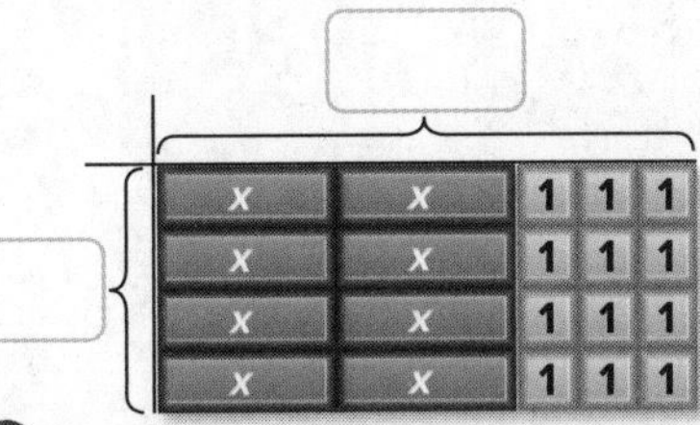

Which MP Mathematical Practices did you use? Shade the circle(s) that applies.

1 Persevere with Problems
2 Reason Abstractly
3 Construct an Argument
4 Model with Mathematics
5 Use Math Tools
6 Attend to Precision
7 Make Use of Structure
8 Use Repeated Reasoning

Essential Question

HOW can you use numbers and symbols to represent mathematical ideas?

Vocabulary

monomial
factor
factored form

Common Core State Standards

Content Standards
7.EE.1, 7.EE.2

MP Mathematical Practices
1, 2, 3, 4

Work Zone

Find the GCF of Monomials

A **monomial** is a number, a variable, or a product of a number and one or more variables.

Monomials	Not Monomials
25, x, $40x$	$x + 4$, $40x + 120$

To **factor** a number means to write it as a product of its factors. A monomial can be factored using the same method you would use to factor a number.

The greatest common factor (GCF) of two monomials is the greatest monomial that is a factor of both.

Examples

Find the GCF of each pair of monomials.

1. **$4x$, $12x$**

$4x = 2 \cdot 2 \cdot x$ Write the prime factorization of $4x$ and $12x$.

$12x = 2 \cdot 2 \cdot 3 \cdot x$ Circle the common factors.

The GCF of $4x$ and $12x$ is $2 \cdot 2 \cdot x$ or $4x$.

2. **$18a$, $20ab$**

$18a = 2 \cdot 3 \cdot 3 \cdot a$ Write the prime factorization of $18a$ and $20ab$.

$20ab = 2 \cdot 2 \cdot 5 \cdot a \cdot b$ Circle the common factors.

The GCF of $18a$ and $20ab$ is $2 \cdot a$ or $2a$.

3. **$12cd$, $36cd$**

$12cd = 2 \cdot 2 \cdot 3 \cdot c \cdot d$ Write the prime factorization of $12cd$ and $36cd$.

$36cd = 2 \cdot 2 \cdot 3 \cdot 3 \cdot c \cdot d$ Circle the common factors.

The GCF of $12cd$ and $36cd$ is $2 \cdot 2 \cdot 3 \cdot c \cdot d$ or $12cd$.

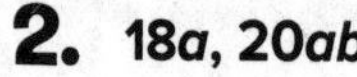

STOP and Reflect

Which of the following is not a factor of 22x? Circle your response.

4 2 11 x

Show your work.

a. ______

b. ______

c. ______

Got It? **Do these problems to find out.**

Find the GCF of each pair of monomials.

a. 12, $28c$ **b.** $25x$, $15xy$ **c.** $42mn$, $14mn$

Factor Linear Expressions

You can use the Distributive Property and the work backward strategy to express a linear expression as a product of its factors. A linear expression is in **factored form** when it is expressed as the product of its factors.

$8x + 4y = 4(2x) + 4(y)$ The GCF of $8x$ and $4y$ is 4.

$= 4(2x + y)$ Distributive Property

Examples

4. Factor $3x + 9$.

Method 1 **Use a model.**

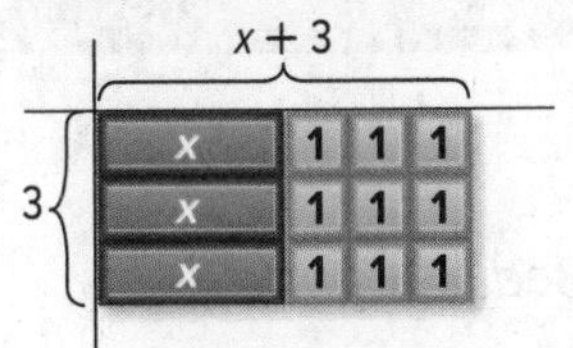

Arrange three x-tiles and nine 1-tiles into equal rows and columns. The rectangle has a width of three 1-tiles, or 3, and a length of one x-tile and three 1-tiles, or $x + 3$.

Method 2 **Use the GCF.**

$3x = 3 \cdot x$ Write the prime factorization of $3x$ and 9.

$9 = 3 \cdot 3$ Circle the common factors.

The GCF of $3x$ and 9 is 3. Write each term as a product of the GCF and its remaining factors.

$3x + 9 = 3(x) + 3(3)$

$= 3(x + 3)$ Distributive Property

So, $3x + 9 = 3(x + 3)$.

5. Factor $12x + 7y$.

Find the GCF of $12x$ and $7y$.

$12x = 2 \cdot 2 \cdot 3 \cdot x$

$7y = 1 \cdot 7 \cdot y$

There are no common factors, so $12x + 7y$ *cannot be factored.*

Factoring Expressions

To check your factored answers, multiply your factors out. You should get your original expression as a result.

Got It? Do these problems to find out.

Factor each expression. If the expression cannot be factored, write *cannot be factored*. Use algebra tiles if needed.

d. $4x - 28$ **e.** $3x + 33y$ **f.** $4x + 35$

d. ________

e. ________

f. ________

Example

6. The drawing of a garden at the right has a total area of $(15x + 18)$ square feet. Find possible dimensions of the garden.

Factor $15x + 18$.

$15x = 3 \cdot 5 \cdot x$ — Write the prime factorization of $15x$ and 18.

$18 = 2 \cdot 3 \cdot 3$ — Circle the common factors.

The GCF of $15x$ and 18 is 3. Write each term as a product of the GCF and its remaining factors.

$15x + 18 = \mathbf{3}(5x) + \mathbf{3}(6)$

$= \mathbf{3}(5x + 6)$ — Distributive Property

So, the possible dimensions are 3 feet by $(5x + 6)$ feet.

Guided Practice

Check

Find the GCF of each pair of monomials. (Examples 1–3)

1. $32x$, 18 ______

2. $27s$, $54st$ ______

3. $18cd$, $30cd$ ______

Factor each expression. If the expression cannot be factored, write cannot be factored. Use algebra tiles if needed. (Examples 4 and 5)

4. $36x + 24$ ______

5. $4x + 9$ ______

6. $14x - 16y$ ______

7. Mr. Phen's monthly income can be represented by the expression $25x + 120$ where x is the number of hours worked. Factor the expression $25x + 120$. (Example 6)

______

8. **Building on the Essential Question** Explain how the GCF is used to factor an expression. Use the term *Distributive Property* in your response.

______

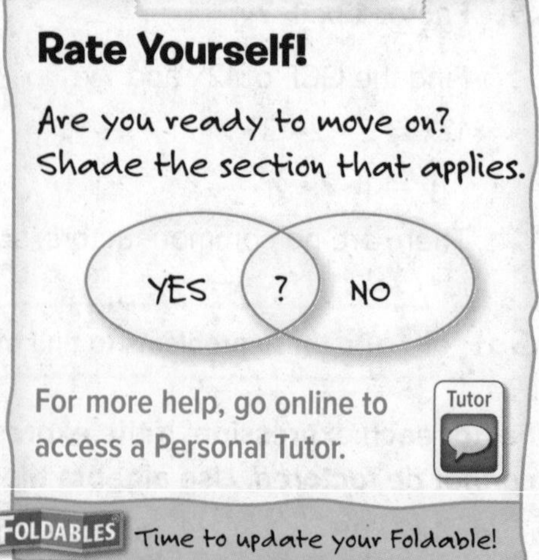

Name ______________________ My Homework ______________

Independent Practice

Go online for Step-by-Step Solutions

Find the GCF of each pair of monomials. (Examples 1–3)

1. 24, 48*m* ______

2. 32*a*, 48*b* ______

3. 36*k*, 144*km* ______

Factor each expression. If the expression cannot be *factored*, write *cannot be factored*. Use algebra tiles if needed. (Examples 4 and 5)

4. $3x + 6$ ______

5. $2x - 15$ ______

6. $12x + 30y$ ______

7. The area of a rectangular dance floor is $(4x - 8)$ square units. Factor $4x - 8$ to find possible dimensions of the dance floor. (Example 6)

____

8. The area of a rectangular porch is $(9x + 18)$ square units. Factor $9x + 18$ to find possible dimensions of the porch. (Example 6)

____

9. Six friends visited a museum to see the new holograms exhibit. The group paid for admission to the museum and \$12 for parking. The total cost of the visit can be represented by the expression $\$6x + \$12$. What expression would represent the cost of the visit for one person?

____

10. The diagram represents a flower border that is 3 feet wide surrounding a rectangular sitting area. Write an expression in factored form that represents the area of the flower border.

____

MP Reason Abstractly Write an expression in factored form to represent the total area of each rectangle.

11.

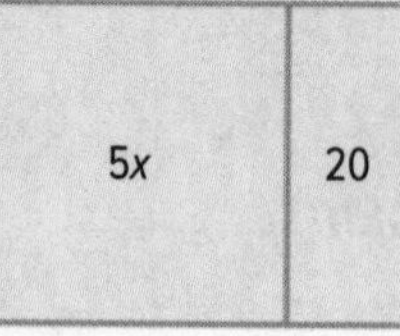

12.

7 | 49x

13.

36 | 20x | 40

14.

18
6x
12

H.O.T. Problems Higher Order Thinking

15. **MP Reason Inductively** Write two monomials whose greatest common factor is $4m$.

16. **MP Find the Error** Jamar is factoring $90x - 15$. Find his mistake and correct it.

17. **MP Persevere with Problems** The area of a rectangle is found using the formula $A = \ell w$, where ℓ is the length and w is the width of the rectangle. Write an expression in factored form that represents the area of the shaded region at the right.

6x
2
4
3y

Name ________________ My Homework ________________

Extra Practice

Find the GCF of each pair of monomials.

18. $63p$, 84 21

$63p = 3 \cdot 3 \cdot 7 \cdot p$

$84 = 2 \cdot 2 \cdot 3 \cdot 7$

Homework Help

The GCF of $63p$ and 84 is $3 \cdot 7$ or 21.

19. $30rs$, $42rs$ $6rs$

$30rs = 2 \cdot 3 \cdot 5 \cdot r \cdot s$

$42rs = 2 \cdot 3 \cdot 7 \cdot r \cdot s$

The GCF of $30rs$ and $42rs$ is $2 \cdot 3 \cdot r \cdot s$ or $6rs$.

20. $60jk$, $45jkm$ ________

21. $40x$, $60x$ ________

22. $54gh$, $72g$ ________

23. $100xy$, $75xyz$ ________

Factor each expression. If the expression cannot be factored, write *cannot be factored*. Use algebra tiles if needed.

24. $5x + 5$ ________

25. $18x + 6$ ________

26. $4x - 7$ ________

27. $10x - 35$ ________

28. $32x + 24y$ ________

29. $30x - 40$ ________

30. James has \$120 in his savings account and plans to save \$$x$ each month for 6 months. The expression \$$6x$ + \$120 represents the total amount in the account after 6 months. Factor the expression $6x + 120$.

____

31. A square scrapbooking page has a perimeter of $(8x + 20)$ inches. What is the length of one side of the scrapbooking page?

____

Copy and Solve **Write an expression in factored form that is equivalent to the given expression. Show your work on a separate piece of paper.**

32. $\frac{1}{2}x + 4$

33. $\frac{2}{3}x + 6$

34. $\frac{3}{4}x - 24$

35. $\frac{5}{6}x - 30$

36. $\frac{2}{5}x + 16$

37. $\frac{3}{8}x + 18$

Power Up! Common Core Test Practice

38. Select the correct terms to fill in the Venn diagram to show the factors of 12 and 18x.

1	9
2	12
3	18
4	x
6	

Factors of 12 | Factors of 18x

What is the GCF of 12 and 18x? Explain how the Venn diagram helped you find the GCF.

39. Which pairs of monomials have a GCF of 4a? Select all that apply.

- ☐ 8a, 18a
- ☐ 16a, 8b
- ☐ 16ab, 12a
- ☐ 28a, 20a

Common Core Spiral Review

Use the Distributive Property to rewrite each expression. 6.EE.1

40. $4(x + 1) =$ ______

41. $3(a + 10) =$ ______

42. $7(2b + 5) =$ ______

43. The letters P, E, M, D, A, and S form PEMDAS. This is a mnemonic device that can be used to help you remember the order of operations. Each letter stands for something. Complete the organizer. 6.EE.3

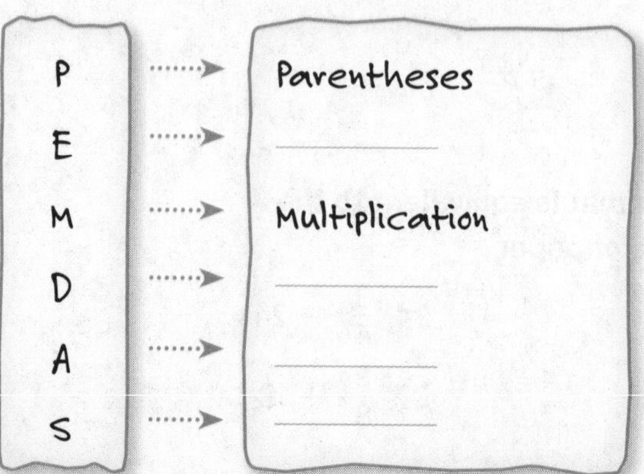

21ST CENTURY CAREER in Animal Conservation

Shark Scientist

Are you fascinated by sharks, especially those that are found around the coasts of the United States? If so, you should consider a career as a shark scientist. Shark scientists use satellite-tracking devices, called tags, to study and track the movements of sharks. By analyzing the data transmitted by the tags, scientists are able to learn more about the biology and ecology of sharks. Their research is helpful in protecting shark populations around the world.

Is This the Career for You?

Are you interested in a career as a shark scientist? Take some of the following courses in high school.

- Algebra
- Calculus
- Physics
- Statistics

Find out how math relates to a career in Animal Conservation.

Tag, You're It!

The *fork length* of a shark is the length from the tip of the snout to the fork of the tail. Use the information on the note cards to solve each problem.

1. Write an expression to represent the total length of a hammerhead shark that has a fork length of *f* feet. ____________

2. Use the expression from Exercise 1 to find the total length of a hammerhead shark that has a fork length of 11.6 feet. ____________

3. Write an expression to represent the average fork length of a tiger shark, given the average fork length s of a sandbar shark. ____________

4. Use the expression from Exercise 3 to find the average fork length of a tiger shark if the average fork length of a sandbar shark is 129 centimeters. ____________

5. Write an expression to find the average fork length of a white shark with a total length of *t* centimeters. ____________

6. The total length of a white shark is 204 centimeters. Use the expression in Exercise 5 to find the approximate fork length of the white shark. ____________

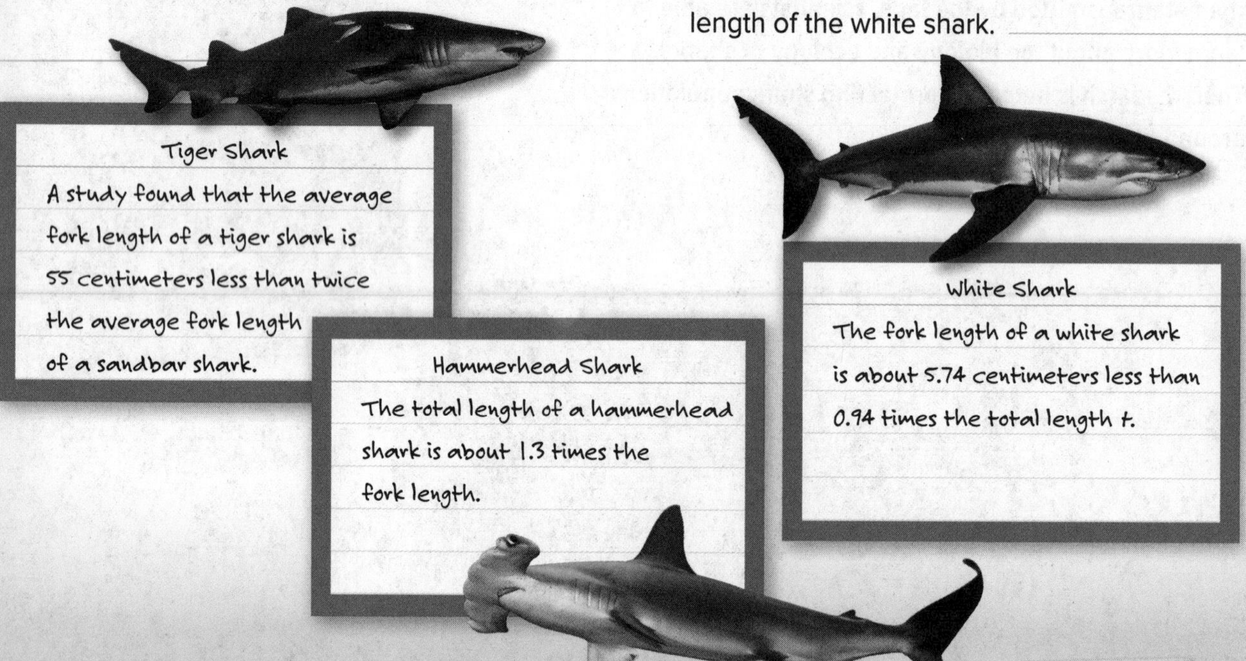

Career Project

It's time to update your career portfolio! Describe the skills that would be necessary for a shark scientist to possess. Determine whether this type of career would be a good fit for you.

List several challenges associated with this career.

-
-
-
-
-

Chapter Review

Vocabulary Check

In the puzzle below, write a vocabulary term for each clue.

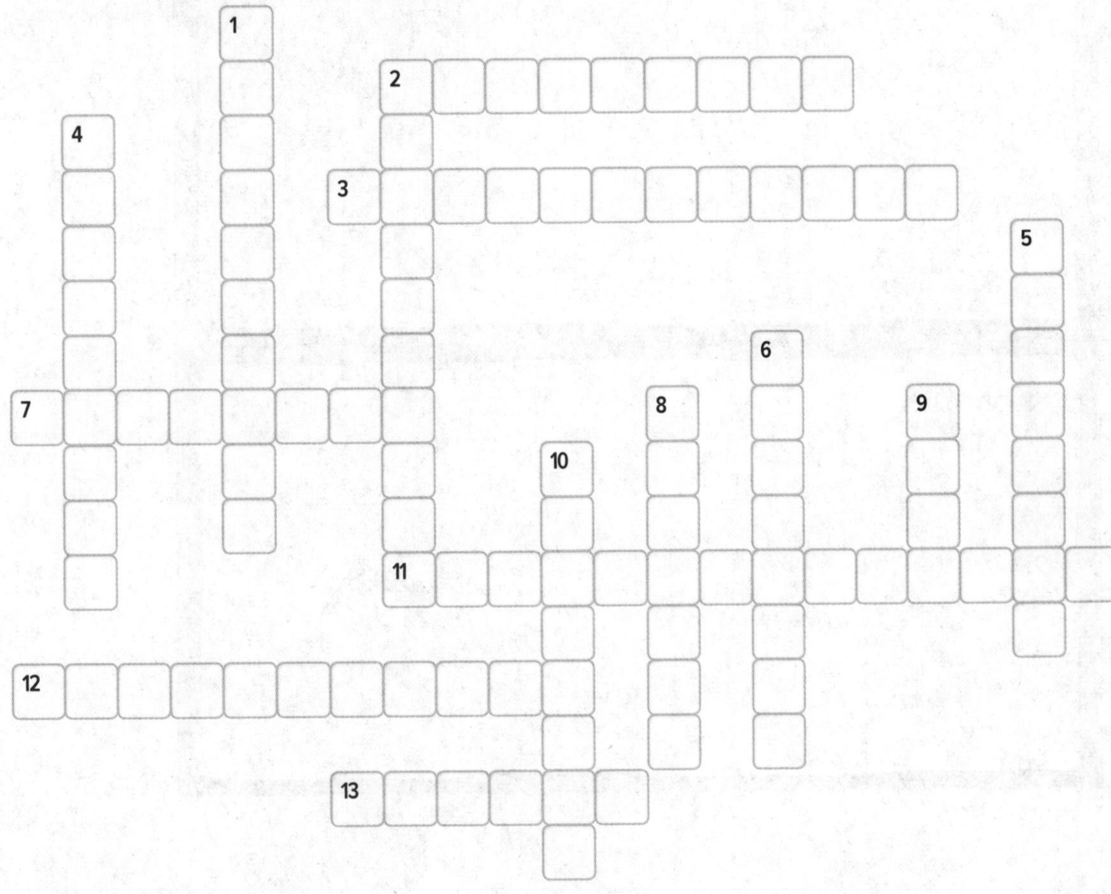

Across

2. a type of expression that contains a variable or variables
3. an algebraic expression that has no like terms and no parentheses is in this form (two words)
7. an ordered list of numbers
11. an example showing a statement is not true
12. the numerical factor of a multiplication expression
13. what is done to a variable to represent an unknown quantity

Down

1. expressions like 4(3 + 2) and 4(3) + 4(2)
2. a sequence in which each term is found by adding the same number
4. terms that include the same variable
5. a letter or symbol
6. a statement that is true for any number or variable
8. a branch of mathematics that uses variables
9. a number in a sequence
10. a term that contains a number only

Key Concept Check

Use Your FOLDABLES

Use your Foldable to help review the chapter.

Tape here

Linear Expressions

Explanation

Explanation

Got it?

Draw a line to match each expression with its equivalent expression.

1. $3 + 1$

2. $4(2 - x)$

3. $3x - 2 - x + 6$

4. $2(x + 2) + (3x + 1)$

5. $3x + 21$

a. $8 - 4x$

b. $5x + 5$

c. $3(x + 7)$

d. $1 + 3$

e. $2x + 4$

CCSS Power Up! Performance Task

Movie Time

The Townsends, a family of five, are going to the local movie theater. The family consists of two adults and three children. Mr. Townsend wants to calculate the cost of the night out. He looks up the admission prices online.

Admission:

Adults — \$10.50
Children — \$6.50

All shows before 6 P.M. $\frac{1}{2}$ price

Before leaving, Mr. Townsend decides that he will get some items at the theater concession stand, a large drink for each person and a large tub of popcorn for everyone to share. He will not know the prices of the items at the concession stand until they arrive.

Write your answers on another piece of paper. Show all of your work to receive full credit.

Part A

Write an expression that represents the cost of the admission prices and the concession stand items based on the available information. Let d represent the cost for a large drink and let p represent the cost of the popcorn. The initial expression must include parentheses. Then simplify the expression by using the Distributive Property and combining like terms.

Part B

Two children from next door join the Townsends. The neighbor children have movie passes and have already eaten, so Mr. Townsend will only need to pay for two more large drinks. At the concession stand, one of the children gives Mr. Townsend a five dollar bill to help pay for the drinks. Write an expression that represents the cost of the drinks for the neighbor children and includes the money given to Mr. Townsend.

Part C

While at the concession stand, Mr. Townsend sees that the large tub of popcorn is \$7.50, and large drinks are \$6 each. Using your answers from Part A and Part B, write an expression that represents the total cost. Then substitute the values for the popcorn and drinks in your expression. What is the total cost for the evening?

Reflect

Answering the Essential Question

Use what you learned about algebraic expressions to complete the graphic organizer. Then answer the chapter's Essential Question below.

When do you use a variable?

Essential Question

HOW can you use numbers and symbols to represent mathematical ideas?

How do you know which operation symbol to use?

Answer the Essential Question. HOW can you use numbers and symbols to represent mathematical ideas?

Chapter 6 Equations and Inequalities

Essential Question

WHAT does it mean to say two quantities are equal?

Common Core State Standards

Content Standards
7.EE.3, 7.EE.4, 7.EE.4a, 7.EE.4b

MP **Mathematical Practices**
1, 2, 3, 4, 5, 7

Math in the Real World

Driving Suppose you live in a state where you must be at least 16 years of age to obtain a driver's license. Circle the statement that represents this age.

DRIVER LICENSE
NAME: Joan Smith
AGE: $a < 16$
$a = 16$
$a \geq 16$
ADDRESS:
1234 Anyplace Dr.

FOLDABLES® Study Organizer

Cut out the Foldable on page FL5 of this book.

Place your Foldable on page 524.

Use the Foldable to help you learn about equations and inequalities.

What Tools Do You Need?

Vocabulary

Addition Property of Equality
Addition Property of Inequality
coefficient
Division Property of Equality
Division Property of Inequality
equation
equivalent equation
inequality
Multiplication Property of Equality
Multiplication Property of Inequality
solution
Subtraction Property of Equality
Subtraction Property of Inequality
two-step equation
two-step inequality

Study Skill: Reading Math

Identify Key Information Have you ever tried to solve a word problem and didn't know where to start. Start by looking for key words in the text and images. Then write the important information in one sentence.

1. Highlight or circle key words in the following real-world problem.

   During a recent Super Bowl, millions of pounds of potato chips and tortilla chips were consumed. The number of pounds of potato chips consumed was 3.1 million pounds more than the number of pounds of tortilla chips. How many pounds of tortilla chips were consumed?

2. Write a sentence that summarizes the information provided. Include information from the text and the image. ______________________

What Do You Already Know?

Place a checkmark below the face that expresses how much you know about each concept. Then scan the chapter to find a definition or example of it.

☹ I have no clue. 😐 I've heard of it. ☺ I know it!

Equations and Inequalities				
Concept	☹	😐	☺	Definition or Example
inequalities				
solving one-step equations				
solving inequalities by addition or subtraction				
solving inequalities by multiplication or division				
solving two-step equations				
solving two-step inequalities				

When Will You Use This?

Here are a few examples of how equations are used in the real world.

Activity 1 Describe a situation when you only had a set amount of money to spend and you needed to buy a certain number of items. Then explain how you determined what you could buy.

Activity 2 Go online at **connectED.mcgraw-hill.com** to read the graphic novel ***Movie Night***. How much does each DVD cost? How much money do they need for popcorn? ______

Are You Ready?

Try the Quick Check below.
Or, take the Online Readiness Quiz.

Quick Review

Common Core Review 6.EE.2a, 6.EE.5

Example 1

Write the phrase as an algebraic expression.

Phrase: five dollars more than Jennifer earned

Variable: Let d represent the number of dollars Jennifer earned.

Expression: $d + 5$

Example 2

Is 3, 4, or 5 the solution of the equation $x + 8 = 12$?

Value of x	$x + 8 = 12$	Are both sides equal?
3	$3 + 8 \stackrel{?}{=} 12$ $11 \neq 12$	no
4	$4 + 8 \stackrel{?}{=} 12$ $12 = 12$	yes ✓
5	$5 + 8 \stackrel{?}{=} 12$ $13 \neq 12$	no

The solution is 4 since replacing x with 4 results in a true sentence.

Quick Check

Words and Symbols **Write the phrase as an algebraic expression.**

1. 3 more runs than the Pirates scored
2. a number decreased by eight
3. ten dollars more than Grace has

One-Step Equations **Identify the solution of each equation from the list given.**

4. $8 + w = 17$; 7, 8, 9 ______
5. $d - 12 = 5$; 16, 17, 18 ______
6. $6 = 3y$; 2, 3, 4 ______
7. $7 \div c = 7$; 0, 1, 2 ______
8. $a + 8 = 23$; 13, 14, 15 ______
9. $10 = 45 - n$; 35, 36, 37 ______

How Did You Do?

Which problems did you answer correctly in the Quick Check? Shade those exercise numbers below.

① ② ③ ④ ⑤ ⑥ ⑦ ⑧ ⑨

Inquiry Lab

Solve One-Step Addition and Subtraction Equations

HOW can bar diagrams or algebra tiles help you solve an equation?

Content Standards
7.EE.4, 7.EE.4a

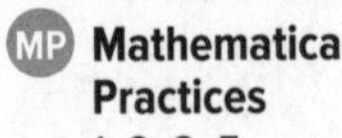

Mathematical Practices
1, 2, 3, 5

In a recent year, 19 of the 50 states had a law banning the use of handheld cell phones while driving a school bus. Determine how many states did *not* have this law.

Hands-On Activity 1

You can represent this situation with an equation.

Step 1 The bar diagram represents the total number of states and the number of states that have passed a cell phone law. Fill in the missing information.

|------------ ☐ states ------------|

states with a law	states that do not have a law
☐ states	?

Step 2 Write an equation from the bar diagram. Let x represent the states that do not have a cell phone law for school bus drivers.

$$19 + x = 50$$

Step 3 Use the *work backward* strategy to solve the equation. Since $19 + x = 50$, $x = 50 - 19$. So, $x =$ ☐.

Check $19 +$ ☐ $= 50$ ✓

So, ☐ states did *not* have a law banning the use of cell phones by bus drivers.

Investigate

Collaborate

Work with a partner to solve each problem.

1. Draw a bar diagram and write an addition equation to represent the following situation. Then solve the equation.

The sum of a number and four is equal to 18.

Equation: ________ Solution: $x =$ ________

2. **Use Math Tools** Jack collects postage stamps. He sold 7 of his stamps and had 29 stamps left. Complete the bar diagram below. Then write and solve a subtraction equation to find the number of stamps Jack had at the beginning.

Equation: ________ Solution: $n =$ ________

So, Jack had ☐ stamps at the beginning.

Analyze and Reflect

Collaborate

3. Suppose Jack sold 15 stamps and had 21 stamps left. How would the bar diagram change?

4. **Reason Abstractly** Suppose Jack had 40 stamps in the beginning and sold 7 of them. How would the bar diagram change? What equation could you write to represent the situation?

Hands-On Activity 2

Solve $x - 3 = -2$ using algebra tiles.

Remember a 1-tile and −1 tile combine to make a *zero pair*. You can add or subtract zero pairs from either side of an equation without changing its value.

Step 1 Model the equation.

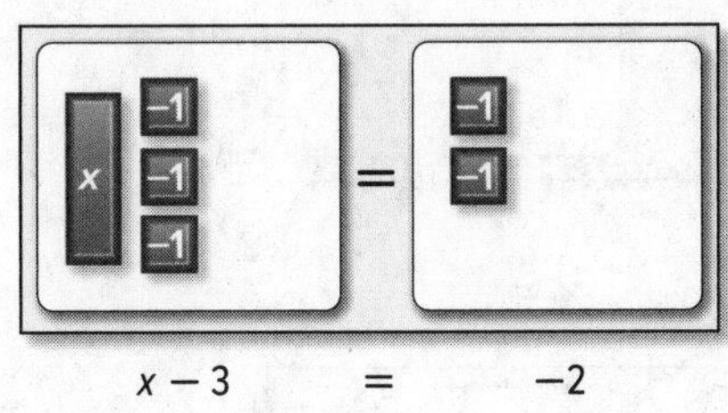

$x - 3 \quad = \quad -2$

Step 2 Add three 1-tiles to the left side of the mat and ______ 1-tiles to the right side of the mat to form zero pairs on each side of the mat.

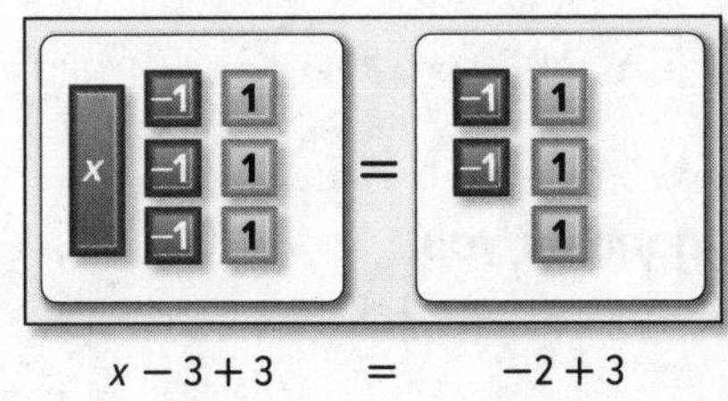

$x - 3 + 3 \quad = \quad -2 + 3$

Step 3 Remove all of the zero pairs from each side. There is ______ 1-tile on the right side of the mat.

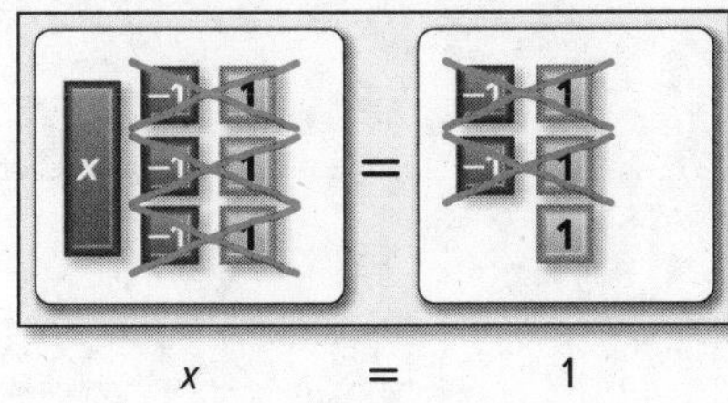

$x \quad = \quad 1$

Therefore, $x =$ ☐.

Check ☐ $- 3 = -2$ ✓

Investigate

Use Math Tools Work with a partner to solve each equation. Use algebra tiles. Show your work using drawings.

5. $x + 4 = 4$ $x =$ ______

6. $-2 = x + 1$ $x =$ ______

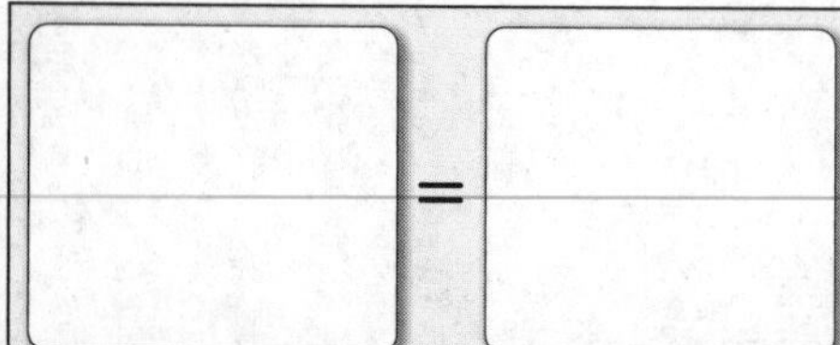

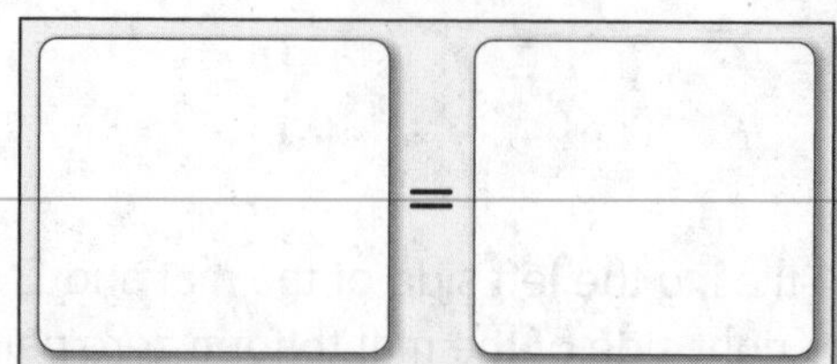

7. $x - 1 = -3$ $x =$ ______

8. $4 = x - 2$ $x =$ ______

Analyze and Reflect

Work with a partner to complete the table. The first one is done for you.

	Equation	Related Equation
	$x + 3 = 4$	$x = 4 - 3$
9.	$6 + x = 10$	
10.	$x + 3 = -1$	
11.	$6 + x = -7$	

Create

12. **Construct an Argument** Write a rule that you can use to solve addition equations without using models or a drawing.

______

______

13. Inquiry HOW can bar diagrams or algebra tiles help you solve an equation?

______

______

Solve One-Step Addition and Subtraction Equations

Vocabulary Start-Up

An **equation** is a sentence stating that two quantities are equal. The value of a variable that makes an equation true is called the **solution** of the equation.

$$\begin{aligned} x + 2 &= 6 \\ -2 &= -2 \\ \hline x &= 4 \end{aligned}$$

The equations $x + 2 = 6$ and $x = 4$ are **equivalent equations** because they have the same solution, 4.

Circle the equations below that are equivalent to $x = 3$. Use algebra tiles if needed.

$x + 3 = 6$	$x + 1 = 6$	$x + 6 = 8$
$x + 3 = 3$	$x + 1 = 4$	$x + 2 = 5$

Essential Question

WHAT does it mean to say two quantities are equal?

Vocabulary

equation
solution
equivalent equation
Subtraction Property of Equality
Addition Property of Equality

Common Core State Standards

Content Standards
7.EE.4, 7.EE.4a

MP Mathematical Practices
1, 2, 3, 4, 5

Real-World Link

Video Games Robyn had some video games, and then she bought 4 more games. Now she has 10 games. This scenario can be described using the equation $x + 4 = 10$.

1. What does x represent in the equation?

2. Write two different equations that are equivalent to $x + 4 = 10$.

Which MP Mathematical Practices did you use? Shade the circle(s) that applies.

① Persevere with Problems	⑤ Use Math Tools
② Reason Abstractly	⑥ Attend to Precision
③ Construct an Argument	⑦ Make Use of Structure
④ Model with Mathematics	⑧ Use Repeated Reasoning

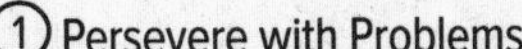
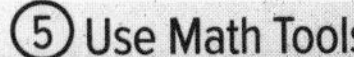

Key Concept

Subtraction Property of Equality

Words The **Subtraction Property of Equality** states that the two sides of an equation remain equal when you subtract the same number from each side.

Symbols If $a = b$, then $a - c = b - c$.

Work Zone

You can use bar diagrams and the *work backward* problem-solving strategy to solve equations arithmetically. Or, you can use the properties of equality to solve equations algebraically.

Examples

1. Solve $x + 6 = 4$. Check your solution.

$x + 6 = 4$ Write the equation.

$-6 = -6$ Subtraction Property of Equality

$x = -2$ Simplify.

Check $x + 6 = 4$ Write the original equation.

$-2 + 6 \stackrel{?}{=} 4$ Replace x with -2.

$4 = 4$ ✓ The sentence is true.

So, the solution is -2.

Solutions

Notice that your new equation, $x = -2$, has the same solution as the original equation, $x + 6 = 4$.

2. Solve $-5 = b + 8$. Check your solution.

$-5 = b + 8$ Write the equation.

$-8 = -8$ Subtraction Property of Equality

$-13 = b$ Simplify.

Check $-5 = b + 8$ Write the original equation.

$-5 \stackrel{?}{=} -13 + 8$ Replace b with -13.

$-5 = -5$ ✓ The sentence is true.

So, the solution is -13.

Got it? **Do these problems to find out.**

Solve each equation. Check your solution.

a. $y + 6 = 9$ **b.** $x + 3 = 1$ **c.** $-3 = a + 4$

Show your work.

a. ________

b. ________

c. ________

Example

3. An angelfish can grow to be 12 inches long. If an angelfish is 8.5 inches longer than a clown fish, how long is a clown fish?

Words	An angelfish is 8.5 inches longer than a clown fish.
Variable	Let c represent the length of the clown fish.
Equation	$12 = c + 8.5$

$$\begin{aligned} 12 &= c + 8.5 \\ -8.5 &= \quad -8.5 \\ \hline 3.5 &= c \end{aligned}$$

Write the equation.
Subtraction Property of Equality
Simplify.

A clown fish is 3.5 inches long.

Solve Arithmetically

You can use a bar diagram to solve an equation arithmetically.

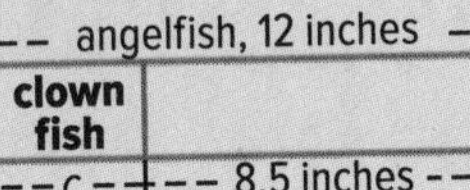

Work backward to solve for c.
$c = 12 - 8.5 = 3.5$

Got it? Do this problem to find out.

d. The highest recorded temperature in Warsaw, Missouri, is 118°F. This is 158° greater than the city's lowest recorded temperature. Find the lowest recorded temperature.

d. ________

Addition Property of Equality

Key Concept

Words The **Addition Property of Equality** states that the two sides of an equation remain equal when you add the same number to each side.

Symbols If $a = b$, then $a + c = b + c$.

Example

4. Solve $x - 2 = 1$. Check your solution.

$$\begin{aligned} x - 2 &= 1 \\ +2 &= +2 \\ \hline x &= 3 \end{aligned}$$

Write the equation.
Addition Property of Equality
Simplify.

The solution is 3. **Check** $3 - 2 = 1$ ✓

Got it? Do these problems to find out.

e. $y - 3 = 4$ **f.** $r - 4 = -2$ **g.** $q - 8 = -9$

e. ________

f. ________

g. ________

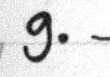

Example

5. A pair of shoes costs $25. This is $14 less than the cost of a pair of jeans. Find the cost of the jeans.

Shoes are $14 less than jeans. Let j represent the cost of jeans.

$25 = j - 14$	Write the equation.
$+14 = \quad +14$	Addition Property of Equality
$39 = j$	Simplify.

The jeans cost $39.

Models

A bar diagram can be used to represent this situation.

jeans, j	
shoes	
$25	$14

$j = 25 + 14 = 39$

Got it? Do this problem to find out.

h. The average lifespan of a tiger is 17 years. This is 3 years less than the average lifespan of a lion. Write and solve an equation to find the average lifespan of a lion.

h. ____________

Guided Practice

Solve each equation. Check your solution. (Examples 1, 2, and 4)

1. $n + 6 = 8$

2. $7 = y + 2$

3. $-7 = c - 6$

4. Orville and Wilbur Wright made the first airplane flights in 1903. Wilbur's flight was 364 feet. This was 120 feet longer than Orville's flight. Write an equation to represent the flights. Use a bar diagram if needed. Then solve to find the length of Orville's flight. (Examples 3 and 5)

5. **Building on the Essential Question** What are two methods for solving a real-world problem that can be represented by an equation?

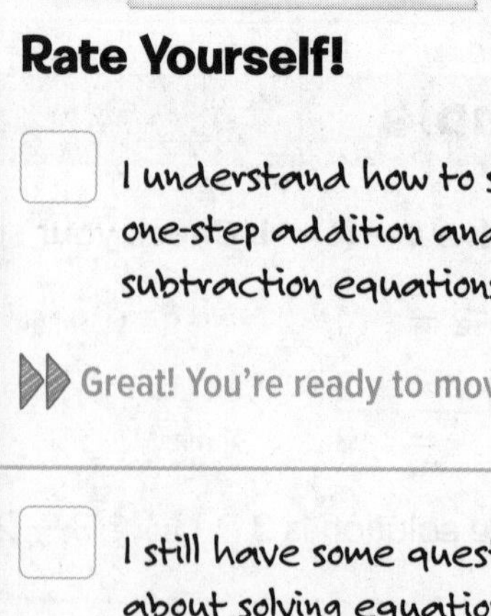

Name ______________________ My Homework ______________________

Independent Practice

Go online for Step-by-Step Solutions

Solve each equation. Check your solution. (Examples 1, 2, and 4)

1. $a + 3 = 10$

2. $y + 5 = -11$

3. $s - 8 = 9$

4. $5 = x + 8$

5. $-2 = p - 1$

6. $14 = s + 7$

Use a bar diagram to solve arithmetically. Then use an equation to solve algebraically. (Examples 3 and 5)

7. Last week Tiffany practiced her bassoon a total of 7 hours. This was 2 hours more than she practiced the previous week. How many hours did Tiffany practice the previous week?

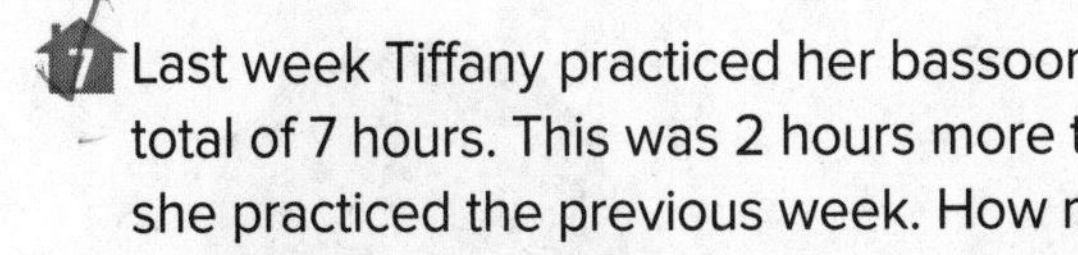

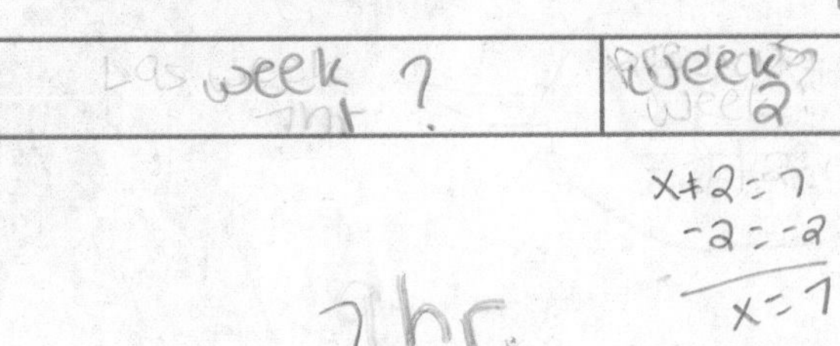

8. In a recent presidential election, Ohio had 18 electoral votes. This is 20 votes less than Texas had. How many electoral votes did Texas have?

9. **Multiple Representations** Use the table to solve.

Tallest Wooden Roller Coasters	Height (feet)	Drop (feet)	Speed (mph)
Colossos	h	159	68
T Express	184	151	65
El Toro	181	176	s
Voyage	163	d	67

a. **Symbols** The difference in speeds of El Toro and T Express is 5 miles per hour. If El Toro has the greater speed, write and solve a subtraction equation to find its speed.

b. **Diagram** Voyage has a drop that is 22 feet less than El Toro. Draw a bar diagram to the right and write an equation to find the height of Voyage.

c. **Words** Let h represent the height of the Colossos roller coaster. Explain why $h - 13 = 184$ and $h - 34 = 163$ are equivalent equations. Then explain the meaning of the solution.

10. The sum of the measures of the angles of a triangle is 180°. Write and solve an equation to find the missing measure.

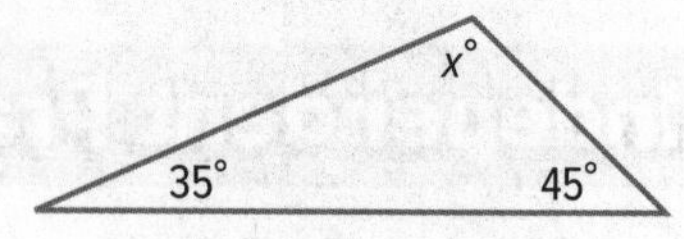

11. The sum of the measures of a quadrilateral is 360°. Write and solve an equation to find the missing measure.

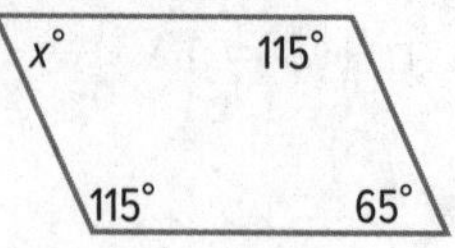

H.O.T. Problems Higher Order Thinking

12. **MP Reason Inductively** Write an addition equation and a subtraction equation that have 10 as a solution.

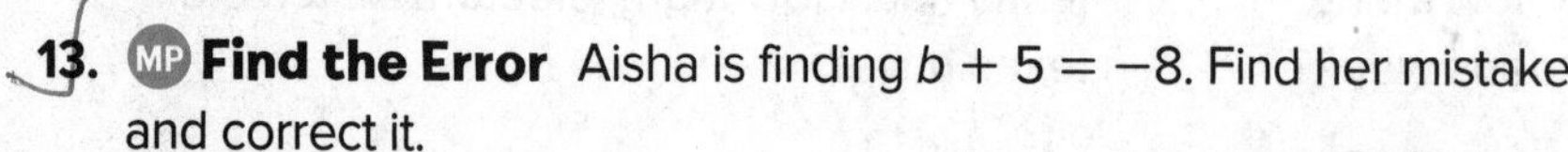

13. **MP Find the Error** Aisha is finding $b + 5 = -8$. Find her mistake and correct it.

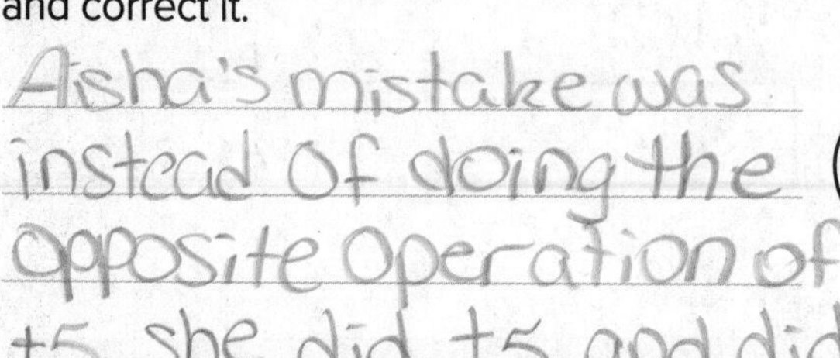

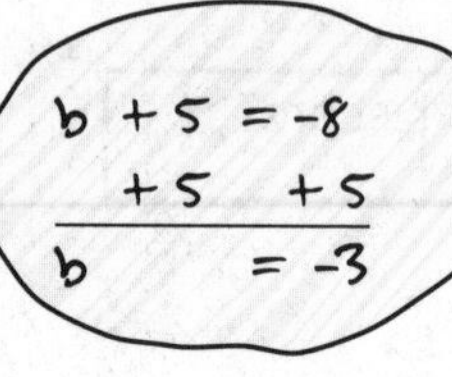

b+5=-8
-5=-5
b=-13

14. **MP Reason Abstractly** Suppose $x + y = 11$ and the value of x increases by 2. If their sum remains the same, what must happen to the value of y? Justify your response

15. **MP Which One Doesn't Belong?** Identify the equation that does not belong with the other three. Explain your reasoning.

$x + 4 = -2$	$x + 5 = -1$	$x + 2 = 8$	$3 - x = 9$

x=-6 x=-6 x=6

3-x=9 does not belong because x does not equal 6 or -6 like the others

16. **MP Reason Inductively** In the equation $x + y = 5$, the value for x is a whole number greater than 2 but less than 6. Find the possible solutions for y.

Name ______________________ My Homework ______________________

Extra Practice

Solve each equation. Check your solution.

17. $r + 6 = -3$

Homework Help

$r + 6 = -3$
$-6 = -6$
$r = -9$

18. $w - 7 = 11$

19. $k + 3 = -9$

20. $-1 = q - 8$

21. $9 = r + 2$

22. $y + 15 = 11$

MP **Use Math Tools** **Use a bar diagram to solve arthimetically. Then use an equation to solve algebraically.**

23. The Miami Heat scored 79 points. This was 13 points less than the Chicago Bulls. How many points did the Chicago Bulls score?

24. Zach is $15\frac{1}{2}$ years old. This is 3 years younger than his brother Lou. How old is Lou?

25. The table shows a golfer's scores for four rounds of a recent U.S. Women's Open. Her total score was even with par. What was her score for the third round?

Round	Score
First	−1
Second	−3
Third	s
Fourth	+2

Copy and Solve **Solve each equation. Check your solution. Show your work on a separate piece of paper.**

26. $a - 3.5 = 14.9$

27. $b + 2.25 = 1$

28. $-\frac{1}{3} = r - \frac{3}{4}$

29. $x - 2.8 = 9.5$

30. $r - 8.5 = -2.1$

31. $z - 9.4 = -3.6$

32. $m + \frac{5}{6} = \frac{11}{12}$

33. $-\frac{5}{6} + c = -\frac{11}{12}$

34. $s - \frac{1}{9} = \frac{5}{18}$

Power Up! Common Core Test Practice

35. The model represents the equation $x - 2 = 5$. Determine if each statement is true or false.

a. To solve the equation, add 2 positive counters to each side of the equation mat. ☐ True ☐ False

b. To solve the equation, add 5 negative to each side of the equation mat. ☐ True ☐ False

c. The value of x is 7. ☐ True ☐ False

36. Britney practiced the piano a total of 7 hours this week. This is 3 hours less than she practiced last week. Select the correct labels to complete the bar diagram that is used to find the number of hours w Britney practiced last week.

this week
last week, w
3 hours
4 hours
7 hours
10 hours

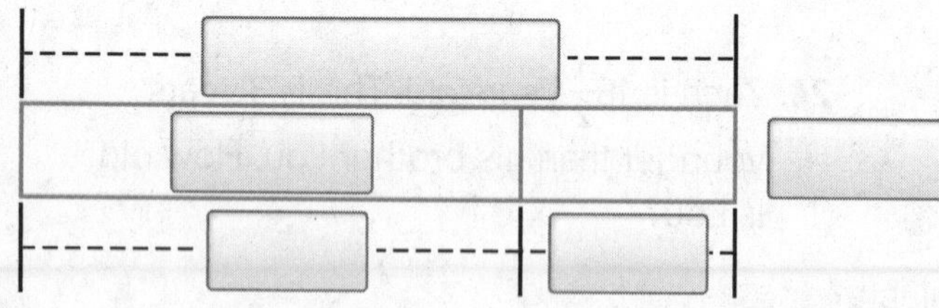

How many hours did Britney practice the piano last week? ________

Common Core Spiral Review

Multiply or divide. 7.NS.2

37. $5(-4) =$ ________

38. $\frac{36}{-9} =$ ________

39. $(-10)(-6) =$ ________

40. $\frac{-42}{-7} =$ ________

41. $(-3)(12) =$ ________

42. $\frac{-54}{2} =$ ________

43. While playing a round of golf, Tina had a score of three under par after the first three holes. Write and solve an equation to find Tina's average score per hole h after three holes. 7.NS.3 ________

44. On Friday morning, the temperature dropped 2 degrees per hour for four hours. Write and solve an equation to find the total number of degrees d the temperature dropped on Friday morning. 7.NS.3 ________

Inquiry Lab

Multiplication Equations with Bar Diagrams

HOW do you know which operation to use when solving an equation?

Content Standards
7.EE.4, 7.EE.4a

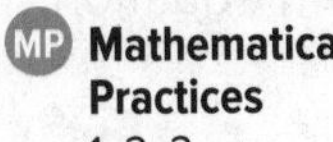
Mathematical Practices
1, 2, 3

Sakiya tutors students to earn money to buy a new Blu-ray™ player that costs \$63. She is able to tutor seven hours in a week. How much should she charge per hour to have enough money by the end of the week?

What do you know? ______________________________

What do you need to find? ______________________________

Hands-On Activity

Step 1 Draw a bar diagram that represents the money Sakiya needs to earn and the number of hours she is available to tutor that week.

\$ ☐

hour 1	hour 2	hour 3	hour 4			

?

Step 2 Write an equation from the bar diagram. Let x represent the amount she should charge each hour.

$$7x = 63$$

Step 3 Use the *work backward* strategy to solve the equation. Since $7x = 63$, $x = 63 \div 7$. So, $x =$ ☐.

Check $7 \times$ ☐ $= 63$ ✓

So, Sakiya should charge ☐ per hour.

Investigate

Work with a partner to solve.

1. The screen on Lin's cell phone allows for 8 lines of text per message. The maximum number of characters for each message is 160. How many characters can each line hold? Complete the bar diagram below and write an equation. Then solve the equation.

Analyze and Reflect

Work with a partner to answer the following question.

2. MP **Make a Conjecture** Refer to Exercise 1. Suppose Lin's cell phone allows 4 lines of text and a maximum of 80 characters for each text message. How would the bar diagram and equation change?

Create

3. MP **Reason Abstractly** Keyani spent $70 for 4 hours of dance classes. How much did she spend per hour of dance class? Draw a bar diagram below and write an equation. Then solve the equation.

4. Inquiry HOW do you know which operation to use when solving an equation?

Lesson 2

Multiplication and Division Equations

Vocabulary Start-Up

The expression $3x$ means *3 times the value of x*. The numerical factor of a multiplication expression like $3x$ is called a **coefficient**. So, 3 is the coefficient of x.

The figure below illustrates the multiplication equation $3x = 6$.

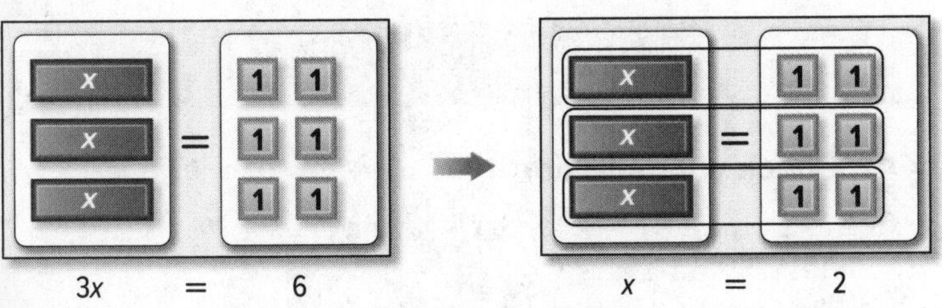

The solution of $3x = 6$ is 2.

Essential Question

WHAT does it mean to say two quantities are equal?

Vocabulary

coefficient
Division Property of Equality
Multiplication Property of Equality

Common Core State Standards

Content Standards
7.EE.4, 7.EE.4a

MP Mathematical Practices
1, 2, 3, 4, 7

Write an equation that represents each of the models below. Identify the coefficient in your equation. Then solve.

1. 

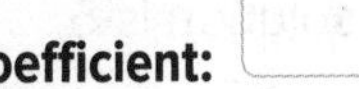

Equation: ______

Coefficient: ☐

Solution: ☐

2. 

Equation: ______

Coefficient: ☐

Solution: ☐

Which MP **Mathematical Practices** did you use? Shade the circle(s) that applies.

① Persevere with Problems
② Reason Abstractly
③ Construct an Argument
④ Model with Mathematics
⑤ Use Math Tools
⑥ Attend to Precision
⑦ Make Use of Structure
⑧ Use Repeated Reasoning

Key Concept

Division Property of Equality

Words The **Division Property of Equality** states that the two sides of an equation remain equal when you divide each side by the same nonzero number.

Symbols If $a = b$ and $c \neq 0$, then $\frac{a}{c} = \frac{b}{c}$.

Work Zone

You can use the Division Property of Equality to solve multiplication equations.

Examples

1. Solve $20 = 4x$. Check your solution.

$20 = 4x$ Write the equation.

$\frac{20}{4} = \frac{4x}{4}$ Division Property of Equality

$5 = x$ Simplify.

Check $20 = 4x$ Write the original equation.

$20 \stackrel{?}{=} 4(5)$ Replace x with 5.

$20 = 20$ ✓ This sentence is true.

So, the solution is 5.

2. Solve $-8y = 24$. Check your solution.

$-8y = 24$ Write the equation.

$\frac{-8y}{-8} = \frac{24}{-8}$ Division Property of Equality

$y = -3$ Simplify.

Check $-8y = 24$ Write the original equation.

$-8(-3) \stackrel{?}{=} 24$ Replace y with -3.

$24 = 24$ ✓ This sentence is true.

So, the solution is -3.

Got it? Do these problems to find out.

Solve each equation. Check your solution.

a. $30 = 6x$ **b.** $-6a = 36$ **c.** $-9d = -72$

Show your work.

a. ____________

b. ____________

c. ____________

Example

3. Lelah sent 574 text messages last week. On average, how many messages did she send each day?

Let m represent the number of messages Lelah sent.

$574 = 7m$	Write the equation. There are 7 days in one week.
$\frac{574}{7} = \frac{7m}{7}$	Division Property of Equality
$82 = m$	Simplify.

Lelah sent 82 messages on average each day.

Solve Arithmetically

You can use a bar diagram to solve an equation arithmetically.

text messages in 1 week, 574						
m	*m*	*m*	*m*	*m*	*m*	*m*

text messages in 1 day

Work backward to solve for *m*.

$m = 574 \div 7 = 82$

Got it? **Do this problem to find out.**

d. Mrs. Acosta's car can travel an average of 24 miles on each gallon of gasoline. Write and solve an equation to find how many gallons of gasoline she will need for a trip of 348 miles.

d. ________

Multiplication Property of Equality

Key Concept

Words The **Multiplication Property of Equality** states that the two sides of an equation remain equal if you multiply each side by the same number.

Symbols If $a = b$, then $ac = bc$.

You can use the Multiplication Property of Equality to solve division equations.

Example

4. Solve $\frac{a}{-4} = -9$.

$\frac{a}{-4} = -9$	Write the equation.
$\frac{a}{-4}(-4) = -9(-4)$	Multiplication Property of Equality
$a = 36$	Simplify.

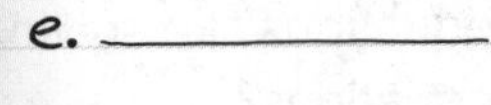

Got it? **Do these problems to find out.**

e. $\frac{y}{-3} = -8$ **f.** $\frac{m}{5} = -7$ **g.** $30 = \frac{b}{-6}$

e. ________

f. ________

g. ________

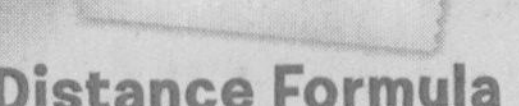

Distance Formula

The distance formula, distance = rate × time, can be written as $d = rt$, $r = \frac{d}{t}$, or $t = \frac{d}{r}$.

Example

5. The distance *d* Tina travels in her car while driving 60 miles per hour for 3 hours is given by the equation $\frac{d}{3} = 60$. How far did she travel?

$\frac{d}{3} = 60$ Write the equation.

$\frac{d}{3}(3) = 60(3)$ Multiplication Property of Equality

$d = 180$ Simplify.

Tina traveled 180 miles.

Guided Practice

Solve each equation. Check your solution. (Examples 1, 2, and 4)

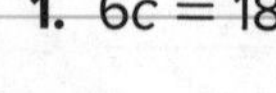

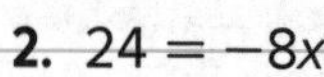

1. $6c = 18$

2. $24 = -8x$

3. $7m = -28$

4. $\frac{p}{9} = 9$

5. $\frac{a}{12} = -3$

6. $\frac{n}{-10} = -4$

7. Antonia earns \$6 per hour helping her grandmother. Write and solve an equation to find how many hours she needs to work to earn \$48. (Example 3) ________

8. A shark can swim at an average speed of 25 miles per hour. At this rate, how far can a shark swim in 2.4 hours? Use $r = \frac{d}{t}$. (Example 5) ________

9. **Building on the Essential Question** How is the process for solving multiplication and division one-step equations like solving one-step addition and subtraction equations?

Rate Yourself!

How confident are you about solving one-step multiplication and division equations? Check the box that applies.

For more help, go online to access a Personal Tutor.

Independent Practice

Go online for Step-by-Step Solutions eHelp

Solve each equation. Check your solution. (Examples 1, 2, and 4)

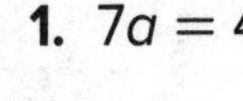

1. $7a = 49$

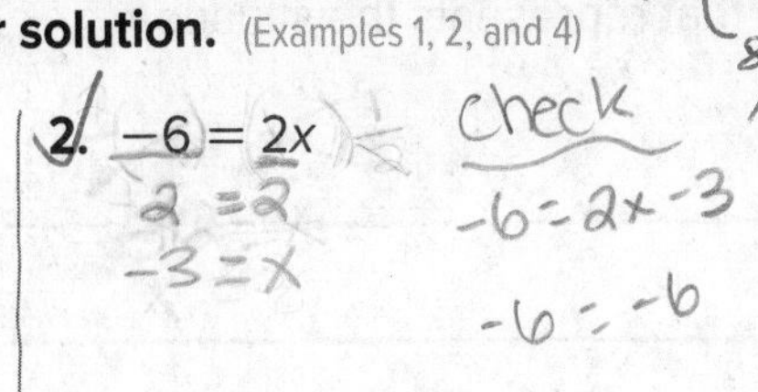

2. $-6 = 2x$

3. $-32 = -4b$

4. $\frac{u}{6} = 9$

5. $-8 = \frac{c}{-10}$

6. $54 = -9d$

7. $-12y = 60$

8. $\frac{r}{20} = -2$

9. $\frac{g}{10} = -9$

10. Brandy wants to buy a digital camera that costs $300. Suppose she saves $15 each week. In how many weeks will she have enough money for the camera? Use a bar diagram to solve arithmetically. Then use an equation to solve algebraically. (Example 3) ______________

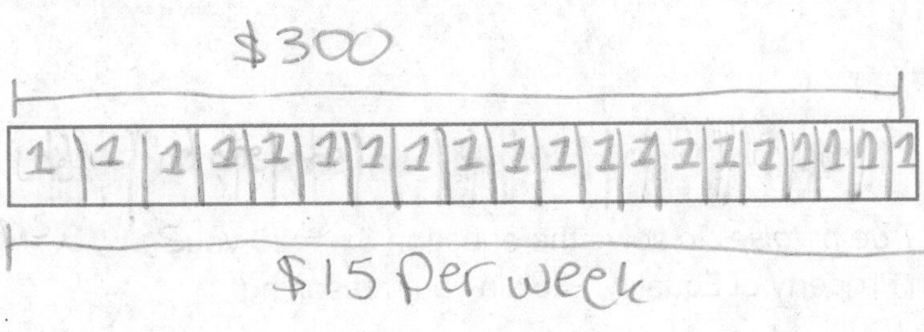

11. A race car can travel at a rate of 205 miles per hour. At this rate, how far would it travel in 3 hours? Use $r = \frac{d}{t}$. Write an equation and then solve. (Example 5)

____

12. A certain hurricane travels at 20.88 kilometers per hour. The distance from Cuba to Key West is 145 kilometers. Write and solve a multiplication equation to find about how long it would take the hurricane to travel from Cuba to Key West.

____

13. MP **Multiple Representations** Kennedy saves $5.50 for each hour she works. She needs to save an additional $44 to buy an E-reader. How many more hours does Kennedy need to work to pay for the E-reader?

a. **Diagram** Draw a bar diagram that represents the situation.

b. **Algebra** Write an equation that represents the situation.

c. **Words** Describe the process you would use to solve your equation. Then solve.

H.O.T. Problems Higher Order Thinking

14. MP **Reason Abstractly** Describe a real-world situation in which you would use a division equation to solve a problem. Write your equation and then solve your problem.

Situation:

Equation: Solution:

15. MP **Identify Structure** *True* or *false*. To solve the equation $5x = 20$ you can use the Multiplication Property of Equality. Explain your reasoning.

False because you have to use the opposite operation. True

16. MP **Persevere with Problems** Solve $3|x| = 12$. Explain your reasoning.

17. MP **Persevere with Problems** Explain how you would solve $\frac{-30}{x} = 6$. Then solve the equation.

First I would multiply -30 by -30 and cancel -30 out & then do the same with 6.

$-30\left(\frac{-30}{x}\right) = (6)-30$

$x = 126$

$x\left(\frac{-30}{x}\right) = (6)$

$-30 = 6x$

-5

Extra Practice

Solve each equation. Check your solution.

18. $-4j = 36$

Homework Help

$-4j = 36$

$\frac{-4j}{-4} = \frac{36}{-4}$

$j = -9$

19. $-4s = -16$

20. $63 = -9d$

21. $\frac{m}{10} = 7$

$\frac{m}{10} = 7$

$\frac{m}{10}(10) = 7(10)$

$m = 70$

22. $\frac{h}{-3} = 12$

23. $\frac{g}{12} = -10$

24. The width of a computer monitor is 1.25 times its height. Find the height of the computer monitor at the right. Use a bar diagram to solve arithmetically. Then use an equation to solve algebraically. ______________

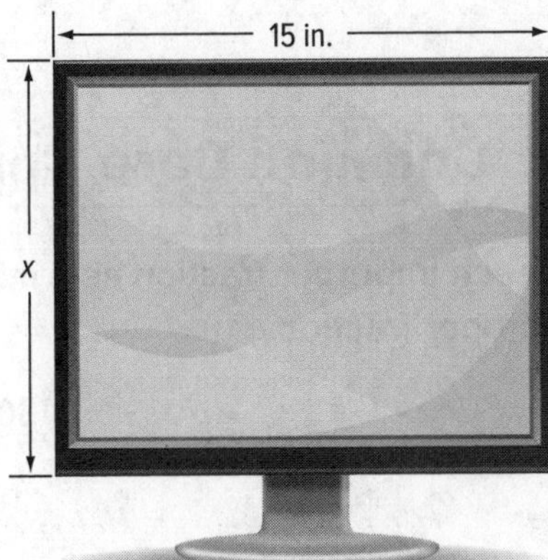

25. A dragonfly, the fastest insect, can fly a distance of 50 feet at a speed of 25 feet per second. Find the time in seconds. Write the equation in the form $d = rt$, then solve.

____

26. MP **Find the Error** Raul is solving $-6x = 72$. Find his mistake and correct it.

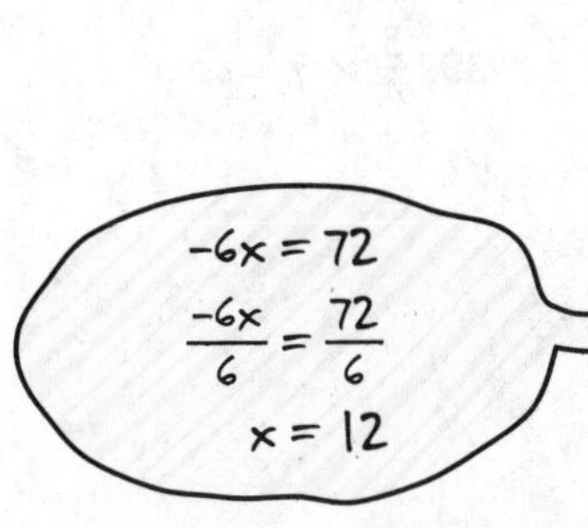

CCSS Power Up! Common Core Test Practice

27. The formula $A = bh$ can be used to find the area A of a parallelogram with base b and height h. The parallelogram shown has an area of 56 square inches.

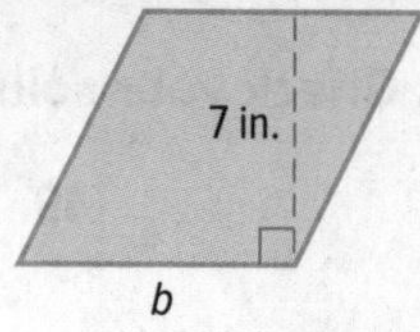

What is the length of the base? [ ]

28. The table shows the prices of different satellite radio plans. Mrs. Freedman paid $99 for m months of satellite radio under Plan A. Fill in each box to write a multiplication equation to represent the situation.

[ ] × [ ] = [ ]

m	16.50
11.99	99
14.35	

Satellite Radio Plans	
Plan	Cost per Month ($)
A	16.50
B	14.35
C	11.99

How many months of service did Mrs. Freedman purchase? [ ]

CCSS Common Core Spiral Review

Write each improper fraction as a mixed number and each mixed number as an improper fraction. 5.NF.3

29. $\frac{10}{3} =$ _______ **30.** $\frac{40}{7} =$ _______ **31.** $\frac{101}{100} =$ _______

32. $2\frac{2}{7} =$ _______ **33.** $3\frac{1}{4} =$ _______ **34.** $10\frac{5}{9} =$ _______

Divide. 6.NS.3

35. $6 \div 1.5 =$ _______ **36.** $3.6 \div 0.4 =$ _______ **37.** $2.73 \div 1.3 =$ _______

Multiply. Write in simplest form. 5.NF.4

38. $\frac{2}{9} \times \frac{7}{5} =$ _______ **39.** $\frac{3}{4} \times 7 =$ _______ **40.** $\frac{5}{8} \times \frac{4}{15} =$ _______

Inquiry Lab

Solve Equations with Rational Coefficients

HOW can you use bar diagrams to solve equations with rational coefficients?

Content Standards
7.EE.4, 7.EE.4a

Mathematical Practices
1, 3

Two thirds of Chen's homeroom class plan to participate in the school talent show. If 16 students from the class plan to participate, how many students are in the homeroom class?

What do you know? ______________________________

____

What do you need to find? ______________________________

____

Hands-On Activity

You can represent the situation above with an equation.

Step 1 Draw a bar diagram that represents the total number of students in the class and how many plan to participate.

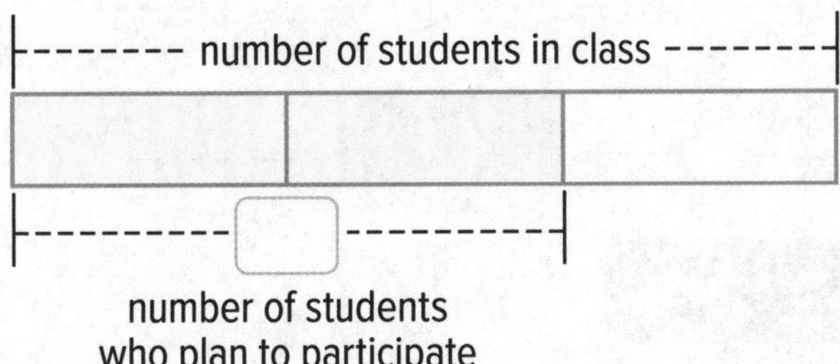

Step 2 Write an equation from the bar diagram. Let c represent the total number of students in the class. ______________________________

Step 3 Find the number of students represented by the sections of the bar. Write that number in each section of the bar in Step 1.

Since each section represents 8 students, there are 8×3 or ☐ students in the class.

Check $\frac{2}{3} \times 24 = \frac{2}{3} \times \frac{24}{1}$

$= \frac{48}{3}$ or 16 ✓

Investigate

Collaborate

Work with a partner to solve the following problem.

1. Eliana is spending $\frac{3}{5}$ of her monthly allowance on a costume for the talent show. She plans to spend $24. Draw a bar diagram to represent the situation. Then write and solve an equation to find the amount of Eliana's monthly allowance.

Equation: ________ Solution: ________

Analyze and Reflect

Collaborate

Work with a partner to answer the following question.

2. **MP Make a Conjecture** Suppose Eliana planned on spending $\frac{3}{4}$ of her monthly allowance on a costume. How would the diagram and equation be different?

Create

On Your Own

3. **MP Model with Mathematics** Write a real-world problem that could be represented by the equation $\frac{2}{3}x = 12$. Then solve the equation.

4. **Inquiry** HOW can you use bar diagrams to solve equations with rational coefficients?

Solve Equations with Rational Coefficients

Real-World Link

Social Networks Three-fourths of the students in Aaliyah's class belong to a social network. There are 15 students in her class that belong to a social network.

1. Create a bar diagram and shade $\frac{3}{4}$, or 0.75, of it.

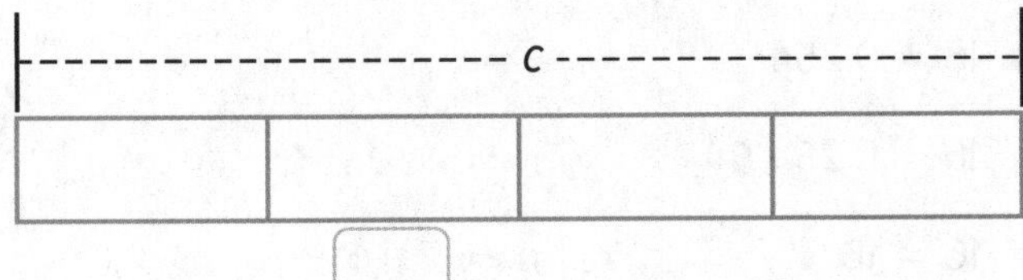

Label 15 along the bottom to show the amount of the bar that represents 15 students.

2. Based on the diagram, circle the equation that can be used to find c, the number of students in Aaliyah's class.

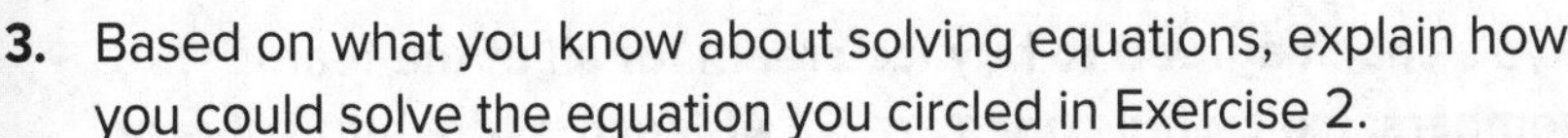

$15c = \frac{3}{4}$ $\qquad$ $0.75c = 15$ $\qquad$ $4c = 15$

3. Based on what you know about solving equations, explain how you could solve the equation you circled in Exercise 2.

4. How many students are in Aaliyah's class?

Essential Question

WHAT does it mean to say two quantities are equal?

Common Core State Standards

Content Standards
7.EE.4, 7.EE.4a

MP Mathematical Practices
1, 2, 3, 4

Which MP Mathematical Practices did you use? Shade the circle(s) that applies.

① Persevere with Problems
② Reason Abstractly
③ Construct an Argument
④ Model with Mathematics
⑤ Use Math Tools
⑥ Attend to Precision
⑦ Make Use of Structure
⑧ Use Repeated Reasoning

Work Zone

Division with Decimals

$$0.25\overline{)16.00} = 64.$$
-150
100
-100
0

a. ____________

b. ____________

c. ____________

d. ____________

Decimal Coefficients

If the coefficient is a decimal, divide each side by the coefficient.

Example

1. Solve 16 = 0.25*n*. Check your solution.

$16 = 0.25n$ — Write the equation.

$\frac{16}{0.25} = \frac{0.25n}{0.25}$ — Division Property of Equality

$64 = n$ — Simplify.

Check $16 = 0.25n$ — Write the original equation.

$16 \stackrel{?}{=} 0.25 \cdot 64$ — Replace n with 64.

$16 = 16$ ✓ — This sentence is true.

The solution is 64.

Got it? **Do these problems to find out.**

a. $6.4 = 0.8m$

b. $-2.8p = 4.2$

c. $-4.7k = -10.81$

Real World

Example

2. Jaya's coach agreed to buy ice cream for all of the team members. Ice cream cones are $2.40 each. Write and solve an equation to find how many cones the coach can buy with $30.

Let n represent the number of cones the coach can buy.

$2.4n = 30$ — Write the equation; $2.40 = 2.4.

$\frac{2.4n}{2.4} = \frac{30}{2.4}$ — Division Property of Equality

$n = 12.5$ — Simplify.

Since the number of ice cream cones must be a whole number, there is enough money for 12 ice cream cones.

Got it? **Do this problem to find out.**

d. Suppose the ice cream cones cost $2.80 each. How many ice cream cones could the coach buy with $42?

Fraction Coefficients

Recall that two numbers with a product of 1 are called multiplicative inverses, or reciprocals. If the coefficient in a multiplication equation is a fraction, multiply each side by the reciprocal of the coefficient.

Examples

3. Solve $\frac{3}{4}x = \frac{12}{20}$.

$\frac{3}{4}x = \frac{12}{20}$ — Write the equation.

$\left(\frac{4}{3}\right) \cdot \frac{3}{4}x = \left(\frac{4}{3}\right) \cdot \frac{12}{20}$ — Multiply each side by the reciprocal of $\frac{3}{4}$, $\frac{4}{3}$.

$\frac{\overset{1}{\cancel{4}}}{\underset{1}{\cancel{3}}} \cdot \frac{\overset{1}{\cancel{3}}}{\underset{1}{\cancel{4}}}x = \frac{\overset{1}{\cancel{4}}}{\underset{1}{\cancel{3}}} \cdot \frac{\overset{4}{\cancel{12}}}{\underset{5}{\cancel{20}}}$ — Divide by common factors.

$x = \frac{4}{5}$ — Simplify. Check the solution.

---

4. Solve $-\frac{7}{9}d = 5$. Check your solution.

$-\frac{7}{9}d = 5$ — Write the equation.

$\left(-\frac{9}{7}\right) \cdot \left(-\frac{7}{9}\right)d = \left(-\frac{9}{7}\right) \cdot 5$ — Multiply each side by the reciprocal of $-\frac{7}{9}$, $-\frac{9}{7}$.

$\left(-\frac{9}{7}\right) \cdot \left(-\frac{7}{9}\right)d = \left(-\frac{9}{7}\right) \cdot \frac{5}{1}$ — Write 5 as $\frac{5}{1}$.

$\left(-\frac{\overset{1}{\cancel{9}}}{\underset{1}{\cancel{7}}}\right) \cdot \left(-\frac{\overset{1}{\cancel{7}}}{\underset{1}{\cancel{9}}}\right)d = \left(-\frac{9}{7}\right) \cdot \frac{5}{1}$ — Divide by common factors.

$d = -\frac{45}{7}$ or $-6\frac{3}{7}$ — Simplify.

Check $-\frac{7}{9}d = 5$ — Write the original equation.

$-\frac{7}{9}\left(-\frac{45}{7}\right) \stackrel{?}{=} 5$ — Replace d with $-\frac{45}{7}$.

$\frac{315}{63} \stackrel{?}{=} 5$ — Simplify.

$5 = 5$ ✓ — This sentence is true.

Got it? Do these problems to find out.

e. $\frac{1}{2}x = 8$

f. $-\frac{3}{4}x = 9$

g. $-\frac{7}{8}x = -\frac{21}{64}$

Fractions as Coefficients

The expression $\frac{3}{4}x$ can be read as $\frac{3}{4}$ of x, $\frac{3}{4}$ multiplied by x, $3x$ divided by 4, or $\frac{x}{4}$ multiplied by 3.

e. ______

f. ______

g. ______

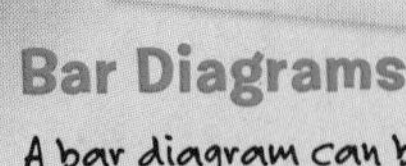

Bar Diagrams

A bar diagram can be used to represent this situation.

$n = 6 \div \frac{2}{3} = \frac{6}{1} \times \frac{3}{2} = 9$

Real World Example

5. Valerie needs $\frac{2}{3}$ yard of fabric to make each hat for the school play. Write and solve an equation to find how many hats she can make with 6 yards of fabric.

Write and solve a multiplication equation. Let n represent the number of hats.

$\frac{2}{3}n = 6$	Write the equation.
$\left(\frac{3}{2}\right) \cdot \frac{2}{3}n = \left(\frac{3}{2}\right) \cdot 6$	Multiply each side by $\frac{3}{2}$.
$n = 9$	Simplify.

So, Valerie can make 9 hats.

Guided Practice

Solve each equation. Check your solution. (Examples 1, 3, and 4)

1. $1.6k = 3.2$

2. $-2.5b = 20.5$

3. $-\frac{1}{2} = -\frac{5}{18}h$

Write and solve an equation. (Examples 2 and 5)

4. The average growth of human hair is 0.5 inch per month. Find how long it takes a human to grow 3 inches of hair.

   Equation: ______ Solution: ______

5. Three fourths of the fruit in a refrigerator are apples. There are 24 apples in the refrigerator. How many pieces of fruit are in the refrigerator?

   Equation: ______ Solution: ______

6. **Building on the Essential Question** What is the process for solving a multiplication equation with a rational coefficient? ______________________

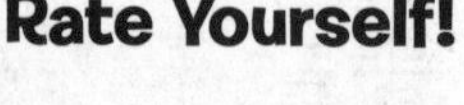

Rate Yourself!

Are you ready to move on? Shade the section that applies.

YES ? NO

For more help, go online to access a Personal Tutor.

Name ______________________ My Homework ______________________

Independent Practice

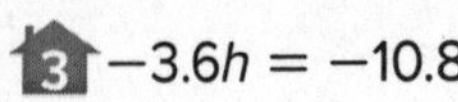

Solve each equation. Check your solution. (Examples 1, 3, and 4)

1. $1.2x = 6$

2. $14.4 = -2.4b$

3 $-3.6h = -10.8$

4. $\frac{2}{5}t = \frac{12}{25}$

5. $-3\frac{1}{3} = -\frac{1}{2}g$

6. $-\frac{7}{9}m = \frac{11}{6}$

7 **Financial Literacy** Dillon deposited $\frac{3}{4}$ of his paycheck into the bank. The deposit slip shows how much he deposited. Write and solve an equation to find the amount of his paycheck. (Example 2)

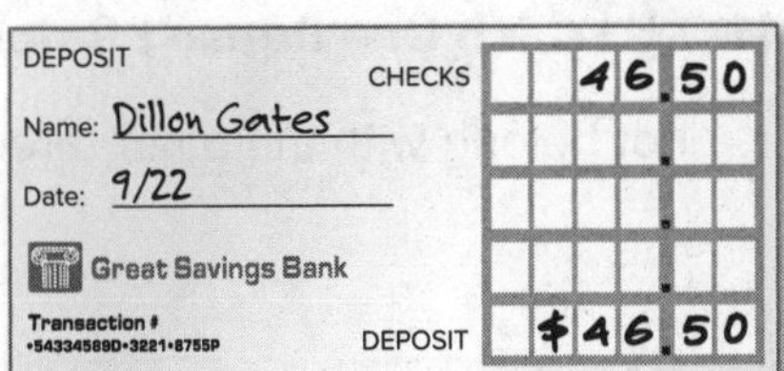

Equation: __________ Solution: __________

8. Twenty-four students brought their permission slips to attend the class field trip to the local art museum. If this represented eight tenths of the class, how many students are in the class? Use a bar diagram to solve arithmetically. Then use an equation to solve algebraically. (Example 5)

Equation: __________ Solution: __________

9. MP **Justify Conclusions** Seventy-five percent, or 15, of the students in Emily's homeroom class are going on a field trip. Two thirds, or 12, of the students in Santiago's homeroom class are going on the field trip. Which class has more students? Justify your answer. ______________________

10. **MP Reason Abstractly** Nora and Ryan are making stuffed animals for a toy drive. The table shows the fabric purchases they made. Who purchased the more expensive fabric?

Purchaser	Amount Purchased (yd)	Amount Paid ($)
Nora	$\frac{2}{3}$	4
Ryan	0.8	6

Explain your reasoning. ____________

H.O.T. Problems Higher Order Thinking

11. **MP Reason Inductively** Complete the statement: If $8 = \frac{m}{4}$, then $m - 12 = ■$. Explain. ____________

12. **MP Which One Doesn't Belong?** Identify the pair of numbers that does not belong with the other three. Explain. ____________

$\frac{9}{6}, \frac{6}{9}$	$4, \frac{1}{4}$	$\frac{3}{5}, 5$	$\frac{2}{7}, \frac{7}{2}$

13. **MP Persevere with Problems** The formula for the area of a trapezoid is $A = \frac{1}{2}h(b_1 + b_2)$, where b_1 and b_2 are both bases and h is the height. Find the value of h in terms of A, b_1, and b_2. Justify your answer.

14. **MP Model with Mathematics** Write a real-world problem that can be represented by the equation $224 = 3.5r$. Then solve the problem and explain the solution.

Name ______________________ My Homework ______________________

Extra Practice

Solve each equation. Check your solution.

15. $0.4d = 2.8$

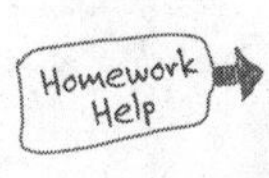

$0.4d = 2.8$

$\frac{0.4d}{0.4} = \frac{2.8d}{0.4}$

$d = 7$

16. $-5w = -24.5$

17. $-22.8 = 6n$

18. $\frac{7}{8}k = \frac{5}{6}$

$\frac{7}{8}k = \frac{5}{6}$

$\left(\frac{8}{7}\right) \cdot \frac{7}{8}k = \left(\frac{8}{7}\right) \cdot \frac{5}{6}$

$k = \frac{40}{42}$ or $\frac{20}{21}$

19. $-6\frac{1}{4} = \frac{3}{5}c$

20. $-\frac{4}{7}v = -8\frac{2}{3}$

21. The Mammoth Cave Discovery Tour includes an elevation change of 140 feet. This is $\frac{7}{15}$ of the elevation change on the Wild Cave Tour. What is the elevation change on the Wild Cave Tour? Use a bar diagram to solve arithmetically. Then use an equation to solve algebraically.

Equation: __________ Solution: __________

22. MP **Model with Mathematics** Refer to the graphic novel frame below. Write and solve an equation to find how many movies they have time to show.

Equation: __________ Solution: __________

Power Up! Common Core Test Practice

23. Which of the following high speed trains are traveling at a rate of 150 miles per hour? Select all that apply.

- ☐ a train that travels 100 miles in $\frac{2}{3}$ hour
- ☐ a train that travels 160 miles in $\frac{5}{6}$ hour
- ☐ a train that travels 125 miles in $\frac{4}{5}$ hour
- ☐ a train that travels 90 miles in $\frac{3}{5}$ hour

24. The table shows the results of a survey. Of those surveyed, 275 students said they prefer pop music.

Music Preference	
Type	**Fraction of Students**
Jazz	$\frac{1}{8}$
Pop	$\frac{5}{8}$
Rap	$\frac{1}{4}$

Write an equation that could be used to find the total number of students *s* who were surveyed. [ ]

How many students were surveyed? [ ]

Common Core Spiral Review

Use the order of operations to evaluate each expression. 6.EE.2c

25. 6 × 4 − 2 = ______

26. 70 − 5 × 4 = ______

27. 18 ÷ 2 − 7 = ______

28. Write *add, divide, multiply,* and *subtract* in the correct order to complete the following sentence. 6.EE.2c
When using the order of operations to evaluate an expression,

always ____________ and ____________ before you ____________

and ____________.

Write and evaluate an expression for each situation. 6.EE.1

29. Used paperback books are $0.25, and hardback books are $0.50. If you buy 3 paperback books and 5 hardback books, how much money do you spend?

Expression: ______________________ Solution: ____________

30. Suppose you order 2 pizzas, 2 garlic breads, and 1 order of BBQ wings. How much change would you receive from $30?

Item	Cost
14" pizza	$8
garlic bread	$2
BBQ wings	$4

Expression: ______________________ Solution: ____________

Inquiry Lab

Solve Two-Step Equations

HOW can a bar diagram or algebra tiles help you solve a real-world problem?

Content Standards
7.EE.4, 7.EE.4a

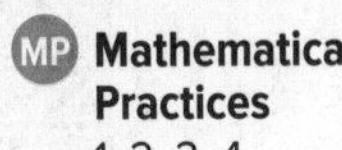

Mathematical Practices
1, 2, 3, 4

Latoya plays basketball and tennis. She has two basketballs and three tennis balls that weigh a total of 48 ounces. Each tennis ball weighs 2 ounces. What is the weight of a basketball?

Hands-On Activity 1

You can use a bar diagram to represent the situation.

Step 1 Draw a bar diagram that represents the total weight.

Step 2 Write an equation that is modeled by the bar diagram.
Let x represent the weight of a basketball.

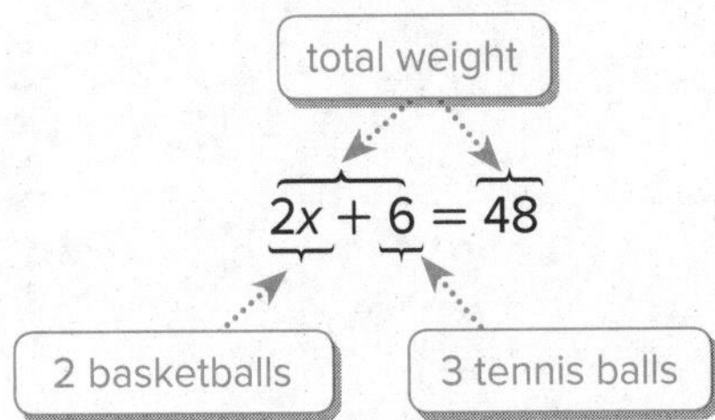

Step 3 Use the bar diagram to solve the equation. Subtract the weight of the tennis balls, ___ ounces, from the total weight, ___ ounces.

The two basketballs together weigh ___ − ___, or ___ ounces.

Divide the weight by ___ to find the weight of one basketball.

So, $x =$ ___. The weight of one basketball is ___ ÷ ___, or ___ ounces.

Check $2 \cdot$ ___ $+ 6 = 48$ ✓

The weight of one basketball is ___ ounces.

Hands-On Activity 2

You can use algebra tiles to model and solve the equation $4x - 2 = 10$.

Step 1 Model the equation.

$4x - 2 = 10$

Step 2 Add ____ 1-tiles to each side of the mat to form zero pairs on the left side.

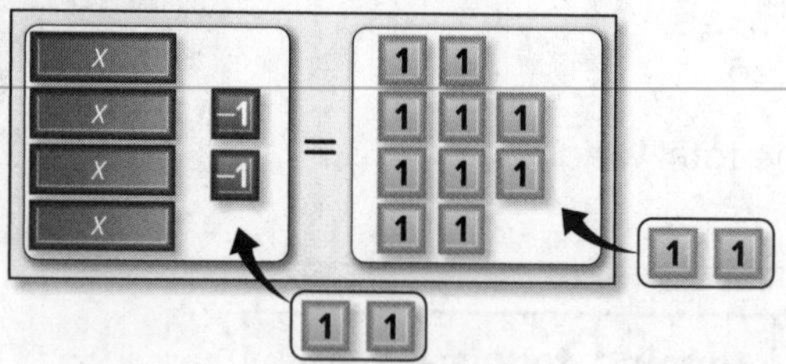

$4x - 2 + 2 = 10 + 2$

Step 3 Remove both zero pairs from the left side so that the variable is by itself.

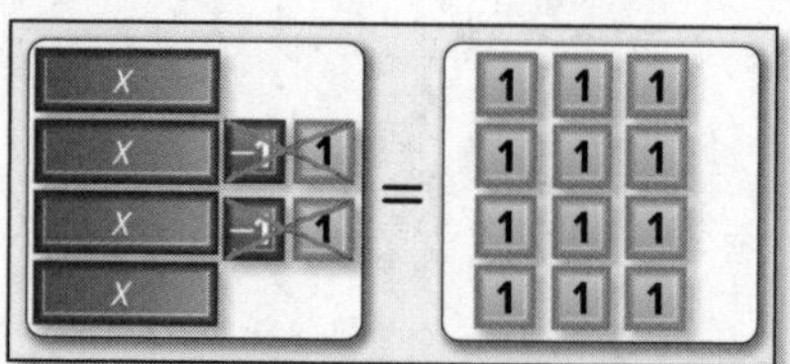

$4x = 12$

Step 4 Divide the remaining tiles into ____ equal groups.

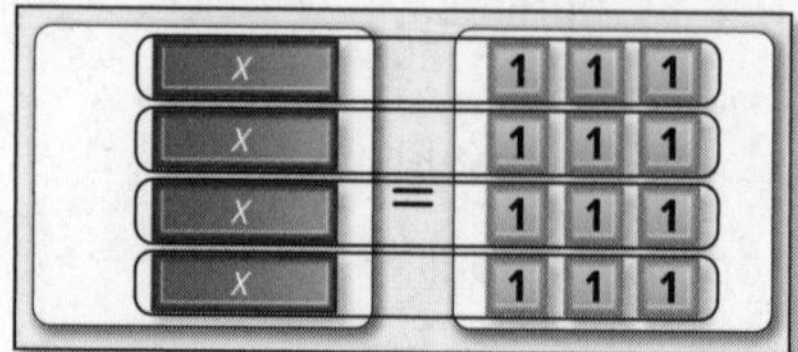

$\frac{4x}{4} = \frac{12}{4}$

So, $x =$ ____.

Check $4 \cdot$ ____ $- 2 = 10$ ✓

Investigate

Work with a partner to solve the following problem.

1. **MP Reason Abstractly** Ryan is saving money to buy a skateboard that costs \$85. He has already saved \$40. He plans to save the same amount each week for three weeks. Draw a bar diagram. Then write an equation. How much should Ryan save each week?

Work with a partner to solve each equation. Use algebra tiles. Show your work using drawings.

2. $2x + 1 = 5$ $x =$ ______

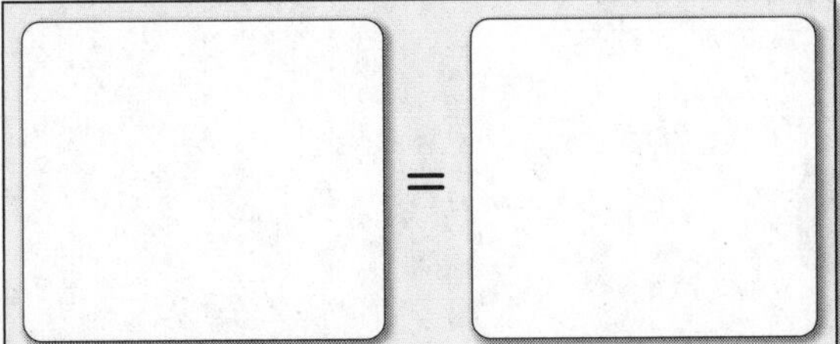

3. $3x + 2 = 11$ $x =$ ______

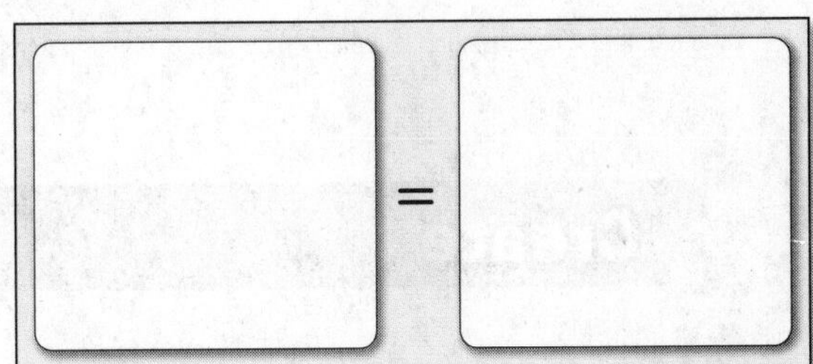

4. $4x + 3 = -5$ $x =$ ______

5. $2x - 1 = 7$ $x =$ ______

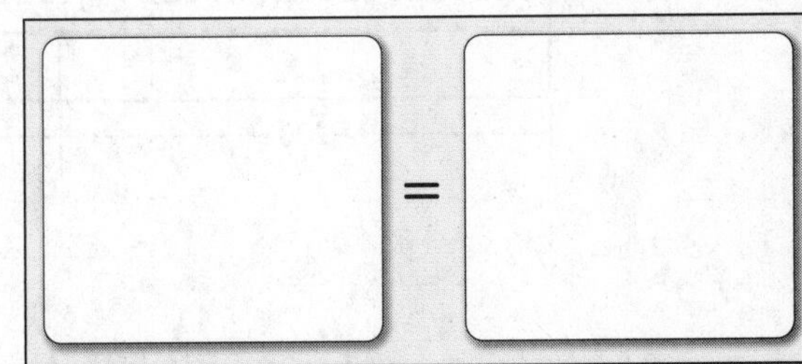

6. $5x - 2 = -7$ $x =$ ______

7. $3x - 4 = 5$ $x =$ ______

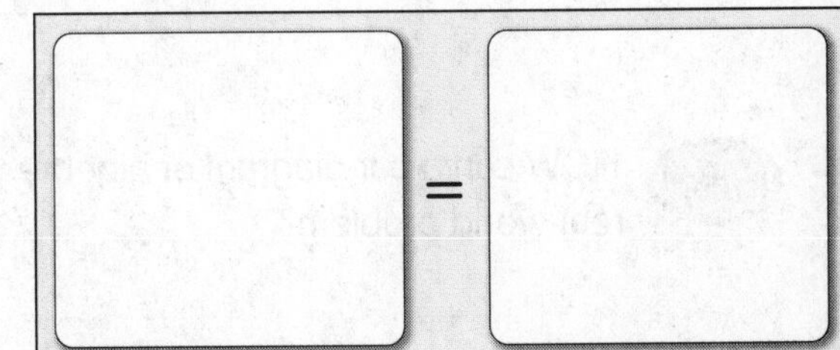

Analyze and Reflect

8. MP **Reason Inductively** Work with a partner. Read the steps to model and solve an equation using algebra tiles. Then circle each correct equation.

Steps to Solve	Choices of Equation		
• Add three 1-tiles to each side of the mat. • Divide tiles into two equal groups.	$2x + 3 = 15$	$3x + 2 = 15$	$2x - 3 = 15$
• Add four 1-tiles to each side of the mat. • Divide tiles into three equal groups.	$3x - 4 = 11$	$3x + 4 = 11$	$4x - 3 = 11$
• Remove seven 1-tiles from each side of the mat. • Divide tiles into three equal groups.	$7x + 3 = 10$	$3x + 7 = 10$	$3x - 7 = 10$
• Add two −1-tiles to each side of the mat. • Remove two zero pairs from the left side of the mat. • Divide tiles into five equal groups.	$5x - 2 = -8$	$5x + 2 = -8$	$2x + 5 = -8$

9. MP **Construct an Argument** What did you observe while choosing the correct equations in the table above?

Create

10. MP **Model with Mathematics** Write a real-world problem and an equation that the bar diagram below could represent. Then solve your problem.

540

200 ? ?

11. Inquiry HOW can a bar diagram or algebra tiles help you solve a real-world problem?

Solve Two-Step Equations

Real-World Link

Balloons A company charges \$2 for each balloon in an arrangement and a \$3 delivery fee. You have \$9 to spend. The equation $2x + 3 = 9$, where x is the number of balloons, represents the situation. Work backward to solve for x.

Start with the amount of money you have to spend. ☐ → Subtract the \$3 delivery fee. ☐ → Since each balloon is \$2, divide by two. ☐

So, you can purchase ☐ balloons.

Check your work by substituting your solution into the equation.

$$2(\square) + 3 \stackrel{?}{=} 9.$$

$$\square + 3 \stackrel{?}{=} 9$$

$$\square = 9$$

1. How many balloons could you have purchased if there was a \$1 delivery charge?

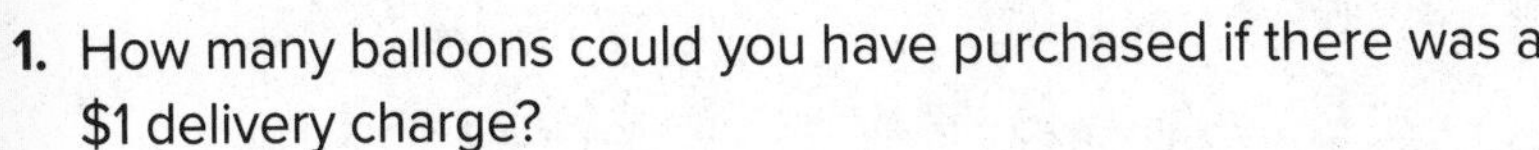

Start with the amount of money you have to spend. ☐ → Subtract the \$1 delivery fee. ☐ → Since each balloon is \$2, divide by two. ☐

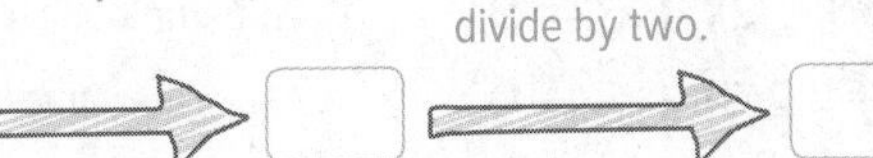

Which MP **Mathematical Practices** did you use? Shade the circle(s) that applies.

① Persevere with Problems
② Reason Abstractly
③ Construct an Argument
④ Model with Mathematics
⑤ Use Math Tools
⑥ Attend to Precision
⑦ Make Use of Structure
⑧ Use Repeated Reasoning

Essential Question

WHAT does it mean to say two quantities are equal?

Vocabulary

two-step equation

Common Core State Standards

Content Standards
7.EE.4, 7.EE.4a

MP **Mathematical Practices**
1, 2, 3, 4

Work Zone

Solve Two-Step Equations

Recall that the *order of operations* ensures that numerical expressions, such as $2 \cdot 5 + 3$, have only one value. To reverse the operations, undo them in reverse order.

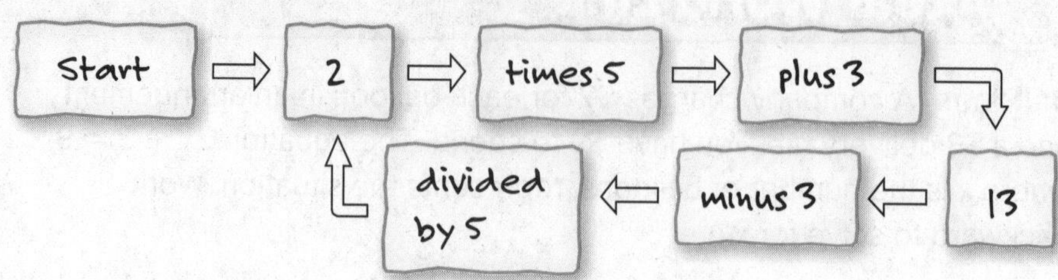

This equation is written as $px + q = r$, where p, q, and r are rational numbers.

A **two-step equation**, such as $2x + 3 = 9$, has two different operations, multiplication and addition. To solve a two-step equation, undo the operations in reverse order of the order of operations.

Step 1 Undo the addition or subtraction first.

Step 2 Undo the multiplication or division.

Examples

Tutor

1. Solve $2x + 3 = 9$. Check your solution.

$2x + 3 = 9$	Write the equation.
$-3 = -3$	Undo the addition first by subtracting 3 from each side.
$2x = 6$	
$\frac{2x}{2} = \frac{6}{2}$	Next, undo the multiplication by dividing each side by 2.
$x = 3$	Simplify.

Check $2x + 3 = 9$	Write the original equation.
$2(3) + 3 \stackrel{?}{=} 9$	Replace x with 3.
$9 = 9$ ✓	The sentence is true.

The solution is 3.

STOP and Reflect

What are the two operations you would perform to solve $3x - 4 = 8$? Write your answer below.

2. **Solve $3x + 2 = 23$. Check your solution.**

$3x + 2 = 23$	Write the equation.
$-2 = -2$	Undo the addition first by subtracting 2 from each side.
$3x = 21$	
$\frac{3x}{3} = \frac{21}{3}$	Division Property of Equality
$x = 7$	Simplify.

Check $3x + 2 = 23$	Write the original equation.
$3(7) + 2 \stackrel{?}{=} 23$	Replace x with 7.
$23 = 23$ ✓	The sentence is true.

The solution is 7.

3. **Solve $-2y - 7 = 3$. Check your solution.**

$-2y - 7 = 3$	Write the equation.
$+7 = +7$	Undo the subtraction first by adding 7 to each side.
$-2y = 10$	
$\frac{-2y}{-2} = \frac{10}{-2}$	Division Property of Equality
$y = -5$	Simplify.
The solution is -5.	Check the solution.

Equations

Remember, solutions of the new equation are also solutions of the original equation.

4. **Solve $4 + \frac{1}{5}r = -1$. Check your solution.**

$4 + \frac{1}{5}r = -1$	Write the equation.
$-4 \quad = -4$	Undo the addition first by subtracting 4 from each side.
$\frac{1}{5}r = -5$	
$5 \cdot \frac{1}{5}r = 5 \cdot (-5)$	Multiplication Property of Equality
$r = -25$	Simplify.
The solution is -25.	Check the solution.

Got it? **Do these problems to find out.**

Solve each equation. Check your solution.

a. $2x + 4 = 10$ **b.** $3x + 5 = 14$ **c.** $5 = 2 + 3x$

d. $4x + 5 = 13$ **e.** $-5s + 8 = -2$ **f.** $-2 + \frac{2}{3}w = 10$

a. ________

b. ________

c. ________

d. ________

e. ________

f. ________

Solve Arithmetically

You can use a bar diagram to solve an equation arithmetically.

$78	
pizza	tickets
$27	$8.50n

Subtract 27 from 78. Then divide by 8.5.

78 − 27 = 51; 51 ÷ 8.5 = 6

Example

5. **Toya had her birthday party at the movies. It cost $27 for pizza and $8.50 per friend for the movie tickets. How many friends did Toya have at her party if she spent $78?**

Words	Cost of pizza	plus	Cost of 1 friend	times	number of friends	equals $78.
Variable	Let n represent the number of friends.					
Equation	27	+	8.50	·	n	= 78

$27 + 8.50n = 78$ Write the equation.

$-27 \quad = -27$ Subtract 27 from each side.

$8.50n = 51$

$\frac{8.50n}{8.50} = \frac{51}{8.50}$ Division Property of Equality

$n = 6$ Simplify.

Toya can have 6 friends at her party.

Guided Practice

Solve each equation. Check your solution. (Examples 1–4)

1. $13 = 1 + 4s$

2. $-3y - 5 = 10$

3. $-7 = 1 + \frac{2}{3}n$

4. Syreeta wants to buy some CDs that each cost $14, and a DVD that costs $23. She has $65. Write and solve an equation to find how many CDs she can buy. (Example 5)

Equation: ______________

Solution: ______________

5. **Building on the Essential Question** When solving an equation, explain why it is important to perform identical operations on each side of the equals sign.

____

____

____

Rate Yourself!

How well do you understand solving two-step equations? Circle the image that applies.

Clear

Somewhat Clear

Not So Clear

For more help, go online to access a Personal Tutor.

Time to update your Foldable!

Name ________________ My Homework ________________

Independent Practice

Go online for Step-by-Step Solutions eHelp 

Solve each equation. Check your solution. (Examples 1–4)

1. $3x + 1 = 10$

2. $-3 + 8n = -5$

3. $4h - 6 = 22$

4. $-8s + 1 = 33$

5. $-4w - 4 = 8$

6. $5 + \frac{1}{7}b = -2$

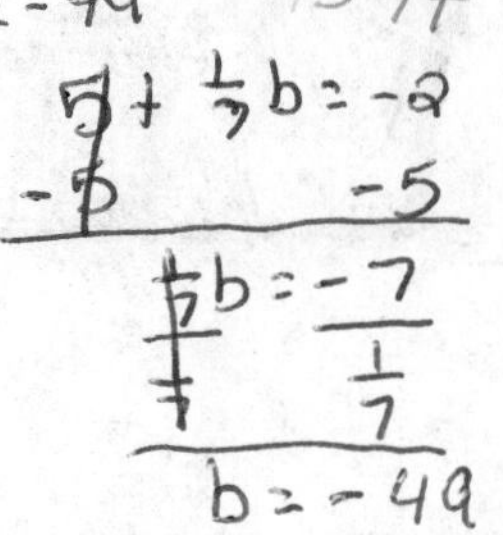

7. **MP Reason Abstractly** Cristiano is saving money to buy a bike that costs \$189. He has saved \$99 so far. He plans on saving \$10 each week. In how many weeks will he have enough money to buy the bike? Use a bar diagram to solve arithmetically. Then use an equation to solve algebraically. (Example 5)

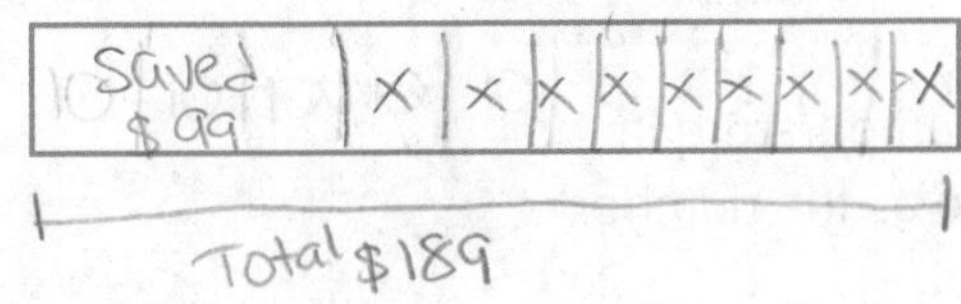

Solve each equation. Check your solution.

8. $2r - 3.1 = 1.7$

9. $4t + 3.5 = 12.5$

10. $8m - 5.5 = 10.1$

11. Temperature is usually measured on the Fahrenheit scale (°F) or the Celsius scale (°C). Use the formula $F = 1.8C + 32$ to convert from one scale to the other.

Alaska Record Low Temperatures (°F) by Month	
January	−80
April	−50
July	16
October	−48

a. Convert the temperature for Alaska's record low in July to Celsius. Round to the nearest degree.

b. Hawaii's record low temperature is −11°C. Find the difference in degrees Fahrenheit between Hawaii's record low temperature and the record low temperature for Alaska in January.

12. **Model with Mathematics** Refer to the graphic novel frame below. Jamar figured that they will spend $39 for popcorn. Each movie cost $19. Write and solve an equation to find how many movies they can purchase.

100 - 39 ÷ 19 = x

H.O.T. Problems Higher Order Thinking

13. **Reason Inductively** Refer to Exercise 11. Is there a temperature in the table at which the number of degrees Celsius is the same as the number of degrees Fahrenheit? If so, find it. If not, explain why not.

14. **Persevere with Problems** Suppose your school is selling magazine subscriptions. Each subscription costs $20. The company pays the school half of the total sales in dollars. The school must also pay a one-time fee of $18. Write and solve an equation to determine the fewest number of subscriptions that can be sold to earn a profit of $200.

22 subscriptions 20+x - 18 = 200 ½(20x) - 18 = 200 10x - 18 = 200

15. **Model with Mathematics** Write a real-world problem that can be represented by the equation $\frac{(12 + 14) \times h}{2} = 52$. Then solve the problem.

Name ______________________ My Homework ______________________

Extra Practice

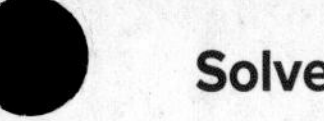

Solve each equation. Check your solution.

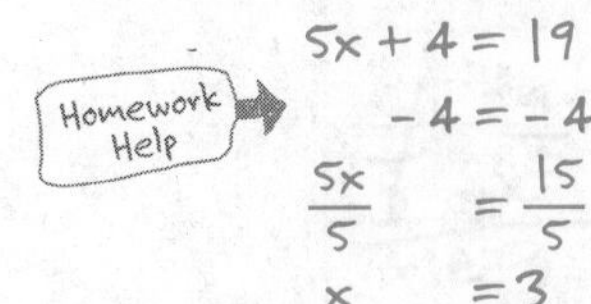

16. $5x + 4 = 19$

17. $6m + 1 = -23$

18. $5 + 4d = 37$

19. $-7y + 3 = -25$

20. $25 + \frac{11}{12}b = 47$

21. $15 - \frac{1}{2}b = -3$

22. It costs $7.50 to enter a petting zoo. Each cup of food to feed the animals is $2.50. If you have $12.50, how many cups can you buy? Use a bar diagram to solve arithmetically. Then use an equation to solve algebraically.

23. MP **Multiple Representations** The perimeter of a rectangle is 48 centimeters. Its length is 16 centimeters. What is the width w?

a. Draw a bar diagram that represents this situation.

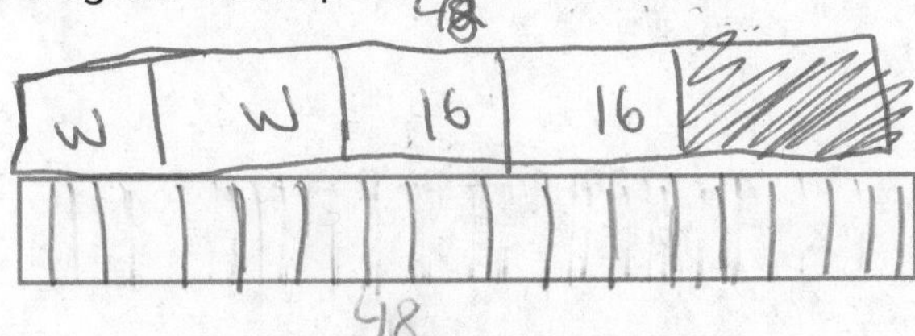

b. Write and solve an equation that represents this situation.

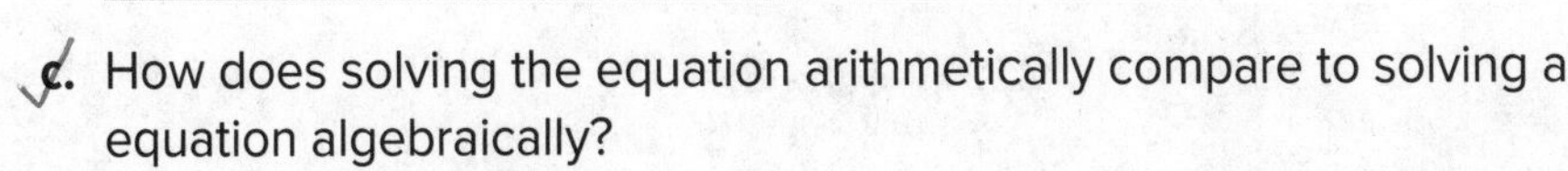

c. How does solving the equation arithmetically compare to solving an equation algebraically?

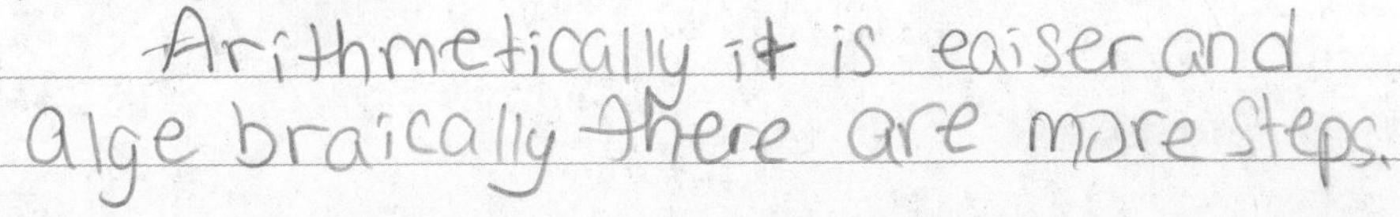

CCSS Power Up! Common Core Test Practice

24. Admission to an amusement park costs $15 and game tickets cost $0.50 each. Craig has $22 to pay for admission and game tickets. Select the correct labels to complete the bar diagram that can be used to find the number of game tickets *t* that Craig can purchase.

admission
game tickets
0.50
15
22
0.50*t*
t

How many game tickets can Craig purchase?

25. A rental car company charges a fixed fee of $30 plus $0.05 per mile. Let *c* represent the total cost of renting a car and driving it *m* miles.

Write an equation that could be used to find the total cost of renting a car and driving it any number of miles.

The Boggs family paid $49.75 for their car rental. How many miles did they drive?

CCSS Common Core Spiral Review

Use the Distributive Property to rewrite each expression. 6.EE.3

26. $2(x + 7) =$

27. $6(10 + n) =$

28. $5(k - 4) =$

Factor each expression. 6.NS.4

29. $5x + 5 \cdot 7 =$

30. $4n + 4 \cdot 2 =$

31. $10t + 10 \cdot 3 =$

32. $7v + 7 \cdot 8 =$

Inquiry Lab

More Two-Step Equations

HOW are equations in $p(x + q) = r$ form different from $px + q = r$ equations?

Content Standards
7.EE.4, 7.EE.4a

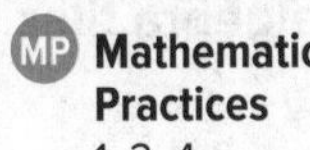

Mathematical Practices
1, 3, 4

Mark has two summer jobs. He babysits and helps with the gardening. He works at each job three days a week and earns a total of \$240. The table shows his earnings each day. How much does he earn each day babysitting?

Job	Daily Earnings (\$)
Babysitting	x
Gardening	30

What do you know? ______________________________

____

What do you need to find? ______________________________

____

Hands-On Activity 1

Step 1 Draw a bar diagram that represents the situation.

\$x + \$30	\$x + \$30	\$x + \$30
earnings each day	earnings each day	earnings each day

Step 2 Write an equation that is modeled by the bar diagram.

$3(\$x + \$30) =$ ☐

From the diagram, you can see that one third of Mark's total earnings is equal to $\$x + \$30$. So, $\$x + \$30 = \frac{\$240}{3}$ or ☐.

Mark earns ☐ – \$30, or ☐ each day babysitting.

Vijay and his brother bought two hamburgers and two lemonades. The hamburgers cost $6 each. They spent a total of $16. How much did each lemonade cost?

Hands-On Activity 2

Use algebra tiles to model the situation described above.

Step 1 Model $2(x + 6) = 16$ using algebra tiles. Use ___ groups of $(x + 6)$ tiles.

$2(x + 6) \quad = \quad 16$

Step 2 Divide the tiles into ___ equal groups on each side of the mat. Remove ___ group from each side.

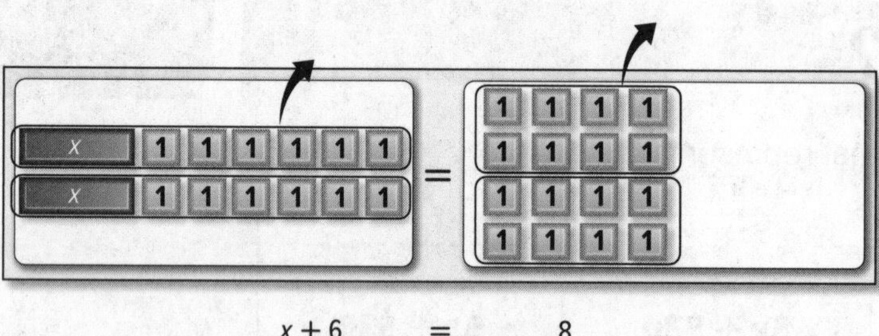

$x + 6 \quad = \quad 8$

Step 3 Remove the same number of 1-tiles from each side.

$x \quad = \quad 2$

So, $x =$ ___. Each lemonade costs ___.

Investigate

Work with a partner to model and solve each equation. Use a bar diagram for Exercises 1 and 2. Use algebra tiles for Exercises 3–6.

1. $3(x + 5) = 21$ $x =$ ______

2. $2(x - 3) = 10$ $x =$ ______

3. $4(x + 1) = 8$ $x =$ ______

4. $3(x + 2) = -12$ $x =$ ______

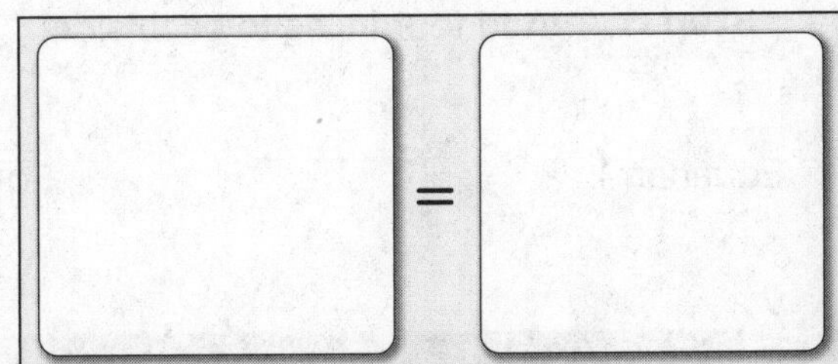

5. $2(x - 1) = 6$ $x =$ ______

6. $3(x - 4) = -3$ $x =$ ______

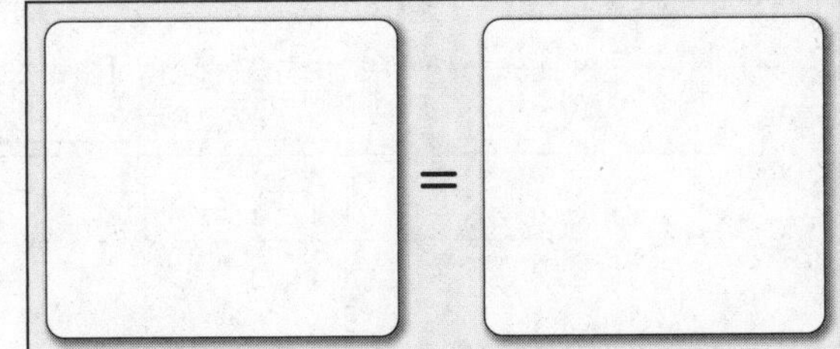

Analyze and Reflect

Collaborate

Work with a partner to write and solve an equation that represents each problem.

7. Refer to Activity 1. If Mark worked four days a week and made \$360, how much did he earn babysitting each day?

8. Refer to Activity 2. If Vijay and his brother spent a total of \$15, how much did each lemonade cost?

9. MP **Reason Inductively** After modeling an equation using algebra tiles, Angelina used the steps shown below to solve the equation. Write two different equations in $p(x + q) = r$ form that Angelina could have solved.

Step 1 Divide the tiles into three equal groups on both sides of the mat.

Step 2 Remove two groups from each side.

Step 3 Add four 1-tiles to each side.

Equation 1: ________ Equation 2: ________

Create

On Your Own

10. MP **Model with Mathematics** Write a real-world problem that can be represented by the equation $4(x + 15) = 140$. Then solve the problem.

11. Inquiry HOW are equations in $p(x + q) = r$ form different from $px + q = r$ equations?

Lesson 5

More Two-Step Equations

Real-World Link

Museums A new exhibit about dinosaurs is being constructed. The exhibit is a rectangle that is 36 feet long. It has a perimeter of 114 feet. Follow the steps to write an equation that can be used to find the width of the museum exhibit.

Essential Question

WHAT does it mean to say two quantities are equal?

Common Core State Standards

Content Standards
7.EE.4, 7.EE.4a

MP Mathematical Practices
1, 2, 3, 4

Step 1 Draw a diagram to help visualize the exhibit.

Label the length and width. Let w represent the width.

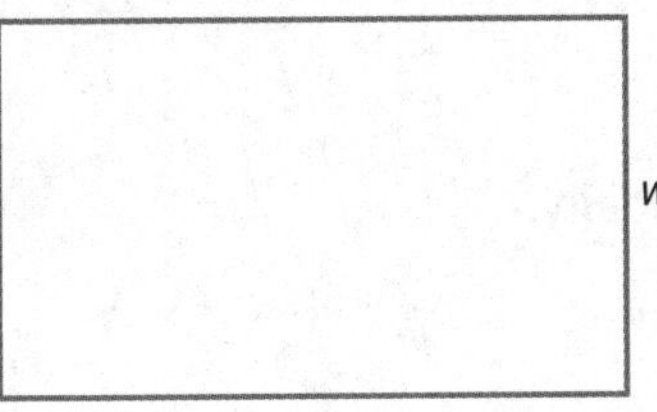

Step 2 Write an expression that represents the sum of the length and width of the exhibit. ________

Step 3 Write an expression that represents twice the sum of the length and width. ________

Step 4 Write an equation that represents the perimeter of the exhibit. ________

Which MP Mathematical Practices did you use? Shade the circle(s) that applies.

① Persevere with Problems
② Reason Abstractly
③ Construct an Argument
④ Model with Mathematics
⑤ Use Math Tools
⑥ Attend to Precision
⑦ Make Use of Structure
⑧ Use Repeated Reasoning

Work Zone

Solve Two-Step Equations

An equation like $2(w + 36) = 114$ is in the form $p(x + q) = r$. It contains two factors, p and $(x + q)$, and is considered a two-step equation. Solve these equations using the properties of equality.

Examples

1. **Solve $3(x + 5) = 45$.**

Method 1 **Solve arithmetically.**

x + 5	x + 5	x + 5

(Total: 45; one part: ?)

Draw a bar diagram. From the diagram, you can see that $x + 5 = 45 \div 3$ or 15. So, $x = 15 - 5$ or 10.

Method 2 **Solve algebraically.**

$3(x + 5) = 45$ Write the equation.

$\frac{3(x + 5)}{3} = \frac{45}{3}$ Division Property of Equality

$x + 5 = 15$ Simplify.

$\underline{-5 = -5}$ Subtraction Property of Equality

$x = 10$ Simplify.

2. **Solve $5(n - 2) = -30$.**

$5(n - 2) = -30$ Write the equation.

$\frac{5(n - 2)}{5} = \frac{-30}{5}$ Division Property of Equality

$n - 2 = -6$ Simplify.

$\underline{+2 = +2}$ Addition Property of Equality

$n = -4$ Simplify. Check the solution.

Check Your Work

Remember to plug your solution back into the original equation to see if it makes a true statement.

a. ____________

b. ____________

c. ____________

Got it? **Do these problems to find out.**

a. $2(x + 4) = 20$ **b.** $3(b - 6) = 12$ **c.** $-7(6 + d) = 49$

Equations with Rational Coefficients

Sometimes the factor p, in $p(x + q)$, will be a fraction or decimal.

Examples

3. Solve $\frac{2}{3}(n + 6) = 10$. Check your solution.

$\frac{2}{3}(n + 6) = 10$	Write the equation.
$\frac{3}{2} \cdot \frac{2}{3}(n + 6) = \frac{3}{2} \cdot 10$	Multiplication Property of Equality
$(n + 6) = \frac{3}{\cancel{2}_1} \cdot \left(\frac{\cancel{10}^5}{1}\right)$	$\frac{2}{3} \cdot \frac{3}{2} = 1$; write 10 as $\frac{10}{1}$.
$n + 6 = 15$	Simplify.
$-6 = -6$	Subtraction Property of Equality
$n = 9$	Simplify.

Check

$\frac{2}{3}(n + 6) = 10$	Write the original equation.
$\frac{2}{3}(9 + 6) \stackrel{?}{=} 10$	Replace n with 9. Is this sentence true?
$10 = 10$ ✓	The sentence is true.

4. Solve $0.2(c - 3) = -10$. Check your solution.

$0.2(c - 3) = -10$	Write the equation.
$\frac{0.2(c - 3)}{0.2} = -\frac{10}{0.2}$	Division Property of Equality
$c - 3 = -50$	Simplify.
$+3 = +3$	Addition Property of Equality
$c = -47$	Simplify.

Check

$0.2(c - 3) = -10$	Write the original equation.
$0.2(-47 - 3) \stackrel{?}{=} -10$	Replace c with -47. Is this sentence true?
$-10 = -10$ ✓	The sentence is true.

Reciprocals

The product of a number and its reciprocal is 1.

Got it? Do these problems to find out.

d. $\frac{1}{4}(d - 3) = -15$ **e.** $0.75(6 + d) = 12$ **f.** $(t + 3)\frac{5}{9} = 40$

d. ______

e. ______

f. ______

STOP and Reflect

Solve the problem in Example 5 arithmetically. How does the arithmetic solution compare to the algebraic solution? Write your answer below.

Example

5. Jamal and two cousins received the same amount of money to go to a movie. Each boy spent \$15. Afterward, the boys had \$30 altogether. Write and solve an equation to find the amount of money each boy received.

Let m represent the amount of money each boy received.

$3(m - 15) = 30$	Write the equation.
$\frac{3(m - 15)}{3} = \frac{30}{3}$	Division Property of Equality
$m - 15 = 10$	Simplify.
$+15 = +15$	Addition Property of Equality
$m = 25$	Simplify.

So, each boy received \$25.

Guided Practice

Solve each equation. Check your solution. (Examples 1–4)

1. $2(p + 7) = 18$

2. $(4 + g)(-11) = 121$

3. $(v + 5)\left(-\frac{1}{9}\right) = 6$

4. $0.8(m - 5) = 10$

5. Mr. Singh had three sheets of stickers. He gave 20 stickers from each sheet to his students and has 12 total stickers left. Write and solve an equation to find how many stickers were originally on each sheet. (Example 5)

Equation: ________

Solution: ________

6. **Building on the Essential Question** What is the difference between $px + q = r$ and $p(x + q) = r$?

1. $(8 \cdot s) + (8 \cdot 3) = 72$

$8s + 24 = 72$

$-24 = -24$

$\frac{8s}{8} = \frac{48}{8}$

$s = 6$

2. $(-7 \cdot z) - (-7 \cdot 6) = -70$

$-7z + 42 = -70$

$-42 = -42$

$\frac{-7z}{-7} = \frac{-112}{-7}$

$z = 16$

3. $(-2 \cdot 8) + (-2 \cdot t) = 12$

$-16 + -2t = 12$

$+16 \qquad +16$

$\frac{-2t}{-2} = \frac{28}{-2}$

$t = -14$

4. $(\frac{8}{11} \cdot n) - (\frac{8}{11} \cdot 10) = 64$

$\frac{11}{8}(\frac{8}{11}n) - (7\frac{3}{11})\frac{11}{8} \quad \frac{80}{11}$

$n = 10$

$n = 98$

5. $(-0.6 \cdot r) + (-0.6 \cdot 0.2) = 1.8$

$-0.6r + -0.12 = 1.8$

$-0.12 = -0.12$

$\frac{-0.6r}{-0.6} = \frac{1.92}{-0.6}$

$r = -3.2$

Name ______________________ My Homework ______________________

Independent Practice

Go online for Step-by-Step Solutions

Solve each equation. Check your solution. (Examples 1–4)

1. $8(s + 3) = 72$

2. $-7(z - 6) = -70$

3. $(t + 8)(-2) = 12$

4. $\frac{8}{11}(n - 10) = 64$

5. $-0.6(r + 0.2) = 1.8$

6. $\left(w - \frac{4}{9}\right)\left(-\frac{2}{3}\right) = -\frac{4}{5}$

7. The length of each side of an equilateral triangle is increased by 5 inches, so the perimeter is now 60 inches. Write and solve an equation to find the original length of each side of the equilateral triangle. (Example 5)

Equation: ______________ Solution: ______________

8. MP **Multiple Representations** Miguel and three of his friends went to the movies. They originally had a total of $40. Each boy had the same amount of money and spent $7.50 on a ticket. How much money did each boy have left after buying his ticket?

a. Model Draw a bar diagram that represents the situation.

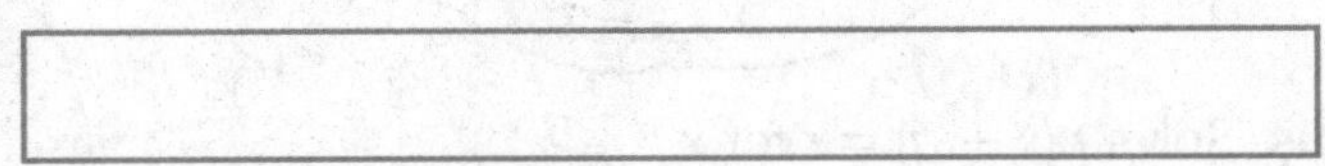

b. Algebra Write and solve an equation that represents the situation.

____

c. Words Explain how you solved your equation.

____

____

d. Words Compare the arithmetic solution and the algebraic solution.

____

____

9. Mrs. Sorenstam bought one ruler, one compass, and one mechanical pencil at the prices shown in the table for each of her 12 students.

Item	Price ($)
compass	1.49
mechanical pencil	0.59
ruler	0.49

a. Suppose Mrs. Sorenstam had 36 cents left after buying the school supplies. Write an equation to find the amount of money Mrs. Sorenstam initially had to spend on each student.

b. Describe a two-step process you could use to solve your equation. Then solve the equation.

H.O.T. Problems Higher Order Thinking

10. **MP Model with Mathematics** Write a real-world situation that can be represented by the equation $2(n + 20) = 110$.

11. **MP Find the Error** Marisol is solving the equation $6(x + 3) = 21$. Find her mistake and correct it.

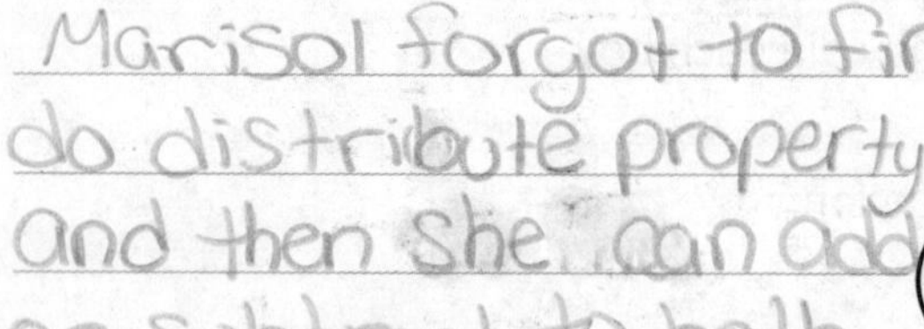

$$6(x + 3) = 21$$
$$-3 = -3$$
$$6x = 18$$
$$x = 3$$

12. **MP Persevere with Problems** Solve $p(x + q) = r$ for x.

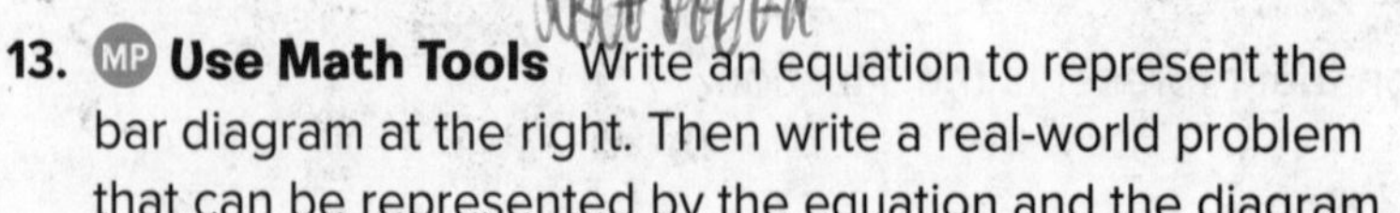

13. **MP Use Math Tools** Write an equation to represent the bar diagram at the right. Then write a real-world problem that can be represented by the equation and the diagram.

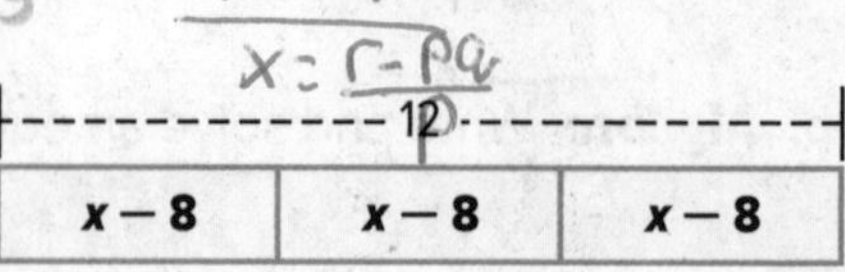

15. $12(x-20)=-48$

$(12\cdot x)-(12\cdot 20)=-48$

$12x-240=-48$

$+240=+240$

$12x=192$

$\frac{12x}{12}=\frac{192}{12}$

$x=16$

16. $-28=7(n+3)$

$-28=(7\cdot n)+(7\cdot 3)$

$-28=7n+21$

$-21=-21$

$-49=7n$

$\frac{-49}{7}=\frac{7n}{7}$

$-7=n$

Mrs. Kolsin problems 1-3

19. $(d-3)\frac{2}{5}=30$

$(\frac{2}{5}\cdot 3)+(\frac{2}{5}\cdot d)=30$

$\frac{6}{5}+\frac{2}{5}d=30$

$-\frac{6}{5} \quad -\frac{6}{5}$

$\frac{5}{2}(\frac{2}{5}d)=(28\frac{4}{5})\frac{5}{2}$

$d=72$

1. $5+2b+2-{}^{-}5=-5$

$2b+5+2+{}^{-}5=-5$

$7b+2+5=-5$

$9b+5=-5$

$-5=-5$

$9b=-10$

$\frac{9b}{9}=\frac{-10}{9}$ $b=1$

$b=-1\frac{1}{9}$

check

$5+2(-1\frac{1}{9})+2-{}^{-}5=-5$

$5+-2\frac{2}{9}+2-{}^{-}5=-5$

$2\frac{7}{9}+2-{}^{-}5=-5$

$4\frac{7}{9}-{}^{-}5=-5$

$9\frac{7}{9}\neq -5$

2. $r+2+2r=7$

$r+2r+2=7$

$3r+2=7$

$-2=-2$

$3r=5$

$\frac{3r}{3}=\frac{5}{3}$

$r=1\frac{2}{3}$ $r=-3$

check

$1\frac{2}{3}+2+2(1\frac{2}{3})=7$

$1\frac{2}{3}+2+3\frac{1}{3}=7$

$3\frac{2}{3}+3\frac{1}{3}=7$

$7=7$

Extra Practice

Solve each equation. Check your solution.

14. $0.25(3 + a) = 0.5$

$0.25(3 + a) = 0.5$

$\frac{0.25(3 + a)}{0.25} = \frac{0.5}{0.25}$

$3 + a = 2$

$a = -1$

15. $12(x - 20) = -48$

16. $-28 = 7(n + 3)$

17. $(t + 9)20 = 140$

18. $\frac{5}{9}(8 + c) = -20$

19. $(d - 3)\frac{2}{5} = 30$

20. **Reason Abstractly** Anne bought a necklace for each of her three sisters. She paid \$7 for each necklace. Suppose she had \$9 left. Write and solve an equation to find how much money Anne had initially to spend on each sister.

Equation: ______

Solution: ______

Solve each equation. Check your solution.

21. $1\frac{3}{5}(t - 6) = -0.4$ ______

22. $\left(x + 5\frac{1}{2}\right)0.75 = \frac{5}{8}$ ______

23. Mr. Gomez bought fruit to make fruit salad. He bought $2\frac{1}{2}$ pounds of apples and spent \$4.50 on apples and oranges. Write and solve an equation to determine the number of pounds of oranges Mr. Gomez bought.

______

Fruit	Price per Pound (\$)
apples	1.20
bananas	0.50
grapes	1.50
oranges	1.20

Power Up! Common Core Test Practice

24. A rectangular classroom is 32 feet long and has a perimeter of 120 feet. Label the drawing with the correct values to represent the situation. Let w represent the width of the classroom.

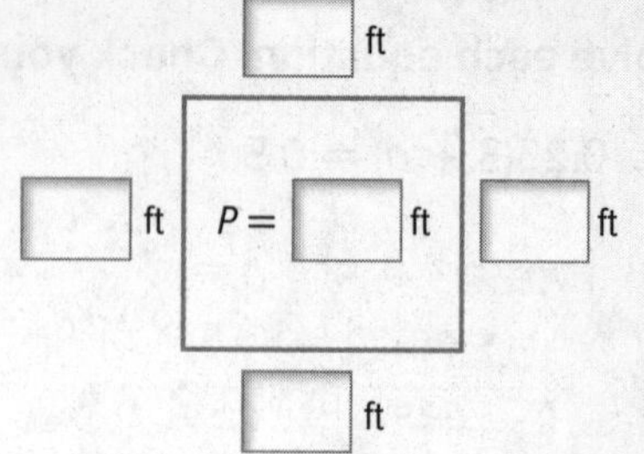

Write an expression that represents the sum of the length and width. ______

Write an expression that represents twice the sum of the length and width. ______

Write an equation you could use to find the perimeter of the classroom. ______

What is the width of the classroom? ______

25. Which of the following are operations that you should use to solve the equation $p(x - q) = r$ for x? Select all that apply.

- ☐ Subtract q from both sides.
- ☐ Multiply both sides by p.
- ☐ Divide both sides by p.
- ☐ Add q to both sides.

Common Core Spiral Review

Solve each equation. 6.EE.6

26. $x + 3 = 5$

27. $x - 2 = -6$

28. $4x = 12$

29. $-6x = -24$

30. $\frac{x}{2} = -1$

31. $\frac{x}{-3} = 1$

Write the number or numbers from the set {−3, −2, −1, 0, 1, 2, 3} that make each statement true. 6.EE.5

32. $4m = 12$ ______

33. $y - 1 = 1$ ______

34. $v > 0$ ______

35. $r \leq 0$ ______

Problem-Solving Investigation
Work Backward

Content Standards
7.EE.3

Mathematical Practices
1, 3, 4

Case #1 Yard Work

Mike earned extra money by doing yard work for his neighbor. Then he spent $5.50 at the convenience store and four times that amount at the bookstore. Now he has $7.75 left.

How much money did Mike have before he went to the convenience store and the bookstore?

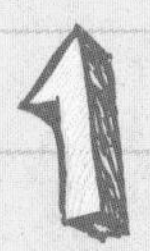

Understand *What are the facts?*

You know Mike has $7.75 left. You need to find the amount before his purchases.

Plan *What is your strategy to solve this problem?*

Start with the end result and work backward.

Solve *How can you apply the strategy?*

He has $7.75 left.

Undo the four times $5.50 spent at the bookstore.
Since $5.50 × 4 is $22, add $7.75 and $22.

$$\begin{array}{r} \$7.75 \\ +\ \$22.00 \\ \hline \$29.75 \end{array}$$

Undo the $5.50 spent at the convenience store.

Add $5.50 and ______.

$$\begin{array}{r} +\ \$5.50 \\ \hline \$35.25 \end{array}$$

So, Mike's starting amount was ______.

Check *Does the answer make sense?*

Assume Mike started with $35.25. He spent $5.50 and $22. He had

$35 − $5.50 − $22 or ______ left. So, $35.25 is correct. ✓

Analyze the Strategy

Construct an Argument Describe how to solve a problem by working backward. ______________________________

____

____

Case #2 Money

Marisa spent $8 on a movie ticket. Then she spent $5 on popcorn and one half of what was left on a drink. She had $2 left.

How much did she have initially?

Understand

Read the problem. What are you being asked to find?

I need to find ______________________.

Underline key words and values. What information do you know?

I know Marisa has ☐ left and that she spent ☐, ☐, and ______________________.

Is there any information that you do *not* need to know?

I do not need to know ______________________.

Plan

Choose a problem-solving strategy.

I will use the ______________________ strategy.

Solve

Use your problem-solving strategy to solve the problem.

Marisa has $2 left.

Undo the half-of-what-was-left amount. Multiply by 2. ____________

Undo the spent $5. Add $5. ____________

Undo the spent $8. Add $8. ____________

So, Marisa had ☐ initially.

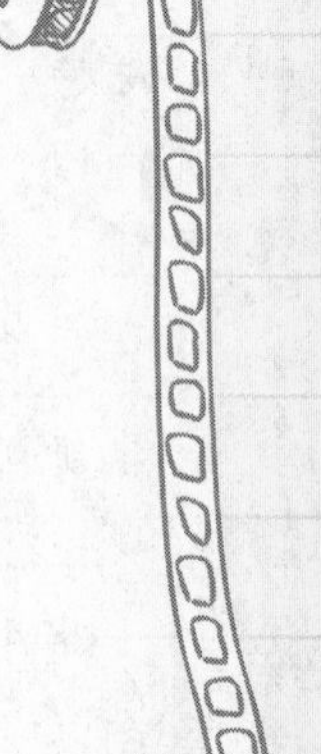

Check

Use information from the problem to check your answer.

Marisa's initial amount: ____________

Amount after spending $8: ____________

Amount after spending $5: ____________

Amount after spending half of what was left: ____________

Work with a small group to solve the following cases. Show your work on a separate piece of paper.

Case #3 Waterfalls

Angel Falls in Venezuela is 3,212 feet high. It is 29 yards higher than 2.5 times the architectural height of the Empire State Building.

Find the architectural height, in feet, of the Empire State Building.

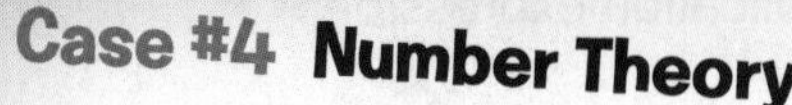

Case #4 Number Theory

Travis works at a kite factory. He checks all the kites before they are packaged. Travis discovered that for every 28 kites that he inspected, 7 kites did not pass: 4 kites did not have tails, and 3 kites had the wrong colors.

Of the 476 kites Travis examined, how many did not have tails and how many had the wrong colors?

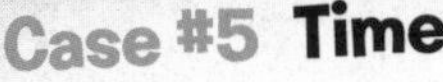

Case #5 Time

Timothy's morning schedule is shown.

At what time does Timothy wake up if he arrives at school at 7:35 A.M.?

Timothy's Schedule	
Activity	**Time**
Wake up	■
Get ready for school — $\frac{3}{4}$ h	■
Walk to school — $\frac{5}{12}$ h	7:35 A.M.

Case #6 Money

Antonio has saved $28 to spend at the arcade.

If he has 5 bills, how many of each kind of bill does he have?

Mid-Chapter Check

Vocabulary Check

1. Define *equation*. Give an example of two equivalent equations. (Lesson 1)

2. Fill in the blank with the correct term. (Lesson 2)

   A ______________ is the numerical factor of a multiplication expression like $3x$.

Skills Check and Problem Solving

Solve each equation. Check your solution. (Lessons 1–5)

3. $21 + m = 33$

4. $a - 5 = -12$

5. $5f = -75$

6. $15 = \frac{b}{15}$

7. $19 = 4p + 5$

8. $3(n - 7) = -30$

9. Cameron has 11 adult Fantail goldfish. This is 7 fewer Fantail goldfish than his friend Julia has. Write and solve a subtraction equation to determine the number of Fantail goldfish g that Julia has. (Lesson 1)

   Equation: ______________ Solution: ______________

10. MP **Persevere with Problems** The pentagon shown is a regular pentagon, so each side has the same length. The perimeter of the pentagon is 22.5 centimeters. What is the value of x? ______

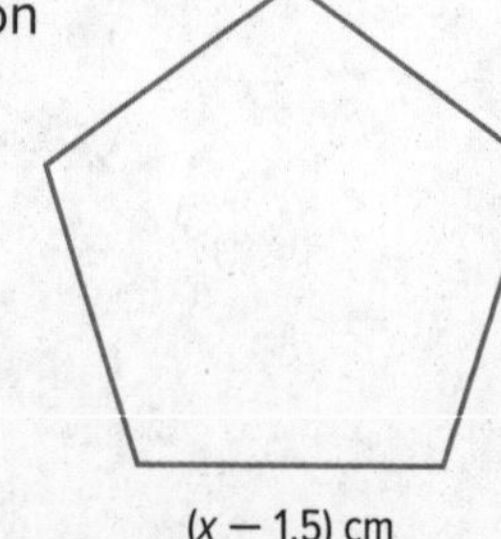

Inquiry Lab

Solve Inequalities

HOW is an inequality like an equation? How is it different?

Content Standards
7.EE.4, 7.EE.4b

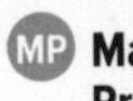

Mathematical Practices
1, 2, 3, 4

Mr. Numkena volunteered to drive Hinto and his friends to the school dance. The car can carry up to 5 people, including the driver. How many friends can ride in the car with Hinto?

What do you know? ______________________________

____

What do you need to find? ______________________________

____

Hands-On Activity 1

The real-world situation described above can be represented by the inequality $x + 2 \leq 5$. Let x represent the friends that can ride with Hinto.

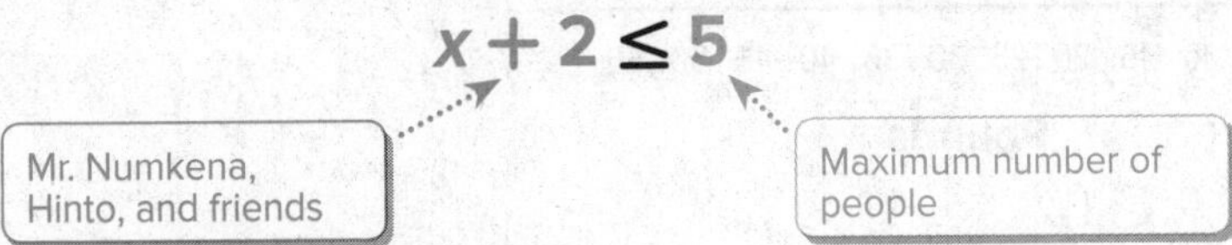

You can use a balance to model and solve the inequality $x + 2 \leq 5$.

Step 1 On one side of a balance, place a paper bag and ☐ cubes to model $x + 2$.

Step 2 On the other side of a balance, place ☐ cubes.

Add one cube to the bag at a time. Then complete the table.

Number of Friends, x	$x + 2$	Less than or equal to 5?
1	3	yes
2		
3		
4		

So, up to ☐ friends can ride with Hinto to the school dance.

An *inequality* is a mathematical sentence that compares quantities. The table shows two examples of inequalities.

Words	Symbols
x is less than two	$x < 2$
x is greater than or equal to four	$x \geq 4$

To solve an inequality means to find values for the variable that make the sentence true. You can use bar diagrams to solve inequalities.

Hands-On Activity 2

An airline charges for checked luggage that weighs more than 50 pounds. Mia's suitcase currently weighs 35 pounds and she still needs to pack her shoes. Find the maximum amount her shoes can weigh so Mia will not be charged a fee.

Step 1 In the bar diagram, write the maximum weight Mia's luggage can be without a fee. Label the weight of Mia's luggage without her shoes.

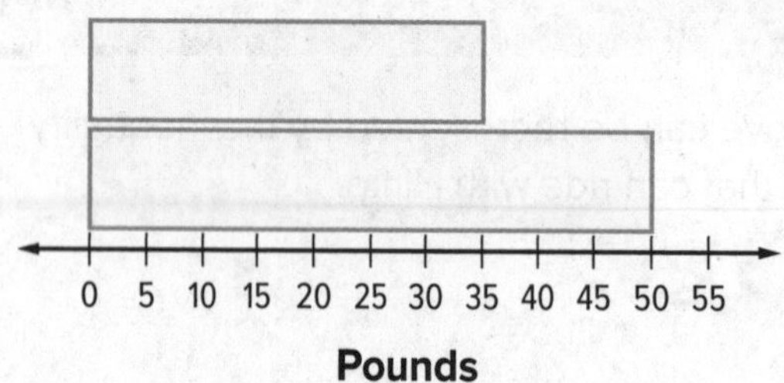

Step 2 In the bar diagram, write an x beside the bar that represents the weight of Mia's luggage.

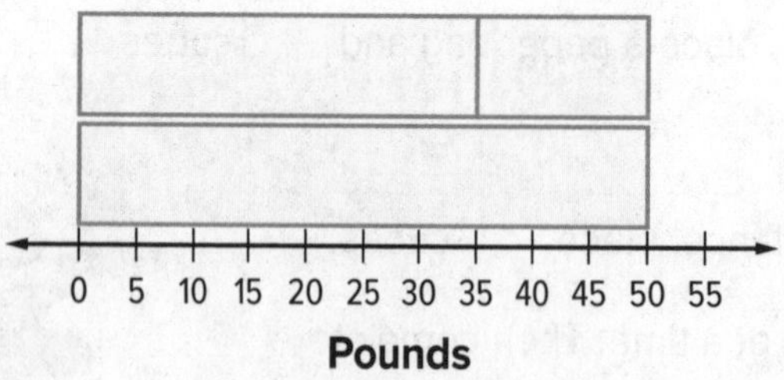

The weight of Mia's suitcase plus the weight of her shoes must be less than or equal to the maximum luggage weight.

This can be written as $35 + x \leq 50$.

Using the bar diagram, Mia's shoes cannot weigh more than $50 - 35$ or ______ pounds.

Work with a partner to solve the following problems.

MP **Reason Inductively** For Exercises 1–3, assume the paper bag is weightless. Write the inequality represented by each balance. Then write the different possible numbers of cubes in the paper bag if the sides of each balance remain unlevel.

1. 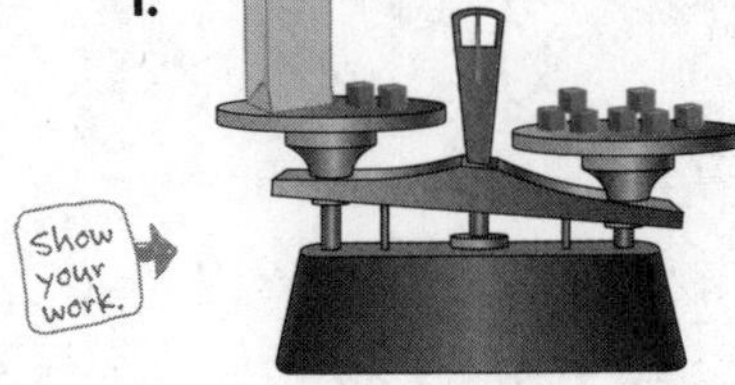

Inequality: ____________

Number of Cubes: ____________

2. 

Inequality: ____________

Number of Cubes: ____________

3. 

Inequality: ____________

Number of Cubes: ____________

4. MP **Reason Abstractly** At an amusement park, roller coaster riders are required to be at least 48 inches tall. Last year, Myron was 42 inches tall. Complete the bar diagram to determine the number of inches x Myron needed to grow this year to be able to ride the roller coaster. Then write an inequality to represent the situation.

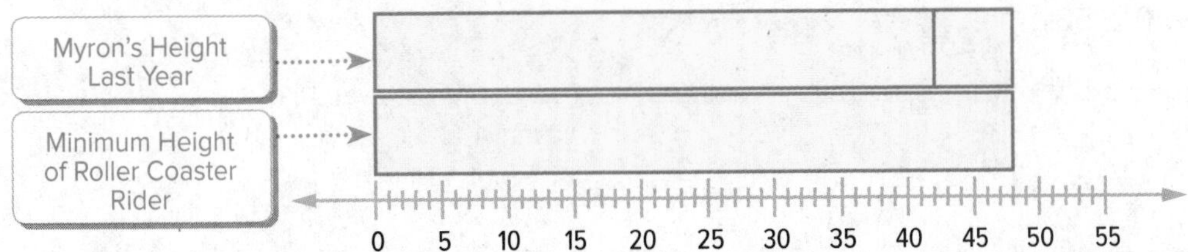

So, Myron needed to grow at least ________ inches.

Inequality: ____________

Analyze and Reflect

Work with a partner to circle the correct inequality for each situation. The first one is done for you.

	Real-World Situation	Inequalities	
	Yolanda wants to score at least 84% on the next history test.	$x \leq 84$	$x \geq 84$ (circled)
5.	To see a certain movie, you must be at least 13 years old.	$n \leq 13$	$n \geq 13$
6.	Kai has $4.99 left on a music download gift card. She has a download costing $1.99 in her online shopping cart. How much money does Kai have left to spend?	$x + 1.99 \leq 4.99$	$x + 1.99 > 4.99$
7.	In some states, teens must be at least 16 years old to obtain a driver's license.	$x < 16$ $x \leq 16$	$x > 16$ $x \geq 16$
8.	The Walter family budgets a maximum amount of $125 per week for groceries. Mr. Walter already spent $40. How much more can the Walter family spend on groceries?	$x + 40 < 125$ $x + 40 \leq 125$	$x + 40 > 125$ $x + 40 \geq 125$
9.	Miles pays $30 for a ticket to an amusement park. He cannot spend more than $50. How much more money can Miles spend at the amusement park?	$x + 30 < 50$ $x + 30 \leq 50$	$x + 30 > 50$ $x + 30 \geq 50$

Create

10. MP **Model with Mathematics** Write a real-world situation that could be represented by $x + 20 \geq 50$.

____

____

11. Inquiry HOW is an inequality like an equation? How is it different?

____

____

____

Lesson 6

Solve Inequalities by Addition or Subtraction

Real-World Link

Mail A first class stamp can be used for letters and packages weighing thirteen ounces or less. Fisher is mailing pictures to his grandmother, and only has a first class stamp. His envelope weighs 2 ounces. Follow the steps to determine how much the pictures can weigh so that Fisher can use the stamp.

Step 1 Let x represent the weight of the pictures. Write and solve an equation to find the maximum weight of the pictures.

weight of the envelope		weight of the pictures		maximum weight of the package
☐	+	x	=	☐

Solve for x.

So, the maximum weight of the pictures is ☐ ounces.

Step 2 Replace the equals sign in your equation with the less than or equal to symbol, $\leq$.

$2 + x$ ☐ 13

Refer to Step 2. Name three possible values of x that will result in a true sentence.

Which MP **Mathematical Practices** did you use? Shade the circle(s) that applies.

① Persevere with Problems
② Reason Abstractly
③ Construct an Argument
④ Model with Mathematics
⑤ Use Math Tools
⑥ Attend to Precision
⑦ Make Use of Structure
⑧ Use Repeated Reasoning

Essential Question

WHAT does it mean to say two quantities are equal?

Vocabulary

Subtraction Property of Inequality
Addition Property of Inequality
inequality

Common Core State Standards

Content Standards
7.EE.4, 7.EE.4b

MP Mathematical Practices
1, 2, 3, 4

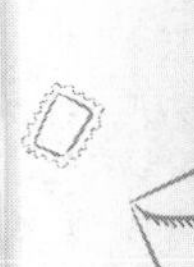
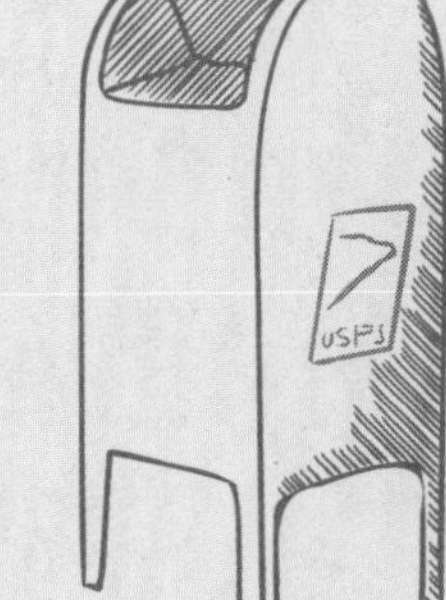

Key Concept

Solve Inequalities

Words You can solve inequalities by using the **Addition Property of Inequalities** and the **Subtraction Property of Inequalities**. When you add or subtract the same number from each side of an inequality, the inequality remains true.

Symbols For all numbers a, b, and c,

1. if $a > b$, then $a + c > b + c$ and $a - c > b - c$.

2. if $a < b$, then $a + c < b + c$ and $a - c < b - c$.

Examples

$$\begin{array}{rr} 2 < & 4 \\ +3 & +3 \\ \hline 5 < & 7 \end{array} \qquad \begin{array}{rr} 6 > & 3 \\ -4 & -4 \\ \hline 2 > & -1 \end{array}$$

Work Zone

An **inequality** is a mathematical sentence that compares quantities. Solving an inequality means finding values for the variable that make the inequality true.

The table below gives some examples of the words you might use when describing different inequalities.

Inequalities				
Words	• is less than • is fewer than	• is greater than • is more than • exceeds	• is less than or equal to • is no more than • is at most	• is greater than or equal to • is no less than • is at least
Symbols	$<$	$>$	$\leq$	$\geq$

Examples

1. **Solve $x + 3 > 10$.**

$x + 3 > 10$ Write the inequality.

$\underline{-3 \quad -3}$ Subtract 3 from each side.

$x > 7$ Simplify.

Therefore, the solution is $x > 7$.

You can check this solution by substituting a number greater than 7 into the original inequality. Try using 8.

Check $x + 3 > 10$ Write the inequality.

$8 + 3 \overset{?}{>} 10$ Replace x with 8. Is this sentence true?

$11 > 10$ This is a true statement. ✓

2. **Solve $-6 \geq n - 5$.**

$-6 \geq n - 5$	Write the inequality.
$+5 \quad +5$	Add 5 to each side.
$-1 \geq n$	Simplify.

The solution is $-1 \geq n$ or $n \leq -1$.

You can check this solution by substituting -1 or a number less than -1 into the original inequality.

Show your work.

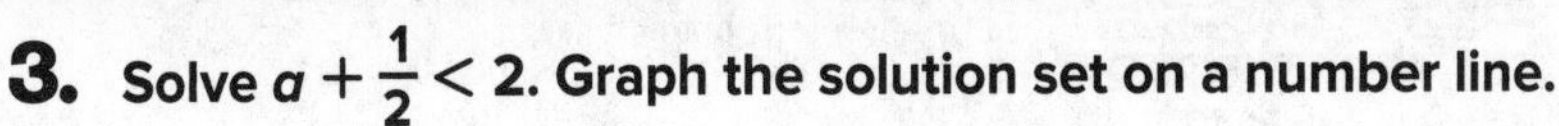

Got it? **Do these problems to find out.**

Solve each inequality.

a. $a - 3 < 8$

b. $0.4 + y \geq 7$

a. ______

b. ______

Example

Tutor

3. **Solve $a + \frac{1}{2} < 2$. Graph the solution set on a number line.**

$a + \frac{1}{2} < 2$	Write the inequality.
$-\frac{1}{2} \quad -\frac{1}{2}$	Subtract $\frac{1}{2}$ from each side.
$a < 1\frac{1}{2}$	Simplify.

The solution is $a < 1\frac{1}{2}$. Check your solution.

Graph the solution.

Place an open dot at $1\frac{1}{2}$. Draw a line and an arrow to the left.

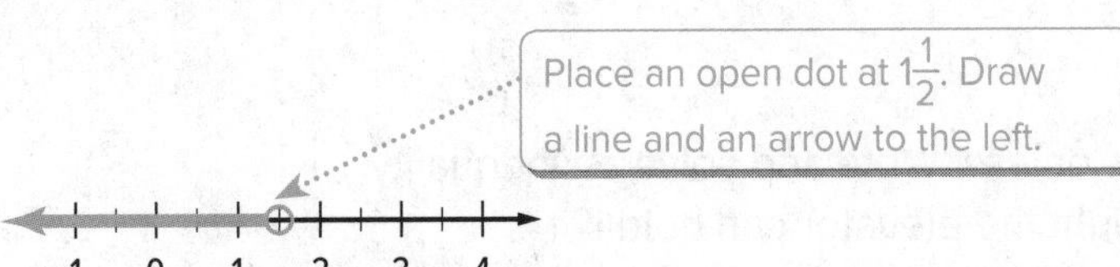

Open and Closed Dots

When graphing inequalities, an open dot is used when the value should not be included in the solution, as with > and < inequalities. A closed dot indicates the value is included in the solution, as with ≤ and ≥ inequalities.

Got it? **Do these problems to find out.**

Solve each inequality. Graph the solution set on the number line provided.

c. $h + 4 > 4$

d. $x - 6 \leq 4$

c. ______

d. ______

Write Inequalities

Inequalities can be used to represent real-world situations. You will want to first identify a variable to represent the unknown value.

Example

4. Dylan has $18 to ride go-karts and play games at the state fair. Suppose the go-karts cost $5.50. Write and solve an inequality to find the most he can spend on games.

Words	Cost of go-kart	plus	cost of games	must be less than or equal to	total amount.
Symbols	Let x = the cost of the games.				
Inequality	5.5	$+$	x	$\leq$	18

$$5.5 + x \leq 18$$ Write the inequality. (5.50 = 5.5)

$$\underline{-5.5 \qquad -5.5}$$ Subtract 5.5 from each side.

$$x \leq 12.5$$ Simplify.

So, the most Dylan can spend on games is $12.50.

Guided Practice

Solve each inequality. Graph the solution set on a number line. (Examples 1–3)

1. $6 + h \geq 12$ ______

2. $14 + t > 5$ ______

3. An elevator can hold 2,800 pounds or less. Write and solve an inequality that describes how much more weight the elevator can hold if it is currently holding 2,375 pounds. Interpret the solution. (Example 4)

4. **Building on the Essential Question** Explain when you would use addition and when you would use subtraction to solve an inequality.

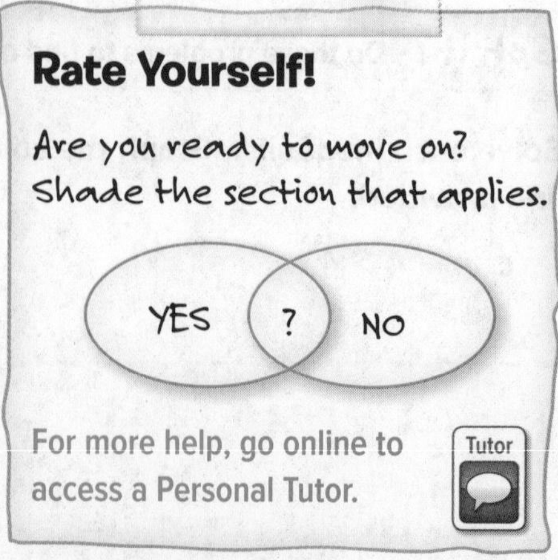

Name ______________________ My Homework ______________________

Independent Practice

Go online for Step-by-Step Solutions

Solve each inequality. (Examples 1 and 2)

1. $h - 16 \leq -24$ ______

2. $y + 6 \geq -13$ ______

3. $-3 < n - 8$ ______

4. $3 \leq m + 1.4$ ______

5. $x + 0.7 > -0.3$ ______

6. $w - 8 \geq 5.6$ ______

Solve each inequality. Graph the solution set on a number line. (Example 3)

7. $m + 5 \geq -1$ ______

8. $-11 > t + 7$ ______

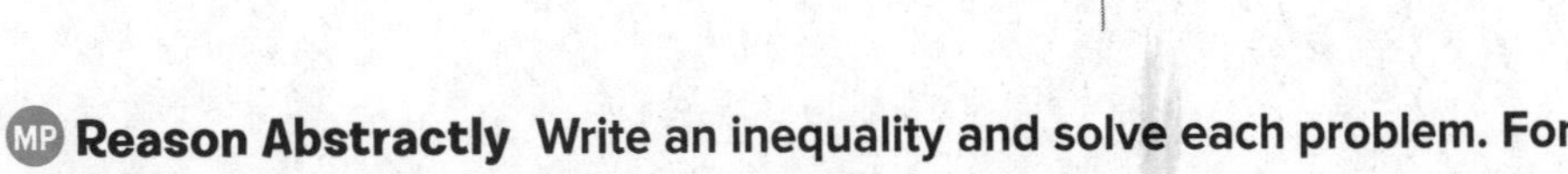

MP **Reason Abstractly Write an inequality and solve each problem. For Exercises 11 and 12, interpret the solution.** (Example 4)

9. Four more than a number is more than 13.

Inequality: ______

Solution: ______

10. The sum of a number and 19 is at least 8.2.

Inequality: ______

Solution: ______

11. The high school soccer team can have no more than 26 players. Write and solve an inequality to determine how many more players can make the team if the coach has already chosen 17 players.

Inequality: ______ Solution: ______

Interpretation: ______

12. Lalo has 1,500 minutes per month on his cell phone plan. How many more minutes can he use if he has already talked for 785 minutes?

Inequality: ______ Solution: ______

Interpretation: ______

13. Refer to the diagram below.

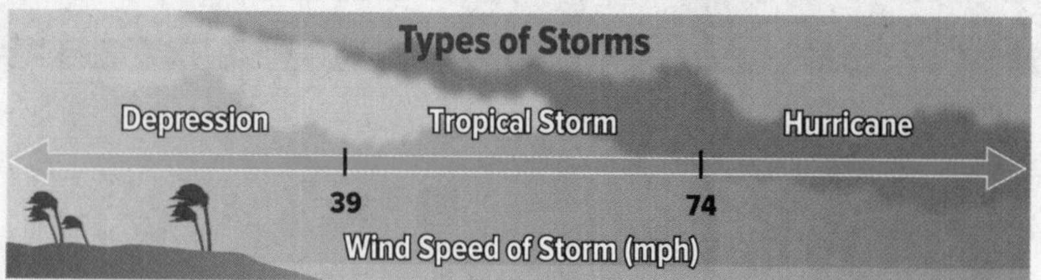

a. A hurricane has winds that are at least 74 miles per hour. Suppose a tropical storm has winds that are 42 miles per hour. Write and solve an inequality to find how much the winds must increase before the storm becomes a hurricane.

Inequality: ______________ Solution: ______________

b. A *major storm* has wind speeds that are at least 110 miles per hour. Write and solve an inequality that describes how much greater these wind speeds are than the slowest hurricane.

Inequality: ______________ Solution: ______________

H.O.T. Problems Higher Order Thinking

14. MP **Reason Inductively** Compare and contrast the solutions of $a - 3 = 15$ and $a - 3 \geq 15$. ______________

15. MP **Model with Mathematics** Write an addition inequality for the solution set graphed below.

____

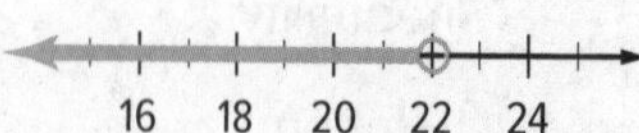

16. MP **Persevere with Problems** Solve $x + b > c$ for x.

____

17. MP **Reason Inductively** Does the graph shown at the right show the solution set of the inequality $x + 3 \geq 2$? If not, explain how you would change the graph to show the actual solution set. ______________

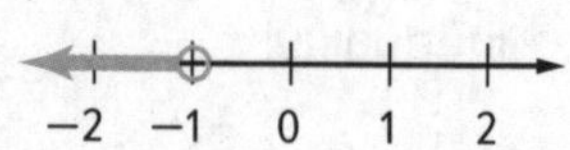

Name ______ My Homework ______

Extra Practice

Solve each inequality.

18. $10 < b - 8$ 18 < b

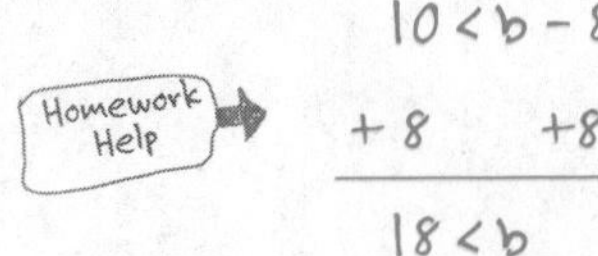

10 < b − 8
+8 +8
18 < b

19. $1.2 + m \leq 5.5$ ______

20. $c - 1\frac{1}{4} > -2\frac{1}{2}$ ______

MP Model with Mathematics **Solve each inequality. Graph the solution set on a number line.**

21. $-21 < a - 16$ ______

22. $t - 6.2 < 4$ ______

Write an inequality and solve each problem.

23. Eight less than a number is less than 10.

Inequality: ______

Solution: ______

24. The difference between a number and $21\frac{1}{2}$ is no more than $14\frac{1}{4}$.

Inequality: ______

Solution: ______

25. There were a total of 125 cars at a car dealership. A salesperson sold 68 of the cars in one month. Write and solve an inequality that describes how many more cars, at most, the salesman has left to sell. Interpret the solution.

Inequality: ______ Solution: ______

Interpretation: ______

Copy and Solve **Solve each inequality. Graph the solution set on a number line. Show your work on a separate sheet of paper.**

26. $n - \frac{1}{5} \leq \frac{3}{10}$

27. $6 > x + 3\frac{1}{3}$

28. $c + 1\frac{1}{4} < 5$

29. $9 \leq m - 2\frac{1}{5}$

30. $\frac{3}{4} + d > 4\frac{1}{2}$

31. $-\frac{7}{8} \leq n + 3\frac{5}{16}$

Power Up! Common Core Test Practice

32. Joaquin can send up to 250 text messages each month. So far this month, he has sent 141 text messages. Let t represent the number of text messages Joaquin can send during the rest of the month.

Write an inequality to model the situation

Solve the inequality for t.

Graph the solution on the number line.

Interpret the solution to the inequality. Explain your reasoning.

33. Which inequality has the solution set shown in the number line below? Select all that apply.

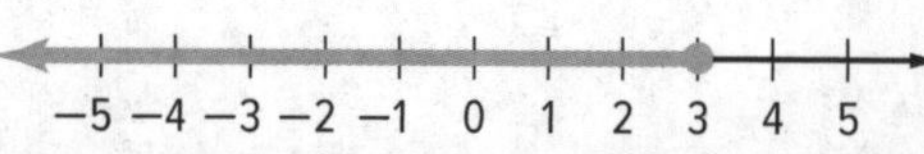

- ☐ $x + 4 \leq 7$
- ☐ $12 > x + 9$
- ☐ $x + 1 \leq 2$
- ☐ $-7 \geq x - 10$

Common Core Spiral Review

Solve each equation. Then graph each solution on the number line below. 6.EE.6

34. $x + 2 = 1$

35. $x - 1 = -5$

36. $2x = 10$

37. $-2x = 4$

38. $\frac{x}{2} = 1$

39. $\frac{x}{-2} = 3$

Lesson 7

Solve Inequalities by Multiplication or Division

Real-World Link

Science An astronaut in a space suit weighs about 300 pounds on Earth, but only 50 pounds on the Moon.

weight on Earth		weight on Moon
300 lb	>	50 lb

1. If the astronaut and space suit each weighed half as much, would the inequality still be true?

$$\frac{300}{2} > \frac{50}{2}$$ Divide each side by 2.

Is the inequality still true? Circle yes or no.

Yes No

2. Is the weight of one astronaut greater on Pluto or Earth? Would the weight of 5 astronauts be greater on Pluto or on Earth? Explain by using an inequality.

Location	Weight of Astronaut (lb)
Earth	300
Moon	50
Pluto	67
Jupiter	796

3. Is the weight of one astronaut greater on Jupiter or on Earth? Would the weight of 5 astronauts be greater on Jupiter or on Earth? Explain by using an inequality.

Essential Question

WHAT does it mean to say two quantities are equal?

Vocabulary

Multiplication Property of Inequality

Division Property of Inequality

Common Core State Standards

Content Standards
7.EE.4, 7.EE.4b

MP Mathematical Practices
1, 2, 3, 4, 7

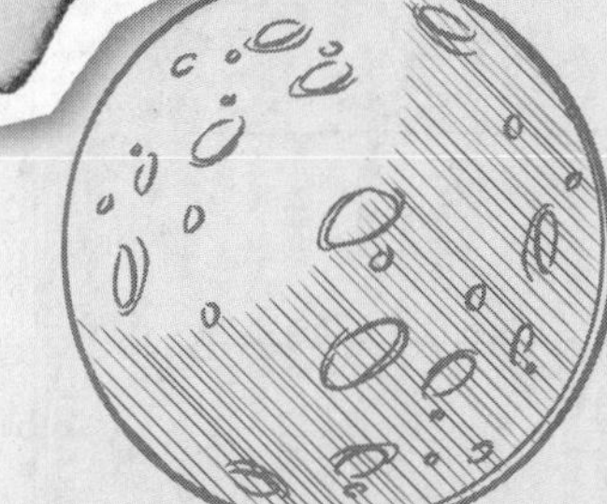

Which MP Mathematical Practices did you use? Shade the circle(s) that applies.

① Persevere with Problems
② Reason Abstractly
③ Construct an Argument
④ Model with Mathematics
⑤ Use Math Tools
⑥ Attend to Precision
⑦ Make Use of Structure
⑧ Use Repeated Reasoning

Key Concept

Multiplication and Division Properties of Inequality, Positive Number

Words The **Multiplication Property of Inequality** and the **Division Property of Inequality** state that an inequality remains true when you multiply or divide each side of an inequality by a positive number.

Symbols For all numbers a, b, and c, where $c > 0$,

1. if $a > b$, then $ac > bc$ and $\frac{a}{c} > \frac{b}{c}$.

2. if $a < b$, then $ac < bc$ and $\frac{a}{c} < \frac{b}{c}$.

These properties are also true for $a \geq b$ and $a \leq b$.

Work Zone

What does the inequality $c > 0$ mean? Explain below.

You can solve inequalities by using the Multiplication Property of Inequality and the Division Property of Inequality.

Examples

1. Solve $8x \leq 40$.

$8x \leq 40$ Write the inequality.

$\frac{8x}{8} \leq \frac{40}{8}$ Divide each side by 8.

$x \leq 5$ Simplify.

The solution is $x \leq 5$. You can check this solution by substituting 5 or a number less than 5 into the inequality.

2. Solve $\frac{d}{2} > 7$.

$\frac{d}{2} > 7$ Write the inequality.

$2\left(\frac{d}{2}\right) > 2(7)$ Multiply each side by 2.

$d > 14$ Simplify.

The solution is $d > 14$. You can check this solution by substituting a number greater than 14 into the inequality.

Got it? **Do these problems to find out.**

a. $4x < 40$

b. $6 \geq \frac{x}{7}$

Show your work.

a. ____________

b. ____________

Multiplication and Division Properties of Inequality, Negative Number

Key Concept

Words When you multiply or divide each side of an inequality by a negative number, the inequality symbol must be reversed for the inequality to remain true.

Symbols For all numbers a, b, and c, where $c < 0$,

1. if $a > b$, then $ac < bc$ and $\frac{a}{c} < \frac{b}{c}$.
2. if $a < b$, then $ac > bc$ and $\frac{a}{c} > \frac{b}{c}$.

Examples

$7 > 1$		$-4 < 16$
$-2(7) < -2(1)$	Reverse the symbols.	$\frac{-4}{-4} > \frac{16}{-4}$
$-14 < -2$		$1 > -4$

These properties are also true for $a \geq b$ and $a \leq b$.

STOP and Reflect

What does the inequality $c < 0$ mean? Expain below.

Examples

Tutor

3. Solve $-2g < 10$. Graph the solution set on a number line.

$-2g < 10$ Write the inequality.

$\frac{-2g}{-2} > \frac{10}{-2}$ Divide each side by −2 and reverse the symbol.

$g > -5$ Simplify.

−7 −6 −5 −4 −3

4. Solve $\frac{x}{-3} \leq 4$. Graph the solution set on a number line.

$\frac{x}{-3} \leq 4$ Write the inequality.

$-3\left(\frac{x}{-3}\right) \geq -3(4)$ Multiply each side by −3 and reverse the symbol.

$x \geq -12$ Simplify.

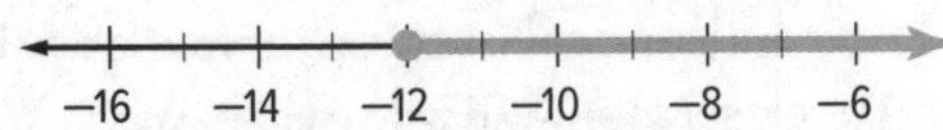

Got it? **Do these problems to find out.**

c. $\frac{k}{-2} < 9$

c. ________

Example

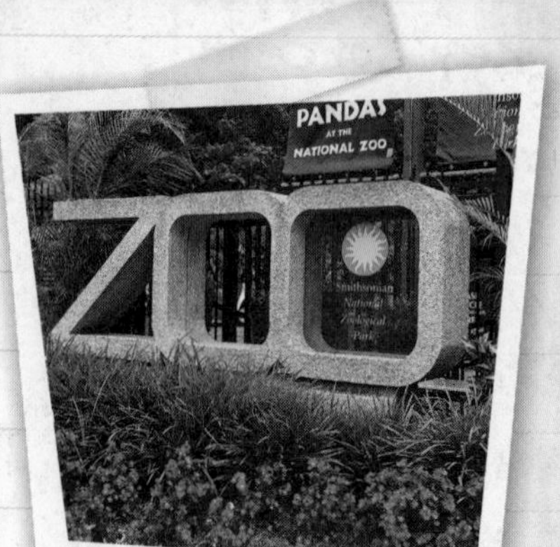

5. **Ling earns \$8 per hour working at the zoo. Write and solve an inequality that can be used to find how many hours she must work in a week to earn at least \$120. Interpret the solution.**

Words	Amount earned per hour	times	number of hours	is at least	amount earned each week.
Variable	Let x represent the number of hours.				
Inequality	8	$\cdot$	x	$\geq$	120

$8x \geq 120$ Write the inequality.

$\frac{8x}{8} \geq \frac{120}{8}$ Divide each side by 8.

$x \geq 15$ Simplify.

So, Ling must work at least 15 hours.

Guided Practice

Solve each inequality. Graph the solution set on a number line. (Examples 1–4)

1. $-3n \leq -22$ ______

2. $\frac{t}{-4} < -11$ ______

3. At a baseball game you can get a single hot dog for \$2. You have \$10 to spend. Write and solve an inequality to find the number of hot dogs you can buy. Interpret the solution. (Example 5) ______

4. **Building on the Essential Question** Explain when you should reverse the inequality symbol when solving an inequality. ______

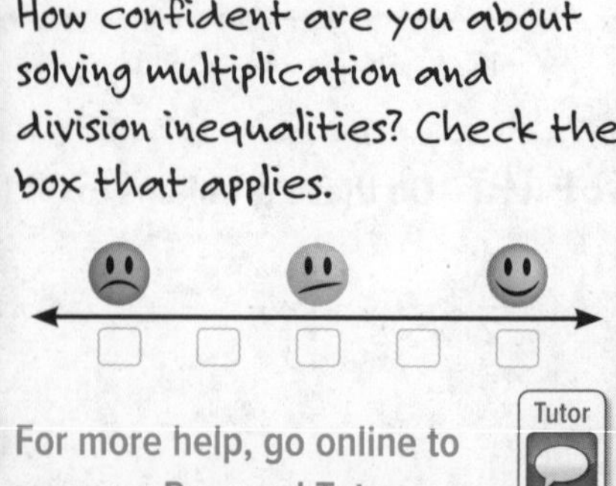

Rate Yourself!

How confident are you about solving multiplication and division inequalities? Check the box that applies.

For more help, go online to access a Personal Tutor.

Name ______________________ My Homework ______________________

Independent Practice

Go online for Step-by-Step Solutions

Solve each inequality. (Examples 1 and 2)

1. $6y < 18$ ________

2. $-3s \geq 33$ ________

3. $60 \leq \frac{m}{3}$ ________

4. $\frac{t}{-2} < 6$ ________

5. $\frac{m}{-14} \leq -4$ ________

6. $-56 \leq -8x$ ________

7. $12n \leq 54$ ________

8. $\frac{h}{9} > \frac{1}{4}$ ________

9. $\frac{w}{-5} \geq 9$ ________

Solve each inequality. Graph the solution set on a number line. (Examples 3 and 4)

10. $4x \geq 36$ ________

11. $20 < 5t$ ________

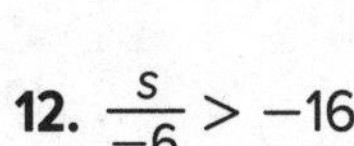

12. $\frac{s}{-6} > -16$ ________

13. $\frac{x}{-4} \geq 8$ ________

14. A pool charges $4 each visit, or you can buy a membership. Write and solve an inequality to find how many times a person should use the pool so that a membership is less expensive than paying each time. Interpret the solution. (Example 5)

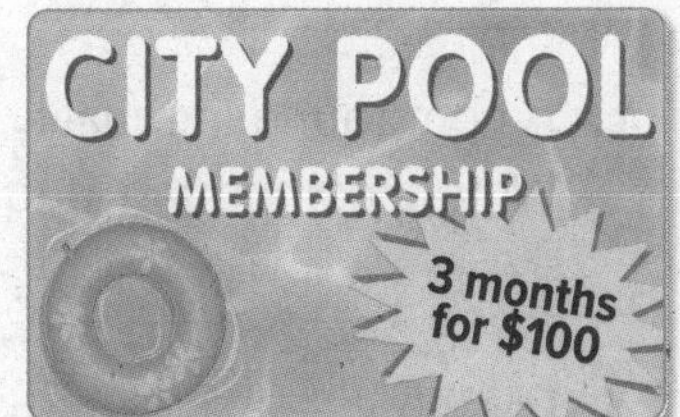

Inequality: ________ Solution: ________

Interpretation: ________________________

15. **MP Reason Inductively** Cross out the inequality that does not belong in the organizer shown at the right. Then explain your reasoning.

$-2x > 12$	$-2 < x + 4$
$\frac{x}{2} < -3$	$-7 > x - 1$

Write an inequality for each sentence. Then solve the inequality.

16. Sixteen is less than eight times a number.

Inequality:

Solution:

17. The product of a number and five is at the most 30.

Inequality:

Solution:

H.O.T. Problems Higher Order Thinking

18. **MP Identify Structure** Write two different inequalities that have the solution $y > 6$. One inequality should be solved using multiplication properties, and the other should be solved using division properties.

19. **MP Persevere with Problems** You score 15, 16, 17, 14, and 19 points out of 20 possible points on five tests. What must you score on the sixth test to have an average of at least 16 points?

20. **MP Reason Inductively** The inequalities $3x > 2$ and $9x > 6$ are equivalent inequalities. Write another inequality that is equivalent to $3x > 2$ and $9x > 6$.

21. **MP Persevere with Problems** Consider the inequalities $b \geq 4$ and $b \leq 13$.

a. Graph each inequality on the number line.

b. Do the solution sets of the two inequalities overlap? If so, what does this overlapping area represent?

c. A compound inequality is an inequality that combines two inequalities. Write a compound inequality for the situation.

d. Look back at the graph of the solutions for both inequalities. Make another graph that shows only the solution of the compound inequality.

Name ______________________ My Homework ______________________

Extra Practice

Solve each inequality.

22. $-10n > -20$ $n < 2$

$-10n > -20$

$\frac{-10n}{-10} < \frac{-20}{-10}$

$n < 2$

23. $-7y < 35$ ______

24. $15 < 3r$ ______

25. $12p \geq -72$ ______

26. $\frac{t}{-7} > 10$ ______

27. $-8 < \frac{y}{5}$ ______

Solve each inequality. Graph the solution set on a number line.

28. $\frac{h}{5} \leq -12$ ______

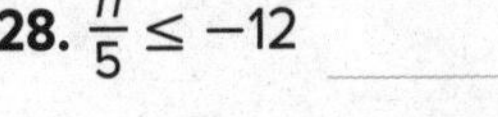

29. $-3w < -39$ ______

30. $15 < 4x$ ______

31. $10 \leq \frac{t}{-2}$ ______

32. MP **Reason Abstractly** Each game at a carnival costs \$0.50, or you can pay \$15 and play an unlimited amount of games. Write and solve an inequality to find how many times you should play a game so that the unlimited game play is less expensive than paying each time. Interpret the solution.

Inequality: ______ Solution: ______

Interpretation: ______

Write an inequality for each sentence. Then solve the inequality.

33. The product of a number and 4 is at least −12.

Inequality: ______

Solution: ______

34. Five times a number is less than −45.

Inequality: ______

Solution: ______

Power Up! Common Core Test Practice

35. Caitlin earns \$7 per hour babysitting. She wants to earn at least \$105 for a camping trip. Determine if each statement is true or false.

a. The inequality $\frac{h}{7} \geq 105$ models how many hours Caitlin must babysit to earn at least \$105. ☐ True ☐ False

b. The inequality $7h \geq 105$ models how many hours Caitlin must babysit to earn at least \$105. ☐ True ☐ False

c. Caitlin must babysit up to 15 hours in order to earn at least \$105. ☐ True ☐ False

36. Soccer balls cost \$24 each at Sports Emporium. Coach Neville can spend at most \$120 on equipment for the soccer team. Let *b* represent the number of soccer balls Coach Neville can buy.

Write an inequality to model the situation.

Solve the inequality for *b*.

Graph the solution on the number line.

How many soccer balls can Coach Neville buy? List all of the possible answers.

Common Core Spiral Review

Solve each equation. Check your solution. 6.EE.6

37. $5k + 6 = 16$

38. $-14 = 2x - 8$

39. $-4n + 3 = 13$

40. $25 = 7m + 4$

41. $10.5 + h = 22.5$

42. $14n - 32 = 22$

Lesson 8

Solve Two-Step Inequalities

Real-World Link

Newspapers Kaitlyn is placing an ad in the local newspaper for a pottery class. The cost of placing an ad is shown in the table.

Service	Cost ($)
10-day ad with 3 lines	38.00
each additional line	9.00

1. Complete the equation to find the total cost c of an ad with 4 or more lines. Use x as the variable.

cost of a 10-day add with only 3 lines		cost of each additional line		total cost
☐	+	☐ x	=	☐

2. How much will it cost to place the ad if it is 5 lines long?

____

3. Suppose Kaitlyn can spend only $50 on the ad. Does she have enough money to place the ad? Circle yes or no.

yes no

If the answer is no, how much more money will Kaitlyn need? Explain. ______________________________

____

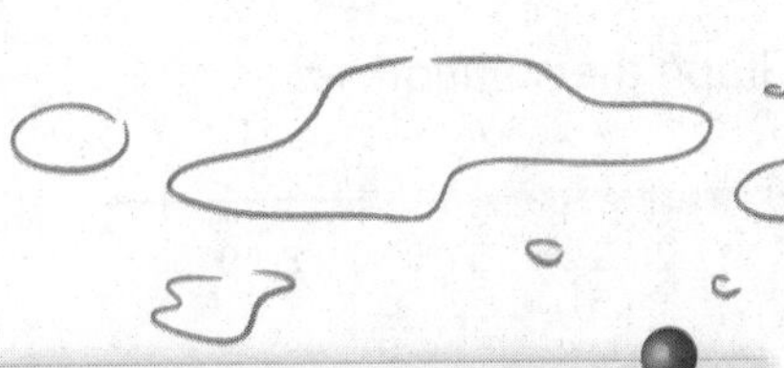

Which MP Mathematical Practices did you use? Shade the circle(s) that applies.

① Persevere with Problems
② Reason Abstractly
③ Construct an Argument
④ Model with Mathematics
⑤ Use Math Tools
⑥ Attend to Precision
⑦ Make Use of Structure
⑧ Use Repeated Reasoning

Essential Question

WHAT does it mean to say two quantities are equal?

Vocabulary

two-step inequality

Common Core State Standards

Content Standards
7.EE.4, 7.EE.4b

MP Mathematical Practices
1, 2, 3, 4, 5

Work Zone

Solve a Two-Step Inequality

A **two-step inequality** is an inequality that contains two operations. To solve a two-step inequality, use inverse operations to undo each operation in reverse order of the order of operations.

Examples

1. Solve $3x + 4 \geq 16$. Graph the solution set on a number line.

$3x + 4 \geq 16$	Write the inequality.
$\underline{-4 \quad -4}$	Subtract 4 from each side
$3x \geq 12$	Simplify.
$\frac{3x}{3} \geq \frac{12}{3}$	Divide each side by 3.
$x \geq 4$	Simplify.

Graph the solution set.

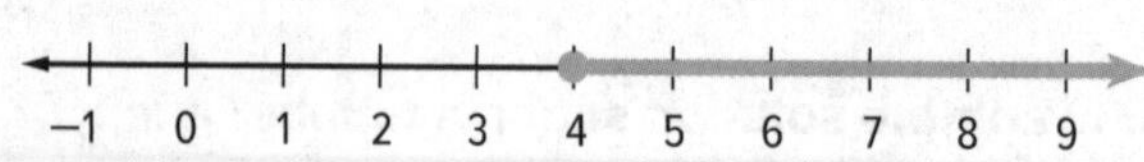

Draw a closed dot at 4 with an arrow to the right.

2. Solve $5 + 4x < 33$. Graph the solution set on a number line.

$5 + 4x < 33$	Write the inequality.
$\underline{-5 \qquad -5}$	Subtract 5 from each side.
$4x < 28$	Simplify.
$\frac{4x}{4} < \frac{28}{4}$	Divide each side by 4.
$x < 7$	Simplify.

Graph the solution set.

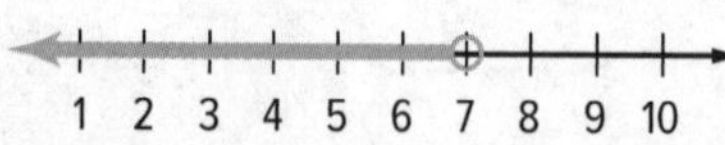

Draw an open dot at 7 with an arrow to the left.

Got it? Do this problem to find out.

a. __________

a. Solve $2x + 8 > 24$. Graph the solution on the number line provided.

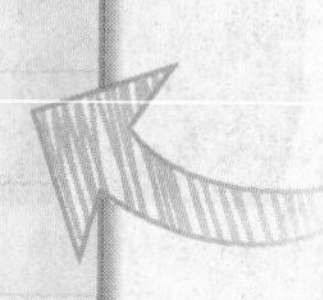

Examples

3. Solve $7 - 2x > 11$. Graph the solution set on a number line.

$7 - 2x > 11$	Write the inequality.
$-7 \quad -7$	Subtract 7 from each side.
$-2x > 4$	Simplify.
$\frac{-2x}{-2} < \frac{4}{-2}$	Divide each side by −2. Reverse inequality symbol.
$x < -2$	Simplify. Check your solution.

Graph the solution set.

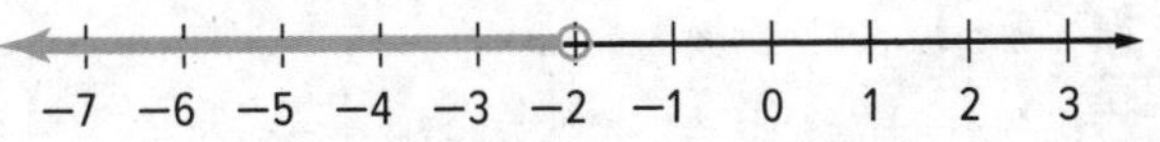

Draw an open dot at −2 with an arrow to the left.

You can check the solution by substituting a number less than −2 into the original inequality. Try using −3.

Check	$7 - 2x > 11$	Write the inequality.
	$7 - 2(-3) \overset{?}{>} 11$	Replace x with −3. Is the sentence true?
	$13 > 11$	This is a true statement. ✓

4. Solve $\frac{x}{2} - 5 < -8$. Graph the solution set on a number line.

$\frac{x}{2} - 5 < -8$	Write the inequality.
$+5 \quad +5$	Add 5 to each side.
$\frac{x}{2} < -3$	Simplify.
$\frac{x}{2}(2) < -3(2)$	Multiply each side by 2.
$x < -6$	Simplify. Check your solution.

Graph the solution set.

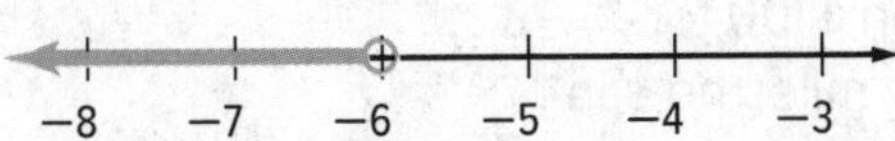

Draw an open dot at −6 with an arrow to the left.

Solving Inequalities

Remember that if multiplying or dividing by a negative number when solving inequalities, reverse the direction of the inequality symbol.

Got it? Do these problems to find out.

Solve each inequality. Graph the solution set on the number line provided.

b. $\frac{x}{2} + 9 \geq 5$

c. $8 - \frac{x}{3} \leq 7$

b. ____________

c. ____________

Example

5. **Halfway through the bowling league season, Stewart has 34 strikes. He averages 2 strikes per game. Write and solve an inequality to find how many more games it will take for Stewart to have at least 61 strikes, the league record. Interpret the solution.**

The number of strikes plus two strikes per game is at least 61. Let g represent the number of games he needs to bowl.

$34 + 2g \geq 61$	Write the inequality.
$-34 \quad\quad -34$	Subtract 34 from each side.
$2g \geq 27$	Simplify.
$\frac{2g}{2} \geq \frac{27}{2}$	Divide each side by 2.
$g \geq 13.5$	Simplify.

Stewart should have at least 61 strikes after 14 more games.

Guided Practice

Solve each inequality. Graph the solution set on a number line. (Examples 1–4)

1. $5x - 7 \geq 43 =$ ______

2. $11 \leq 7 + \frac{x}{5}$ ______

3. **Financial Literacy** A rental car company charges $45 plus $0.20 per mile to rent a car. Mr. Lawrence does not want to spend more than $100 for his rental car. Write and solve an inequality to find how many miles he can drive and not spend more than $100. Interpret the solution. (Example 5)

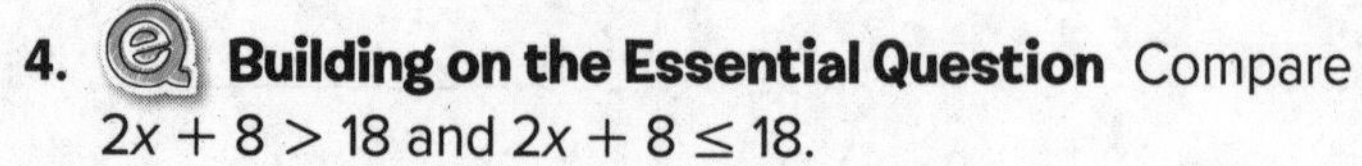

4. **Building on the Essential Question** Compare $2x + 8 > 18$ and $2x + 8 \leq 18$.

Name ______________________ My Homework ______________________

Independent Practice

Go online for Step-by-Step Solutions

Solve each inequality. Graph the solution set on a number line. (Examples 1–4)

1. $6x + 14 \geq 20$

Show your work.

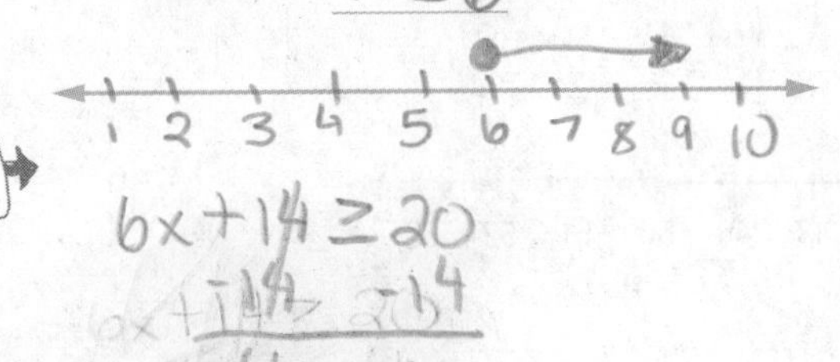

2. $4x - 13 < 11$

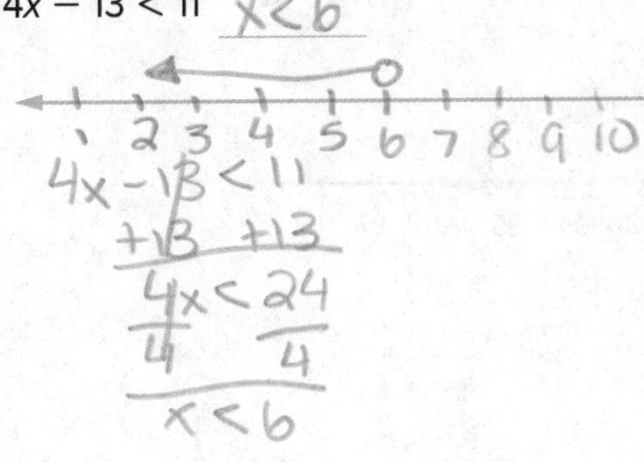

3. $-20 > -2x + 4$

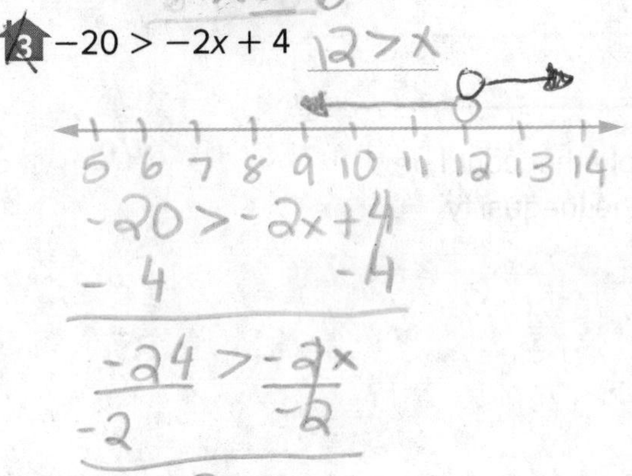

4. $\frac{x}{13} + 3 \geq 4$

5. Tyler needs at least $205 for a new video game system. He has already saved $30. He earns $7 an hour at his job. Write and solve an inequality to find how many hours he will need to work to buy the system. Interpret the solution. (Example 5)

Inequality: ______________ Solution: ______________

Interpretation: ______________________________

MP Reason Abstractly Write and solve an inequality for each sentence.

6. Three times a number increased by four is less than -62.

7. The quotient of a number and -5 increased by one is at most 7.

8. The quotient of a number and 3 minus two is at least -12.

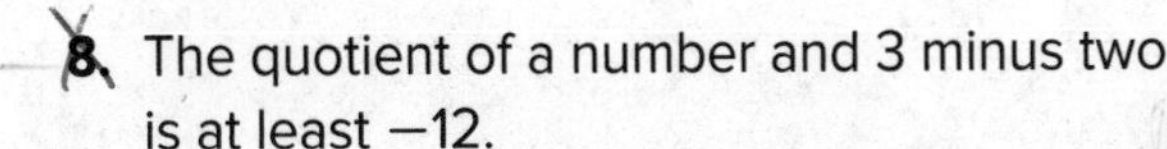

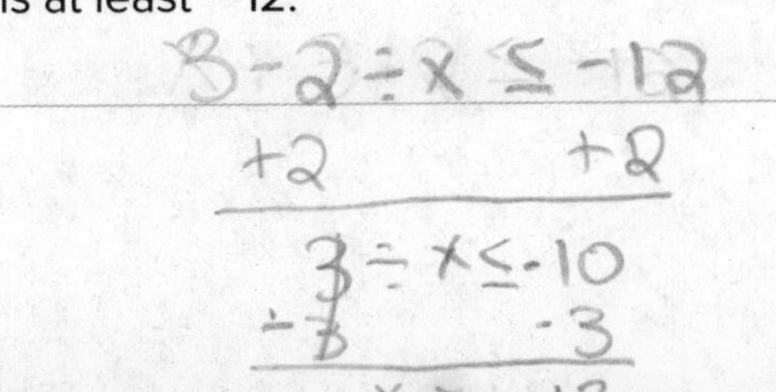

9. The product of -2 and a number minus six is greater than -18.

Write a two-step inequality that could be represented by each number line.

10. 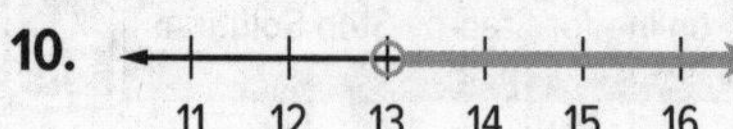

11. 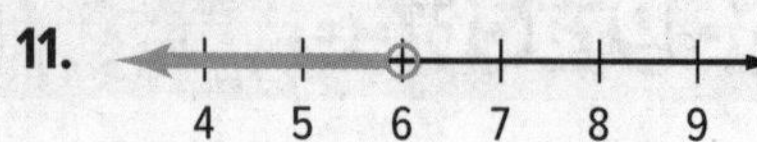

12. 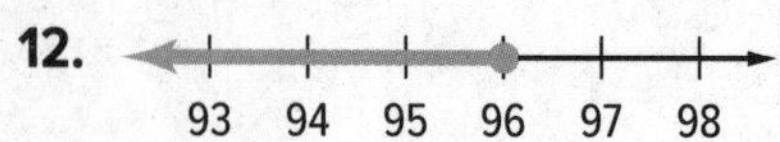

13. 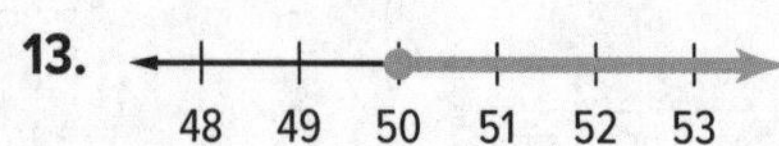

H.O.T. Problems Higher Order Thinking

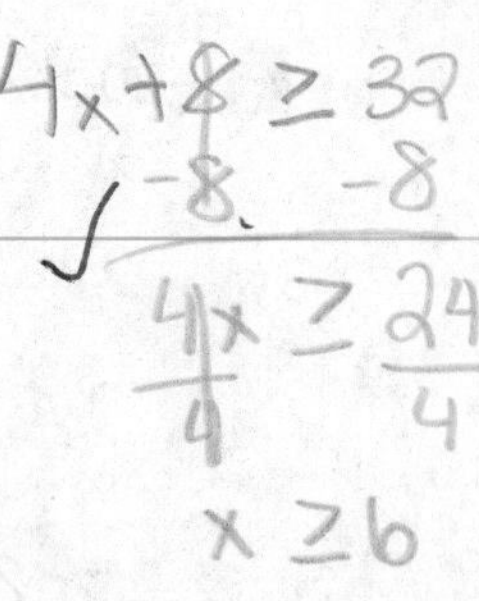

14. **MP Model with Mathematics** Write a real-world example that could be solved by using the inequality $4x + 8 \geq 32$. Then solve the inequality.

Joe wants to buy a game for \$32. He had already saved \$8. If Joe earns \$4 for every hr he does a chore, at least how many hr will it take him to buy the game.

15. **MP Persevere with Problems** In five games, you score 16, 12, 15, 13, and 17 points. Write and solve an inequality to determine how many points you must score in the sixth game to have an average of at least 15 points.

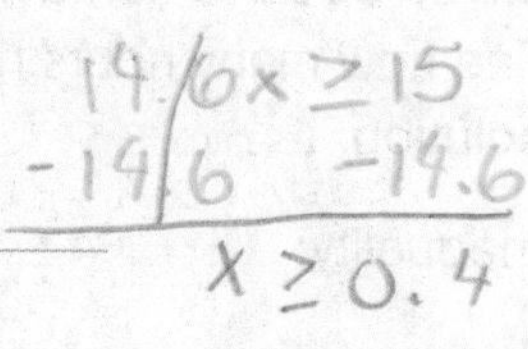

$14.6x \geq 15$

$\frac{73+x}{6} \geq 15$

16. **MP Use Math Tools** Solve $-x + 6 > -(2x + 4)$. Then graph the solution set on the number line.

Solution: ______

17. **MP Model with Mathematics** Write and solve a real-world problem that can be represented by the inequality $4(x - 2.8) \leq 45$.

Extra Practice

Solve each inequality. Graph the solution set on a number line.

18. $4x - 15 \leq 5$ $x \leq 5$

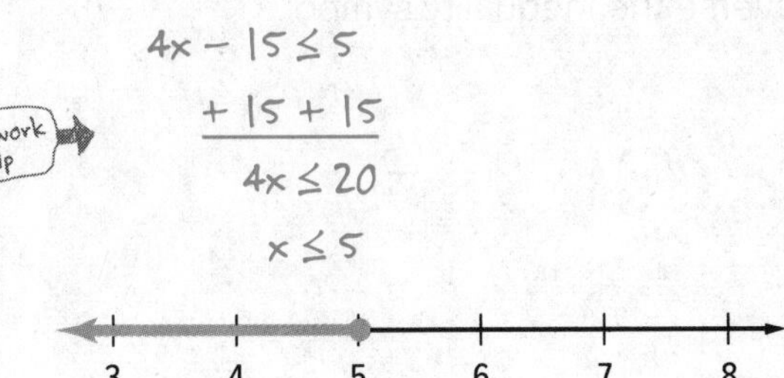

Homework Help

19. $-73 \geq 15 + 11x$ ______

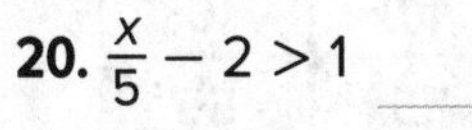

20. $\frac{x}{5} - 2 > 1$ ______

21. $9 \leq \frac{x}{14} + 6$ ______

22. Catie is starting a babysitting business. She spent \$26 to make signs to advertise. She charges an initial fee of \$5 and then \$3 for each hour of service. Write and solve an inequality to find the number of hours she will have to babysit to make a profit. Interpret the solution.

Inequality: $x(5+3) > 26$ Solution: $x > 4$

Interpretation: Catie will have to babysit for 4 or more hr. to make a profit.

23. **MP Reason Abstractly** As a salesperson, Audrey earns \$75 per week plus \$5 per sale. This week, she wants her pay to be at least \$125. Write and solve an inequality for the number of sales Audrey needs to make. Interpret the solution.

Inequality: $5n + 75 \geq 125$ Solution: $n \geq 10$

Interpretation: Audrey will have to make at least 10 or more sales

24. Elijah and his sister went to the movies. They had \$34 altogether and spent \$9.50 per ticket. Elijah and his sister bought the same snacks. Write and solve an inequality for the amount that each person spent on snacks. Interpret the solution.

Inequality: $9.50 - n < 34$ Solution: $n > -24.5$

Interpretation: Elijah & his sisters spent more than

CCSS Power Up! Common Core Test Practice

25. Which of the following are operations that you should use to solve $-2x - 5 < 7$ for x? Select all that apply.

- ☐ Subtract 7 from both sides.
- ☐ Add 5 to both sides.
- ☐ Divide both sides by -2.
- ☐ Reverse the inequality symbol.

26. The table shows the cost of renting a jet ski.

Rental Period	Cost ($)
First hour	$55
Each additional 15-minutes	$10

Jeremy can spend no more than $105 on a jet ski rental. Let x represent the number of additional 15-minute increments. Fill in each box to write an inequality to represent the situation.

☐ + ☐ ≤ ☐

$10x$	10
$55x$	55
$105x$	105
$<$	$\leq$
$>$	$\geq$

What is the greatest length of time Jeremy can rent the jet ski?

☐

CCSS Common Core Spiral Review

Solve and graph each inequality. 7.EE.4b

27. $n + 1 > -2$

Solution: ______

28. $-2y > 12$

Solution: ______

29. $\frac{t}{-1} > -2$

Solution: ______

Solve each equation. Check your solution. 7.EE.4a

30. $5y + 6 = 46$ ______

31. $-4k - 1 = 47$ ______

32. $5 = 8m + 1$ ______

33. Michael's dad is 30 years of age. He is 2 years more than four times Michael's age m. Write and solve a two-step equation to determine Michael's age. 6.EE.7

Equation: ______ Solution: ______

21ST CENTURY CAREER in Veterinary Medicine

Veterinary Technician

If you love being around animals, enjoy working with your hands, and are good at analyzing problems, a challenging career in veterinary medicine might be a perfect fit for you. Veterinary technicians help veterinarians by helping to diagnose and treat medical conditions. They may work in private clinics, animal hospitals, zoos, aquariums, or wildlife rehabilitation centers.

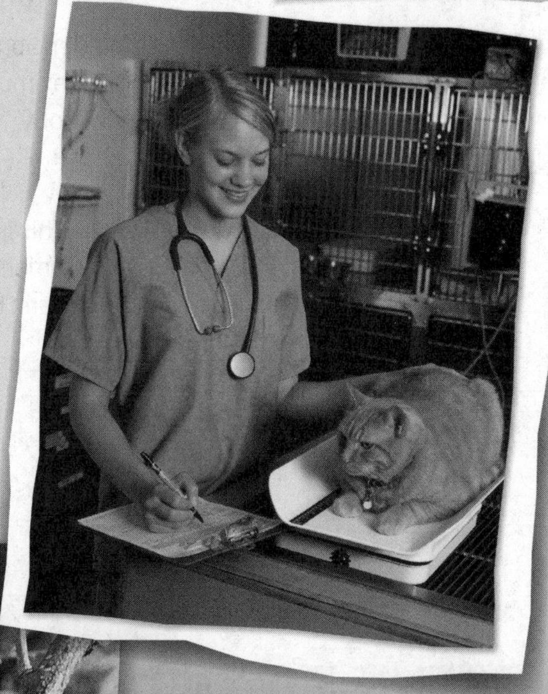

Is This the Career for You?

Are you interested in a career as a veterinary technician? Take some of the following courses in high school.

- Algebra
- Animal Science
- Biology
- Chemistry
- Veterinary Assisting

Find out how math relates to a career in Veterinary Medicine.

MP Vet Techs Don't Monkey Around

For each problem, use the information in the tables to write an equation. Then solve the equation.

1. The minimum tail length of an emperor tamarin is 1.6 inches greater than that of a golden lion tamarin. What is the minimum tail length of a golden lion tamarin? ____________

2. The minimum body length of a golden lion tamarin is 5.3 inches less than the maximum body length. What is the maximum body length? ____________

3. Tamarins live an average of 15 years. This is 13 years less than the years that one tamarin in captivity lived. How long did the tamarin in captivity live? ____________

4. The maximum weight of a golden lion tamarin is about 1.97 times the maximum weight of an emperor tamarin. What is the maximum weight of an emperor tamarin? Round to the nearest tenth. ____________

5. For an emperor tamarin, the maximum total length, including the body and tail, is 27 inches. What is the maximum body length of an emperor tamarin? ____________

Golden Lion Tamarin Monkeys		
Measure	**Minimum**	**Maximum**
Body length	7.9 in.	ℓ
Tail length	t	15.7 in.
Weight	12.7 oz	28 oz

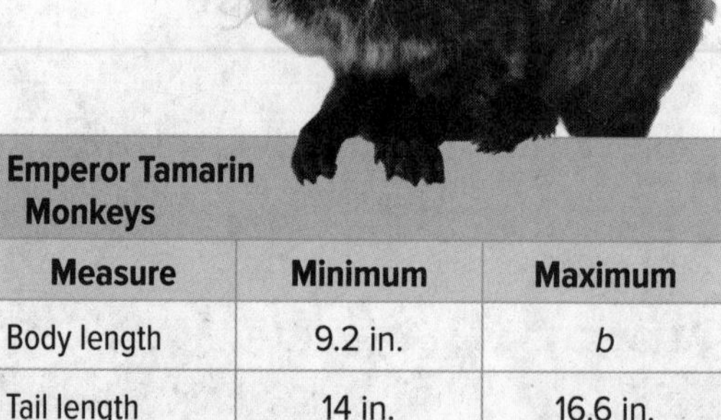

Emperor Tamarin Monkeys		
Measure	**Minimum**	**Maximum**
Body length	9.2 in.	b
Tail length	14 in.	16.6 in.
Weight	10.7 oz	w

MP Career Project

It's time to update your career portfolio! Go to the Occupational Outlook Handbook online and research a career as a veterinary technician. Include brief descriptions of the work environment, education and training requirements, and the job outlook.

Do you think you would enjoy a career as a veterinary technician? Why or why not?

-
-
-
-
-

Chapter Review

Vocabulary Check

Unscramble each of the clue words.

TOW-SETP

[ ][ ][ ][–][ ][ ][ ][ ] — box 2 is numbered 7

PYORERPT

[ ][ ][ ][ ][ ][ ][ ][ ] — box 8 is numbered 8

DODTIINA

[ ][ ][ ][ ][ ][ ][ ][ ] — box 4 is numbered 6

NIIOSDIV

[ ][ ][ ][ ][ ][ ][ ][ ]

LABVIERA

[ ][ ][ ][ ][ ][ ][ ][ ] — box 7 is numbered 5

AILEYQUITN

[ ][ ][ ][ ][ ][ ][ ][ ][ ][ ] — box 4 is numbered 2

BISTAUTORNC

[ ][ ][ ][ ][ ][ ][ ][ ][ ][ ][ ] — box 2 is numbered 3

NUATIEQO

[ ][ ][ ][ ][ ][ ][ ][ ] — box 1 is numbered 1

TIULINTICPOLMA

[ ][ ][ ][ ][ ][ ][ ][ ][ ][ ][ ][ ][ ][ ] — box 10 is numbered 4

Use the numbered letters to find another vocabulary term from this chapter.

[ ][ ][ ][ ][ ][ ][ ][ ]

1 2 3 4 5 6 7 8

Key Concept Check

Use Your FOLDABLES

Use your Foldable to help review the chapter.

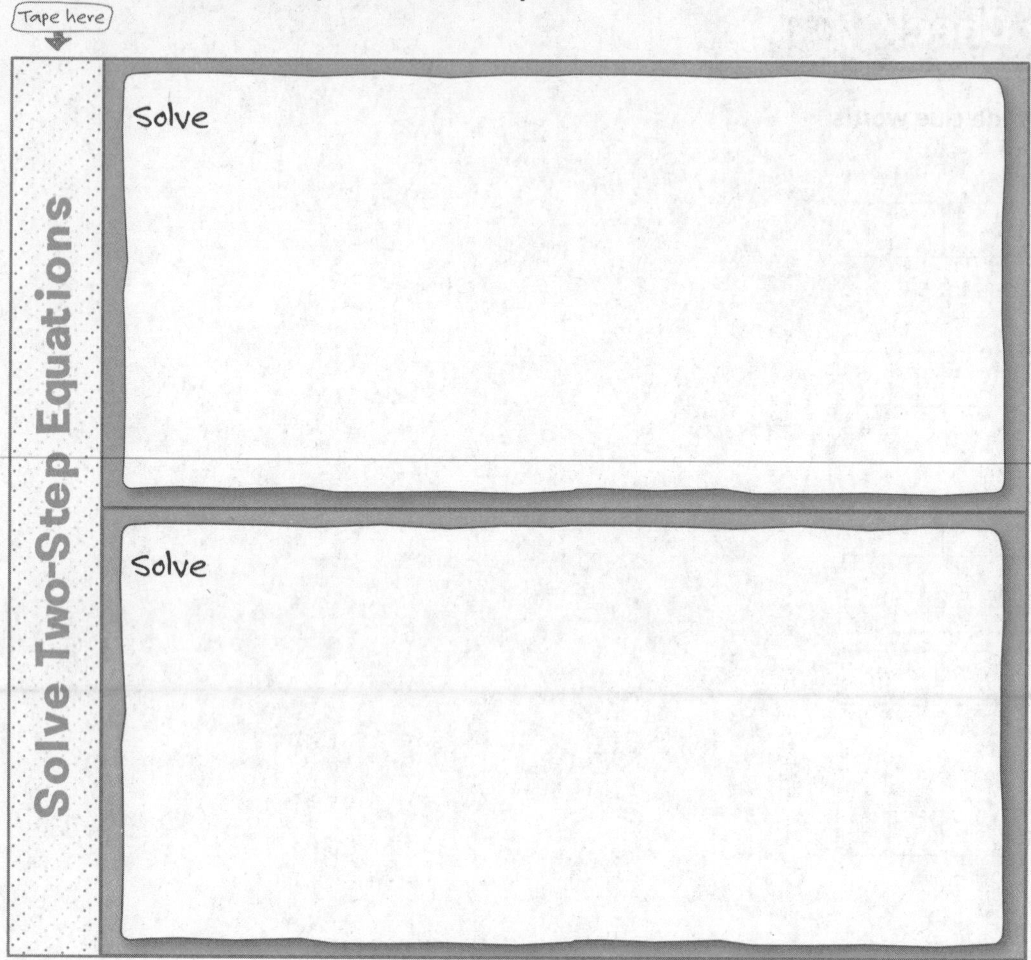

Got it?

Match each phrase with the correct term.

1. the value of a variable that makes an equation true
2. the numerical factor in a multiplication expression
3. equations that have the same solution
4. a sentence stating that two quantities are equal

a. equivalent equations

b. equation

c. Addition Property of Equality

d. coefficient

e. formula

f. solution

Fall Reading

Gordon's English teacher assigned a book to be read by October 31st. The students may select a book from the table, and Gordon chose *City Streets*.

Book	Number of Pages
City Streets	387
Life and Time	411
Myopia	435

Write your answers on another piece of paper. Show all of your work to receive full credit.

Part A

By October 19th, Gordon had read 35 pages. Starting on October 20th, he decides to read the same number of pages each day until he finishes the book on October 30th. Write and solve an equation to represent the situation. Let p represent the number of pages read per day. How many pages does Gordon read per day?

Part B

Gordon's friend, Kendrick, selected *Myopia*. He read eight pages in class on October 19th and begins reading again on October 23rd. He needs to read at least 350 pages by the end of the day on October 28th. Write and solve an inequality to represent this situation and graph the solution on a number line. Let p represent the number of pages read per day. How many pages must Kendrick read per day to accomplish his goal?

Reflect

Answering the Essential Question

Use what you learned about equations and inequalities to complete the graphic organizer.

When do you use an equals sign?

Essential Question

WHAT does it mean to say two quantities are equal?

When do you use an inequality symbol?

Answer the Essential Question. WHAT does it mean to say two quantities are equal?

UNIT PROJECT

Watch

Stand Up and Be Counted The U.S. Census is used to determine the number of U.S. House of Representative members that each state is assigned. In this project you will:

- **Collaborate** with your classmates as you research Census data and the U.S. House of Representatives.
- **Share** the results of your research in a creative way.
- **Reflect** on how you can communicate mathematical ideas effectively.

Collaborate

Go Online Work with your group to research and complete each activity. You will use your results in the Share section on the following page.

1. Explore the official U.S. Census web site to find the 2010 state populations. There will be interactive maps that display this information. Write down a few facts you find interesting.

2. Create a table that displays the population and the number of U.S. Representatives for your state and three other states. Then create a line plot for the number of U.S. Representatives.

3. Write an equation that uses any state's population x and its number of U.S. Representative members y to describes the number of people per U.S. Representative z.

4. Use your equation from Exercise 3 to determine the approximate number of people per U.S. Representative for the four states you chose. Interpret the results.

5. Look at the 2000 and 2010 census. How did the population of your state and states in your region change? Did the population change affect the number of U.S. Representatives assigned?

6. States can be categorized by population size and density. Write at least two inequalities that compare the states using these categories.

With your group, decide on a way to share what you have learned about the U.S. House of Representatives and state populations. Some suggestions are listed below, but you can also think of other creative ways to present your information. Remember to show how you used mathematics to complete each of the activities in this project!

- Act as a Census representative and create a presentation to encourage people to participate in the census and explain why it is important.
- Write a letter or email to your Representative about what you learned in this project and how it can be used to improve your community.

Check out the note on the right to connect this project with other subjects.

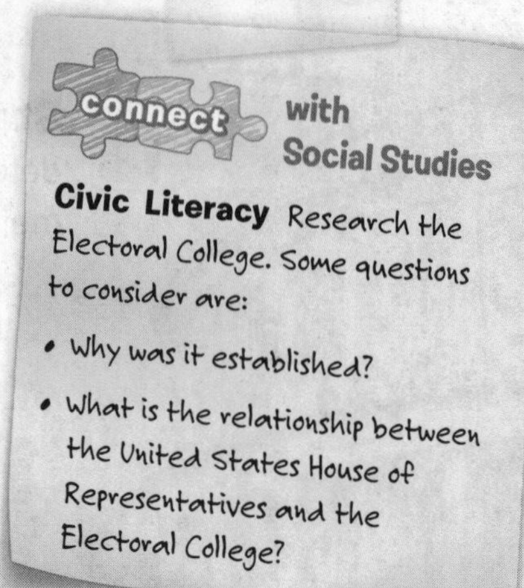

7. **Answer the Essential Question** How can you communicate mathematical ideas effectively?

   a. How did you use what you learned about expressions to help you communicate mathematical ideas effectively in this project?

   b. How did you use what you learned about equations and inequalities to help you communicate mathematical ideas effectively in this project?

UNIT 4

CCSS Geometry

Essential Question

HOW can you use different measurements to solve real-life problems?

Chapter 7
Geometric Figures

Geometric shapes can be drawn freehand, with a ruler and protractor, or using technology. In this chapter, you will draw two- and three-dimensional figures. You will also solve problems involving scale drawings of geometric figures.

Chapter 8
Measure Figures

Real-life problems involving area, surface area, and volume can be solved by using formulas. In this chapter, you will use formulas to find the area and circumference of a circle and to find the surface area and volume of prisms and pyramids.

Unit Project Preview

Turn Over a New Leaf Leaves serve an important purpose for plants and trees, and there are scientific reasons why they are flat.

Make a sketch of a leaf. Estimate the area of the leaf in square units.

At the end of Chapter 8, you'll complete a project to investigate the relationship between the volume and surface area of a leaf. So, put on your hiking shoes, and don't forget to bring along your measuring tools. You're about to go on a nature hike!

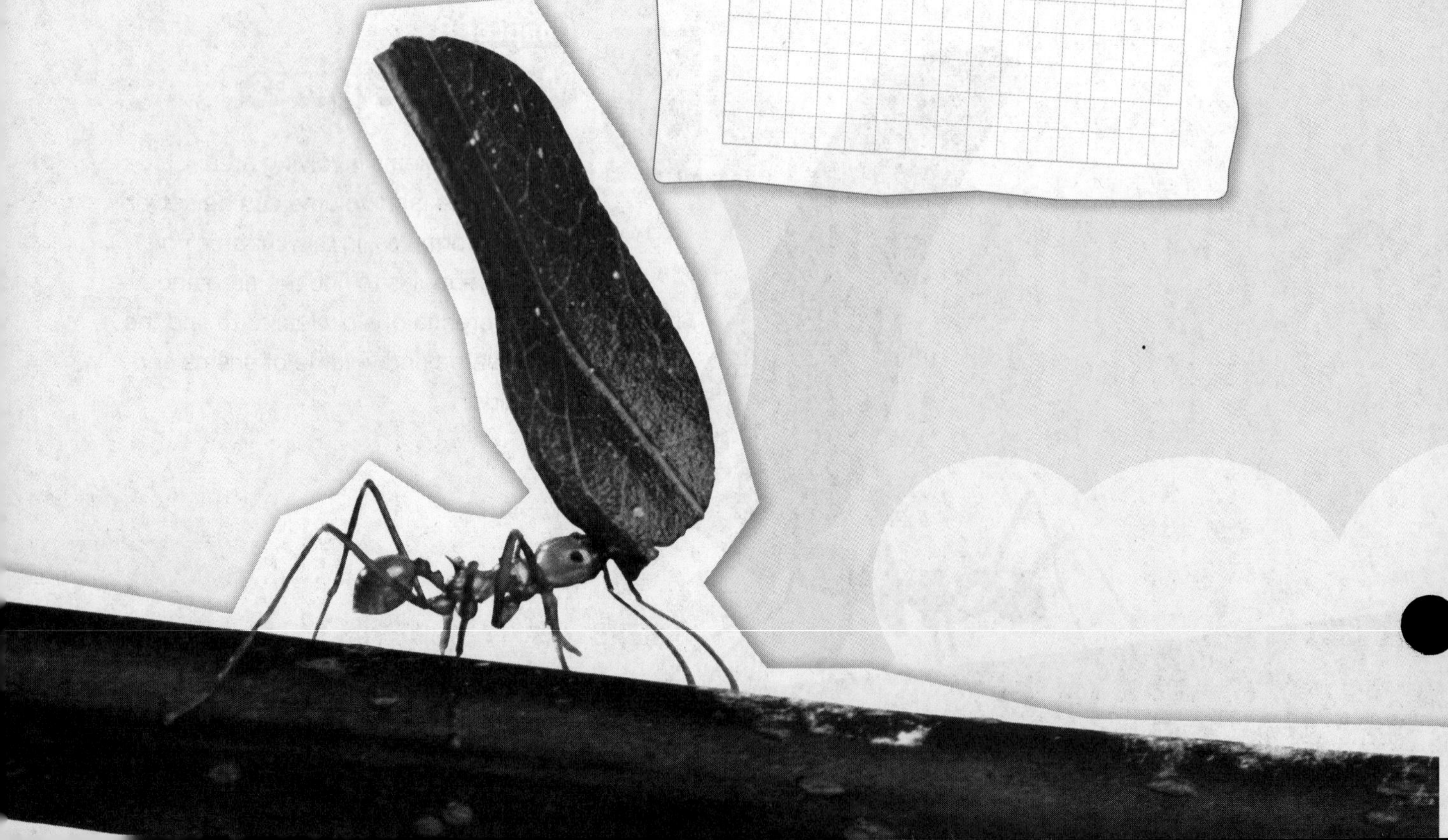

Chapter 7
Geometric Figures

Essential Question

HOW does geometry help us describe real-world objects?

Common Core State Standards

Content Standards
7.G.1, 7.G.2, 7.G.3, 7.G.5

MP Mathematical Practices
1, 2, 3, 4, 5, 6, 7, 8

Math in the Real World

Robots that could be programmed and digitally operated were invented by George Devol in 1954.

The actual length of the robot's arm is 15 inches. A drawing of the robot is $\frac{1}{5}$ the size of the actual robot. Fill in the blank below with the correct measurement for the robot's arm.

FOLDABLES® Study Organizer

1 Cut out the Foldable on page FL7 of this book.

2 Place your Foldable on page 604.

3 Use the Foldable throughout this chapter to help you learn about geometric figures.

What Tools Do You Need?

Vocabulary

acute angle
acute triangle
adjacent angles
base
complementary angles
cone
congruent
congruent segments
coplanar
cross section
cylinder
diagonal
edge
equilateral triangle
face
isosceles triangle
obtuse angle
obtuse triangle
plane
polyhedron
prism
pyramid
right angle
right triangle
scale
scale drawing
scale factor
scale model
scalene triangle
skew line
straight angle
supplementary angles
triangle
vertex
vertical angles

Study Skill: Reading Math

The Language of Mathematics Many of the words you use in math and science are also used in everyday language, such as the leg of a person and the leg of a right triangle.

Usage	Example
Some words are used in science and in mathematics, but the meanings are different.	$x + 4 = -2$ $x = -6$ solution (125, 100, 75, 50)
Some words are used only in mathematics.	hypotenuse

Explain how the everyday meaning of *face* is different than its mathematical meaning.

Everyday meaning: ____________________

____

Mathematical meaning: ____________________

____

What Do You Already Know?

Read each statement. Decide whether you agree (A) or disagree (D). Place a checkmark in the appropriate column and then justify your reasoning.

Geometric Figures			
Statement	A	D	Why?
The common endpoint shared by the two sides of an angle is called the origin.			
A pair of adjacent supplementary angles forms a right angle.			
The sum of the measures of the angles in any triangle is 180°.			
A scale drawing/model represents an object that is too large or too small to be drawn at actual size.			
An isosceles triangle has three congruent sides.			
A cylinder has two parallel congruent circular bases.			

When Will You Use This?

Here are a few examples of how scale drawings are used in the real world.

Activity 1 Use the Internet to find a map of the area where you live. Find the map distance, in inches, between your school and your home. Describe a method you could use to find the actual distance once you know the map distance.

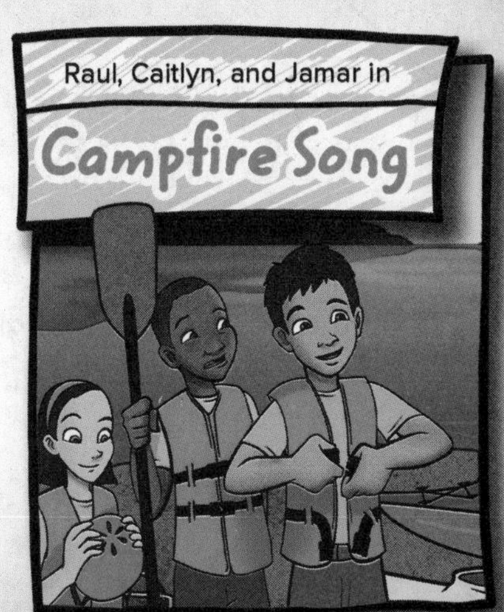

Activity 2 Go online at **connectED.mcgraw-hill.com** to read the graphic novel ***Campfire Song***. On the map, how many units long is the line connecting the cabin to the lake?

Try the Quick Check below.
Or, take the Online Readiness Quiz.

CCSS **Quick Review**

Common Core Review 4.MD.6, 6.G.1

Example 1

Use a protractor to measure angle *ABC*.

Align the center of the protractor with the vertex of the angle.

Make sure one ray of the angle passes through zero on the protractor.

Read the measure on the protractor where the other ray crosses the protractor.

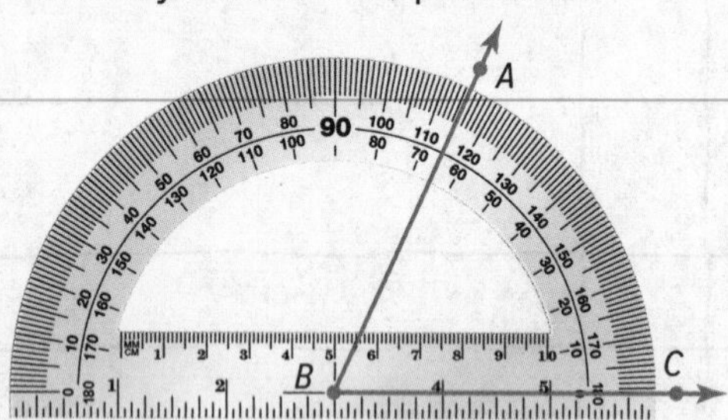

The angle measures 65°.

Example 2

Find the area of the triangle.

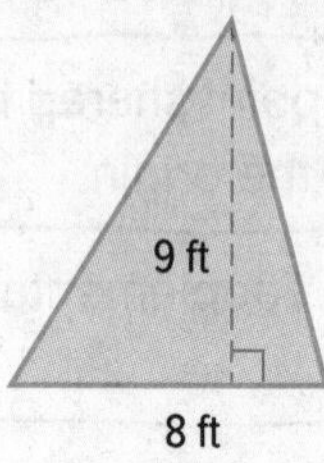

$A = \frac{1}{2}bh$ Area of a triangle

$A = \frac{1}{2}(8 \cdot 9)$ Replace *b* with 8 and *h* with 9.

$A = 36$ Simplify.

The area of the triangle is 36 square feet.

Quick Check

Angle Measures Use a protractor to measure each angle.

1. 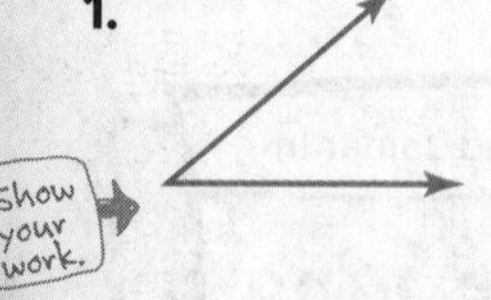

2. 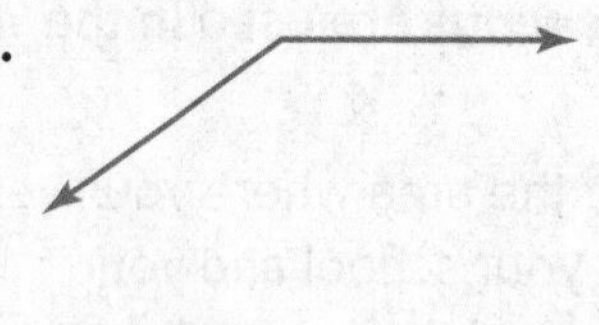

3. 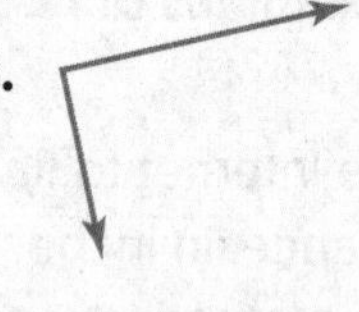

Area Find the area of each triangle.

4. 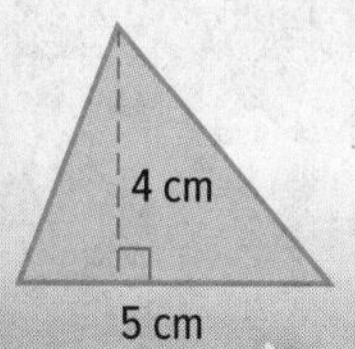

5. base: 3.2 yd
height: 4.2 yd

How Did You Do?

Which problems did you answer correctly in the Quick Check? Shade those exercise numbers below.

① ② ③ ④ ⑤

Lesson 1
Classify Angles

Vocabulary Start-Up 

An angle is formed by two rays that share a common endpoint. The **vertex** is the point where the two rays meet.

Complete the table by drawing the hands of a clock to represent each angle.

Type of Angle			
Right	**Acute**	**Obtuse**	**Straight**
exactly 90°	less than 90°	greater than 90°	exactly 180°

Essential Question

HOW does geometry help us describe real-world objects?

Vocabulary

vertex
right angle
acute angle
obtuse angle
straight angle
vertical angles
congruent
adjacent angles

Math Symbols

∠
≅

Common Core State Standards

Content Standards
7.G.5

MP Mathematical Practices
1, 3, 4, 7

Real-World Link

The angle formed by a bike ramp is shown.

1. What type of angle is formed?

2. Estimate the measure of the angle.

Which **MP Mathematical Practices** did you use? Shade the circle(s) that applies.

① Persevere with Problems
② Reason Abstractly
③ Construct an Argument
④ Model with Mathematics
⑤ Use Math Tools
⑥ Attend to Precision
⑦ Make Use of Structure
⑧ Use Repeated Reasoning

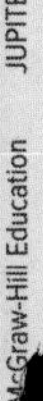

Key Concept

Name and Identify Angles

Work Zone

Words	Models	Symbols
Two angles are **vertical** if they are opposite angles formed by the intersection of two lines. Vertical angles are **congruent** or have the same measure.	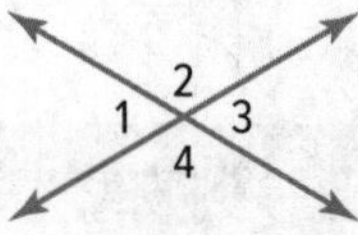 ∠1 and ∠3, ∠2 and ∠4	∠1 ≅ ∠3 ∠2 ≅ ∠4
Two angles are **adjacent** if they share a common vertex, a common side, and do not overlap.	(model: 1, 2, 3, 4)	Adjacent angle pairs are ∠1 and ∠2, ∠2 and ∠3, ∠3 and ∠4, and ∠4 and ∠1.

You can name an angle by its vertex and by its points.

Example

Tutor

Symbols
The symbol for angle is ∠.
The symbol ≅ means is congruent to.

1. Name the angle shown at the right. Then classify it as *acute, right, obtuse,* or *straight*.

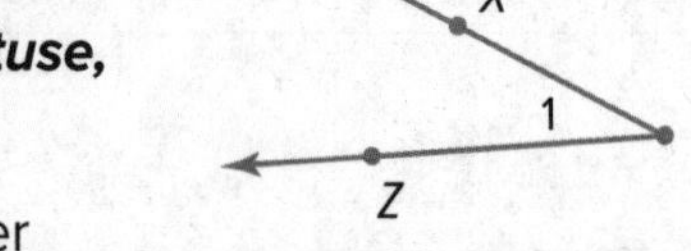

- Use the vertex as the middle letter and a point from each side, ∠XYZ or ∠ZYX.
- Use the vertex only, ∠Y.
- Use a number, ∠1.

Since the angle is less than 90°, it is an acute angle.

Got it? Do these problems to find out.

Name each angle in four ways. Then classify each angle as *acute, right, obtuse,* or *straight*.

a. 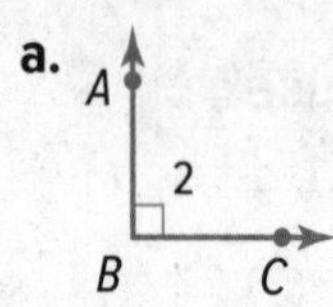

b. 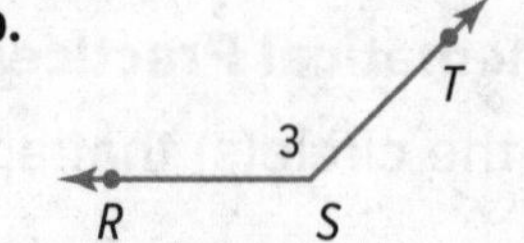

c. 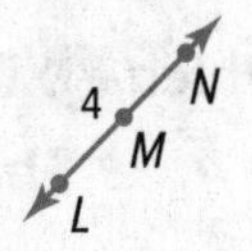

a. ________

b. ________

c. ________

Example

2. Identify a pair of vertical angles and adjacent angles in the diagram at the right. Justify your response.

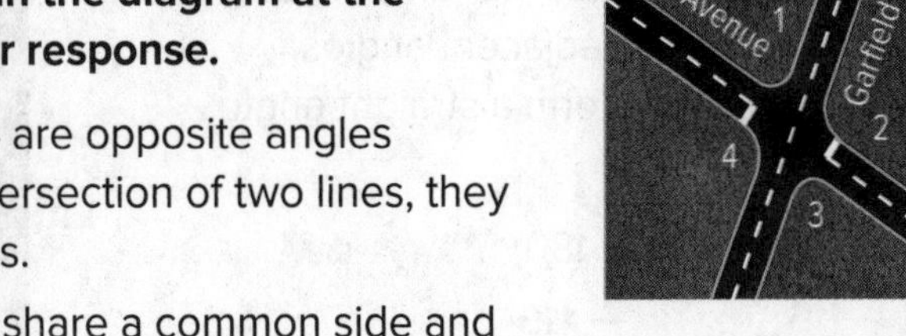

Since $\angle 2$ and $\angle 4$ are opposite angles formed by the intersection of two lines, they are vertical angles.

Since $\angle 1$ and $\angle 2$ share a common side and vertex, and they do not overlap, they are adjacent angles.

Got it? Do this problem to find out.

d. Refer to the diagram in Example 2. Identify different pairs of vertical and adjacent angles. Justify your response.

d. ___________

Find a Missing Measure

You can use what you learned about vertical and adjacent angles to find the value of a missing measure.

Example

3. What is the value of *x* in the figure?

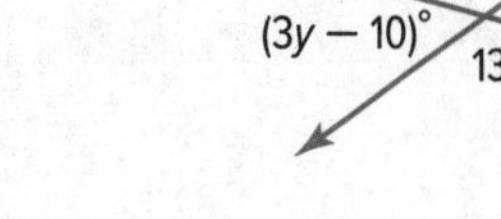

The angle labeled $(2x + 2)°$ and the angle labeled 130° are vertical angles.

Since vertical angles are congruent, $(2x + 2)°$ equals 130°.

$2x + 2 = 130$ Write the equation.

$-2 = -2$ Subtract 2 from each side.

$\frac{2x}{2} = \frac{128}{2}$ Divide each side by 2.

$x = 64$

So, the value of *x* is 64.

Got it? Do this problem to find out.

e. What is the value of *y* in the figure in Example 2?

e. ___________

Real World Example

4. What is the value of *x* shown in the sidewalk?

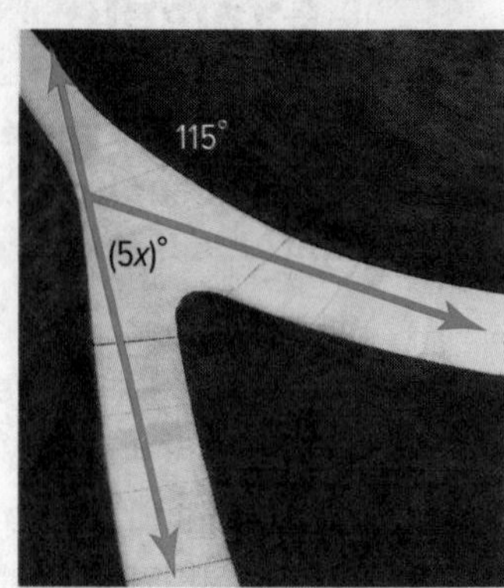

The angle labeled 115° and the angle labeled $5x$ are adjacent angles. Together they form a straight angle or 180°.

$$115 + 5x = 180$$ Write the equation.

$$-115 \quad = -115$$ Subtract 115 from each side.

$$\frac{5x}{5} = \frac{65}{5}$$ Divide each side by 5.

$$x = 13$$

So, the value of x is 13.

Guided Practice

1. Name the angle below in four ways. Then classify it as *acute, right, obtuse,* or *straight.* (Example 1)

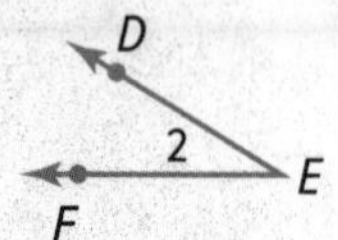

2. Find the value of x in the figure. (Examples 3–4)

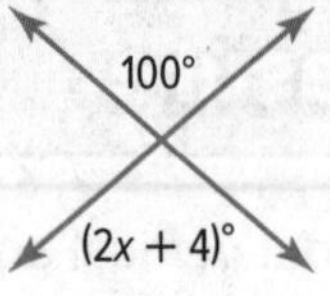

3. Identify a pair of vertical angles and adjacent angles on the railroad crossing sign. Justify your response. (Example 2)

4. **Building on the Essential Question** Describe the differences between vertical and adjacent angles.

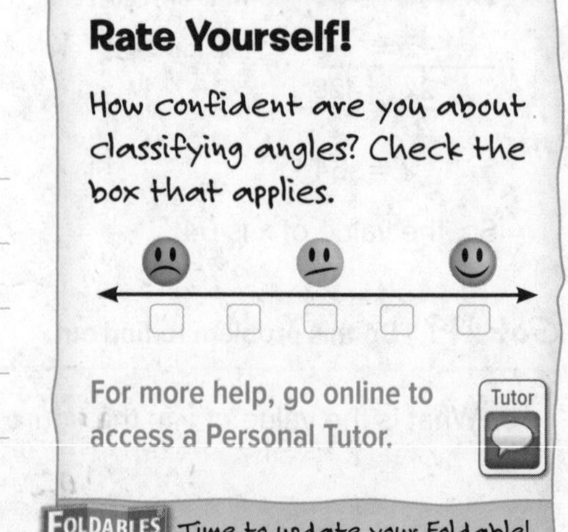

Name ______________________ My Homework ______________________

Independent Practice

Go online for Step-by-Step Solutions

Name each angle in four ways. Then classify the angle as *acute, right, obtuse,* or *straight*. (Example 1)

1. 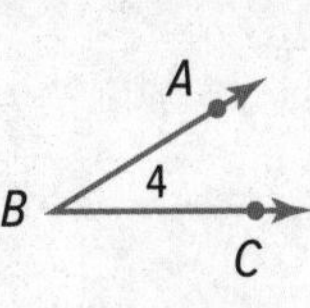

<ABC, <CBA, <B, <4
Acute

2. D, 5, E, F

<DEF, <FED, <E, <5
Right

3. 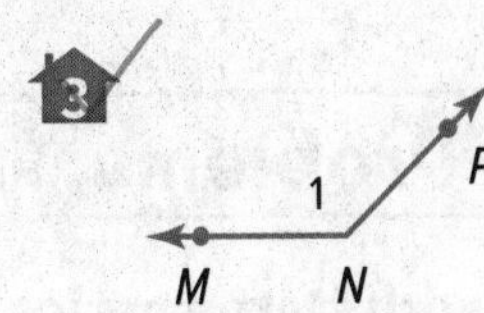

<MNP, <PNM, <N, <1
Obtuse

Identify Structure Refer to the diagram at the right. Identify each angle pair as *adjacent, vertical,* or *neither*. (Example 2)

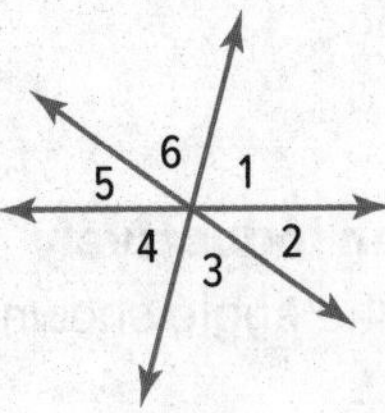

4. ∠2 and ∠5 vertical

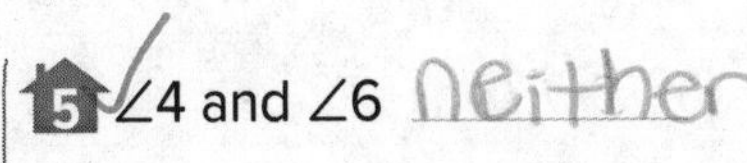

5. ∠4 and ∠6 neither

6. ∠3 and ∠4 adjacent

7. ∠5 and ∠6 adjacent

8. ∠1 and ∠3 neither

9. ∠1 and ∠4 vertical

10. What is the value of *x* in the figure at the right? (Examples 3 and 4) 37

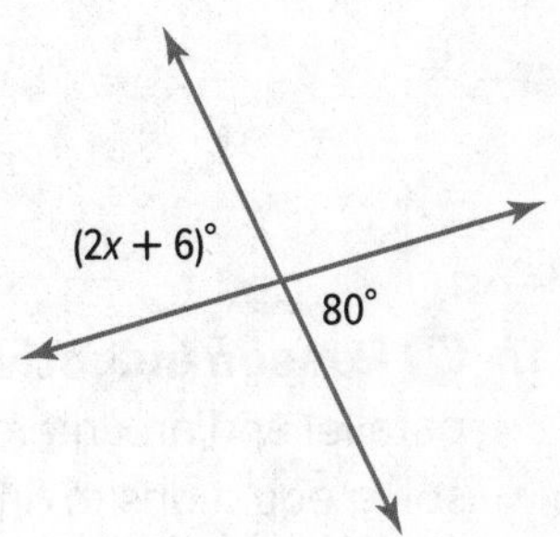

$2x+6=80$
$-6 \quad -6$
$\frac{2x}{2}=\frac{74}{2}$
$x=37$

11. What is the value of *x* in the figure at the right? (Examples 3 and 4) 11

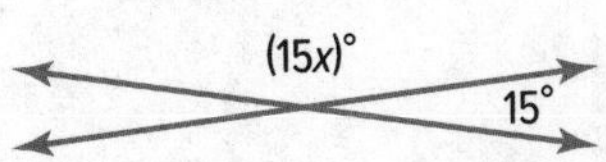

$15x+15=180$
$-15 \quad -15$
$\frac{15x}{15}=\frac{165}{15}$
$x=11$

12. Angles ABC and DBE are vertical angles. If the measure of $\angle ABC$ is 40°, what is the measure of $\angle ABD$?

H.O.T. Problems Higher Order Thinking

13. **MP Model with Mathematics** Draw examples of angles that represent real-world objects. Be sure to include at least three of the following angles: acute, right, obtuse, straight, vertical, and adjacent. Verify by measuring the angles.

14. **MP Reason Inductively** Explain how you can use a protractor to measure the angle shown. Find the measure of the angle.

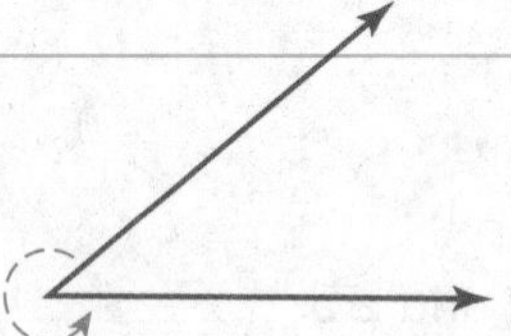

MP Persevere with Problems Determine whether each statement is *true* or *false*. If the statement is true, provide a diagram to support it. If the statement is false, explain why.

15. A pair of obtuse angles can also be vertical angles.

16. A pair of straight angles can also be adjacent angles.

Show your work.

$2x+8$ | $5x-10$ | $3x+42$ | $x+34$
$2(26)+8$ | $5(26)-10$ | $3(26)+42$ | $26+34$
$52+8$ | $130-10$ | $78+42$ | $60°$
$60°$ | $120°$ | $120°$

17. **MP Reason Inductively** Lines ℓ and k shown at the right are parallel and are intersected by line j. Explain how you can write and solve equations to find the measure of each angle. Then find the measure of each angle.

j
$(2x + 8)°$ $(5x - 10)°$ k
$(3x + 42)°$ $(x + 34)°$ ℓ

To set up the equation I would add $(2x+8)°$ with $(5x-10)°$ because there are both adjacent and equal it to 180° because a line is 180°. I would do the same for line ℓ. After finding wha x equals plug it back in the equation.

$3x+42+x+34=180$
$4x+76=180$
$-76\ -76$
$\frac{4x}{4}=\frac{104}{4}$ $x=26$

$2x+8+5x-10=180$
$7x-2=180$
$+2\ +2$
$\frac{7x}{7}=\frac{182}{7}$ $x=26$

Name ______ My Homework ______

Extra Practice

Name each angle in four ways. Then classify the angle as *acute, right, obtuse,* or *straight*.

18. 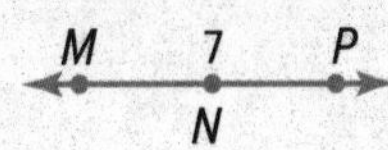

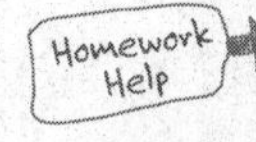

∠MNP, ∠PNM, ∠N, ∠7;
straight

19.  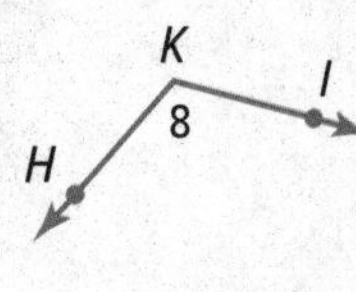

20. 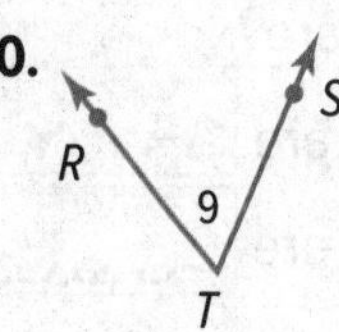

21. The corner where the states of Utah, Arizona, New Mexico, and Colorado meet is called the Four Corners.

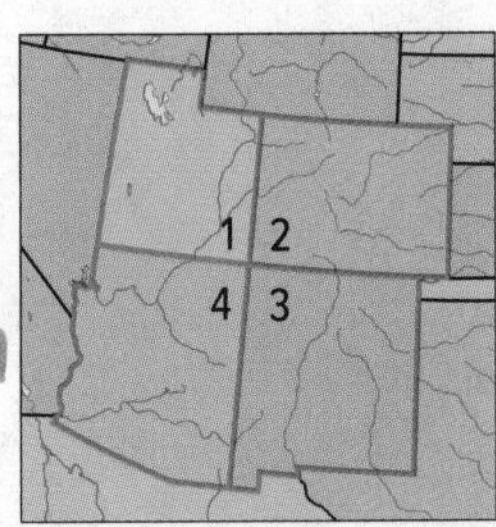

a. Identify a pair of vertical angles. Justify your response.

A pair of vertical angles are ∠1 & ∠2. These are vertical angles because they are opposite angles formed by the intersection of 2 lines.

b. Identify a pair of adjacent angles. Justify your response.

A pair of adjacent angles are ∠4 & ∠1. These are adjacent angles because they share a common vertex, a common side, do not overlap.

22. What is the value of x in the figure at the right?

76° 30°

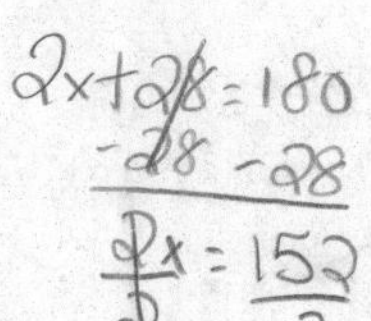

2x+28=180
-28 -28
2x=152
x=76

2x+38=

23. What is the value of x in the figure at the right?

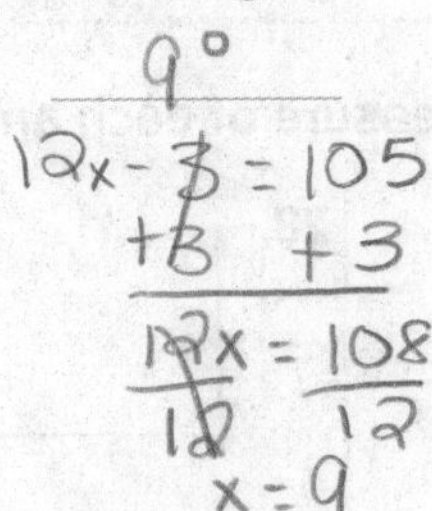
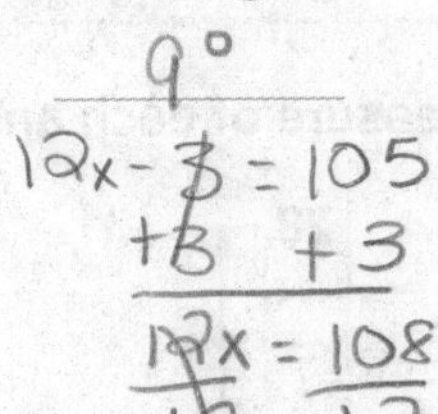

9°

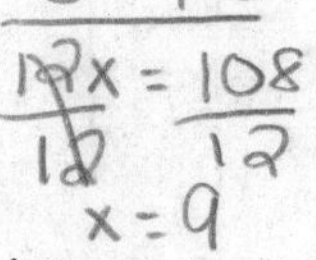

12x-3=105
+3 +3
12x=108
x=9

24. **MP Identify Structure** The John Hancock Center in Chicago is shown at the right. Classify each pair of angles.

a. ∠1 and ∠2 adjacent

b. ∠2 and ∠4 vertical

c. ∠3 and ∠4 adjacent

d. ∠1 and ∠3 vertical

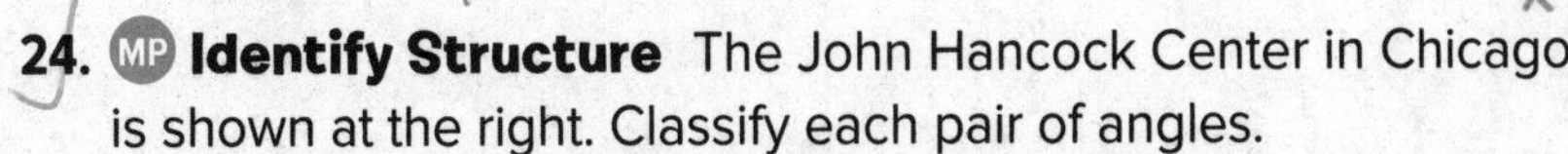

e. If the measure of ∠2 is 66°, what are the measures of the other angles? Angle 1 is 114 Angle 3 is 114°
Angle 4 is 66

Power Up! Common Core Test Practice

25. Refer to the figure at the right.

Fill in each box to make a true statement.

a. ∠1 and ∠4 are [vertical] angles.

b. ∠3 and ∠4 are [adjacent] angles.

c. ∠2 and ∠4 are [adjacent] angles.

d. ∠2 and ∠3 are [vertical] angles.

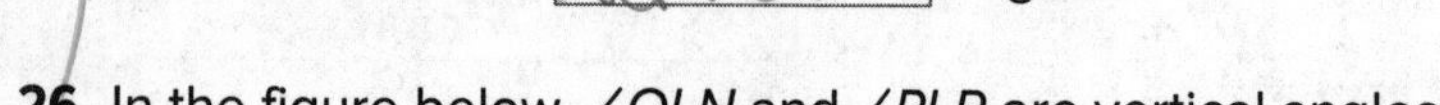

26. In the figure below, ∠*OLN* and ∠*PLR* are vertical angles.

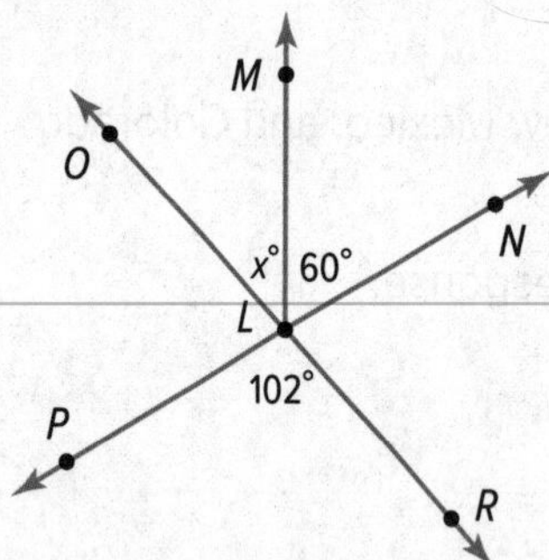

Select values to complete the equation to find the measure of ∠*MLO*.

x°	30°	60°
90°	102°	180°

[x°] + [60°] = [102°]

What is the measure of ∠*MLO*? [42°]

Common Core Spiral Review

Use a protractor to find the measure of each angle. 4.MD.6

27.

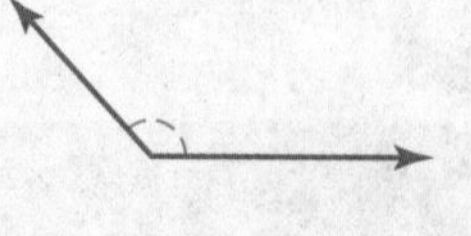

28.

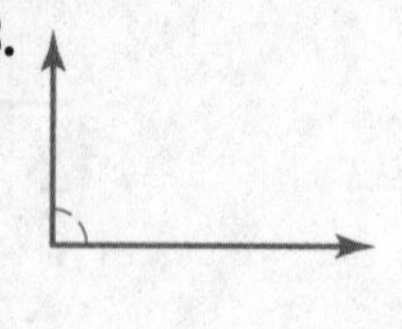

29.

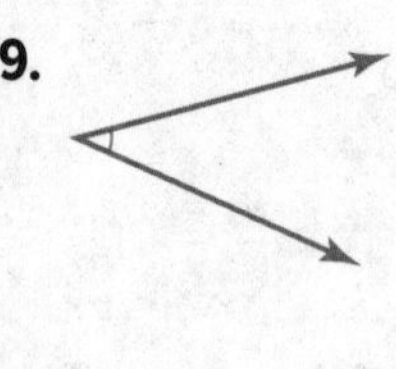

30. Name the line segment at the right in two ways. 5.G.4

31. What is the name for a quadrilateral with all right angles and opposite sides that are parallel and congruent? 5.G.3

Complementary and Supplementary Angles

Real-World Link

Watch

Bridges Engineers use angles to construct bridges. The Golden Gate Bridge is created by combining angles as shown.

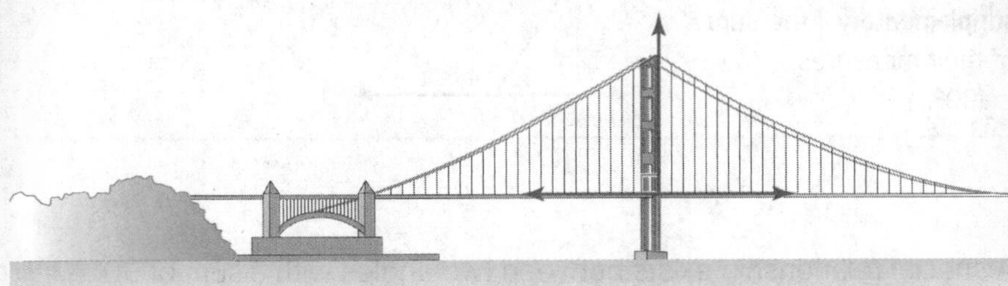

1. What types of angles make up the two angles marked in the drawing of the bridge? ___________

2. What is the sum of the two angles marked in the drawing of the bridge? ________

3. In the space below, draw a figure that contains two angles that have a sum of 90°.

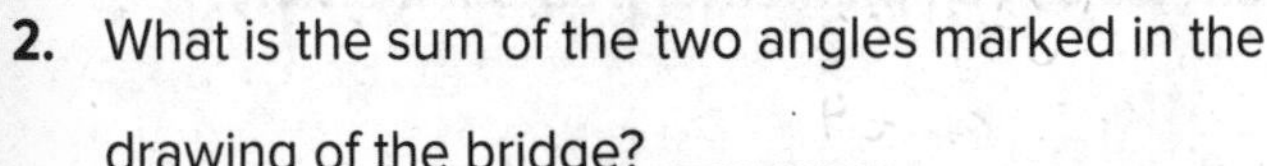

Essential Question

HOW does geometry help us describe real-world objects?

Vocab

Vocabulary

complementary angles
supplementary angles

Math Symbols

$m\angle 1$

Common Core State Standards

Content Standards
7.G.5

MP Mathematical Practices
1, 3, 4, 7

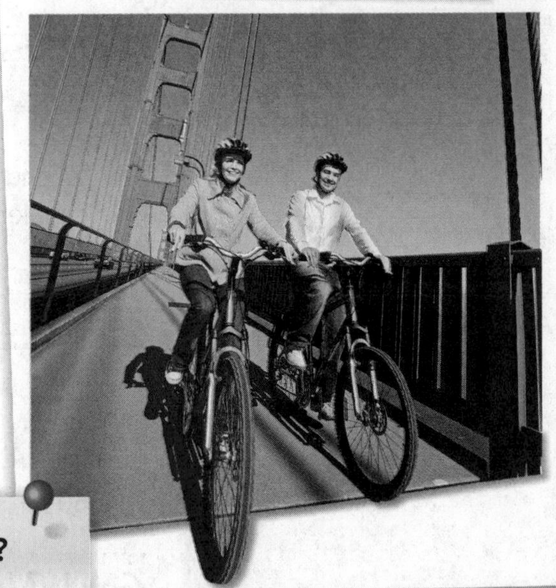

Which MP Mathematical Practices did you use? Shade the circle(s) that applies.

① Persevere with Problems
② Reason Abstractly
③ Construct an Argument
④ Model with Mathematics
⑤ Use Math Tools
⑥ Attend to Precision
⑦ Make Use of Structure
⑧ Use Repeated Reasoning

Key Concept

Pairs of Angles

Words	Models	Symbols
Two angles are **complementary** if the sum of their measures is 90°.	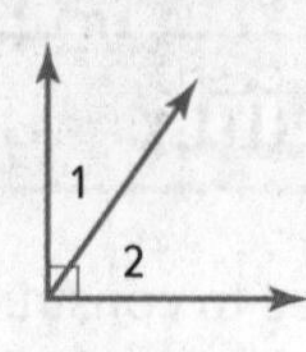	$m\angle 1 + m\angle 2 = 90°$
Two angles are **supplementary** if the sum of their measures is 180°.	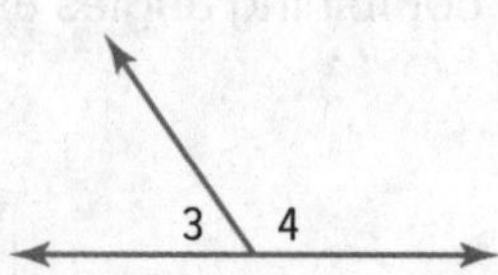	$m\angle 3 + m\angle 4 = 180°$

Work Zone

A special relationship exists between two angles with a sum of 90°. A special relationship also exists between two angles with a sum of 180°. The symbol $m\angle 1$ means *the measure of angle 1.*

Examples

Identify each pair of angles as *complementary, supplementary,* or *neither*.

1.

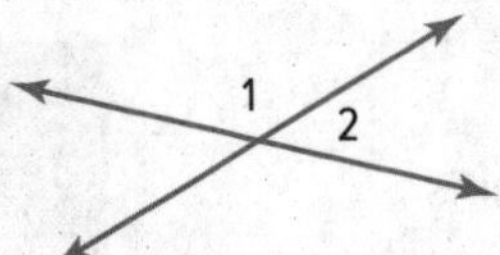

$\angle 1$ and $\angle 2$ form a straight angle. So, the angles are supplementary.

Adjacent
As shown in Example 2, angles do not need to be adjacent to be complementary or supplementary angles.

2.

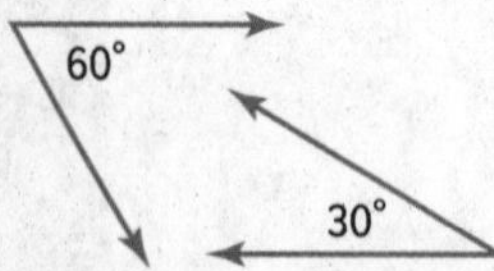

$60° + 30° = 90°$ The angles are complementary.

Got it? Do these problems to find out.

a.

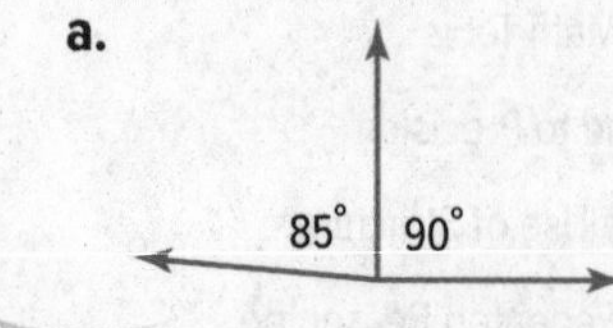

b.

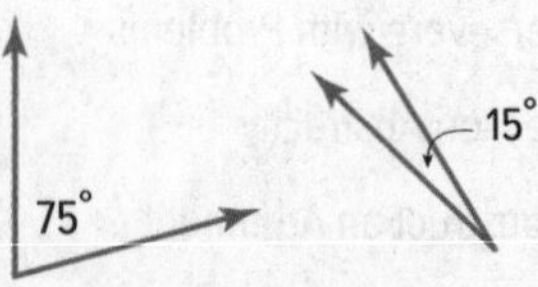

Show your work.

a. ____________

b. ____________

Find a Missing Measure

You can use angle relationships to find missing measures.

Examples

3. Find the value of x.

Since the two angles form a right angle, they are complementary.

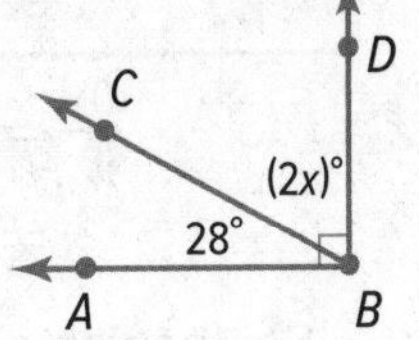

Words	The sum of the measures of $\angle ABC$ and $\angle CBD$	is	90°.
Variable	Let $2x$ represent the measure of $\angle CBD$.		
Equation	$28 + 2x$	$=$	90

$$28 + 2x = 90 \quad \text{Write the equation.}$$
$$-28 \quad = -28 \quad \text{Subtract 28 from each side.}$$
$$\frac{2x}{2} = \frac{62}{2} \quad \text{Divide each side by 2.}$$
$$x = 31$$

So, the value of x is 31.

4. The angles shown are supplementary. Find the value of x.

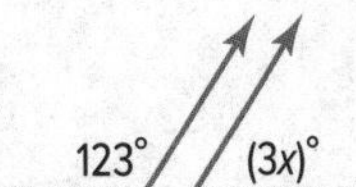

$$123 + 3x = 180 \quad \text{Write the equation.}$$
$$-123 \quad = -123 \quad \text{Subtract 123 from each side.}$$
$$\frac{3x}{3} = \frac{57}{3} \quad \text{Divide each side by 3.}$$
$$x = 19$$

So, the value of x is 19.

Circle true or false.
The sum of two angles that are supplementary is 180°.

True False

Got it? **Do this problem to find out.**

c. Find the value of x.

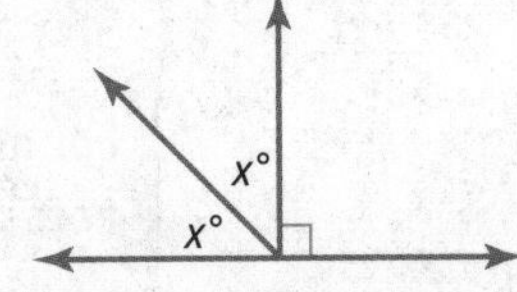

c. ____________

Example

5. The picture shows a support brace for a gate. Find the value of *x*.

The angle labeled 80° and the angle labeled $10x$ are supplementary angles.

$80 + 10x = 180$		Write the equation.
$-80 \quad = -80$		Subtract 80 from each side.
$\frac{10x}{10} = \frac{100}{10}$		Divide each side by 10.
$x = 10$		

So, the value of x is 10.

Got it? Do this problem to find out.

d. ______

d. A pair of scissors forms the angles shown. What is the value of x?

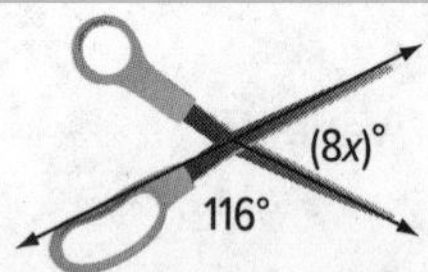

Guided Practice

Identify each pair of angles as *complementary, supplementary,* or *neither*.
(Examples 1 and 2)

1.

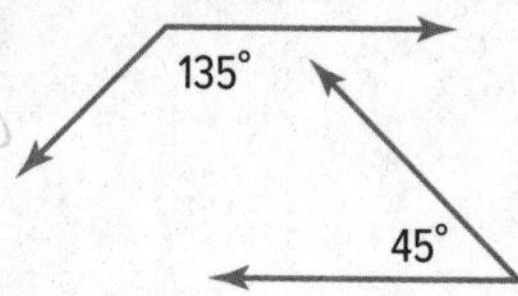

______

2.

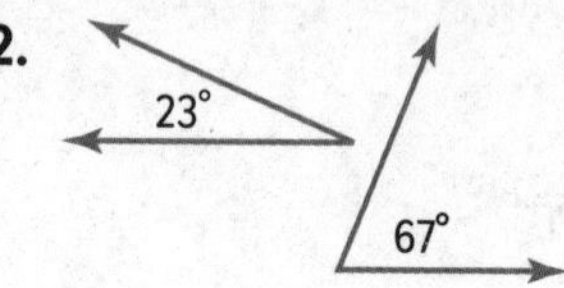

______

3. Find the value of x. (Examples 3–5)

______

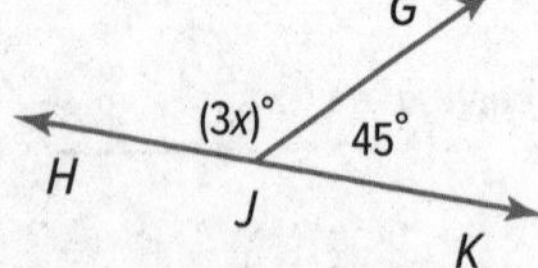

4. **Building on the Essential Question** How are vertical, adjacent, complementary, and supplementary angles related? ______

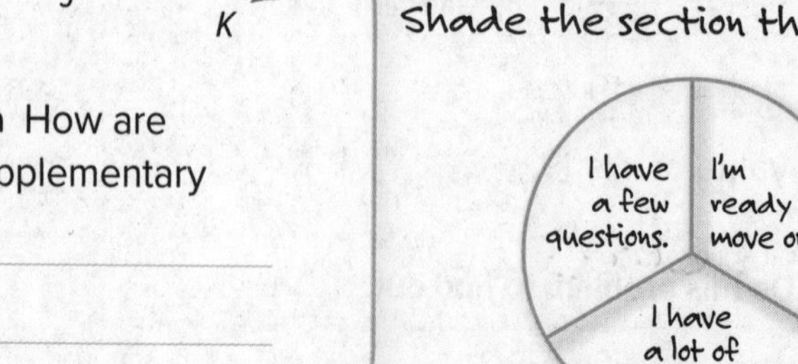

Rate Yourself!

Are you ready to move on? Shade the section that applies.

I have a few questions.
I'm ready to move on.
I have a lot of questions.

For more help, go online to access a Personal Tutor.

Independent Practice

Go online for Step-by-Step Solutions 

Identify each pair of angles as *complementary, supplementary,* or *neither*.
(Examples 1 and 2)

1. 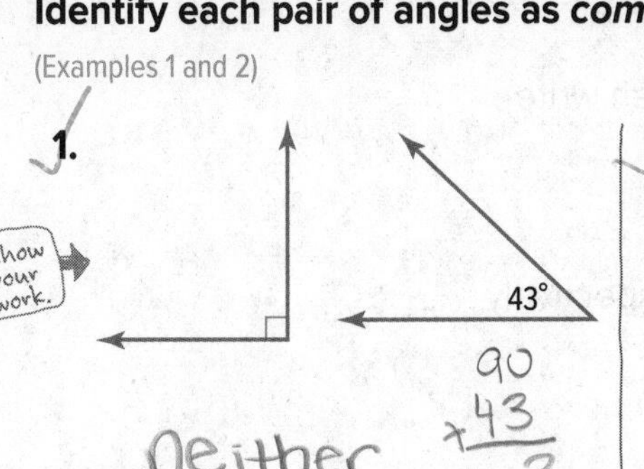

2. 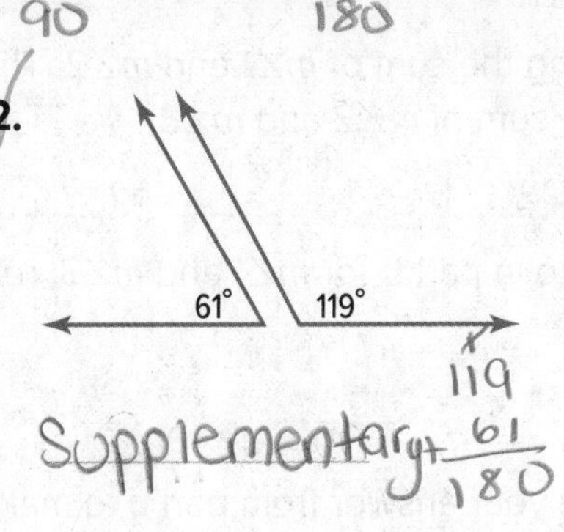

3 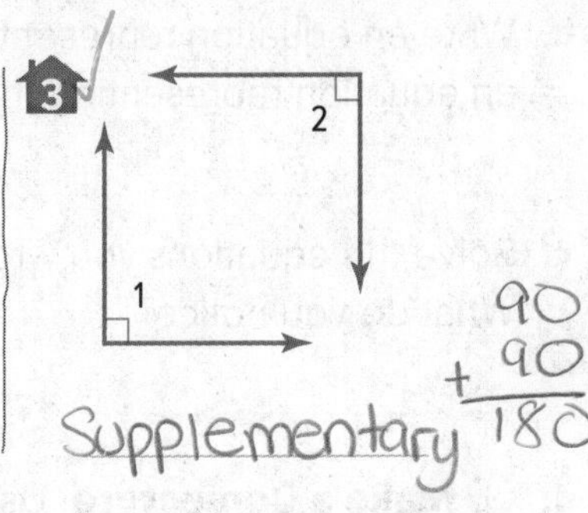

Find the measure of *x* in each figure. (Examples 3 and 4)

4. 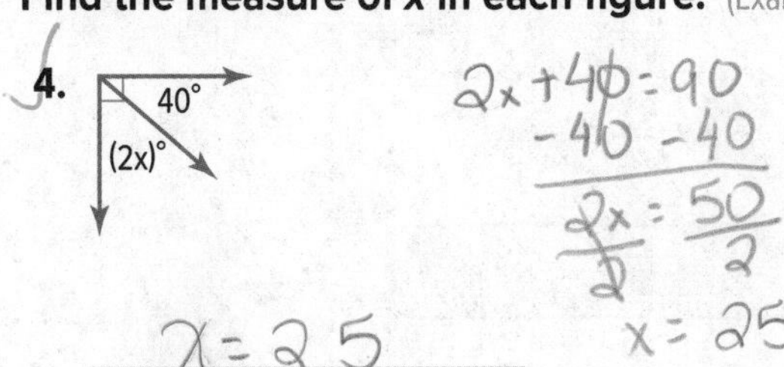

5. 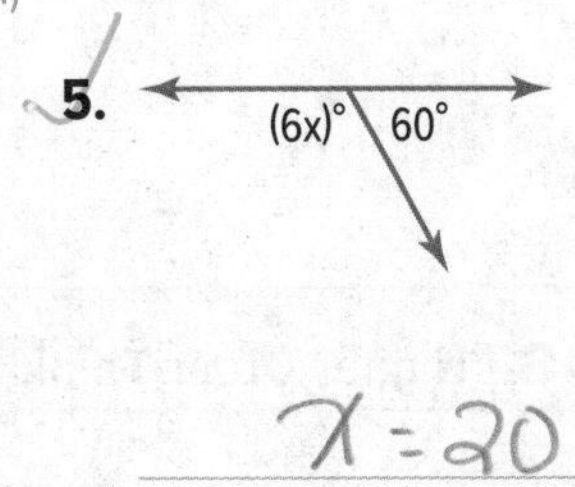

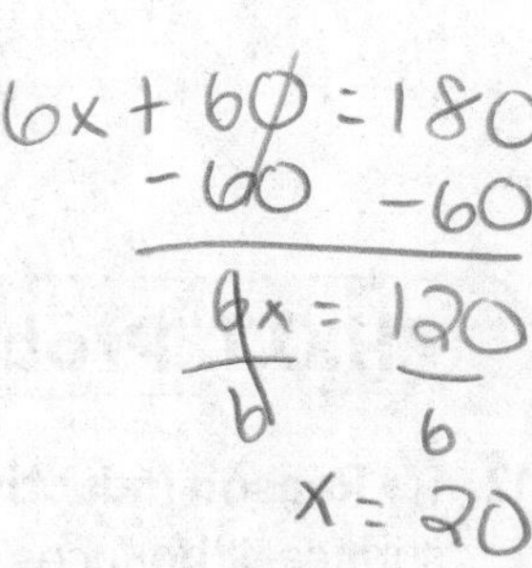

6. $\angle A$ and $\angle B$ are complementary angles. The measure of $\angle B$ is $(4x)°$, and the measure of $\angle A$ is 50°. What is the value of x? (Example 5)

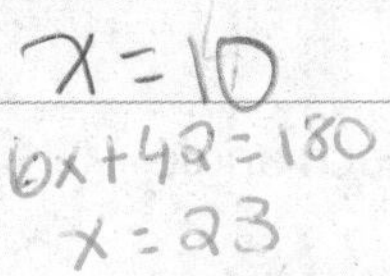

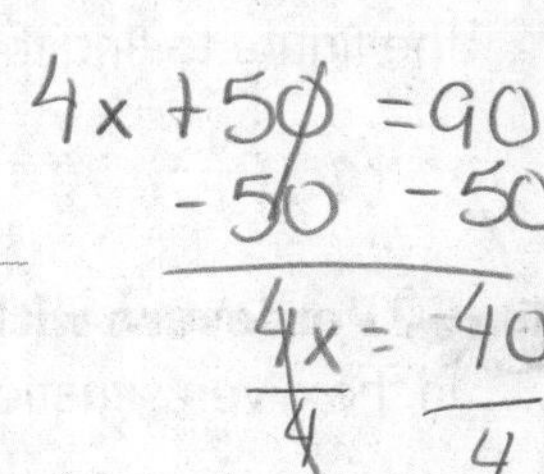

7 A skateboard ramp forms a 42° angle as shown.

Find the value of x. (Example 5)

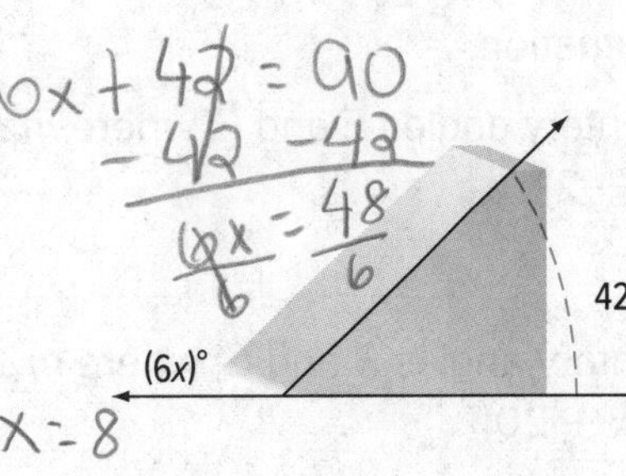

Use the figure at the right to name the following.

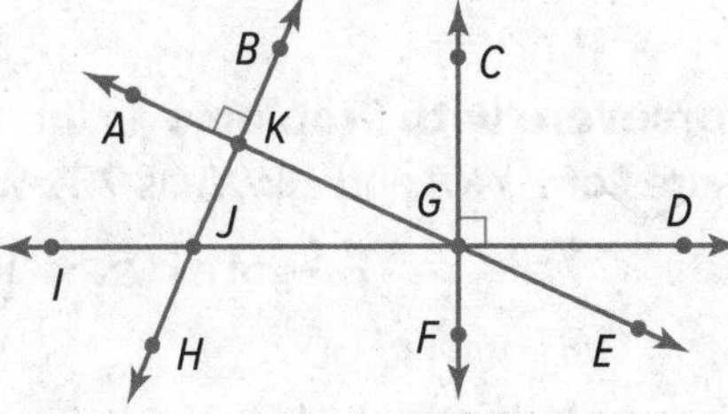

8. a pair of supplementary angles

9. a pair of complementary angles

10. a pair of vertical angles

11. Use the figure at the right.

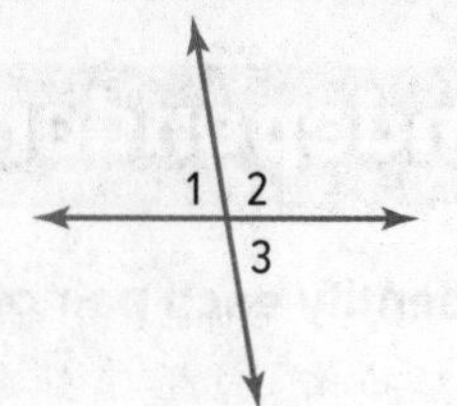

a. Are ∠1 and ∠2 vertical angles, adjacent angles, or neither? ∠2 and ∠3? ∠1 and ∠3?

b. Write an equation representing the sum of $m\angle 1$ and $m\angle 2$. Then write an equation representing the sum of $m\angle 2$ and $m\angle 3$.

c. Solve the equations you wrote in part **b** for $m\angle 1$ and $m\angle 3$, respectively. What do you notice?

d. MP **Make a Conjecture** Use your answer from part **c** to make a conjecture as to the relationship between vertical angles.

H.O.T. Problems Higher Order Thinking

12. MP **Reason Inductively** When a basketball hits a hard, level surface, it bounces off at the same angle at which it hits. Use the figure to find the angle at which the ball hit the floor.

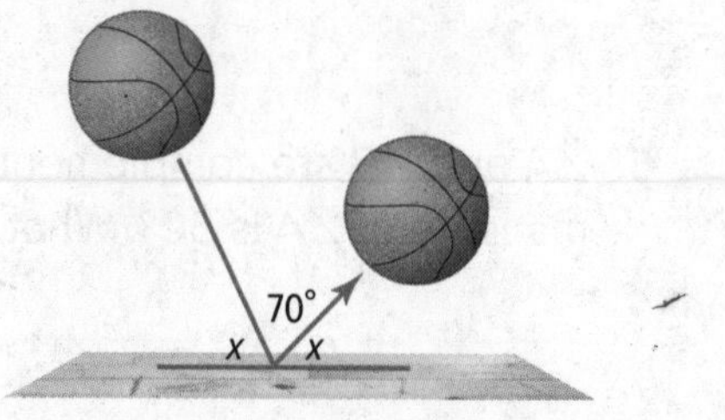

13. MP **Persevere with Problems** Find the measure of each angle in the given situation.

a. complementary angles E and F, where $m\angle E = (x - 10)°$ and $m\angle F = (x + 2)°$

∠E = 39 ∠F = 51

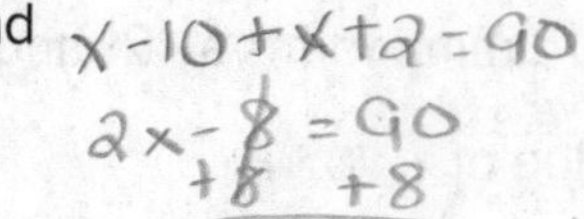

b. supplementary angles B and C, where $m\angle B = (2x - 40)°$ and $m\angle C = (2x + 20)°$

∠B = 60 ∠C = 120

14. MP **Persevere with Problems** In the figure shown, the sum of the measures of ∠YXZ and ∠WXV is 75°. What is the measure of ∠ZXW?

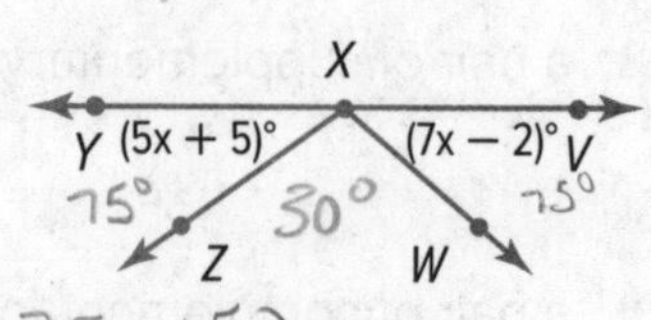

30° 105°

15. MP **Reason Inductively** Is the statement below *always*, *sometimes*, or *never* true? Explain.

If two angles are right angles, they must be supplementary.

Always, because one right angle equals to 90° and a Supplementary angle is 2 angles sum that equals 180°

Name _______________ My Homework

Extra Practice

Identify each pair of angles as *complementary, supplementary,* or *neither*.

16. 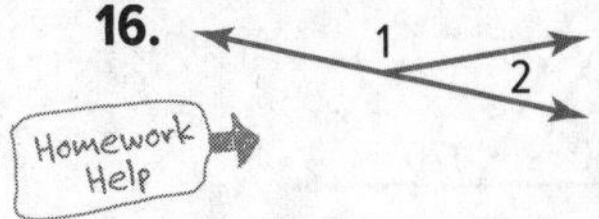

∠1 and ∠2 form a straight angle. So, the angles are supplementary.

17. 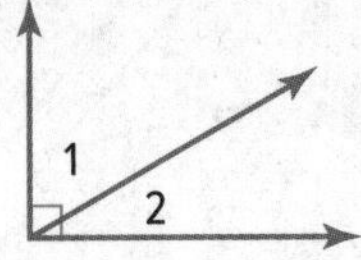

18. 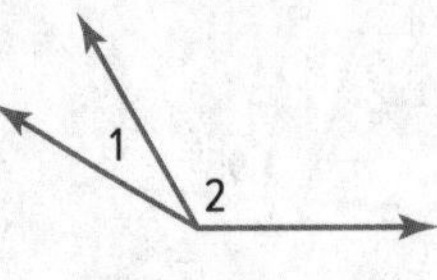

19. ∠*J* and ∠*K* are supplementary. The measure of ∠*J* is $(9x)°$ and the measure of ∠*K* is 45°. What is the value of *x*?

x = 15

$9x + 45 = 180$
$-45 \quad -45$
$\frac{9x}{9} = \frac{135}{9}$ x = 15

20. ∠*C* and ∠*D* are complementary. The measure of ∠*C* is $(4x)°$ and the measure of ∠*D* is 26°. What is the value of *x*?

x = 16

$4x + 26 = 90$
$-26 \quad -26$
$\frac{4x}{4} = \frac{64}{4}$ x = 16

MP Identify Structure Determine whether each statement is *always, sometimes,* or *never* true. Explain your reasoning.

21. Two obtuse angles are supplementary.

Sometimes, because an obtuse angle is more then 90° but can also be less then 180°. Never, both are greater then 90

22. Two vertical angles are complementary.

Sometimes, because the 2 angles can either be an acute or obtuse. we do not know the degree of the vertical angles.

23. **MP Multiple Representations** Line *a* passes through (1, 4) and (−4, −1). Line *b* passes through (−3, 4) and (2, −1).

a. **Graphs** Graph each line on the same coordinate plane.

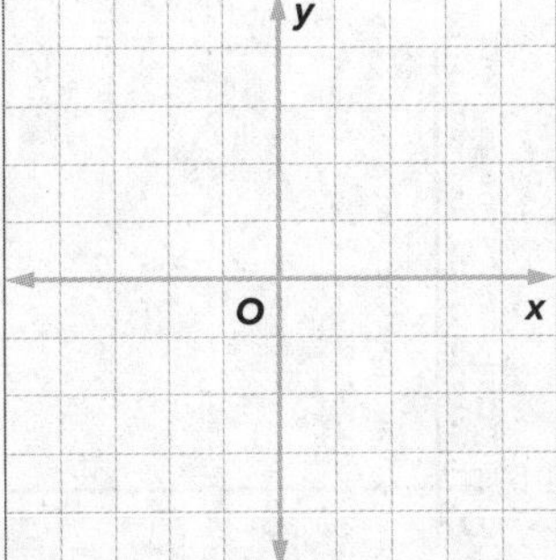

b. **Words** Describe the lines.

c. **Numbers** What is the slope of each line?

Power Up! Common Core Test Practice

24. Which angle pairs below are supplementary? Select all that apply.

- ☐ 75° and 95°
- ☐ 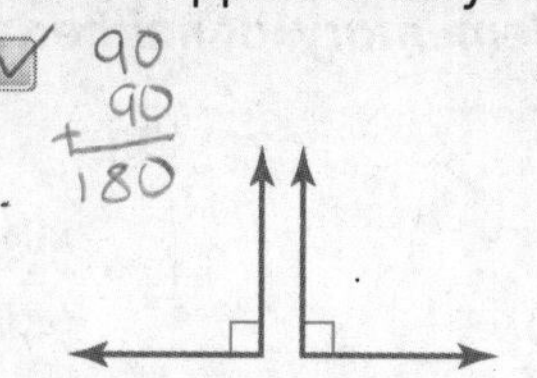
- ☐ 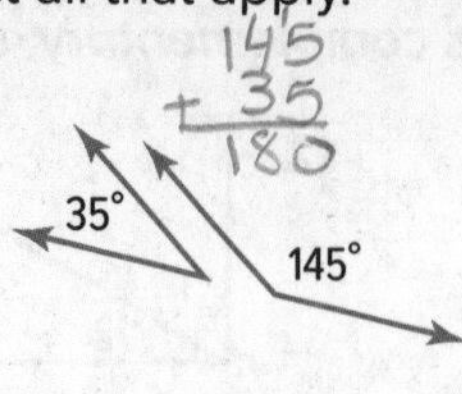

- ☐ 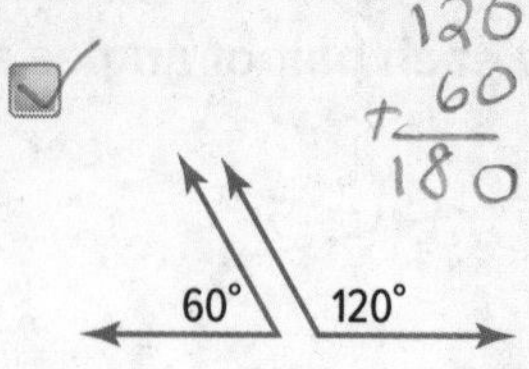

25. The angle at which the light ray hits the water is equal to the angle at which the light ray is reflected from the water.

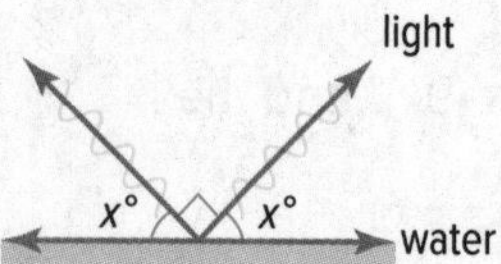

Select values to complete the equation below to find the value of x.

2 × ☐ + ☐ = ☐

x°
45°
60°
90°
180°

What is the measure of the angle at which the light ray hits the water?

☐

What is the measure of the angle at which the light ray is reflected from the water?

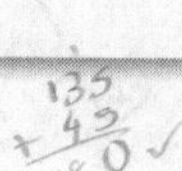

Common Core Spiral Review

Graph each figure with the given vertices on the coordinate plane. Then classify each figure. 6.G.3

26. (1, 3), (1, 6), (5, 5), and (5, 3)

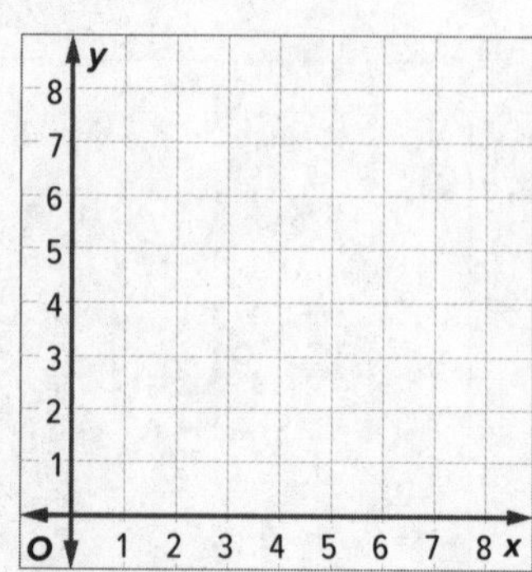

27. (1, 2), (5, 2), (5, 6), and (1, 6)

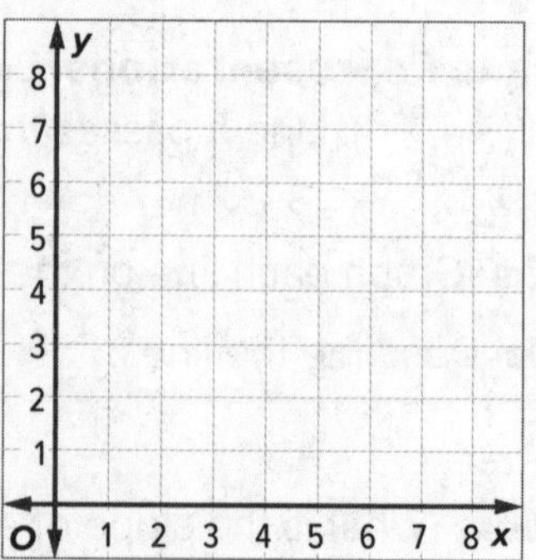

Inquiry Lab
Create Triangles

WHAT do you notice about the measures of the sides or the measures of the angles that form triangles?

Content Standards
7.G.2

Mathematical Practices
1, 3, 5, 8

Dennis has a sailboat. The sail on his boat is in the shape of a triangle with side lengths of 6 feet, 8 feet, and 10 feet. These dimensions work to form a triangle, but not just any three lengths form a triangle. Complete the Activity below to determine which side lengths form triangles.

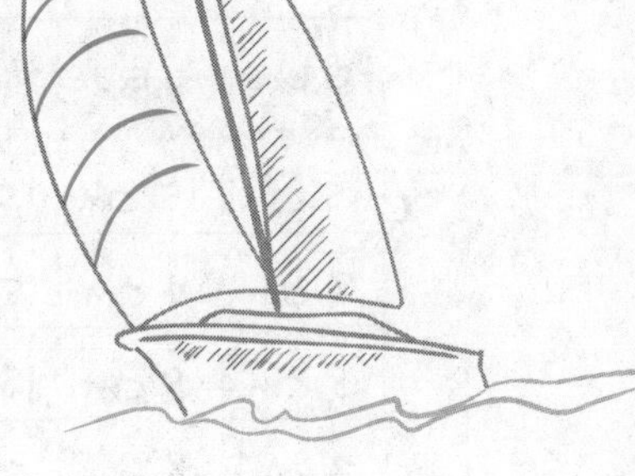

Hands-On Activity 1

Step 1 Measure and cut several plastic straws into lengths that equal 3, 4, 4, 5, 8, 8, 8, 13, 15, 15, 15, and 15 centimeters.

Step 2 Arrange three of the pieces that each measure 15 centimeters to see if you can form a triangle.

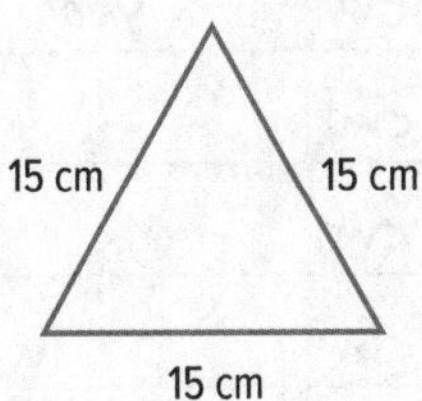

So, you can form a triangle with side lengths of 15 centimeters, 15 centimeters, and 15 centimeters.

Step 3 Continue using pieces of straw to try to form triangles using the different combinations of side lengths given. Determine whether or not the lengths form a triangle. Complete the table.

Side 1	Side 2	Side 3	Do the sides form a triangle?
15 cm	15 cm	15 cm	yes
3 cm	4 cm	5 cm	
8 cm	8 cm	13 cm	
3 cm	4 cm	8 cm	
4 cm	4 cm	5 cm	
8 cm	3 cm	15 cm	
4 cm	8 cm	15 cm	

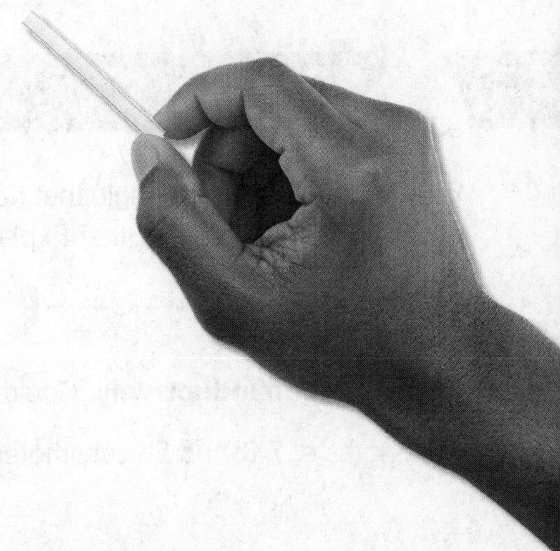

Investigate

Work with a partner. Try to create triangles using the given side lengths. Circle *yes* if you can make a triangle or *no* if you cannot.

1. 5 cm, 8 cm, 15 cm

   Yes or No

2. 13 cm, 8 cm, 15 cm

   Yes or No

3. 13 cm, 4 cm, 4 cm

   Yes or No

Analyze and Reflect

Work with a partner.

4. The table below contains the dimensions you used in Step 3 of the Activity. Transfer your results from the Investigation into the fourth column and then complete the fifth column.

Side 1	Side 2	Side 3	Do the sides form a triangle?	Is Side 1 + Side 2 greater than or less than Side 3?
15 cm	15 cm	15 cm	yes	greater than
3 cm	4 cm	5 cm		
8 cm	8 cm	13 cm		
3 cm	4 cm	8 cm		
4 cm	4 cm	5 cm		
8 cm	3 cm	15 cm		
4 cm	8 cm	15 cm		

5. What do you notice about the figures with a Side 1 and Side 2 sum that is less than the length of Side 3? ______

Create

6. Can you create a triangle that has the same shape as the triangle in the Activity, but different side lengths? Explain.

______

7. MP **Reason Inductively** Could you form a triangle using the side lengths of 7, 8, and 25 centimeters? Explain. ______

Hands-On Activity 2

Use angles of different sizes to determine which ones form a triangle.

Step 1 Draw two sets of angles measuring 30°, 45°, 60°, and 90° on different pieces of patty paper. Extend the rays of each angle to the edges of the patty paper.

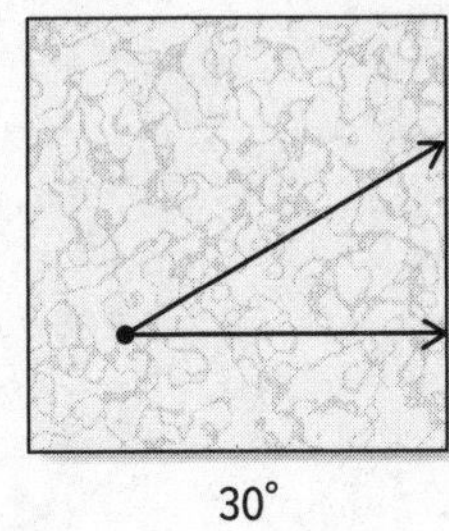

30°

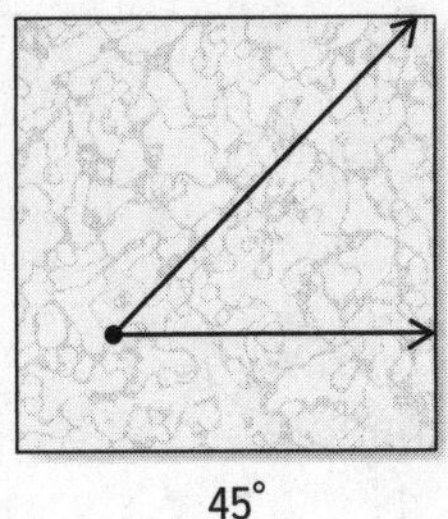

45°

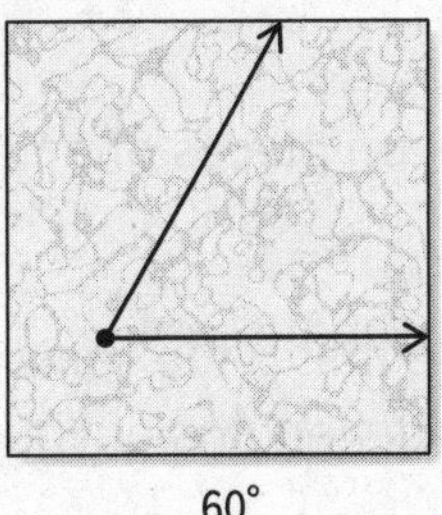

60°

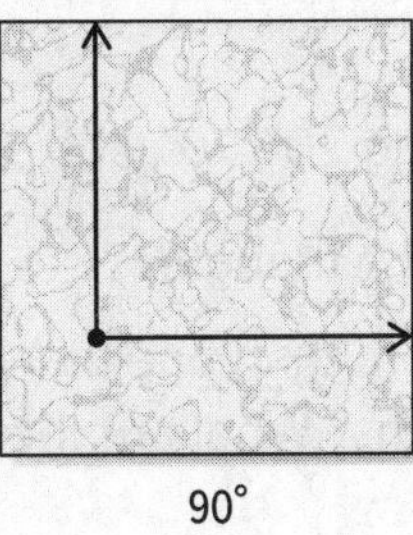

90°

Step 2 Try to form a triangle with one 90° angle and two 45° angles.

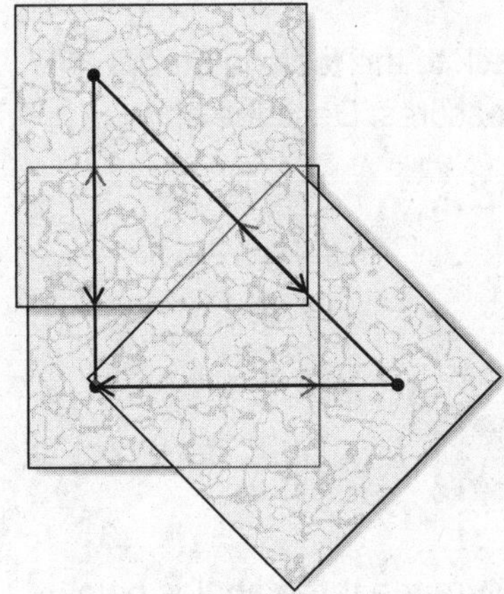

So, a 90° angle and two 45° angles form a triangle.

Step 3 Try to form triangles using the angle measures that are given in the table. Fill in *yes* or *no* in the fourth column of the table.

Angle 1	Angle 2	Angle 3	Do the angles form a triangle?
90°	45°	45°	yes
30°	60°	90°	
30°	45°	60°	
30°	30°	60°	

Investigate

Work with a partner.

8. Draw another 60° angle on a piece of patty paper. Describe the angles and side lengths of the figure you form using three 60° angles.

Show your work.

9. Draw angles measuring 20°, 70°, and 90° on pieces of patty paper.

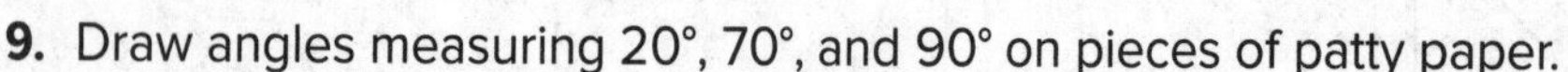

a. Do the angles form a triangle?

b. Can you create more than one triangle that is the same shape with different side lengths? What are the side lengths of your triangle?

Analyze and Reflect

Collaborate

10. MP **Identify Repeated Reasoning** Refer back to the table in Step 3 of Activity 2. Compare the sum of the angle measures. Describe any patterns that are found.

Create

On Your Own

11. MP **Use Math Tools** Use a protractor to measure the three angles below. Would you be able to form a triangle from these angles? Explain.

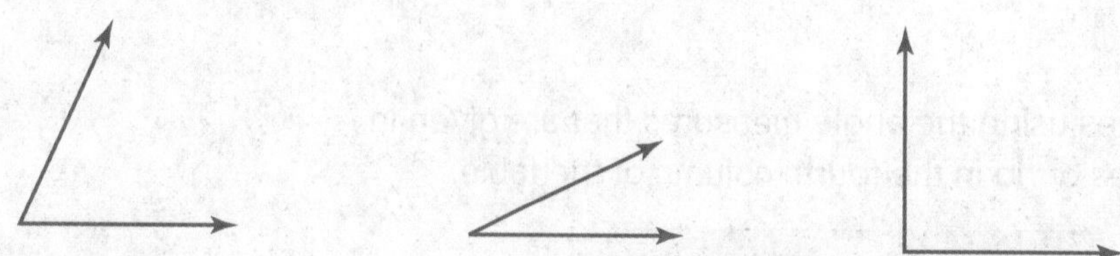

12. Inquiry WHAT do you notice about the measures of the sides or the measures of the angles that form triangles?

Lesson 3

Triangles

Real-World Link

Ramps Julia practices jumping on a ski ramp. The front of the ramp is a triangle like the one shown below.

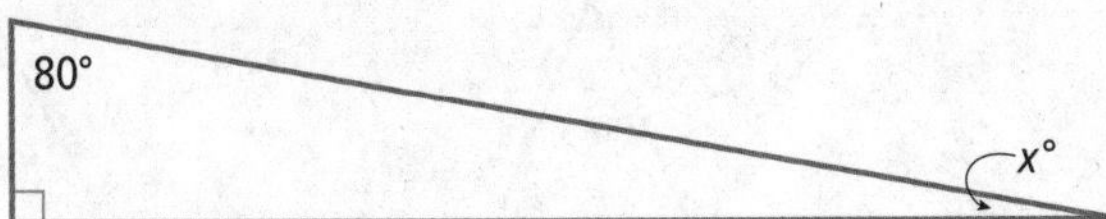

1. Draw an X through the type of angle that is not shown in the triangle.

   right   acute   obtuse

2. Measure the unknown angle. Describe the relationship between the 80° angle and the unknown angle.

3. Draw a triangle with one obtuse angle.

4. Is it possible to draw a triangle with two obtuse angles? Explain.

Essential Question

HOW does geometry help us describe real-world objects?

Vocabulary

acute triangle
right triangle
obtuse triangle
scalene triangle
isosceles triangle
equilateral triangle
triangle
congruent segments

Math Symbols

$\triangle$

Common Core State Standards

Content Standards
7.G.2

MP Mathematical Practices
1, 2, 3, 4

Which MP Mathematical Practices did you use? Shade the circle(s) that applies.

① Persevere with Problems
② Reason Abstractly
③ Construct an Argument
④ Model with Mathematics
⑤ Use Math Tools
⑥ Attend to Precision
⑦ Make Use of Structure
⑧ Use Repeated Reasoning

Key Concept

Classify Triangles

all acute angles

acute triangle

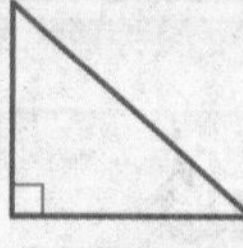
1 right angle

right triangle

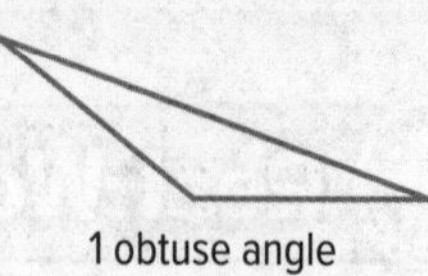
1 obtuse angle

obtuse triangle

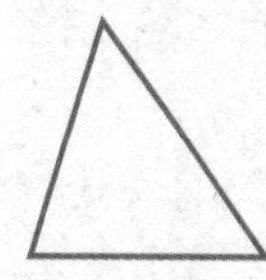
no congruent sides

scalene triangle

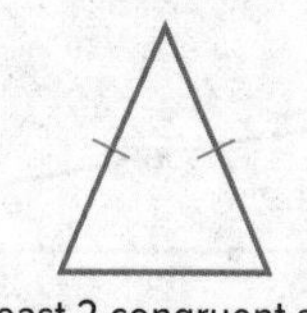
at least 2 congruent sides

isosceles triangle

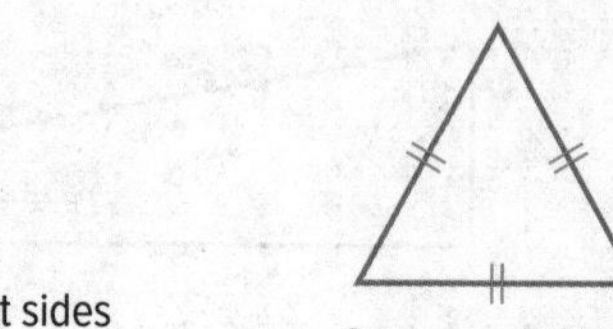
3 congruent sides

equilateral triangle

Work Zone

Congruent Segments

The tick marks on the sides of the triangle indicate that those sides are congruent.

A **triangle** is a figure with three sides and three angles. The symbol for triangle is $\triangle$.

Every triangle has at least two acute angles. One way you can classify a triangle is by using the third angle. Another way to classify triangles is by their sides. Sides with the same length are **congruent segments**.

Example

1. Draw a triangle with one obtuse angle and no congruent sides. Then classify the triangle.

Draw an obtuse angle. The two segments of the angle should have different lengths.

Connect the two segments to form a triangle.

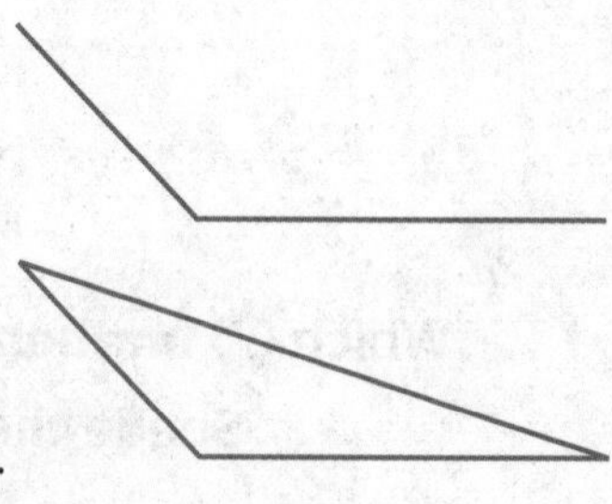

The triangle is an obtuse scalene triangle.

Got it? Do this problem to find out.

Draw a triangle that satisfies the set of conditions below. Then classify the triangle.

a. a triangle with one right angle and two congruent sides

a. ______

Example

2. Classify the triangle on the house by its angles and by its sides.

The triangle has one obtuse angle and two congruent sides. So, it is an obtuse isosceles triangle.

STOP and Reflect

How would you classify a triangle with a right angle and two congruent sides?

Got it? **Do this problem to find out.**

b. Classify the triangle shown by its angles and by its sides.

b. ________

Angles of a Triangle

Key Concept

Words The sum of the measures of the angles of a triangle is 180°.

Model

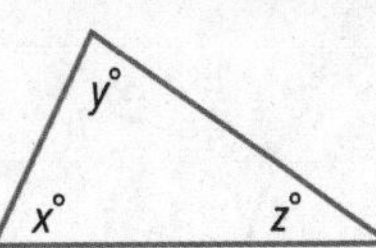

Algebra $x + y + z = 180$

You can write and solve an equation to find the missing angle measure of a triangle.

Example

3. Find $m\angle Z$.

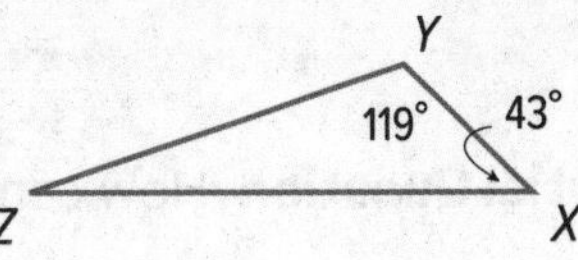

The sum of the angle measures in a triangle is 180°.

$m\angle Z + 43° + 119° = 180°$ Write the equation.

$m\angle Z + 162° = 180°$ Simplify.

$\mathbf{-162° = -162°}$ Subtract 162° from each side.

$m\angle Z = 18°$

So, $m\angle Z$ is 18°.

Got it? **Do this problem to find out.**

c. In $\triangle ABC$, if $m\angle A = 25°$ and $m\angle B = 108°$, what is $m\angle C$?

c. ________

Example

4. The Alabama state flag is shown. What is the missing measure in the triangle?

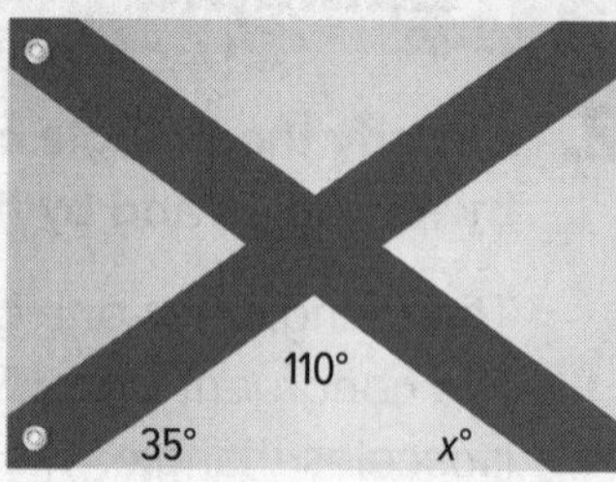

To find the missing measure, write and solve an equation.

$x + 110 + 35 = 180$	The sum of the measures is 180.
$x + 145 = 180$	Simplify.
$-145 = -145$	Subtract 145 from each side.
$x = 35$	

The missing measure is 35°.

Guided Practice

1. Draw a triangle with three acute angles and two congruent sides. Classify the triangle. (Examples 1 and 2) ______

2. Find $m\angle T$ in $\triangle RST$ if $m\angle R = 37°$ and $m\angle S = 55°$. (Example 3) ______

3. A triangle is used in the game of pool to rack the pool balls. Find the missing measure of the triangle. (Example 4)

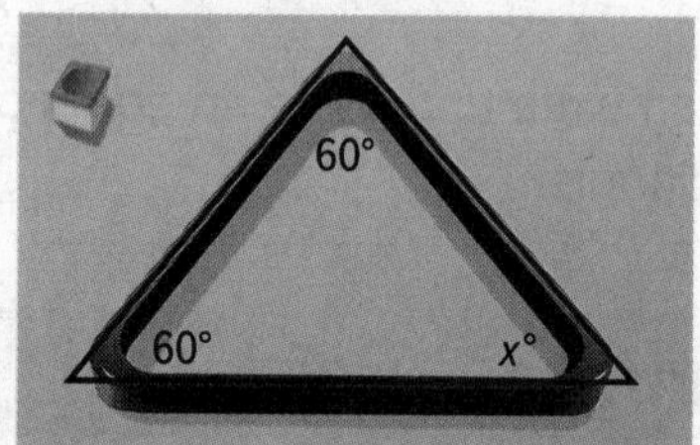

4. **Building on the Essential Question** How can triangles be classified?

Rate Yourself!

Are you ready to move on? Shade the section that applies.

YES ? NO

For more help, go online to access a Personal Tutor.

FOLDABLES Time to update your Foldable!

Independent Practice

Go online for Step-by-Step Solutions

Draw a triangle that satisfies each set of conditions. Then classify the triangle. (Example 1)

1. a triangle with three acute angles and three congruent sides ______

2. a triangle with one right angle and no congruent sides ______

Classify the marked triangle by its angles and by its sides. (Example 2)

3. 

Acute Equilateral

4. 

Acute Equilateral

5. 

~~Acute isosceles~~ Obtuse isosceles

Find the value of *x*. (Examples 3 and 4)

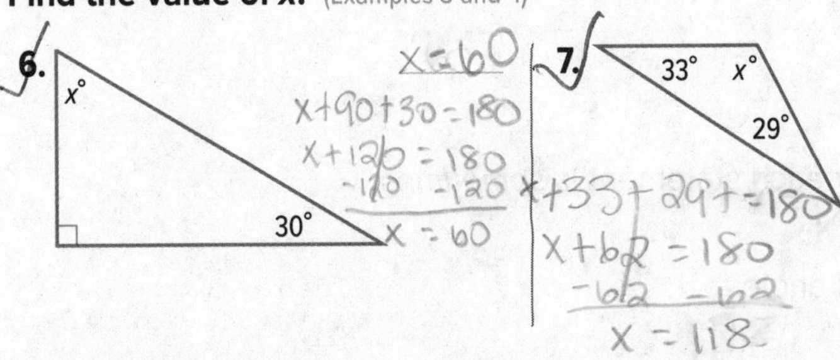

8. 21° x° 132°

x=27

x+132+21=180

x+153=180

-153 -153

x=27

9. MP **Model with Mathematics** Refer to the graphic novel below. Classify the triangle formed by the cabin, ropes course, and mess hall by its angles and sides.

Acute ~~Scalene~~ isosceles

10. Triangle *ABC* is formed by two parallel lines and two other intersecting lines. Find the measure of each angle *A*, *B*, and *C* of the triangle.

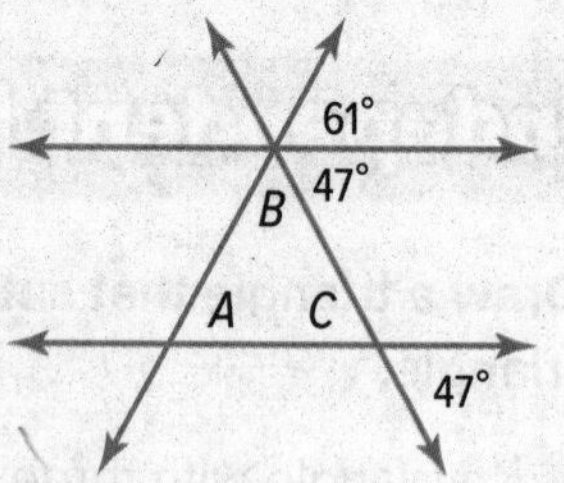

H.O.T. Problems Higher Order Thinking

11. **MP Persevere with Problems** Apply what you know about triangles to write and solve equations to find the missing angle measures in the figure.

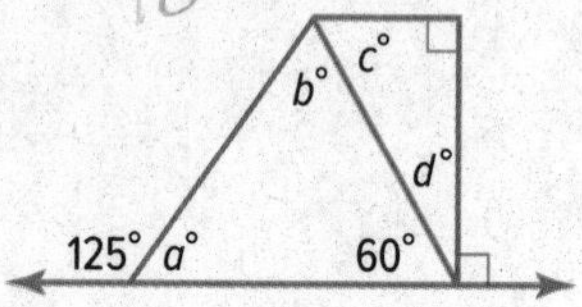

12. **MP Model with Mathematics** Draw an acute scalene triangle. Describe the angles and sides of the triangle.

Show your work.

13. **MP Justify Conclusions** Determine whether each statement is *sometimes*, *always*, or *never* true. Justify your answer.

a. It is possible for a triangle to have two right angles.

b. It is possible for a triangle to have two obtuse angles.

14. **MP Reason Inductively** Miguel says that an equilateral triangle is sometimes an obtuse triangle. Jane says that an equilateral triangle is always an acute triangle. Is either of them correct? Explain your reasoning.

Name ________________ My Homework ________________

Extra Practice

Classify the marked triangle in each object by its angles and by its sides.

15. 

The triangle has all acute angles and two congruent sides. It is an acute isosceles triangle.

16. 

____

17. 

____

Draw a triangle that satisfies each set of conditions. Then classify the triangle.

18. a triangle with three acute angles and no congruent sides ________________

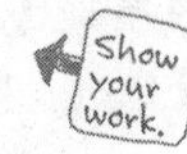

19. a triangle with one obtuse angle and two congruent sides ________________

Find the value of *x*.

20. 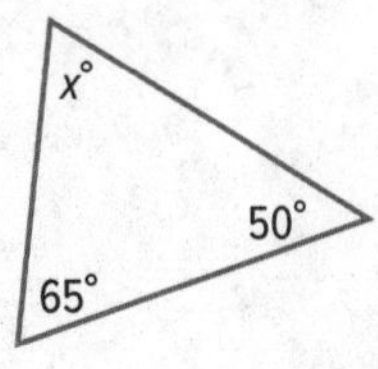

____

21. 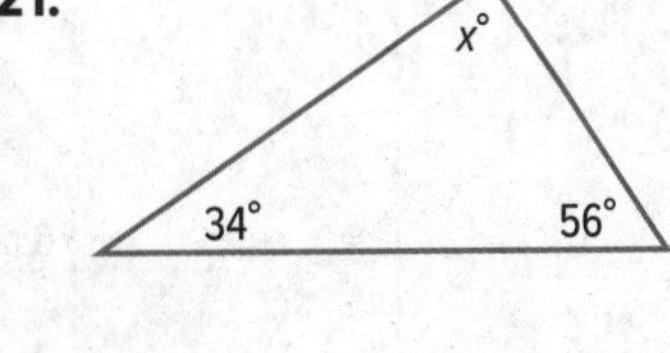

____

22. 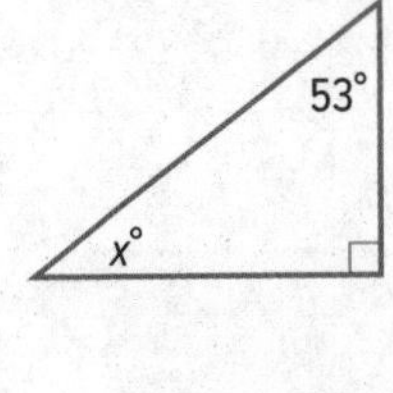

____

23. Find $m\angle Q$ in $\triangle QRS$ if $m\angle R = 25°$ and $m\angle S = 102°$. ________________

MP Reason Abstractly **Find the value of *x* in each triangle.**

24. 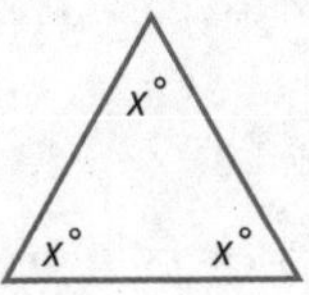

____

25. 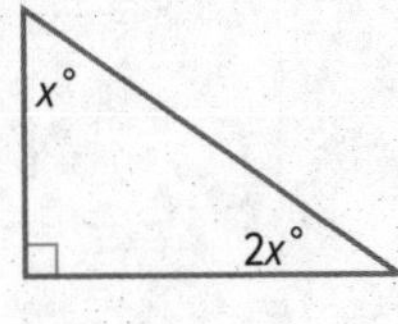

____

26. 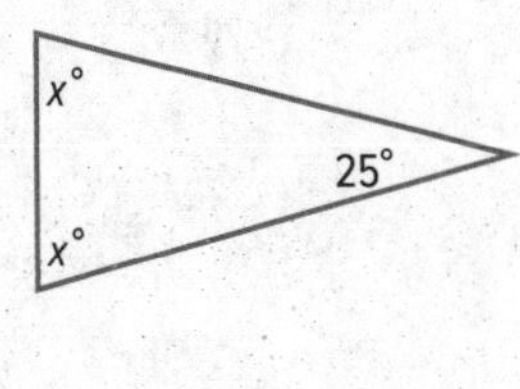

____

Power Up! Common Core Test Practice

27. Refer to the figure shown. Determine if each statement is true or false.

R, S, T; 30°, 30°

a. To find $m\angle R$, subtract 30° from 90°. ☐ True ☐ False

b. The measure of $\angle R$ is 120°. ☐ True ☐ False

c. Triangle RST is an acute triangle. ☐ True ☐ False

28. In a right triangle, the measure of one of the angles is 43°. Sketch a diagram to represent this situation.

What is the measure of the other angle? ☐

Common Core Spiral Review

Find the area of each figure. 6.G.1

29. 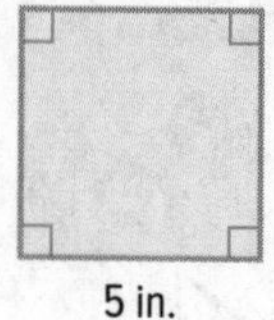

______

30. 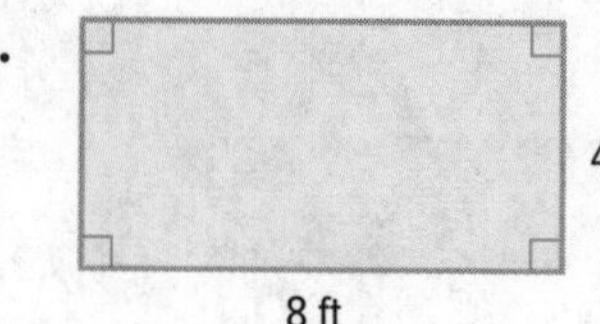

______

31. 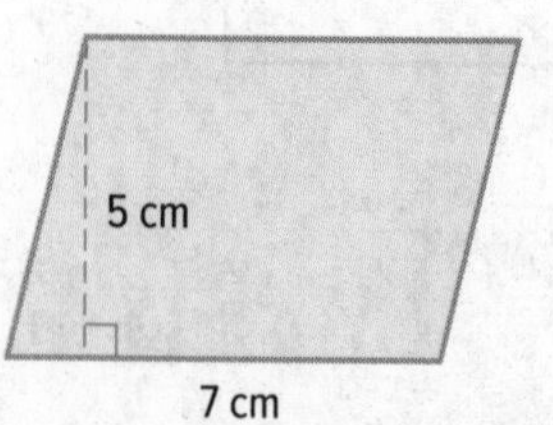

______

32. 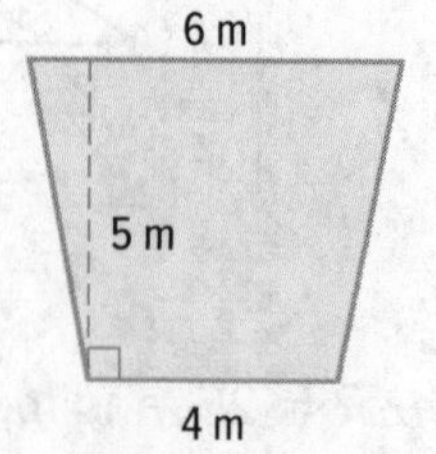

______

33. 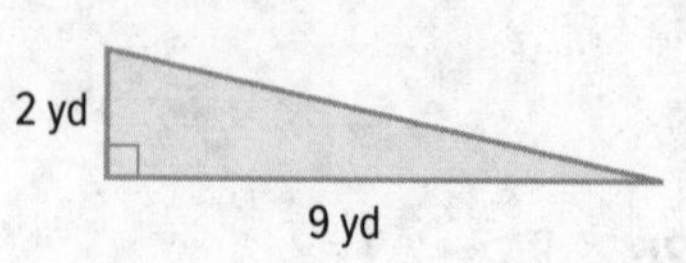

______

34. 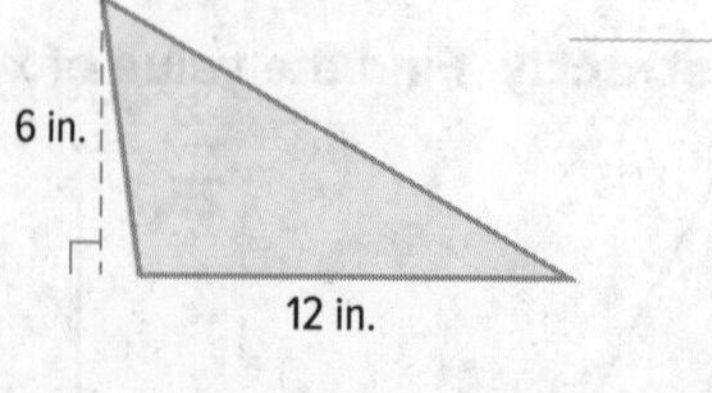

______

Inquiry Lab

Draw Triangles

HOW can you use technology to draw geometric shapes?

Content Standards
7.G.2

Mathematical Practices
1, 3, 5

The Spirit Club is selling triangular-shaped pennants for Homecoming. Teresa is making a poster to advertise the pennants. She wants to use a computer program to draw a model of the pennant.

Hands-On Activity 1

You can use dynamic geometry software such as The Geometer's Sketchpad® to draw triangles given three angle measures. In this investigation, you will draw a triangle with angle measures of 30°, 60°, and 90°.

Step 1 First, click on **Edit**. Go to **Preferences**. Change the angle precision from *hundredths* to *units*. Next, use the **Straightedge (segment)** tool. Click and drag three times to create a triangle like the one shown.

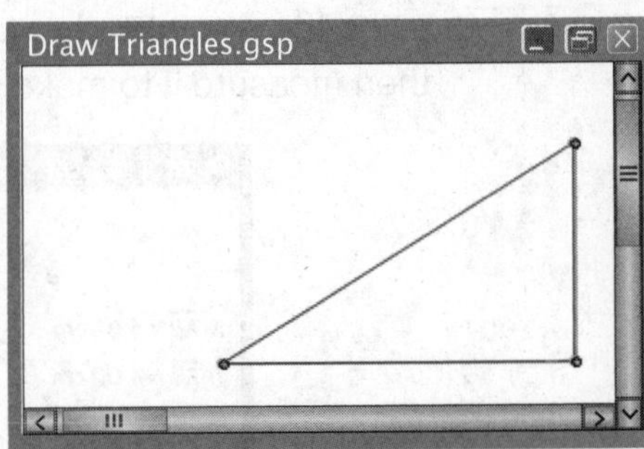

Step 2 Using the **Selection Arrow**, click on each of the vertex points *A*, *B*, and *C*. Then select **Measure** and **Angle**. Labels will automatically be assigned to the vertices. You found that the measure of $\angle ABC$ is ______.

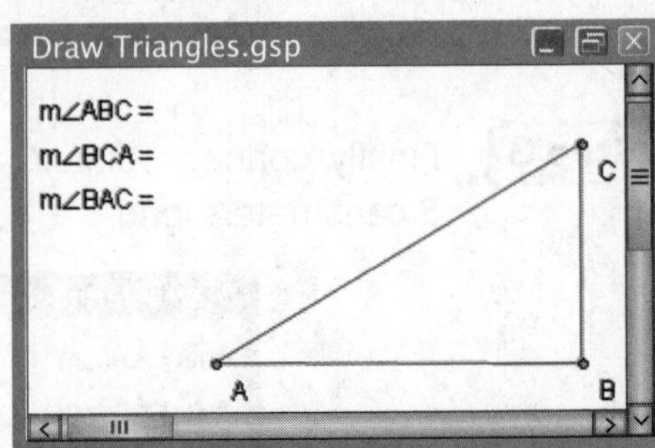

Step 3 Click on points *B*, *C*, and *A*. Click **Measure** and **Angle** again. Repeat for points *B*, *A*, and *C*. The angle measures should be displayed on your screen.

Step 4 If the angles do not measure 30°, 60°, and 90°, use the **Selection Arrow** to move the vertices. Click and drag one or more points so that the angles move.

Hands-On Activity 2

You can also use The Geometer's Sketchpad® to draw triangles given three side measures. In this Activity, you will draw a triangle with side measures of 3 centimeters, 4 centimeters, and 5 centimeters.

Step 1 First, click on **Edit**. Go to **Preferences**. Check that the distance precision is set to hundredths. Using the **Straightedge (segment)** tool, click and drag to create a line segment with endpoints *A* and *B*. Use the **Selection Arrow** to select the segment. Click on **Measure** and **Length**. Then drag one of the endpoints so that the line segment measures 5 centimeters.

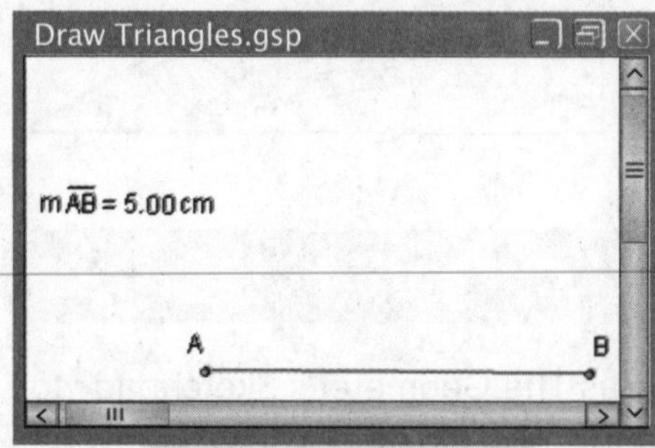

Step 2 Next, create a line segment from point *A* that is 4 centimeters long using the **Straightedge (segment)** tool. Draw the segment first and then measure it to make sure it is 4 centimeters.

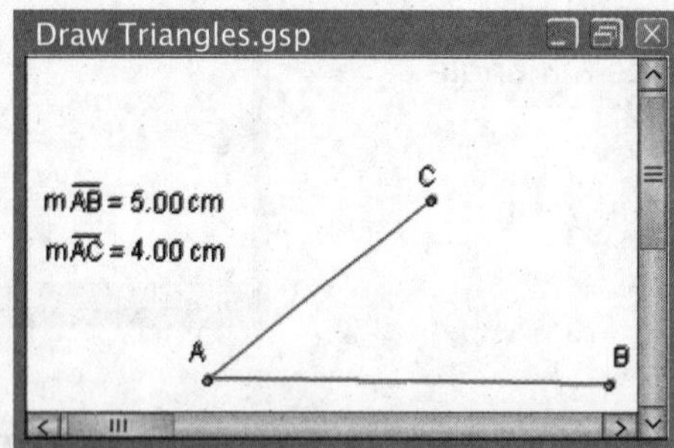

Step 3 Finally, connect points *C* and *B* with a line segment that is 3 centimeters long.

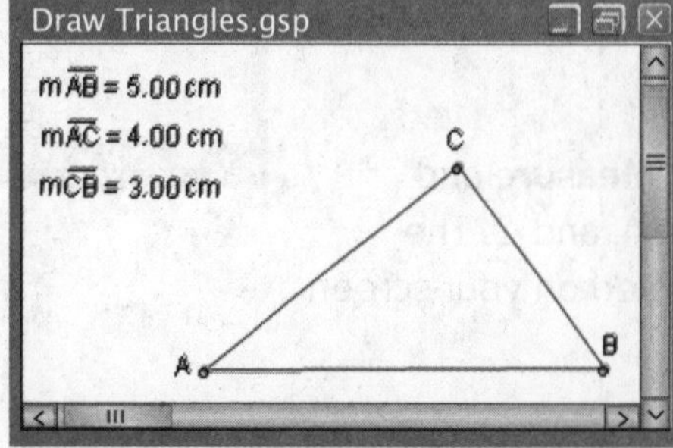

You have created a triangle with side lengths of 3 centimeters, 4 centimeters, and 5 centimeters.

Investigate

MP Use Math Tools Work with a partner to construct each triangle. Once you have constructed a triangle, draw the text and image that appears on your display.

1. $\angle ABC = 90°$

   $\angle BCA = 70°$

   $\angle BAC = 20°$

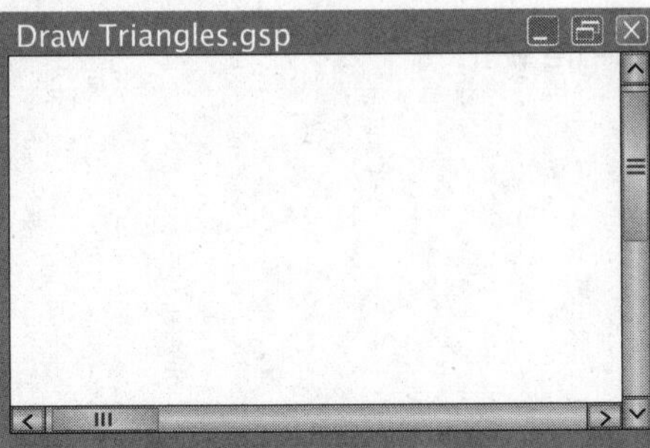

2. $\angle ABC = 90°$

   $\angle BCA = 45°$

   $\angle BAC = 45°$

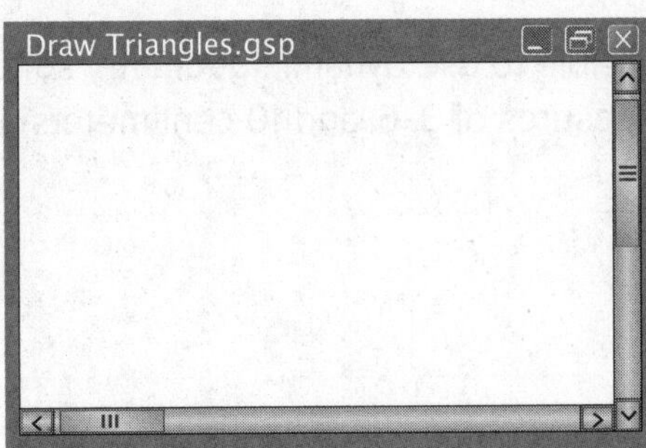

3. Explain the steps you would take to create a triangle if you were given the measures of all three angles.

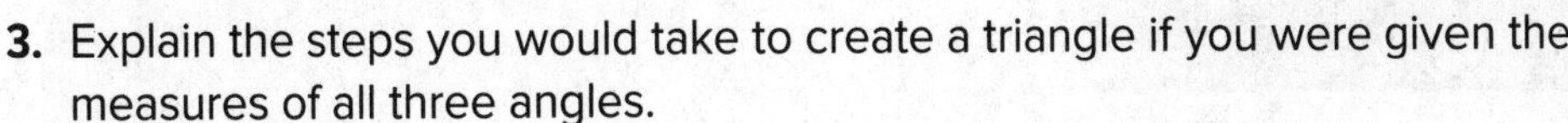

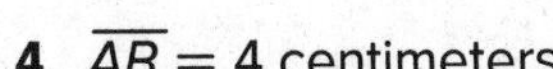

4. $\overline{AB} = 4$ centimeters

   $\overline{AC} = 6$ centimeters

   $\overline{CB} = 9$ centimeters

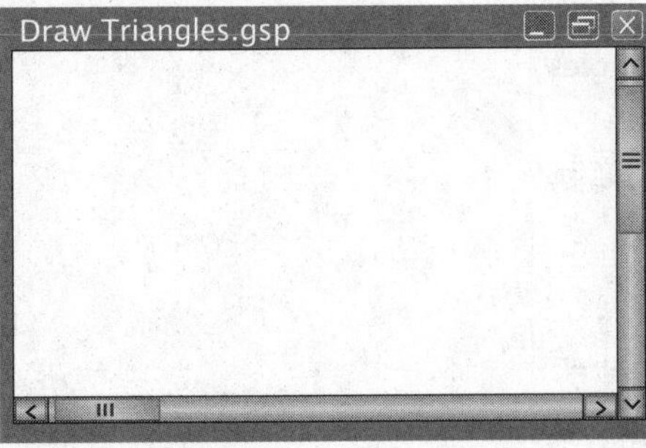

5. $\overline{AB} = 2$ centimeters

   $\overline{AC} = 5$ centimeters

   $\overline{CB} = 4$ centimeters

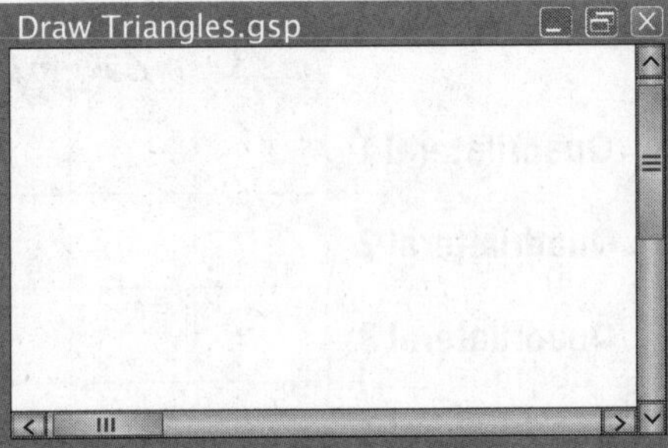

6. MP **Justify Conclusions** Explain the steps you would take to create a triangle if you were given the lengths of all three sides.

Analyze and Reflect

Collaborate

Work with a partner to answer each of the following questions.

7. Is it possible to use dynamic geometry software to draw a triangle with angles of 50°, 65°, and 70°? Explain.

8. Is it possible to use dynamic geometry software to draw a triangle with side measures of 3, 6, and 10 centimeters? Explain.

Create

9. MP **Reason Inductively** You know the rule to find the sum of the interior angles of a triangle. Does a similar rule exist for the sum of the interior angles of a quadrilateral? Use dynamic geometry software to draw four different quadrilaterals and complete the table below to find out. (*Hint:* Do not draw more than one square or rectangle.)

	$m\angle 1$	$m\angle 2$	$m\angle 3$	$m\angle 4$	Sum of Angles
Quadrilateral 1					
Quadrilateral 2					
Quadrilateral 3					
Quadrilateral 4					

10. Inquiry HOW can you use technology to draw geometric shapes?

Problem-Solving Investigation
Make a Model

Content Standards
7.G.1

Mathematical Practices
1, 4, 6

Case #1 Science Project

Jordan is making a model of Mount Saint Helens for a science project. The height of the actual volcano is about 2,500 meters. She uses a scale of 250 meters equals 1 centimeter.

What is the height of the volcano in Jordan's model?

Understand *What are the facts?*

- The height of the actual volcano is about 2,500 meters.
- The scale for her model is 250 meters = 1 centimeter.

Plan *What is your strategy to solve this problem?*

Draw a model that represents the actual volcano and Jordan's volcano to help you visualize the problem.

Solve *How can you apply the strategy?*

The scale is 250 meters = 1 centimeter.
Write and solve a proportion using the scale.

$$\frac{250 \text{ m}}{1 \text{ cm}} = \frac{\square \text{ m}}{x \text{ cm}}$$

$$250 \cdot x = 1 \cdot \square$$

$$x = \square \text{ cm}$$

x

2,500 m

So, Jordan's model has a height of ______________.

Check *Does the answer make sense?*

Multiply the height of the model by 250 to see if it matches the actual height.

Analyze the Strategy

Be Precise The height of Mount Saint Helens is about 8,500 feet. What scale could Jordan use to represent the model in the U.S. Customary System?

____

Case #2 Portraits

Alicia created a portrait that is 10 inches wide by 13 inches long. She wants to put it in a frame that is $2\frac{1}{4}$ inches wide on each side.

What is the area of the framed portrait?

Understand

Read the problem. What are you being asked to find?

I need to find ______________________.

2 Plan

What is your strategy to solve this problem?

I will use the ______________________ strategy.

3 Solve

How can you apply the strategy?

I will ______________________.

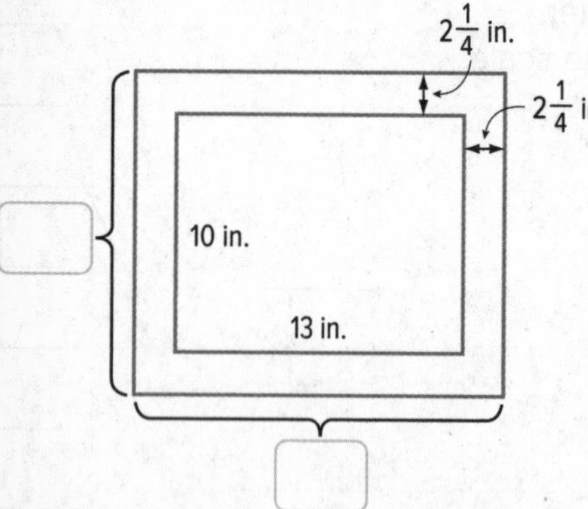

The inner rectangle is the portrait and the outer rectangle is the frame.

Label the combined length and width of the portrait and the frame.

The area of the framed portrait is ______________________.

Check

Estimate the product of the length and width of the framed portrait to determine if your answer is reasonable.

____

____

Work with a small group to solve the following cases. Show your work on a separate piece of paper.

Case #3 Tables

Members of Student Council are setting up tables end-to-end for an awards banquet. Each table will seat one person on each side.

How many square tables will they need to put together for 32 people? Explain your method.

Case #4 Tile

The diagram shows the design of a tile border around a rectangular swimming pool that measures 10.5 meters by 6 meters. Each tile is a square measuring 1.5 meter on each side.

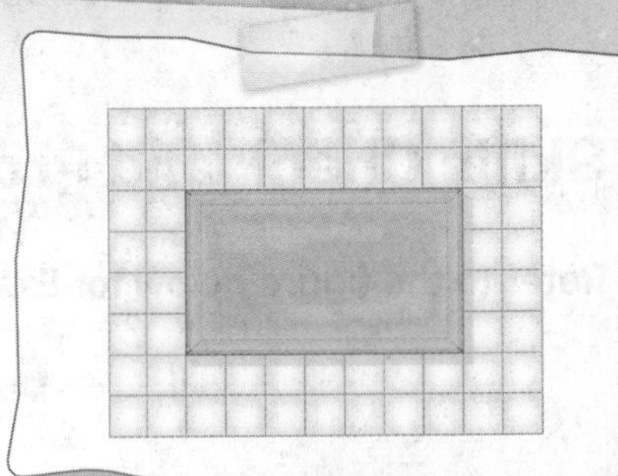

Explain a method you could use to find the area of just the tile border. Then solve.

Case #5 Classes

Faith, Sarah, and Guadalupe take French, Spanish, and German. No person's language class begins with the same letter as their first name. Sarah's best friend takes French.

Which language does each person take?

Case #6 Money

Corri received money for her birthday. She let Lakisha borrow $4.50 and spent half of the remaining money. The next day she received $10 from her uncle. After spending $12.50 at the movies, she still had $7.75 left.

How much money did she receive for her birthday?

Mid-Chapter Check

Vocabulary Check

1. MP **Be Precise** Define *complementary angles*. Give an example of two angles that would be complementary. (Lesson 2)

____

____

2. Fill in the blank in the sentence below with the correct term. (Lesson 3)

A ______________ triangle is made up of one right angle and no congruent sides.

Skills Check and Problem Solving

Refer to the figure below for Exercises 3–5. (Lessons 1 and 2)

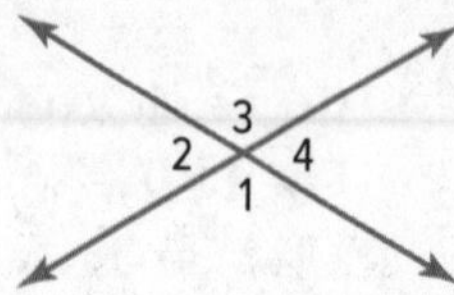

3. Identify a pair of vertical angles.

____

4. Identify a pair of supplementary angles.

____

5. Suppose $m\angle 1 = 127°$. Find the measures of the other angles.

____

6. What is $m\angle A$ in $\triangle ABC$ if $m\angle B = 35°$ and $m\angle C = 92°$? ______________

7. MP **Reason Inductively** Classify the triangle that satisfies each set of conditions.

   a. one right angle and two congruent sides ______________

   b. one obtuse angle and no congruent sides ______________

   c. three acute angles and three congruent sides ______________

Inquiry Lab

Investigate Online Maps and Scale Drawings

HOW is the zoom feature of an online map like the scale of a drawing?

Content Standards
7.G.1

Mathematical Practices
1, 3, 4, 5

Maps and blueprints are *scale drawings* of the locations and buildings they represent. Unlike maps printed on paper, online map services allow users the opportunity to view a location from different distances.

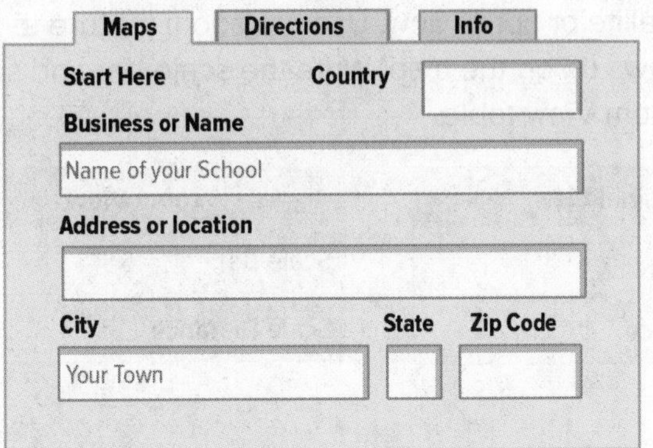

Hands-On Activity 1

Step 1 Use the online map service provided to you by your teacher. Locate your school on a map.

Step 2 Measure the length of the scale bar in centimeters on the online map. Find the scale distance of the map. Write these values in the Original View table in Step 4.

Step 3 Click on the satellite or aerial view. Use the zoom feature to zoom in until your school shows up on the map.

Step 4 Measure the length of the scale bar in centimeters. Find the new scale distance for the map. Write these values in the Zoom View table.

Original View	
Scale Bar	
Scale Distance	

Zoom View	
Scale Bar	
Scale Distance	

What happens when you use the zoom feature?

____

Describe the appearance of the map as you zoomed in.

____

Investigate

MP Use Math Tools **Work with a partner to answer the following questions about using an online map service.**

1. Locate the local public library on the map. Write the scale bar and scale distance values in the Original View table below Exercise 2.

2. Click on the satellite or aerial view. Use the zoom feature to zoom in until the building shows up on the map. Write the scale bar and scale distance values in the Zoom View table.

Original View	
Scale Bar	
Scale Distance	

Zoom View	
Scale Bar	
Scale Distance	

Analyze and Reflect

Work with a partner to answer the following questions about using an online map.

3. Refer to Activity 1. Write a ratio $\frac{\text{scale bar}}{\text{scale distance}}$ for the original view and the zoom view.

Original View: ________ Zoom View: ________

4. How many times bigger is the zoom view?

5. How does zooming in affect the scale on the map?

6. When using the zoom feature on an online map, what changes and what stays the same?

Hands-On Activity 2

The diagram shown represents a garden. The scale is 1 centimeter = 30 meters. That means that each square on the grid measures 1 centimeter by 1 centimeter or 30 meters by 30 meters.

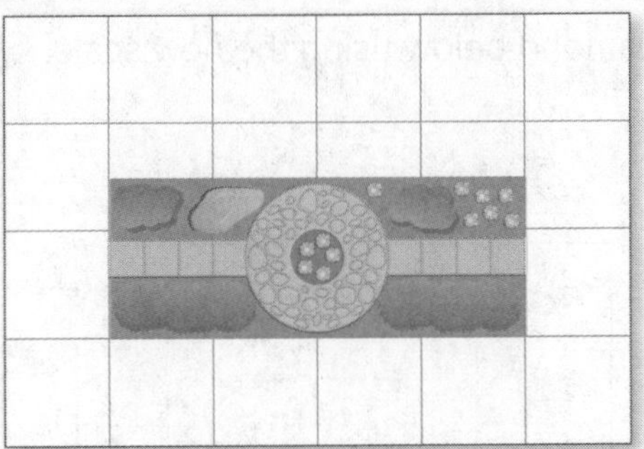

Step 1 Write the length and width of the drawing of the garden.

Length: ______ centimeters Width: ______ centimeters

Step 2 Use the scale to find the dimensions of the garden.

Length: ______ meters Width: ______ meters

Step 3 On the grid below, draw the garden so that the scale is 1 centimeter = 10 meters. Write the dimensions of your drawing.

Length: ______ centimeters Width: ______ centimeters

Step 4 Use the scale on your drawing to compute the dimensions of the garden. How do the dimensions compare to the dimensions in Step 2?

Length: ______ meters Width: ______ meters

____

Investigate

Work with a partner to answer the following questions about reproducing a scale drawing.

7. Recreate the drawing of the baseball diamond below using the new scale.

current scale: 1 unit = 15 ft
new scale: 1 unit = 30 ft

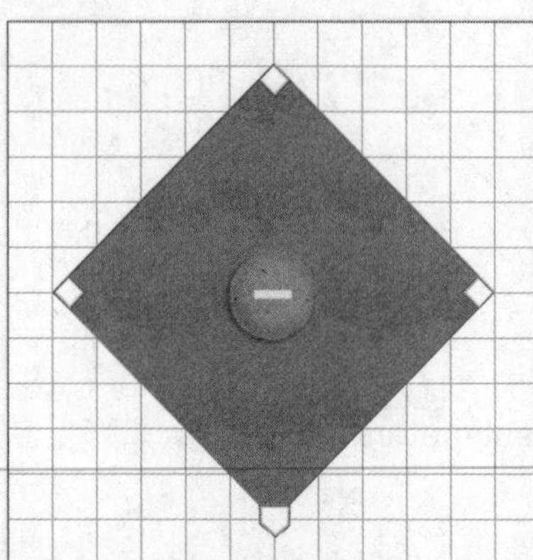
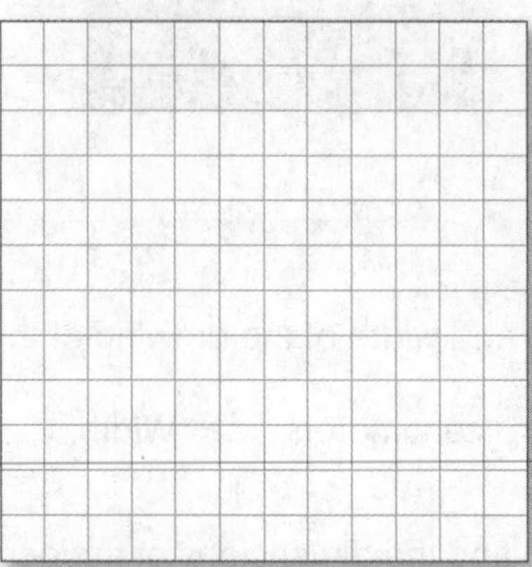

8. A drawing of the Statue of Liberty is 3 inches tall. The scale is 1 inch = 50 feet. How tall would the drawing be if the scale were 0.5 inch = 100 feet? __________

Analyze and Reflect

9. MP **Reason Inductively** The triangle shown in the drawing has an area of 40 square feet. What is the scale of the drawing? __________

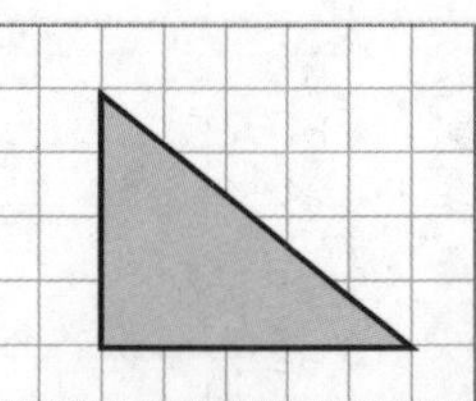

Create

10. MP **Model with Mathematics** Using a separate piece of grid paper, create a map of your classroom or a room in your home. Identify the scale you used.

____

11. Inquiry How is the zoom feature of an online map like the scale of a drawing?

____

____

Lesson 4

Scale Drawings

Real-World Link

Room Model Architects make detailed drawings of rooms and buildings. Conner made a drawing of a bedroom. Follow the steps below to make a model of a room of your choosing.

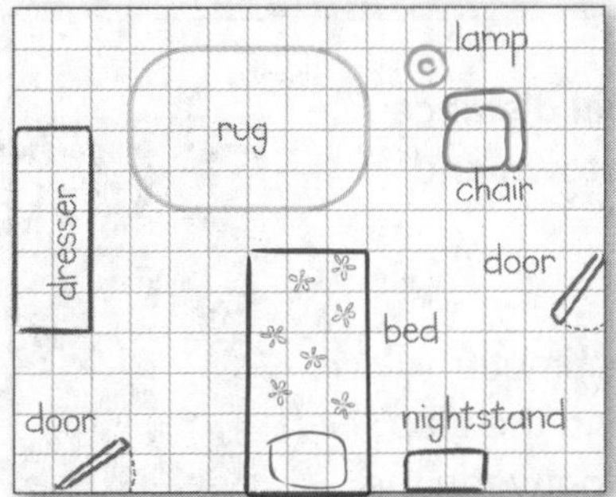

Step 1 Measure the length of three objects in the room. Record each length to the nearest $\frac{1}{2}$ foot in the table below.

Object	Length (ft)	Length (units)

Step 2 Let 1 unit represent 2 feet. So, 4 units = 8 feet. Convert all your measurements to units. Record these values.

Step 3 On grid paper, make a drawing of your room like the one shown.

Which MP Mathematical Practices did you use? Shade the circle(s) that applies.

① Persevere with Problems
② Reason Abstractly
③ Construct an Argument
④ Model with Mathematics
⑤ Use Math Tools
⑥ Attend to Precision
⑦ Make Use of Structure
⑧ Use Repeated Reasoning

Essential Question

HOW does geometry help us describe real-world objects?

Vocabulary

scale drawing
scale model
scale
scale factor

Common Core State Standards

Content Standards
7.G.1

MP Mathematical Practices
1, 2, 3, 4, 5

Work Zone

Use a Scale Drawing or a Scale Model

Scale drawings and **scale models** are used to represent objects that are too large or too small to be drawn or built at actual size. The **scale** gives the ratio that compares the measurements of the drawing or model to the measurements of the real object. The measurements on a drawing or model are proportional to the measurements on the actual object.

Example

1. What is the actual distance between Hagerstown and Annapolis?

Step 1 Use a centimeter ruler to find the map distance between the two cities. The map distance is about 4 centimeters.

Step 2 Write and solve a proportion using the scale. Let d represent the actual distance between the cities.

	Scale		Length	
map → actual →	$\frac{1 \text{ centimeter}}{24 \text{ miles}}$	=	$\frac{4 \text{ centimeter}}{d \text{ miles}}$	← map ← actual

$1 \times d = 24 \times 4$ Cross products

$d = 96$ Simplify.

The distance between the cities is about 96 miles.

Scale

A map scale can be written in different ways, including the following:

1 cm = 20 mi

1 cm : 20 mi

$\frac{1 \text{ cm}}{20 \text{ mi}}$

Got it? **Do this problem to find out.**

a. ____________

a. On the map of Arkansas shown, find the actual distance between Clarksville and Little Rock. Use a ruler to measure.

Example

2. A graphic artist is creating an advertisement for this cell phone. If she uses a scale of 5 inches = 1 inch, what is the length of the cell phone on the advertisement?

Write a proportion using the scale. Let a represent the length of the advertisement cell phone.

	Scale		Length	
advertisement →	5 inches	=	a inches	← advertisement
actual →	1 inch		4 inches	← actual

$5 \cdot 4 = 1 \cdot a$ Cross products

$20 = a$ Simplify.

The length of the cell phone on the advertisement is 20 inches long.

Scale

The scale is the ratio of the drawing/model measure to the actual measure. It is not always the ratio of a smaller measure to a larger measure.

Got it? **Do this problem to find out.**

b. A scooter is $3\frac{1}{2}$ feet long. Find the length of a scale model of the scooter if the scale is 1 inch = $\frac{3}{4}$ feet.

b. ________

Find a Scale Factor

A scale written as a ratio without units in simplest form is called the **scale factor.**

Example

3. Find the scale factor of a model sailboat if the scale is 1 inch = 6 feet.

$\frac{1 \text{ inch}}{6 \text{ feet}} = \frac{1 \text{ inch}}{72 \text{ inches}}$ Convert 6 feet to inches.

$= \frac{1}{72}$ Divide out the common units.

The scale factor is $\frac{1}{72}$.

Got it? **Do this problem to find out.**

c. What is the scale factor of a model car if the scale is 1 inch = 2 feet?

c. ________

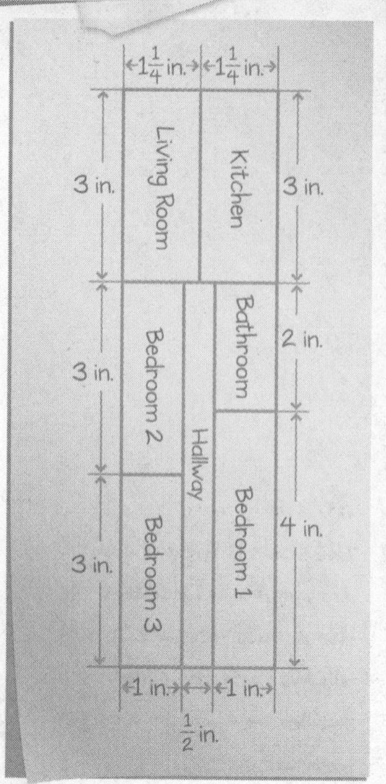

Example

4. A floor plan for a home is shown at the left where $\frac{1}{2}$ inch represents 3 feet of the actual home. What is the actual area of bedroom 1?

Length of Bedroom 1.

$\frac{\frac{1}{2} \text{ in.}}{3 \text{ ft}} = \frac{4 \text{ in.}}{w}$ ← **floor plan** ← **actual**

$\frac{1}{2}w = 12$ Find cross products.

$w = 24$ Divide each side by $\frac{1}{2}$.

Width of Bedroom 1.

$\frac{\frac{1}{2} \text{ in.}}{3 \text{ ft}} = \frac{1 \text{ in.}}{x}$ ← **floor plan** ← **actual**

$\frac{1}{2}x = 3$ Find cross products.

$x = 6$ Divide each side by $\frac{1}{2}$.

So, the area of bedroom 1 is 24 × 6 or 144 square feet.

Got it? Do this problem to find out.

Show your work.

d. What is the actual area of bedroom 3?

d. ______

Guided Practice

1. On a map, the distance from Akron to Cleveland measures 2 centimeters. What is the actual distance if the scale of the map shows that 1 centimeter is equal to 30 kilometers? (Example 1)

2. An engineer makes a model of a bridge using a scale of 1 inch = 3 yards. The length of the actual bridge is 50 yards. What is the length of the model? (Example 2)

3. Julie is constructing a scale model of her room. The rectangular room is $10\frac{1}{4}$ inches by 8 inches. If 1 inch represents 2 feet of the actual room, what is the scale factor and the actual area of the room? (Examples 3 and 4)

4. **Building on the Essential Question** Explain how you could use a map to estimate the actual distance between Miami, Florida, and Atlanta, Georgia.

Rate Yourself!

How well do you understand scale drawings? Circle the image that applies.

For more help, go online to access a Personal Tutor.

Name ________________ My Homework ________________

Independent Practice

Go online for Step-by-Step Solutions

MP Use Math Tools Find the actual distance between each pair of locations in South Carolina. Use a ruler to measure. (Example 1)

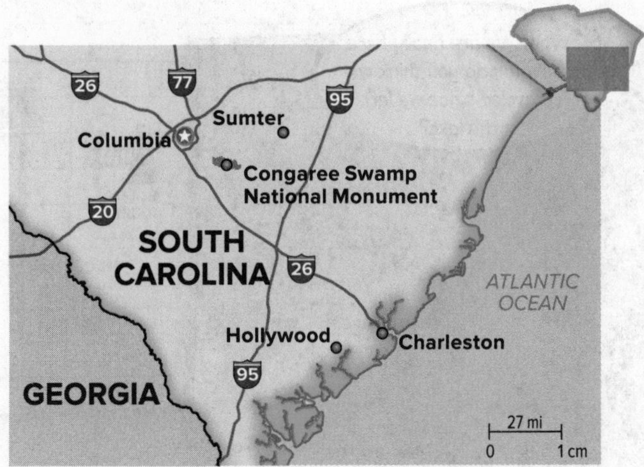

1. Columbia and Charleston ________

2. Hollywood and Sumter ________

Find the length of each model. Then find the scale factor. (Examples 2 and 3)

3.

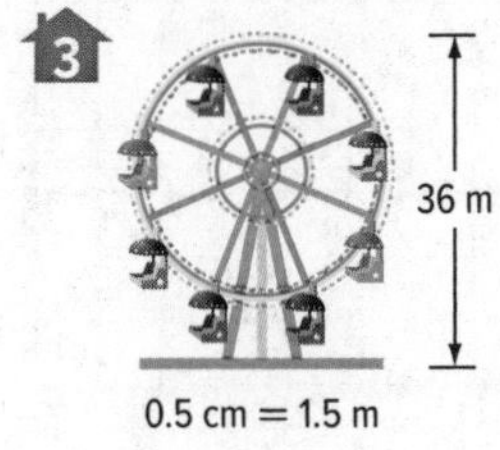

4.

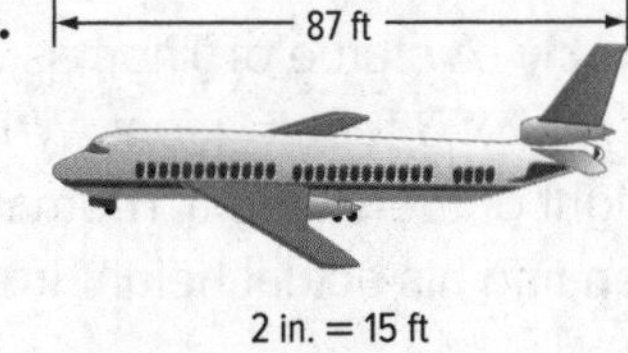

5. A model of an apartment is shown where $\frac{1}{4}$ inch represents 3 feet in the actual apartment. Find the actual area of the master bedroom. (Example 4)

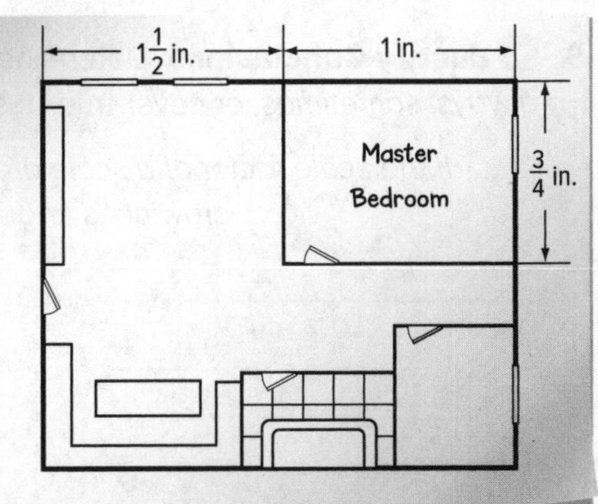

6. **MP Model with Mathematics** Refer to the graphic novel frames below. The scale on the map shows that 1 centimeter is equal to 75 yards. If the red line represents the path they took, how far have Raul, Caitlyn, and Jamar traveled since they left the lake? Each square on the map is 1 centimeter long.

H.O.T. Problems Higher Order Thinking

7. **MP Model with Mathematics** On the grid paper, create a scale drawing of a room in your home. Include the scale that you used. ____________

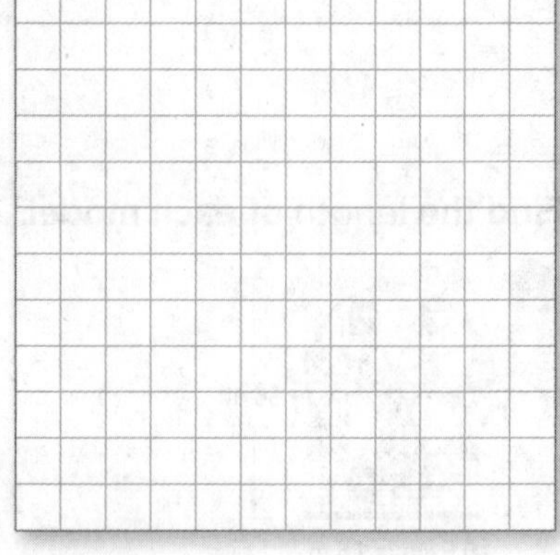

8. **MP Reason Abstractly** A statue of Thomas Jefferson was made using a scale of 3 feet = 1 foot. Write an expression to represent the height of the statue if Thomas Jefferson is *x* feet in height. Then find his actual height if the height of the statue is 19 feet.

9. **MP Justify Conclusions** Determine whether the following statement is *always*, *sometimes*, or *never* true. Justify your reasoning.

If the scale factor of a scale drawing is greater than one, the scale drawing is larger than the actual object.

Name ______________________ My Homework ______________________

Extra Practice

Use Math Tools **Find the actual distance between each pair of cities in New Mexico. Use a ruler to measure.**

10. Carlsbad and Artesia 50 km

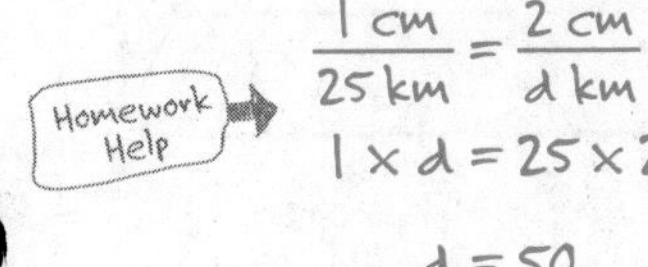

11. Hobbs and Eunice ________

12. Artesia and Eunice ________

13. Lovington and Carlsbad ________

14. Find the length of the model. Then find the scale factor. The length of an actual bird is shown at the right.

____

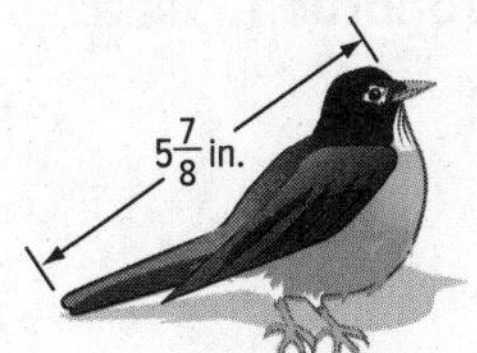

1 in. = 0.5 in.

Copy and Solve **Show your work on a separate piece of paper.**

15. A model of a tree is made using a scale of 1 inch = 25 feet. What is the height of the actual tree if the height of the model is $4\frac{3}{8}$ inches?

16. A map of Bakersfield has a scale of 1 inch = 5 miles. If the city is $5\frac{1}{5}$ inches across on the map, what is the actual distance across the city?

17. Tyson is creating a scale drawing of the area of his school. The rectangular drawing shows the length as 20 inches and the width as 19 inches. The drawing uses a scale of 1 inch = 3 feet. What is the actual area of the school in square feet?

Power Up! Common Core Test Practice

18. A landscape designer created a scale drawing of a bench that will be in a garden as shown. The actual width of the bench is 6 feet, and the actual height is 3 feet. Fill in each box to complete the following statements.

a. The scale of the drawing is ☐ inch(es) = ☐ feet.

b. The height of the scale drawing is ☐ inch(es).

19. A scale drawing of a doctor's office is shown. What are the actual dimensions of the doctor's office? Explain how you found your answer.

Common Core Spiral Review

20. A carpenter sawed a piece of wood into 3 pieces. The ratio of wood pieces is 1:3:6. The longest piece is 2.5 feet longer than the shortest piece. Use the *draw a diagram* strategy to find the length of the original piece. **6.RP.1**

Solve each proportion. 7.RP.2C

21. $\frac{2}{5} = \frac{b}{25}$

22. $\frac{3}{7} = \frac{a}{49}$

23. $\frac{2}{9} = \frac{x}{99}$

24. Dante has 60 baseball cards. This is at least six more than three times as many cards as Anna. Write and solve an inequality to represent the situation. **7.EE.4b**

Inquiry Lab

Scale Drawings

WHAT happens to the size of a scale drawing when it is reproduced using a different scale?

Content Standards
7.G.1

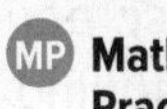

Mathematical Practices
1, 3, 5

The owner of the miniature golf course wants to create a sign with an image of the 18th hole on it. Use the dimensions shown to create a scale drawing using the Geometer's Sketchpad®. Use the scale 1 centimeter = 3 meters.

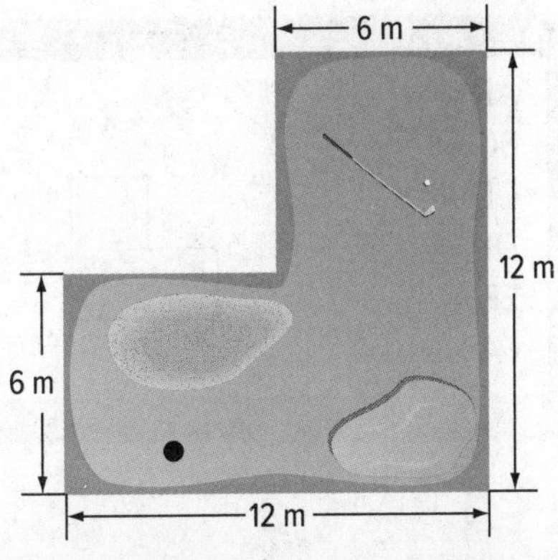

Hands-On Activity

Step 1 Determine the length the 6 meter side and the 12 meter side will be in the drawing.

Scale	Length	Scale	Length
$\frac{1\text{ cm}}{3\text{ m}} =$	$\frac{x\text{ cm}}{6\text{ m}}$	$\frac{1\text{ cm}}{3\text{ m}} =$	$\frac{x\text{ cm}}{12\text{ m}}$

$1 \cdot 6 = 3 \cdot x$ $\qquad$ $1 \cdot 12 = 3 \cdot x$

$x = \square$ $\qquad$ $x = \square$

So, the 6 meter side will be ☐ centimeters and the 12 meter side will be ☐ centimeters in the drawing.

Step 2 Create the drawing using a dynamic geometry software. Then fill in the correct length for each line segment.

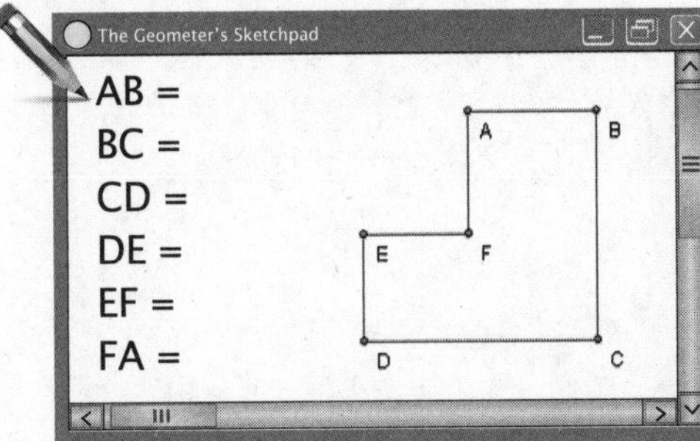

Investigate

Collaborate

Work with a partner. Use a dynamic geometry software.

1. MP **Use Math Tools** The owner wants a different size image of the 18th hole to place on the scorecards. Use the scale 1 centimeter = 6 meters. Fill in the new lengths of the line segments and draw the new scale drawing on the screen below. (*Hint:* You don't have to redraw the figure. Try clicking and dragging on the sides of your first drawing to adjust the side lengths.)

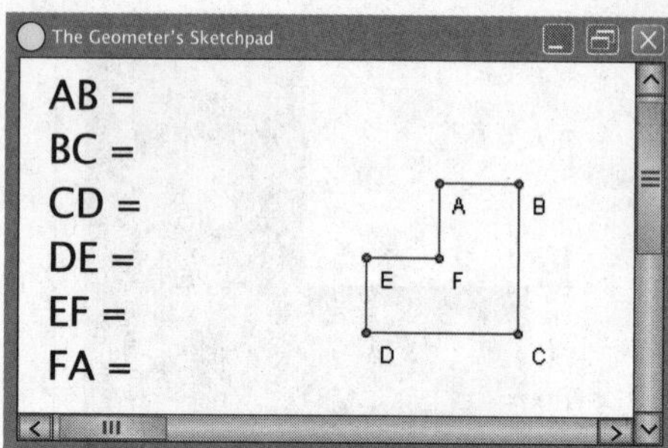

Analyze and Reflect

Collaborate

2. What happened to the size of the scale drawing when the scale changed from 1 centimeter = 3 meters to 1 centimeter = 6 meters?

Create

On Your Own

3. MP **Reason Inductively** Suppose you drew the miniature golf hole again at the scale 1 centimeter = 2 meters. Would the size of your drawing be larger or smaller than the drawing in the Activity? Explain.

4. Inquiry WHAT happens to the size of a scale drawing when it is reproduced using a different scale?

Draw Three-Dimensional Figures

Real-World Link

New York City In art class, Rasheed studied buildings known for their unusual architecture. He studied the Flatiron Building shown.

Three-dimensional figures, such as the Flatiron Building, have length, width, and height. They can be viewed from different perspectives, including the *side* view and the *top* view.

1. What is the two-dimensional figure that makes up the side view?

2. What is the two-dimensional figure that makes up the top view?

3. Sketch the side view of the Flatiron Building.

4. Sketch the top view of the Flatiron Building.

5. The top view, side view, and front view of a three-dimensional figure are shown below. Sketch the figure.

front

Which MP Mathematical Practices did you use? Shade the circle(s) that applies.

① Persevere with Problems
② Reason Abstractly
③ Construct an Argument
④ Model with Mathematics
⑤ Use Math Tools
⑥ Attend to Precision
⑦ Make Use of Structure
⑧ Use Repeated Reasoning

Essential Question

HOW does geometry help us describe real-world objects?

Common Core State Standards

Content Standards
Preparation for 7.G.3

MP Mathematical Practices
1, 3, 4

Work Zone

Draw a Three-Dimensional Figure

You can draw different views of three-dimensional figures. The most common views drawn are the top, side, and front views.

The top, side, and front views of a three-dimensional figure can be used to draw a corner view of the figure.

Examples

Tutor

1. Draw a top, a side, and a front view of the figure at the right.

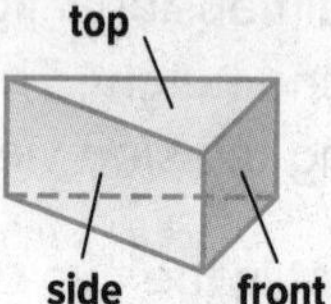

The top view is a triangle.

The side and front view are rectangles.

top **side** **front**

2. Draw a top, a side, and a front view of the figure at the right.

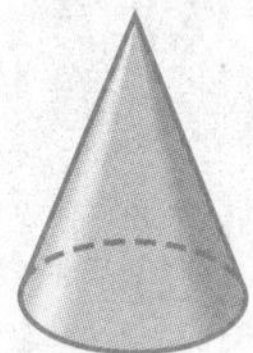

The top view is a circle.

The side and front view are triangles.

top side front

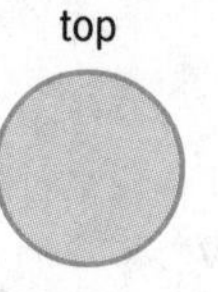

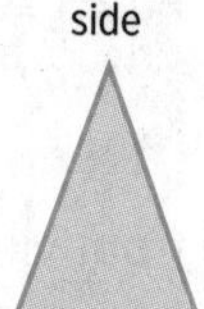

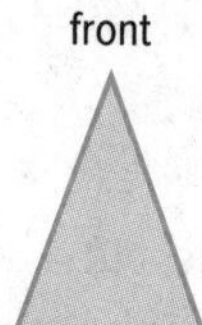

Got it? Do this problem to find out.

a. Draw a top, a side, and a front view of the figure at the right.

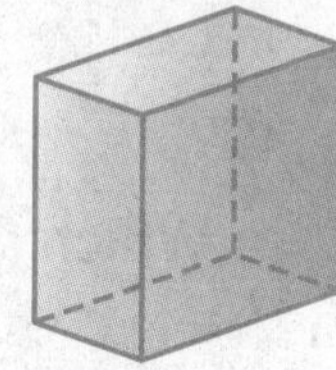

a. ______

Example

3. Draw a top, a side, and a front view of the video console shown.

The top view is a rectangle.

The side and front views are also rectangles.

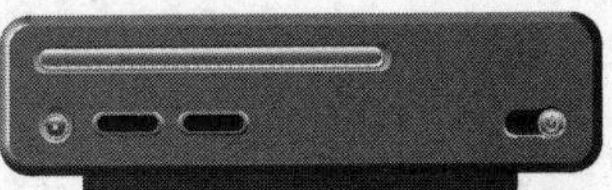

Got it? **Do this problem to find out.**

b. Draw a top, a side, and a front view of the tent shown.

Example

4. Draw a corner view of the three-dimensional figure whose top, side, and front views are shown.

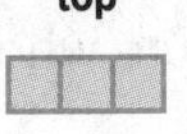

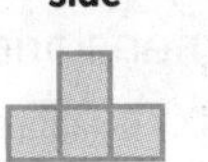

Step 1 Use the top view to draw the base of the figure, a 1-by-3 rectangle.

Step 2 Add edges to make the base a solid figure.

Step 3 Use the side and front views to complete the figure.

Got it? **Do this problem to find out.**

c. Draw a corner view of the three-dimensional figure whose top, side, and front views are shown.

Plane Figures

In geometry, three-dimensional figures are solids and two-dimensional figures such as triangles, circles, and squares are plane figures.

Show your work.

b. ______________

c. ______________

Example

Tutor

5. Draw a corner view of the three-dimensional figure whose top view, side view, and front view are shown.

top side front

Step 1 Use the top view to draw the base of the figure, a 2-by-4 rectangle.

Step 2 Add edges to make the base a solid figure.

Step 3 Use the side and front views to complete the figure.

top

front side

Guided Practice

1. Draw a top, a side, and a front view of the figure. (Examples 1–3)

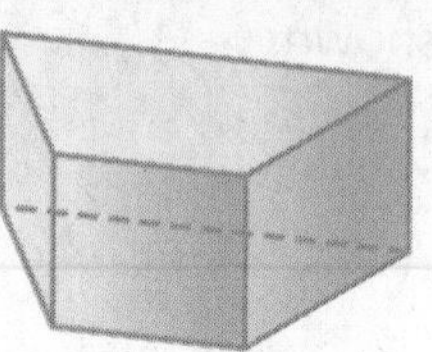

2. Draw a corner view of the three-dimensional figure whose top view, side view, and front view are shown. (Examples 4–5)

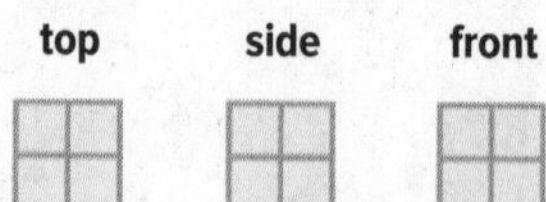

3. **Building on the Essential Question** How does drawing the different views of a three-dimensional figure help you have a better understanding of the figure?

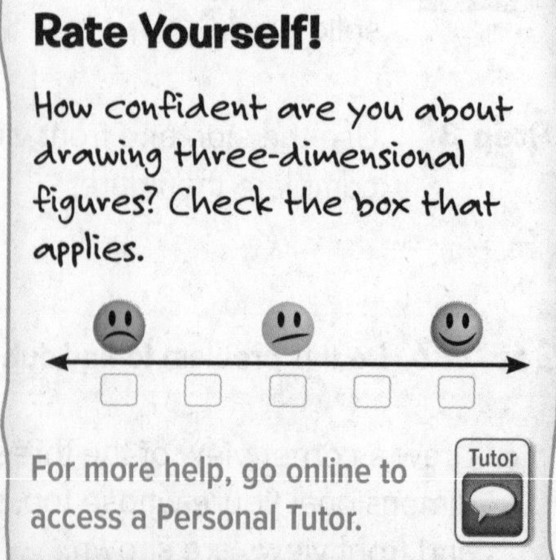

Name ______________________ My Homework ______________________

Independent Practice

Go online for Step-by-Step Solutions

Draw a top, a side, and a front view of each figure. (Examples 1–2)

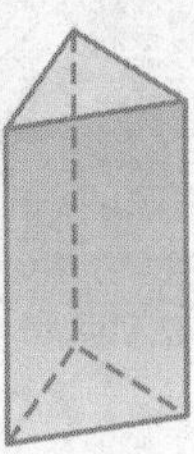

2.

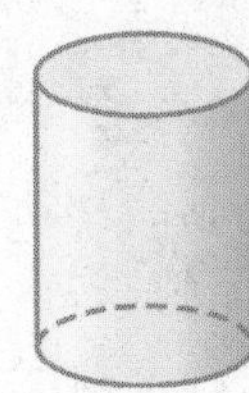

3 Draw a top, a side, and a front view of the eraser shown. (Example 3)

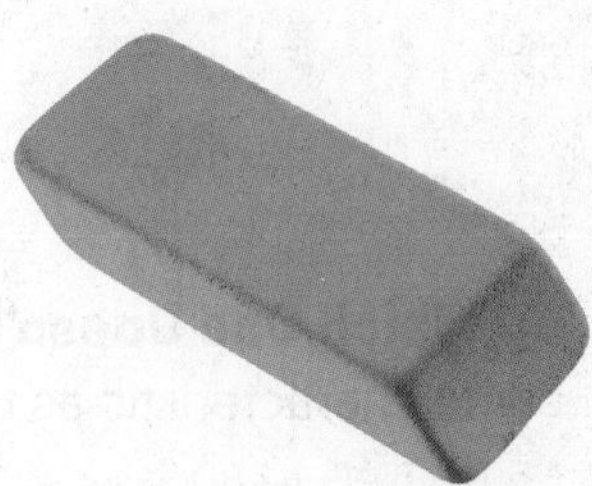

Draw a corner view of each three-dimensional figure whose top view, side view, and front view are shown. (Examples 4–5)

4. **top** **side** **front**

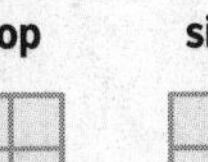

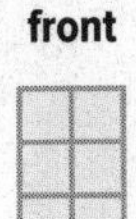

5. **top** **side** **front**

6. Name a real-world object that has a top view of a triangle, and a side and front view that are each rectangles. ______________________

7. **Model with Mathematics** The Quetzalcoatl pyramid in Mexico is shown. Use the photo to sketch views from the top, side, and front of the pyramid.

Show your work.

H.O.T. Problems Higher Order Thinking

8. **Model with Mathematics** Choose an object in your classroom or in your home. Sketch any view of the object. Choose among a top, a side, or a front view.

9. **Which One Doesn't Belong?** Identify the figure that does not have the same characteristic as the other three. Explain your reasoning.

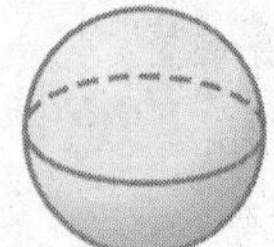 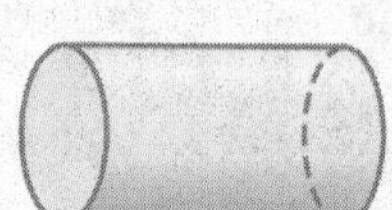 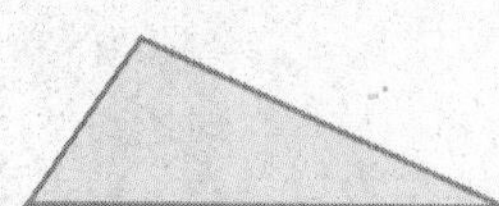 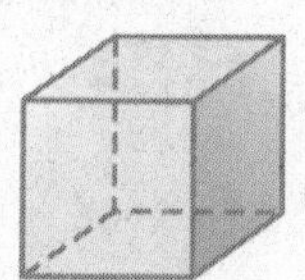

10. **Persevere with Problems** Draw a three-dimensional figure in which the front and top views each have a line of symmetry but the side view does not.

11. **Reason Inductively** Determine whether each statement is *always*, *sometimes*, or *never* true.

a. The bases of a cylinder have different radii. ________

b. Two planes intersect in a single point. ________

c. Three planes do not intersect in a point. ________

Extra Practice

Draw a top, a side, and a front view of each figure.

12.

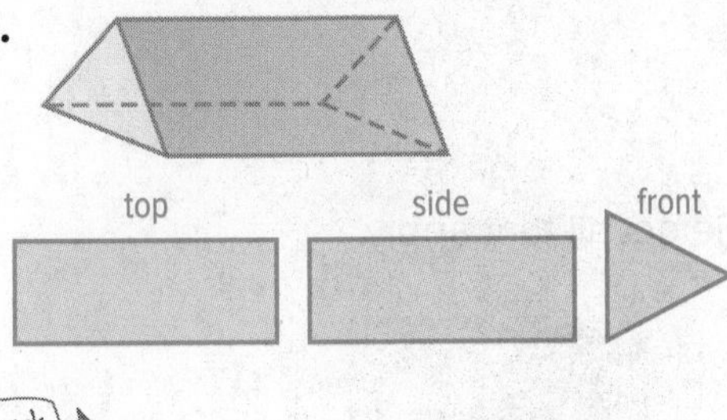

13.

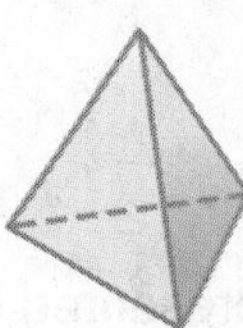

Draw a corner view of each three-dimensional figure whose top view, side view, and and front view are shown.

14. **top** **side** **front**

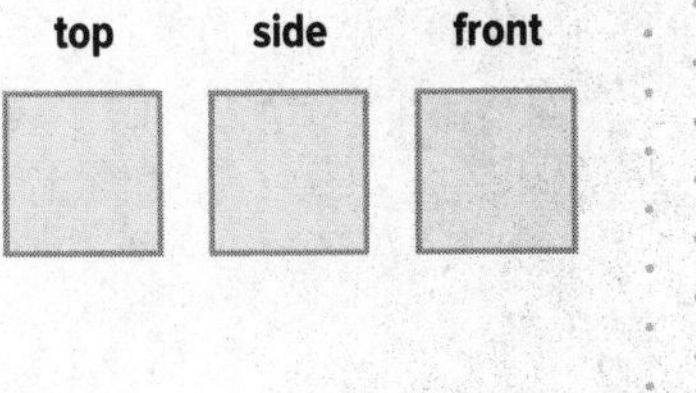

15. **top** **side** **front**

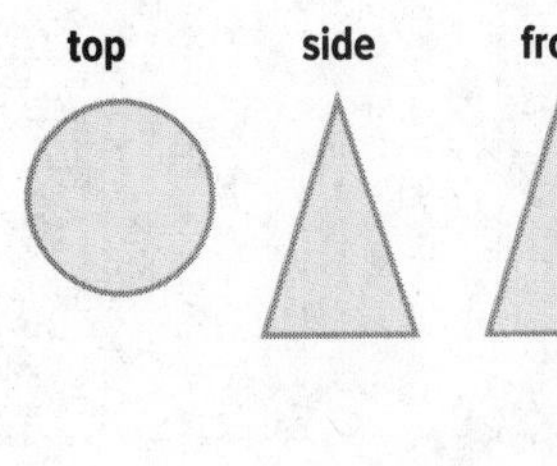

Draw a top, a side, and a front view of each figure.

16.

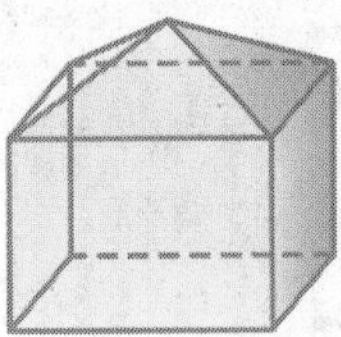

17.

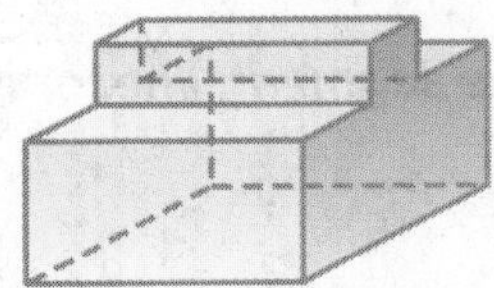

18. MP **Find the Error** Raul drew the side, top, and front view of the figure shown at the right. Find his mistake and correct it.

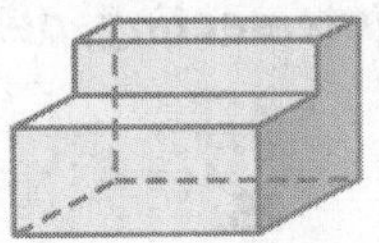

side

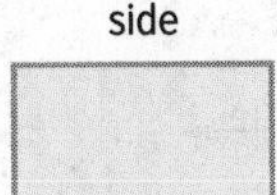

top

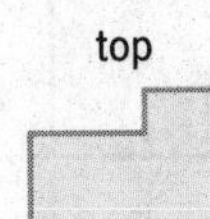

front

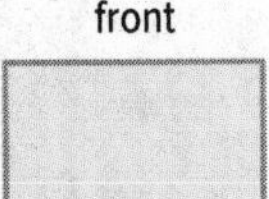

Power Up! Common Core Test Practice

19. The top, side, and front view of a figure made of cubes are shown.

top

front

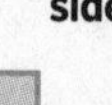

side

Which of the following can be represented by these views? Select all that apply.

20. Draw the front, top, and side views of the three-dimensional figure shown at the right.

Common Core Spiral Review

Identify each figure as a *line segment*, *line*, or *ray*. Then name each figure using symbols. 5.G.3

21.

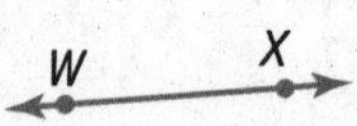

22.

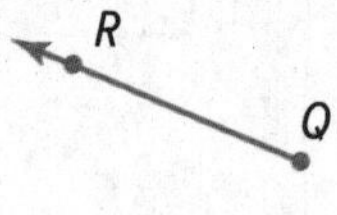

23.

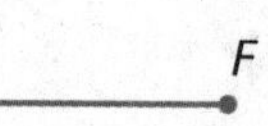

Describe each pair of lines as *intersecting*, *perpendicular*, or *parallel*. Choose the most specific term. 5.G.3

24.

25.

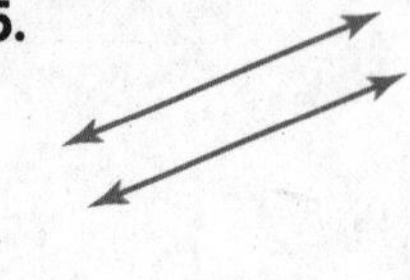

26.

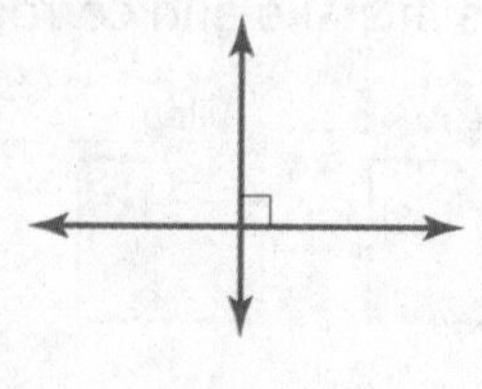

Lesson 6

Cross Sections

Vocabulary Start-Up

A **prism** is a three-dimensional figure with at least two parallel, congruent faces called **bases** that are polygons. A **pyramid** is a three-dimensional figure with one base that is a polygon. Its other faces are triangles.

Write *prism* or *pyramid* on the line below each figure.

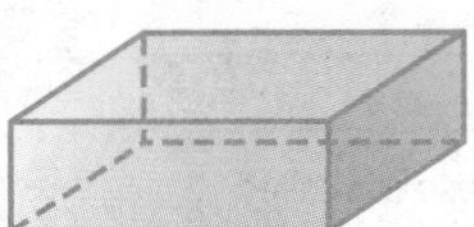

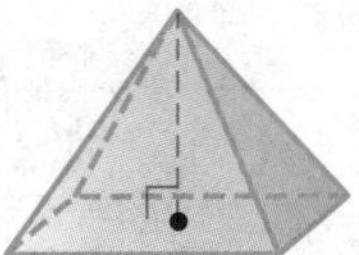

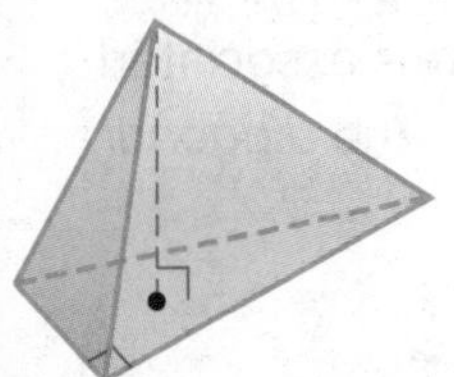

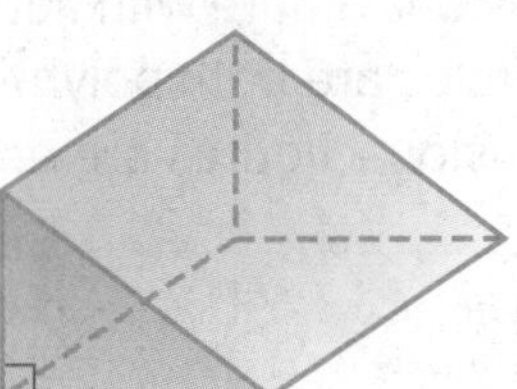

Real-World Link

The Rock and Roll Hall of Fame is shown below. Is the shape of the building a *prism* or *pyramid*? Explain.

Essential Question

HOW does geometry help us describe real-world objects?

Vocabulary

prism
bases
pyramid
plane
coplanar
parallel
polyhedron
edge
face
vertex
diagonal
skew lines
cylinder
cone
cross section

Common Core State Standards

Content Standards
7.G.3

MP **Mathematical Practices**
1, 3, 4

Which MP Mathematical Practices did you use? Shade the circle(s) that applies.

1. Persevere with Problems
2. Reason Abstractly
3. Construct an Argument
4. Model with Mathematics
5. Use Math Tools
6. Attend to Precision
7. Make Use of Structure
8. Use Repeated Reasoning

Work Zone

Identify Three-Dimensional Figures

A **plane** is a flat surface that goes on forever in all directions. The figure at the right shows rectangle *ABCD*. Line segments *AB* and *DC* are **coplanar** because they lie in the same plane. They are also **parallel** because they will never intersect, no matter how far they are extended.

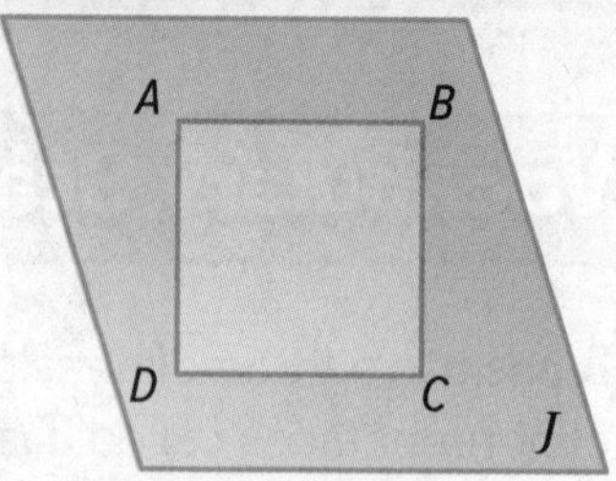

Just as two lines in a plane can intersect or be parallel, there are different ways that planes may be related in space.

Intersect in a Line — ℓ

Intersect at a Point — A

No Intersection — These are called *parallel planes*.

Intersecting planes can form three-dimensional figures. A **polyhedron** is a three-dimensional figure with flat surfaces that are polygons. Prisms and pyramids are both polyhedrons. Some terms associated with three-dimensional figures are *edge, face, vertex*, and *diagonal*.

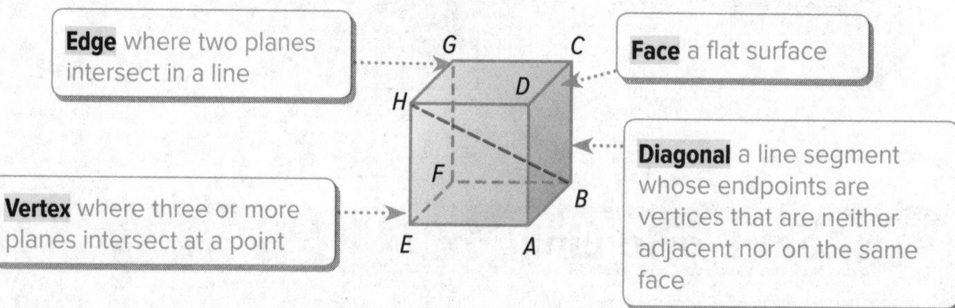

Notice in the figure above, $\overline{GC}$ and $\overline{DA}$ do not intersect. These segments are not parallel because they do not lie in the same plane. Lines that do not intersect and are not coplanar are called **skew lines**.

There are also solids that are not polyhedrons. A **cylinder** is a three-dimensional figure with two parallel congruent circular bases connected by a curved surface. A **cone** has one circular base connected by a curved side to a single vertex.

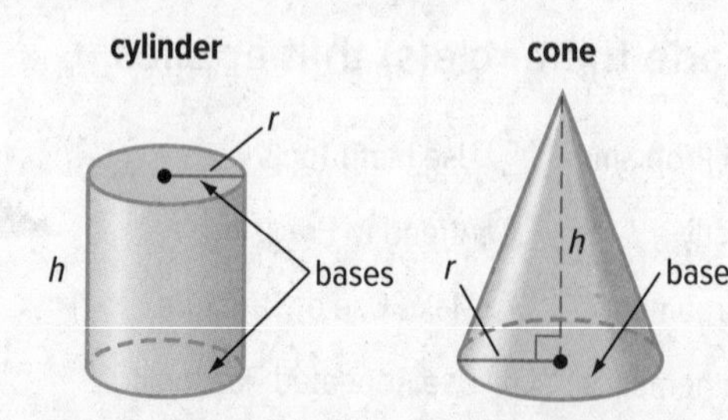

Polygons

The table below lists some common names of polygons.

Sides	Name
5	pentagon
6	hexagon
7	heptagon
8	octagon
9	nonagon
10	decagon

Examples

Identify the figure. Name the bases, faces, edges, and vertices. Then, identify a pair of skew lines.

1.

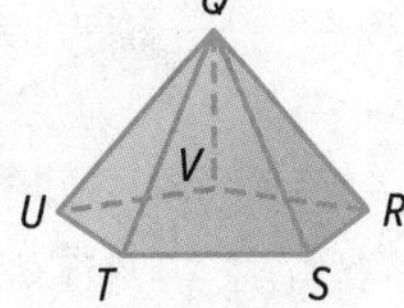

The figure has one base that is a pentagon, so it is a pentagonal pyramid.

base: *RSTUV*

faces: *RSTUV, QVR, QRS, QST, QTU, QUV*

edges: $\overline{QR}, \overline{QS}, \overline{QT}, \overline{QU}, \overline{QV}, \overline{VR}, \overline{RS}, \overline{ST}, \overline{TU}, \overline{UV}$

vertices: *Q, R, S, T, U, V*

skew lines: $\overline{QV}$ and $\overline{TS}$

2.

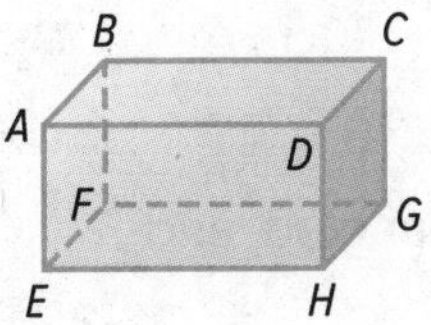

The figure has rectangular bases that are parallel and congruent, so it is a rectangular prism.

bases: *ABCD* and *EFGH*, *ABFE* and *DCGH*, *ADHE* and *BCGF*

faces: *ABCD, EFGH, ABFE, DCGH, ADHE, BCGF*

edges: $\overline{AB}, \overline{BC}, \overline{CD}, \overline{AD}, \overline{EF}, \overline{FG}, \overline{GH}, \overline{EH}, \overline{AE}, \overline{BF}, \overline{CG}, \overline{DH}$

vertices: *A, B, C, D, E, F, G, H*

skew lines: $\overline{AE}$ and $\overline{FG}$

Got it? **Do this problem to find out.**

a.

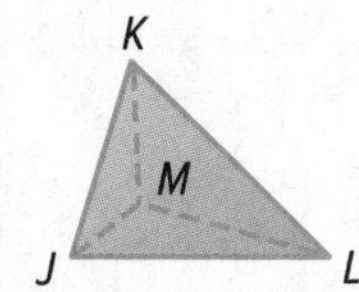

figure name: ____________

base: ____________

faces: ____________

edges: ____________

vertices: ____________

skew lines : ____________

Common Error

In the drawing of a rectangular prism, the bases do not have to be on the top and bottom. Any two parallel rectangles are bases. In a triangular pyramid, any face is a base.

Identify Cross Sections

The intersection of a solid and a plane is called a **cross section** of the solid.

Example

3. **Describe the shape resulting from a vertical, angled, and horizontal cross section of a square pyramid.**

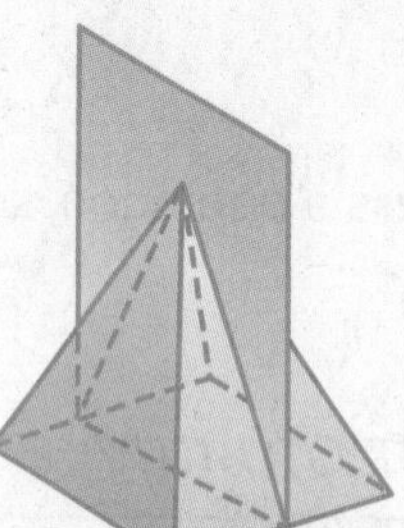

Vertical Slice

The cross section is a triangle.

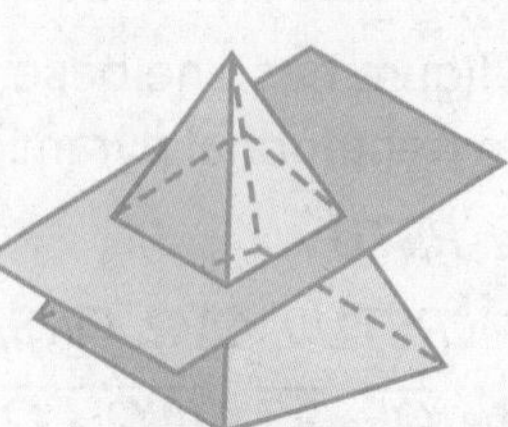

Angled Slice

The cross section is a trapezoid.

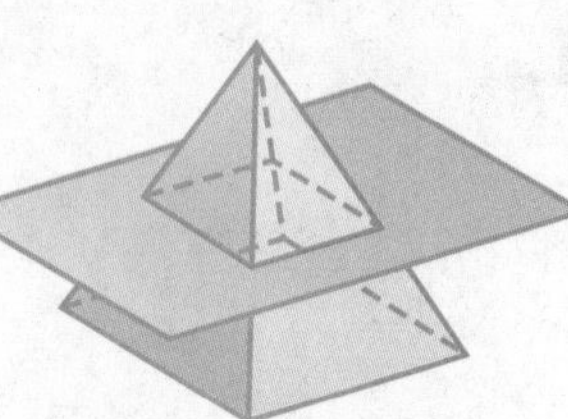

Horizontal Slice

The cross section is a square.

Got it? **Do this problem to find out.**

b. ______

b. Describe the shape resulting from a vertical, angled, and horizontal cross section of a cylinder.

Guided Practice

1. Identify the figure. Then name the bases, faces, edges, and vertices. Then, identify a pair of skew lines. (Examples 1–2)

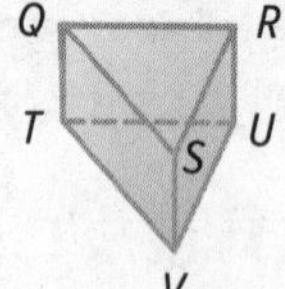

figure name: ______

bases: ______

faces: ______

edges: ______

vertices: ______

skew lines: ______

2. Describe the shape resulting from the cross section shown. (Example 3) ______

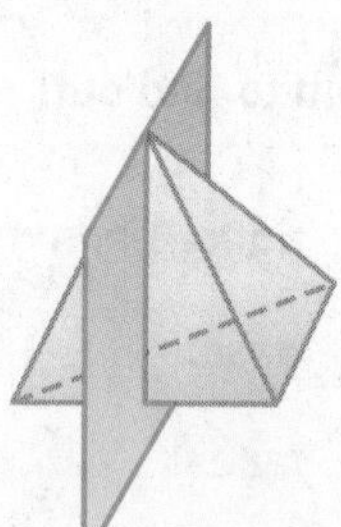

3. **Building on the Essential Question** How can knowing the shape of the base of a three-dimensional figure help you name the figure?

______

______

______

Rate Yourself!

Are you ready to move on? Shade the section that applies.

YES ? NO

For more help, go online to access a Personal Tutor.

Name ______________________ My Homework ______________________

Independent Practice

Go online for Step-by-Step Solutions 

Identify each figure. Name the bases, faces, edges, and vertices. Then, identify a pair of skew lines. (Examples 1–2)

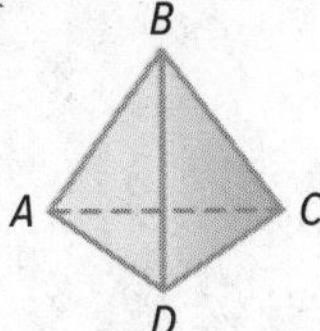

figure name: ______________________

bases: ______________________

faces: ______________________

edges: ______________________

vertices: ______________________

skew lines: ______________________

2.

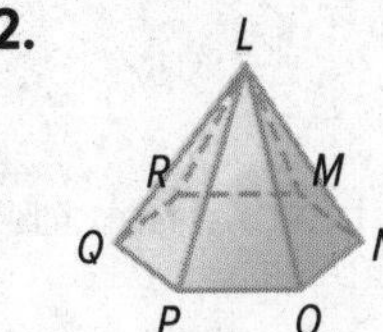

figure name: ______________________

bases: ______________________

faces: ______________________

edges: ______________________

vertices: ______________________

skew lines ______________________

Describe the shape resulting from each cross section. (Example 3)

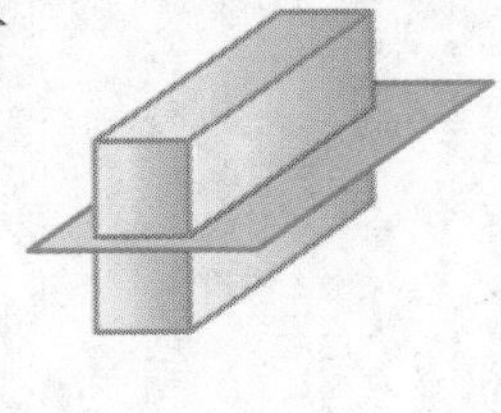

____

4.

____

5.

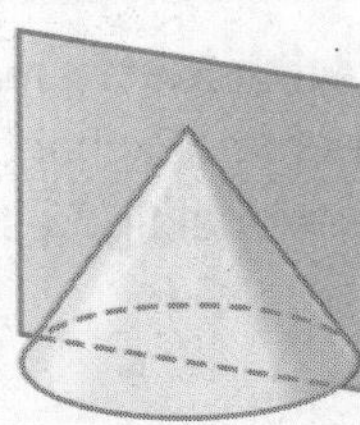

____

6. A basketball is shaped like a *sphere*.

a. Draw a basketball with a vertical, angled, and horizontal slice.

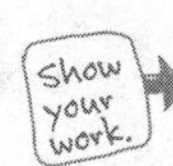

b. Describe the cross section made by each slice.

____

c. Is the basketball a polyhedron? Explain.

____

7. MP **Use a Counterexample** State whether the following conjecture is *true* or *false*. If *false*, provide a counterexample.

Two planes in three-dimensional space can intersect at one point.

8. Draw and label a hexagonal prism. Then identify each of the following.

a. parallel planes

b. skew lines

c. intersecting planes

Show your work.

H.O.T. Problems Higher Order Thinking

9. MP **Model with Mathematics** Draw the cross sections of a polyhedron, cylinder, or cone. Exchange papers with another student. Identify the three-dimensional figures represented by the cross sections.

Show your work.

MP **Persevere with Problems Determine whether each statement is *always*, *sometimes*, or *never* true. Explain your reasoning.**

10. A pyramid has parallel faces.

11. A prism has 2 bases and 4 faces.

12. A parallelogram cannot be a cross section of a triangular prism.

13. A pyramid has a rectangular base.

Extra Practice

Identify each figure. Then name the bases, faces, edges, and vertices. Then identify a pair of skew lines.

14. 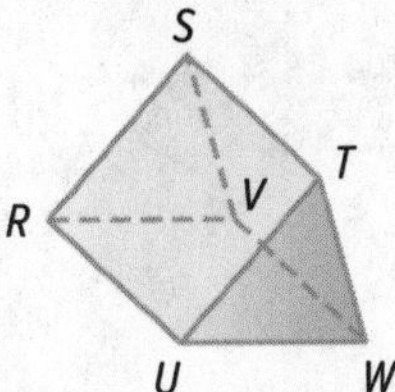

figure name: triangular prism

Homework Help

bases: RSV and UTW

faces: RSV, UTW, RSTU, SVWT, VRUW

edges: $\overline{RS}$, $\overline{SV}$, $\overline{RV}$, $\overline{UT}$, $\overline{TW}$, $\overline{UW}$, $\overline{RU}$, $\overline{VW}$, $\overline{ST}$

vertices: R, S, T, U, V, W

skew lines: Sample answer: $\overline{TU}$ and $\overline{WV}$

15. 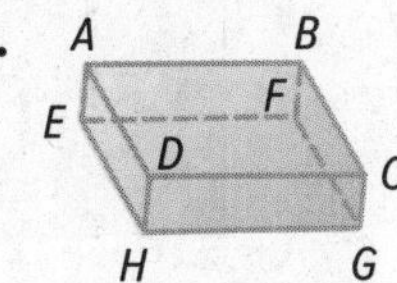

figure name: ______

bases: ______

faces: ______

edges: ______

vertices: ______

skew lines: ______

Describe the shape resulting from each cross section. (Example 4)

16. 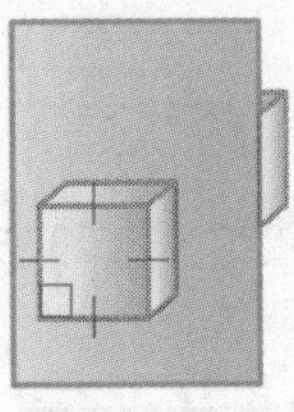

______

17. 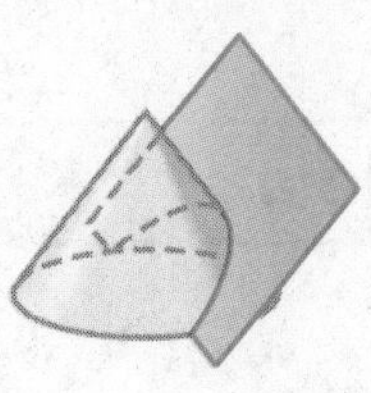

______

18. 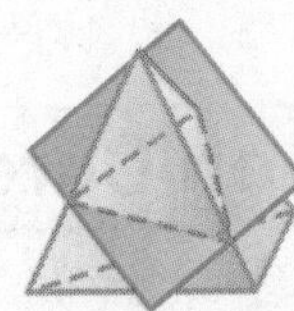

______

19. MP **Find the Error** Hannah is identifying the figure at the right. Find her mistake and correct it.

______

CCSS Power Up! Common Core Test Practice

20. The figure shown is a square pyramid. Which of the following are cross sections of the pyramid? Select all that apply.

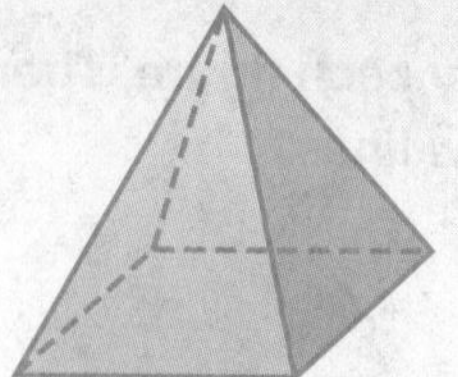

☐ 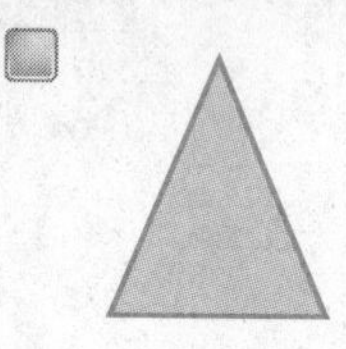

☐ 

☐ 

☐ 

21. Match each number of faces, edges, and vertices to the correct solid figure.

Figure 1
Figure 2
Figure 3

Figure 1

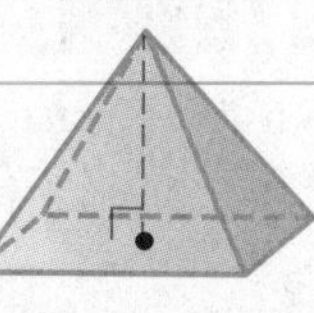

Figure 2

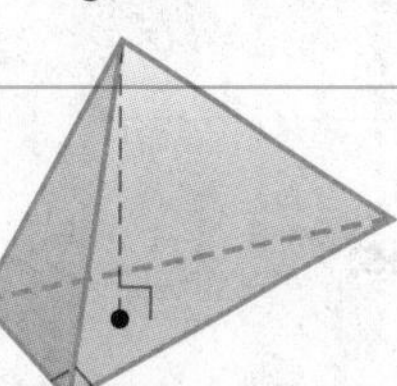

Figure 3

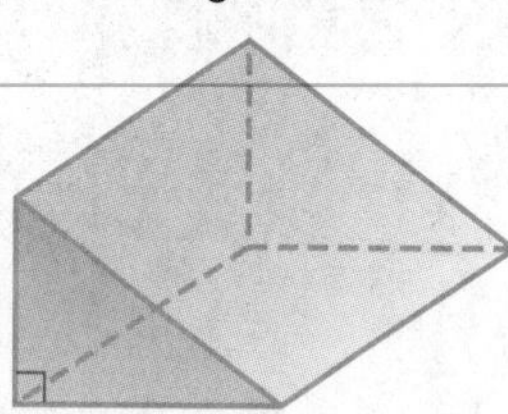

a. 4 faces, 6 edges, 4 vertices ______

b. 5 faces, 8 edges, 5 vertices ______

c. 5 faces, 9 edges, 6 vertices ______

CCSS Common Core Spiral Review

Name each polygon. 5.G.3

22. 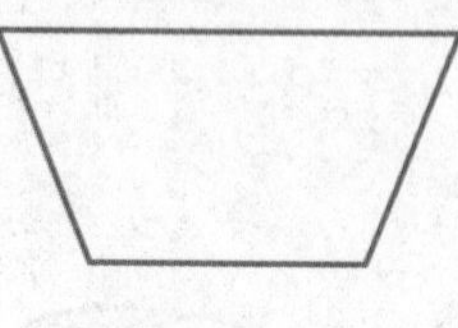

______

23. 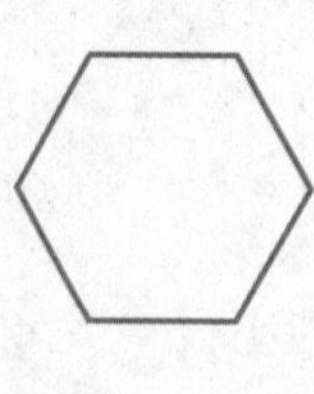

______

24. 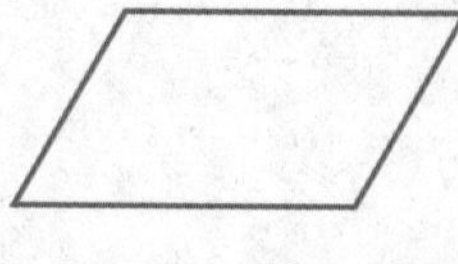

______

25. Find the measure of the missing angle of the polygon. 5.G.3 ______

90° 90°

$x°$

 Need more practice? Download more Extra Practice at **connectED.mcgraw-hill.com.**

21ST CENTURY CAREER in Design Engineering

Roller Coaster Designer

If you have a passion for amusement parks, a great imagination, and enjoy building things, you might want to consider a career in roller coaster design. Roller coaster designers combine creativity, engineering, mathematics, and physics to develop rides that are both exciting and safe. In order to analyze data and make precise calculations, a roller coaster designer must have a solid background in high school math and science.

Is This the Career for You?

Are you interested in a career as a roller coaster designer? Take some of the following courses in high school to get started in the right direction.

- Algebra
- Calculus
- Geometry
- Physics
- Trigonometry

Turn the page to find out how math relates to a career in Design Engineering.

A Thrilling Ride

Use the information in the table to solve each problem.

1. In a scale drawing of SheiKra, a designer uses a scale of 1 inch = 16 feet. What is the height of the roller coaster in the drawing? ______

2. On a model of Montu, the height of the loop is 13 inches. What is the scale? ______

3. In a scale drawing of Montu, the height of the roller coaster is 10 inches. What is the scale factor? ______

4. SheiKra has a hill that goes through a tunnel. On a model of the roller coaster, the hill is 23 inches tall and the scale is 1 inch = 6 feet. What is the actual height of the tunnel hill? ______

5. An engineer is building a model of SheiKra. She wants the model to be about 32 inches high. Choose an appropriate scale for the model. Then use it to find the loop height of the model. ______

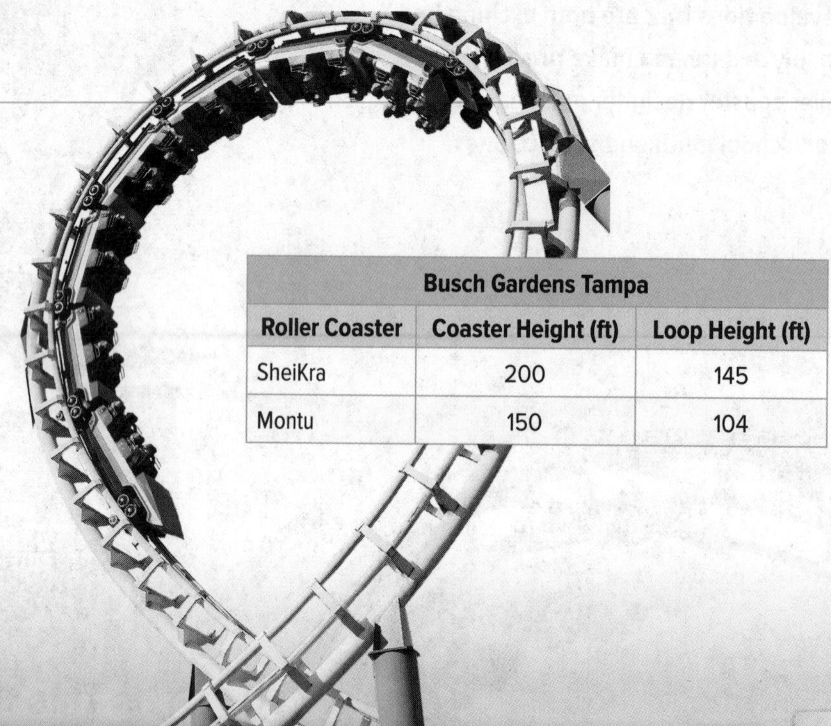

Busch Gardens Tampa		
Roller Coaster	**Coaster Height (ft)**	**Loop Height (ft)**
SheiKra	200	145
Montu	150	104

Career Project

It's time to update your career portfolio! Describe a roller coaster that you, as a roller coaster designer, would create. Include the height and angle of the tallest drop, the total length, maximum speed, number of loops and tunnels, and color scheme. Be sure to include the name of your roller coaster.

______

______

______

______

What problem-solving skills might you use as a roller coaster designer?

- ______
- ______
- ______
- ______
- ______

Chapter Review

Vocabulary Check 

Complete the crossword puzzle using the vocabulary list at the beginning of the chapter.

Across

2. a three-dimensional figure with two parallel, congruent bases that are polygons
4. a triangle with an angle greater than 90 degrees and less than 180 degrees
8. a three-dimensional figure with two parallel congruent circular bases connected by a curved surface
11. a triangle with three congruent sides
12. segments with the same length
14. used to represent an object that is too large to be built at actual size (2 words)
15. two angles with a sum of 90 degrees

Down

1. angles that share a common vertex, a common side, and do not overlap
3. two angles with a sum of 180 degrees
5. a figure with three sides and three angles
6. the ratio that compares the measurements of a model and the real object
7. opposite angles that are formed by the intersection of two lines
8. a three-dimensional figure with one circular base connected by a curved side to a single vertex
9. an angle less than 90 degrees
10. where two rays meet to form an angle
13. a 90 degree angle

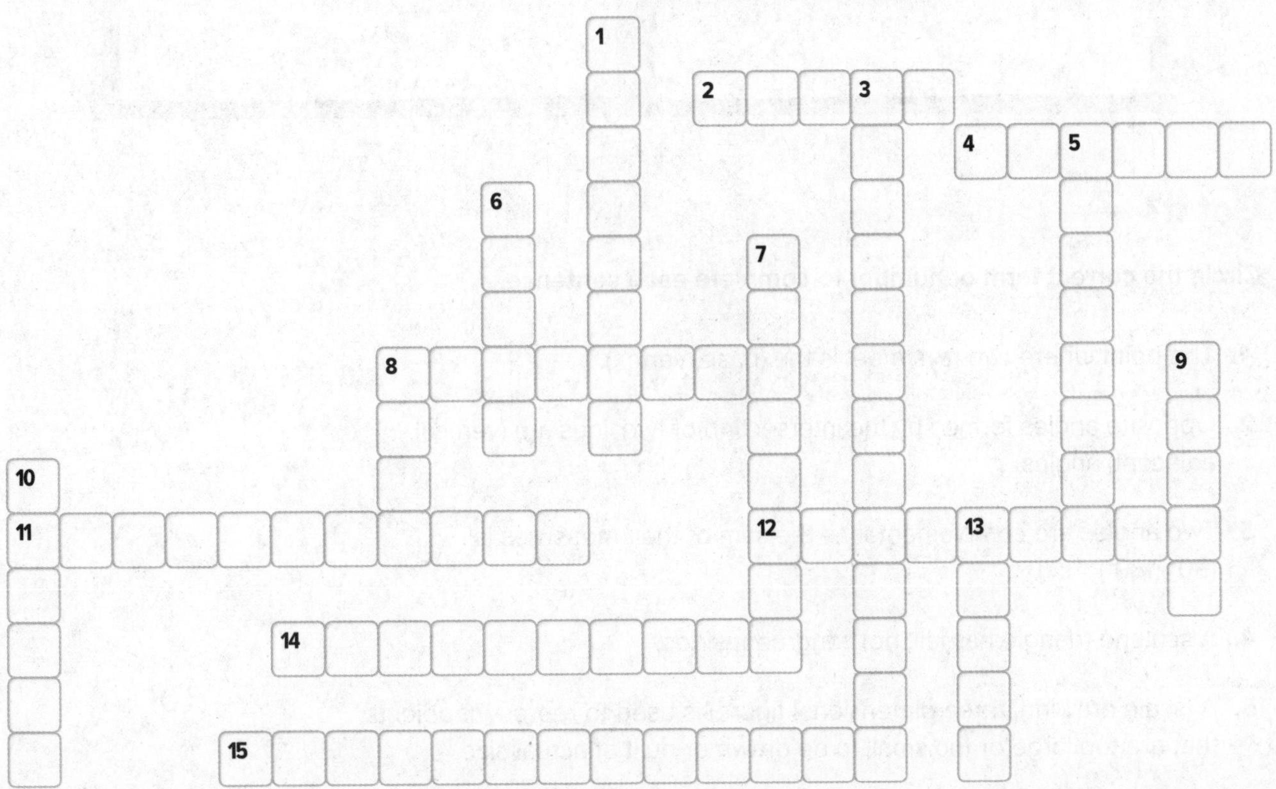

Key Concept Check

Use Your FOLDABLES

Use your Foldable to help review the chapter.

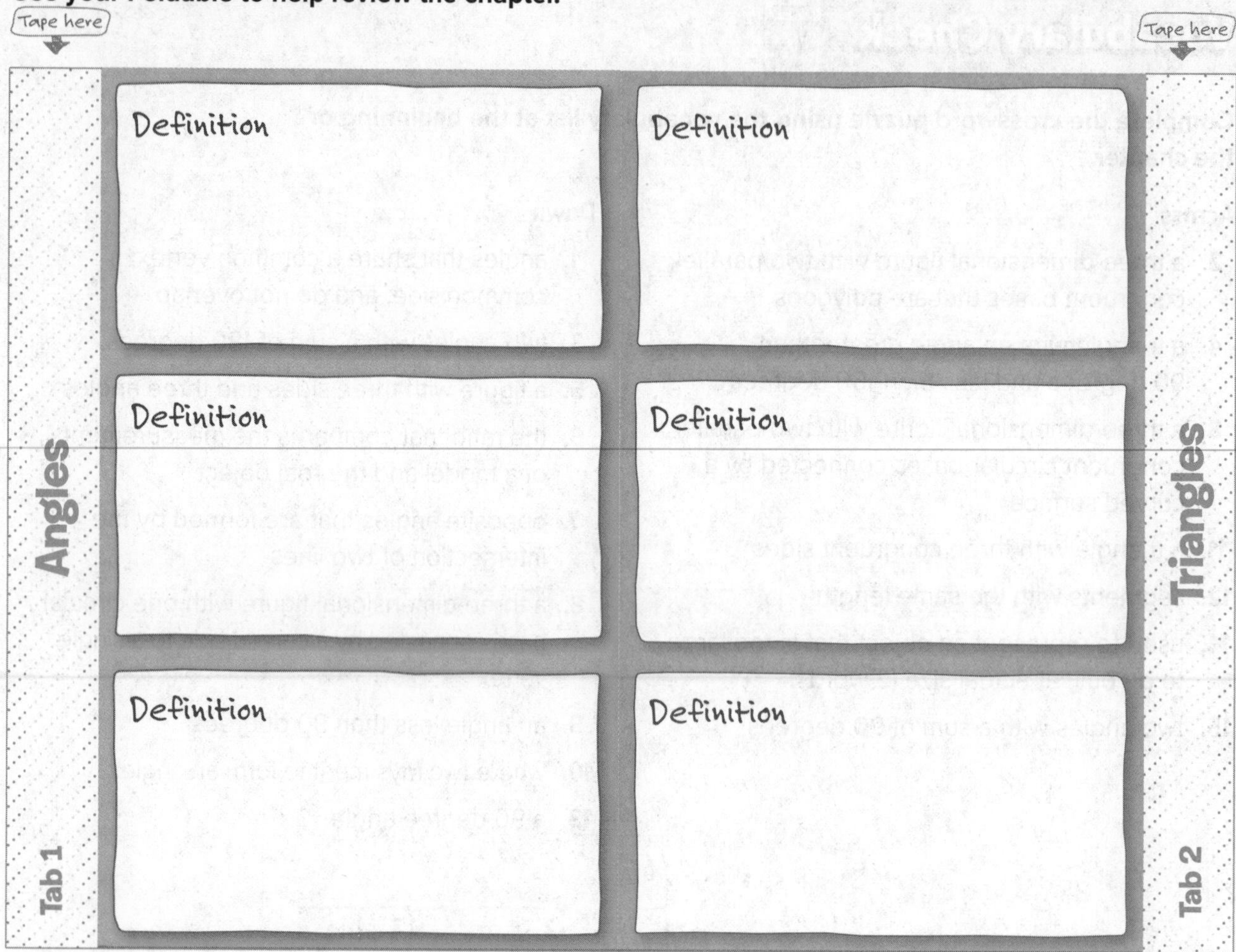

Got it?

Circle the correct term or number to complete each sentence.

1. The point where two rays meet is the (base, vertex).
2. Opposite angles formed by the intersection of two lines are (vertical, adjacent) angles.
3. Two angles are complementary if the sum of their measures is (90°, 180°).
4. A scalene triangle has (all, no) congruent sides.
5. A (scale drawing, three-dimensional figure) is used to represent objects that are too large or too small to be drawn or built at actual size.

CCSS Power Up! Performance Task

Stacking Triangles

Thompson has drawn two different triangles.

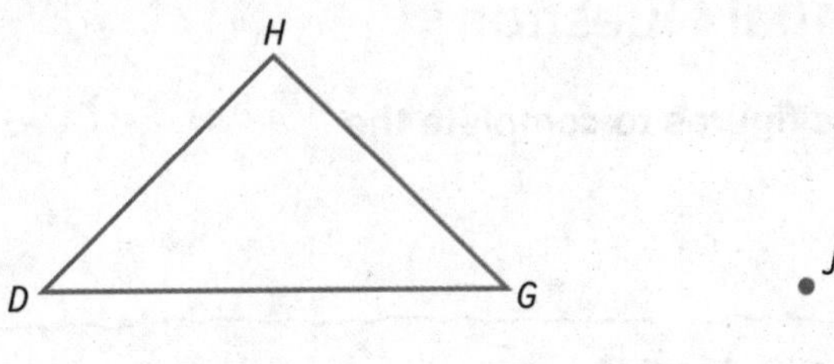

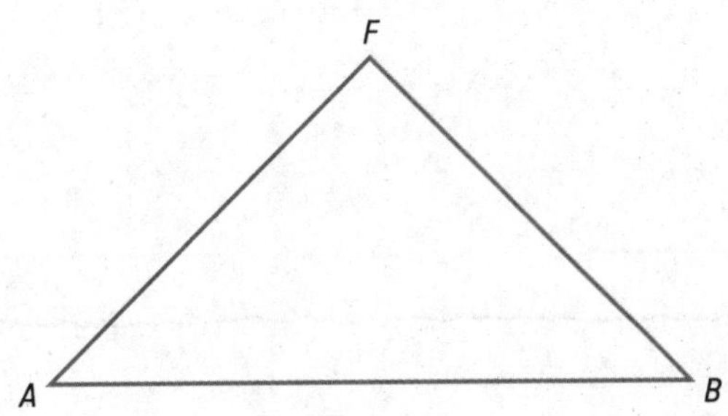

Write your answers on another piece of paper. Show all of your work to receive full credit.

Part A

Triangle *DHG* is a scale drawing of triangle *AFB*. Use a ruler. Measure and label the side lengths of triangle *AFB* and triangle *DHG*. What scale factor was used to make triangle *DHG*?

Part B

Use a scale factor of 2 to make a new scale drawing of triangle *DHG*. Label the new triangle *XYZ*.

Part C

Extend line *DG* though point *J* and extend line *HG* through point *K*. Find the measure of angle *HGJ* and angle *JGK*. Name a pair of complementary or supplementary angles. Justify your response.

Reflect

Answering the Essential Question

Use what you learned about geometric figures to complete the graphic organizer.

How do polygons help us describe real-world objects?

Essential Question

HOW does geometry help us describe real-world objects?

How do polyhedrons help us describe real-world objects?

Answer the Essential Question. HOW does geometry help us describe real-world objects?

Chapter 8
Measure Figures

Essential Question

How do measurements help you describe real-world objects?

Common Core State Standards

Content Standards
7.G.4, 7.G.6

MP Mathematical Practices
1, 2, 3, 4, 5, 6, 8

Math in the Real World

Soccer is a sport that is played on a rectangular field. The dimensions of a regulation size soccer field are 100 yards long and 60 yards wide.

What is the area of the soccer field shown?

$A =$ ______ square yards

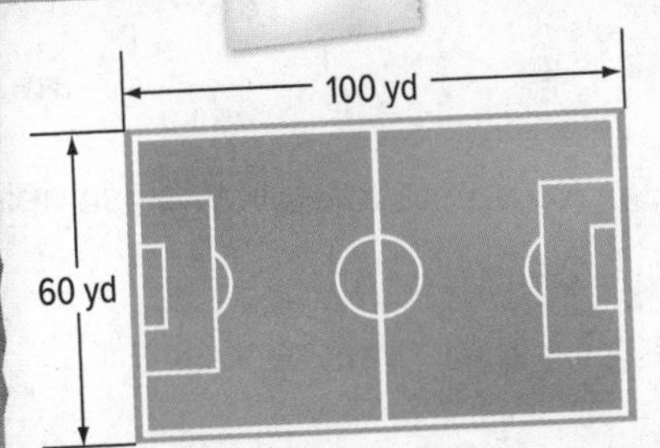

1. Cut out the Foldable on page FL9 of this book.
2. Place your Foldable on page 700.
3. Use the Foldable throughout this chapter to help you learn about measuring figures.

What Tools Do You Need?

Vocabulary

center	lateral face	semicircle
circle	lateral surface area	slant height
circumference	pi	surface area
composite figure	radius	volume
diameter	regular pyramid	

Study Skill: Studying Math

Power Notes *Power notes* are similar to lesson outlines, but they are simpler to organize. Power notes use the numbers 1, 2, 3, and so on. You can have more than one detail under each power. You can even add drawings or examples to your power notes.

Power 1: This is the main idea.
Power 2: This provides details about the main idea.
Power 3: This provides details about Power 2.
and so on...

Complete the following sample of power notes for this chapter.

1: Circles

2: Circumference

3: ______________________________

3: ______________________________

2: Area

3: ______________________________

What Do You Already Know?

Place a checkmark below the face that expresses how much you know about each concept. Then scan the chapter to find a definition or example of it.

☹ I have no clue. 😐 I've heard of it. ☺ I know it!

Integers				
Concept	☹	😐	☺	Definition or Example
area of a circle				
area of composite figures				
lateral surface area				
pi (π)				
total surface area				
volume				

When Will You Use This?

Here are a few examples of how volume and surface area are used in the real world.

Activity 1 When wrapping a present, how do you determine how much paper to use? Describe a method that you could use to make sure that you cut the right size piece of wrapping paper.

Activity 2 Go online at **connectED.mcgraw-hill.com** to read the graphic novel ***The Dunk Tank***. How many square feet of metal sheeting do they have to build the dunk tank?

Are You Ready?

Try the Quick Check below.
Or, take the Online Readiness Quiz.

Quick Review

Common Core Review 6.G.1

Example 1

Find the area of the rectangle.

10 m

4 m

$A = \ell w$ Area of a rectangle

$A = (10)(4)$ Replace ℓ with 10 and w with 4.

$A = 40$ Simplify.

The area of the rectangle is 40 square meters.

Example 2

Find the area of the triangle.

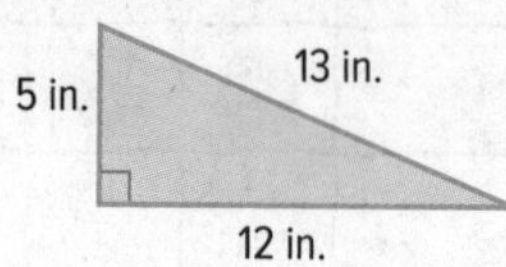

$A = \frac{1}{2}bh$ Area of a triangle

$A = \frac{1}{2}(12)(5)$ Replace b with 12 and h with 5.

$A = \frac{1}{2}(60)$ Multiply.

$A = 30$ Simplify.

The area of the triangle is 30 square inches.

Quick Check

Area Find the area of each figure.

1. 14 m

3 m

$A =$ ______

2. 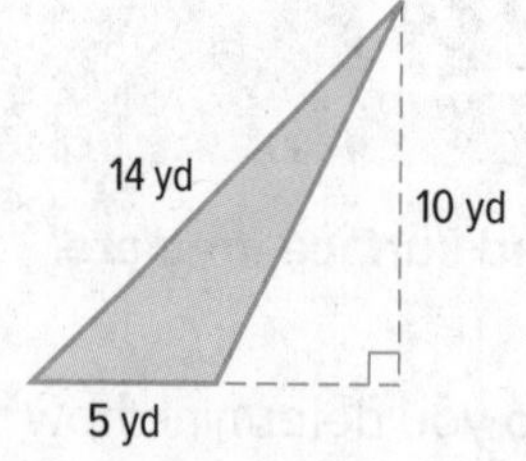

$A =$ ______

3. 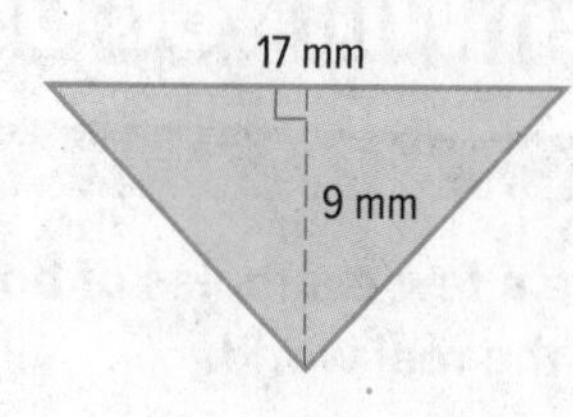

$A =$ ______

4. Anita's yard is in the shape of a triangle. It has a height of 35 feet and a base of 50 feet. What is the area of the yard?

______

How Did You Do?

Which problems did you answer correctly in the Quick Check? Shade those exercise numbers below.

① ② ③ ④

Inquiry Lab

Circumference

HOW is the circumference of a circle related to its diameter?

Content Standards
7.G.4

Mathematical Practices
1, 3, 6

The distance around a flying disc, or its *circumference*, is 37.7 centimeters. The distance across the disc through its center, or its *diameter*, is 12 centimeters. How is the circumference of a circular object, such as a flying disc, related to its diameter?

Hands-On Activity

Step 1 Cut a piece of string the length of the circumference of a circular object such as a jar lid. Use a centimeter ruler to measure the length of the string to the nearest tenth of a centimeter. Record this measurement in the table below.

Object	Circumference (C)	Diameter (d)	$\frac{C}{d}$
Disc	37.7 cm	12 cm	

Step 2 Measure and record the diameter of the lid.

Step 3 Use a calculator to find the ratio of the circumference of the flying disc to its diameter. Repeat for the circular object you measured in Steps 1 and 2. Round answers to the nearest hundredth.

Step 4 Repeat Steps 1 through 3 for other circular objects.

Describe the ratios $\frac{C}{d}$ you found. Identify the number that is closest to the value of each ratio. ______________________

____

Write a rule in the form $\frac{C}{d} \approx$ ■, where ■ is the number you identified in the question above. ______________________

Investigate

MP **Make a Prediction** Work with a partner. Measure the diameter of two different circular objects. Predict each circumference. Check your predictions by measuring. Then find each ratio $\frac{C}{d}$. Record your values to the nearest hundredth in the table below.

	Object	Diameter (d)	Predicted Circumference	Measured Circumference (C)	Ratio $\frac{C}{d}$
1.					
2.					

Analyze and Reflect

3. MP **Reason Inductively** How do the ratios $\frac{C}{d}$ in the table compare to the ones in the Activity? Identify the number that is closest to the value of all of the ratios.

Create

4. MP **Reason Abstractly** Write a formula in the form $\frac{C}{d} = ■$, that gives the approximate ratio of the circumference C of a circle to its diameter d, where ■ is the number you identified in Exercise 3.

5. MP **Reason Abstractly** Multiply both sides of your formula by d to write an equivalent formula in the form $C = ■ \times d$ *that gives the approximate circumference* C if you know the diameter d of a circle.

6. MP **Reason Abstractly** The radius of a circle is one half of its diameter. Write a formula that relates the circumference C of a circle to its radius r.

7. Inquiry HOW is the circumference of a circle related to its diameter?

Lesson 1
Circumference

Vocabulary Start-Up

A **circle** is the set of all points in a plane that are the same distance from a point, called the **center**. The **circumference** is the distance around a circle. The **diameter** is the distance across a circle through its center. The **radius** is the distance from the center to any point on the circle.

Fill in each box with one of the following terms: *center*, *diameter*, and *radius*.

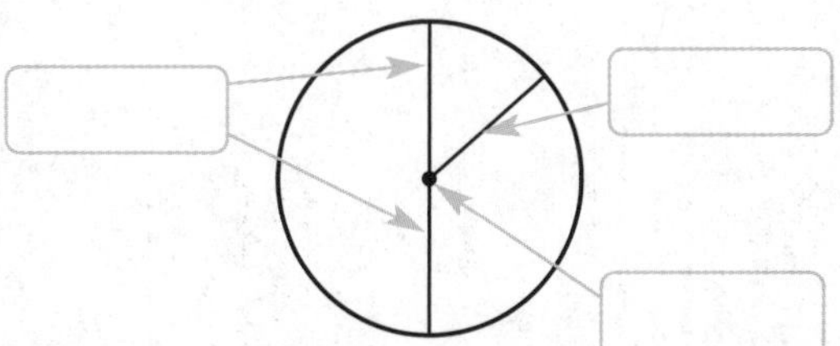

Real-World Link

1. The table shows the approximate measurements of two sizes of hula hoops.

Size	Radius (in.)	Diameter (in.)	Circumference (in.)
child	14	28	88
adult	20	40	126

a. Describe the relationship between the diameter and radius of each hula hoop. __________

b. Describe the relationship between the circumference and diameter of each hula hoop. __________

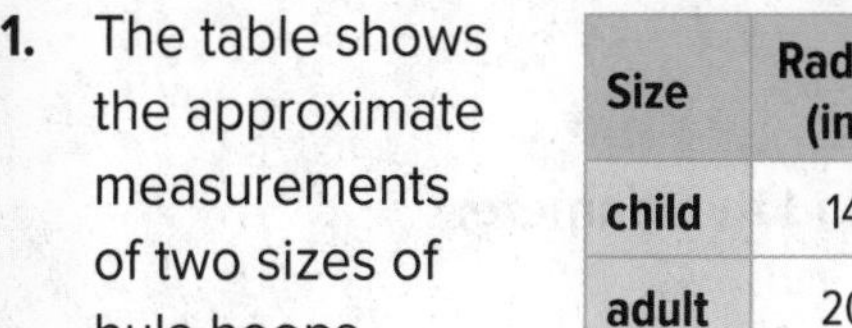

Which MP **Mathematical Practices** did you use? Shade the circle(s) that applies.

1. Persevere with Problems
2. Reason Abstractly
3. Construct an Argument
4. Model with Mathematics
5. Use Math Tools
6. Attend to Precision
7. Make Use of Structure
8. Use Repeated Reasoning

Essential Question

HOW do measurements help you describe real-world objects?

Vocabulary

circle
center
circumference
diameter
radius
pi π

Common Core State Standards

Content Standards
7.G.4

MP **Mathematical Practices**
1, 3, 4, 6, 8

Key Concept

Radius and Diameter

Words The diameter d of a circle is twice its radius r. The radius r of a circle is half of its diameter d.

Symbols $d = 2r$ $r = \frac{d}{2}$

Work Zone

Examples

Tutor

1. The diameter of a circle is 14 inches. Find the radius.

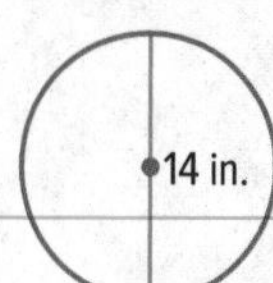

$r = \frac{d}{2}$ Radius of circle

$r = \frac{14}{2}$ Replace d with 14.

$r = 7$ Divide.

The radius is 7 inches.

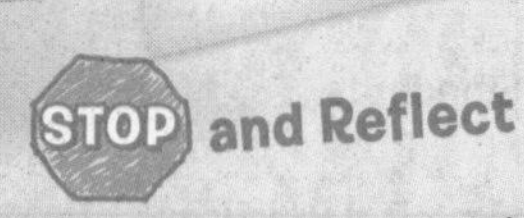

STOP and Reflect

The diameter of a circle is 36 inches. Circle the radius.

72 in. 18 in.

2. The radius of a circle is 8 feet. Find the diameter.

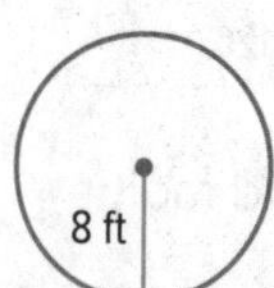

$d = 2r$ Diameter of circle

$d = 2 \cdot 8$ Replace r with 8.

$d = 16$ Multiply.

The diameter is 16 feet.

Show your work.

a. ___

b. ___

c. ___

d. ___

Got it? Do these problems to find out.

Find the radius or diameter of each circle with the given dimension.

a. $d = 23$ cm **b.** $r = 3$ in.

c. $d = 16$ yd **d.** $r = 5.2$

Circumference

Key Concept

Words The circumference of a circle is equal to π times its diameter or π times twice its radius.

Model

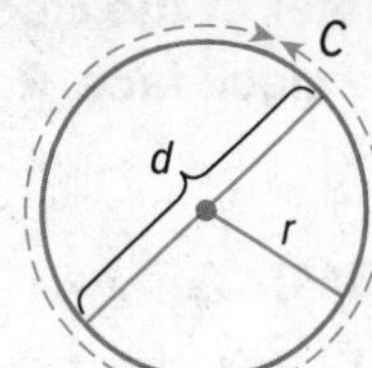

Symbols $C = \pi d$ or $C = 2\pi r$

In the Inquiry Lab, you learned that $\frac{C}{d} \approx 3$. The exact ratio is represented by the Greek letter **π (pi)**. The value of π is 3.1415926... . The decimal never ends, but it is often approximated as 3.14.

Another approximation for π is $\frac{22}{7}$. Use this value when the radius or diameter is a multiple of 7 or has a multiple of 7 in its numerator if the radius is a fraction.

Estimation

To estimate the circumference of a circle, you can use 3 for π since $\pi \approx 3$.

Example

3. Find the circumference of a circle with a radius of 21 inches.

Since 21 is a multiple of 7, use $\frac{22}{7}$ for π.

$C = 2\pi r$ Circumference of a circle

$C \approx 2 \cdot \frac{22}{7} \cdot 21$ Replace π with $\frac{22}{7}$ and r with 21.

$C \approx 2 \cdot \frac{22}{\cancel{7}_1} \cdot \frac{\cancel{21}^3}{1}$ Divide by the GCF, 7.

$C \approx 132$ Simplify.

The circumference of the circle is about 132 inches.

Got it? **Do these problems to find out.**

Find the circumference of each circle. Use $\frac{22}{7}$ for π.

e.

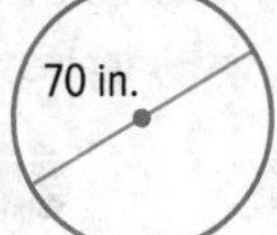

f.

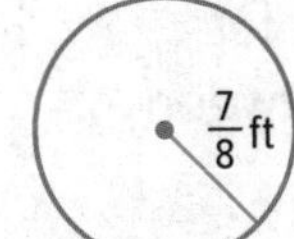

e. ______

f. ______

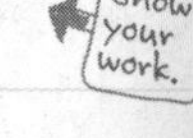

Example

4. Big Ben is a famous clock tower in London, England. The diameter of the clock face is 23 feet. Find the circumference of the clock face. Round to the nearest tenth.

$C = \pi d$	Circumference of a circle
$C \approx$ **3.14(23)**	Replace π with 3.14 and d with 23.
$C \approx 72.2$	Multiply.

So, the distance around the clock is about 72.2 feet.

Got it? Do this problem to find out.

Show your work.

g. A circular fence is being placed to surround a tree. The diameter of the fence is 4 feet. How much fencing is used? Use 3.14 for π. Round to the nearest tenth if necessary.

g. ________

Guided Practice

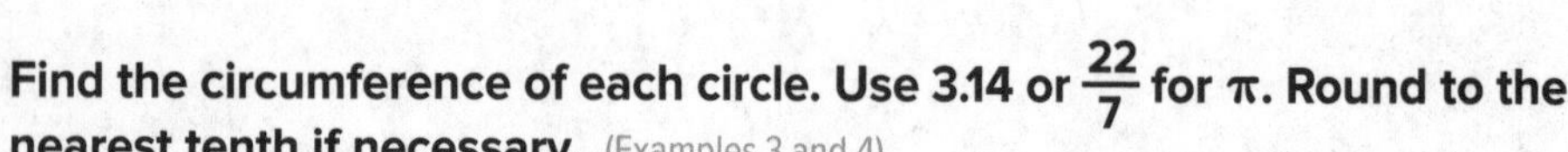

Find the radius or diameter of each circle with the given dimension. (Examples 1 and 2)

1. $d = 3$ m ______
2. $r = 14$ ft ______
3. $d = 20$ in. ______

Show your work.

Find the circumference of each circle. Use 3.14 or $\frac{22}{7}$ for π. Round to the nearest tenth if necessary. (Examples 3 and 4)

4. ______

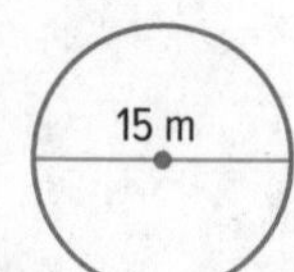

5. ______

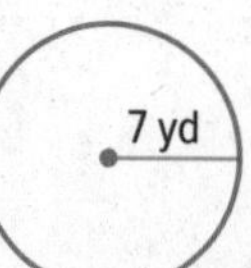

6. **Building on the Essential Question** A circle has a circumference of about 16.3 meters and a diameter of about 5.2 meters. What is the relationship between the circumference and diameter of this circle?

Rate Yourself!

How confident are you about finding the circumference? Check the box that applies.

For more help, go online to access a Personal Tutor.

Name ______________________ My Homework ______________________

Independent Practice

Go online for Step-by-Step Solutions

Find the radius or diameter of each circle with the given dimensions. (Examples 1 and 2)

1. $d = 5$ mm ________

2. $d = 24$ ft ________

3. $r = 17$ cm ________

Find the circumference of each circle. Use 3.14 or $\frac{22}{7}$ for π. Round to the nearest tenth if necessary. (Example 3)

4.

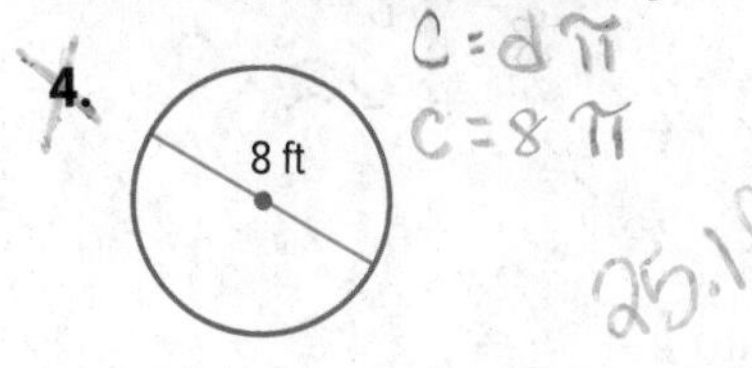

____

5.

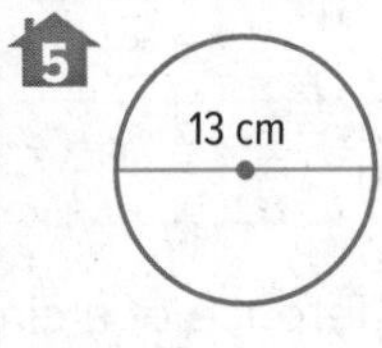

____

6.

3.5 mi

____

7. The largest tree in the world by volume is in Sequoia National Park. The diameter at the base is 36 feet. If a person with outstretched arms can reach 6 feet, how many people would it take to reach around the base of the tree? (Example 4)

____

8. The Belknap shield volcano is located in the Cascade Range in Oregon. The volcano is circular and has a diameter of 5 miles. What is the circumference of this volcano. Round your answer to the nearest tenth? (Example 4)

____

9. MP **Be Precise** Refer to the circle at the right.

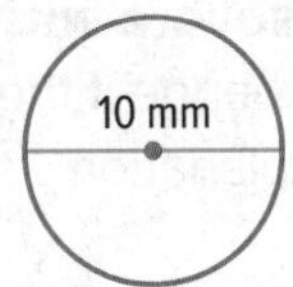

a. Find the circumference of the circle. Use 3 as the estimate of π.

____

b. Find the circumference of the circle using 3.14 for π.

____

c. Another estimate of π is 3.14159. Find the circumference using this estimate.

____

d. What do you notice about the estimate used for π and the circumference of the circle?

____

____

Copy and Solve **For Exercises 10–14, show your work on a separate piece of paper.**

Find the diameter given each circumference. Use 3.14 for π.

10. a satellite dish with a circumference of 957.7 meters

11. a basketball hoop with a circumference of 56.52 inches

12. a nickel with a circumference of about 65.94 millimeters

Find the distance around each figure. Use 3.14 for π.

13.

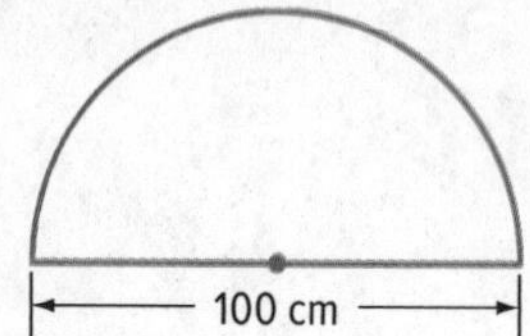

14.

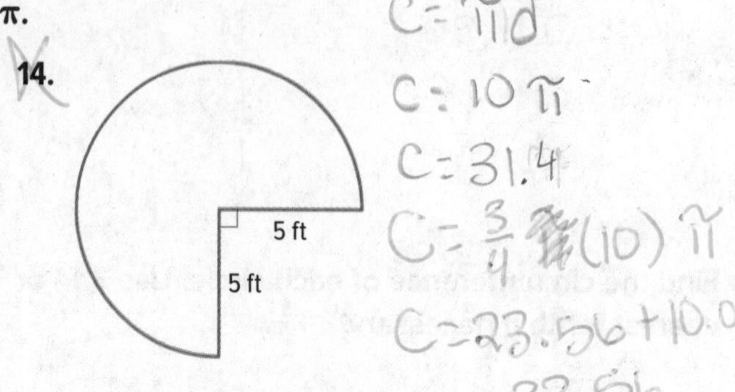

H.O.T. Problems Higher Order Thinking

15. **MP Justify Conclusions** Determine if the circumference of a circle with a radius of 4 feet will be greater or less than 24 feet. Explain.

It will be greater then 25 feet because when you slove in terms of π you get 8π and that equals 25ft when rounded.

C = 2πr
C = 2(4)π
C = 8π
C = 25ft

16. **MP Model with Mathematics** Draw and label a circle that has a diameter more than 5 inches, but less than 10 inches. Estimate its circumference and then find its circumference using a calculator. Compare your results.

17. **MP Persevere with Problems** Analyze how the circumference of a circle would change if the diameter was doubled. Provide an example to support your explanation.

18. **MP Justify Conclusions** Determine whether the relationship between the circumference of a circle and its diameter is a direct variation. If so, identify the constant of proportionality. Justify your response.

Name ______________________ My Homework ______________________

Extra Practice

Find the radius or diameter of each circle with the given dimensions.

19. $d = 7$ in. 3.5 in.

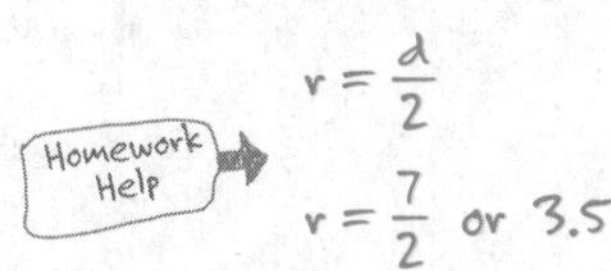

20. $d = 30$ m ______

21. $r = 36$ ft ______

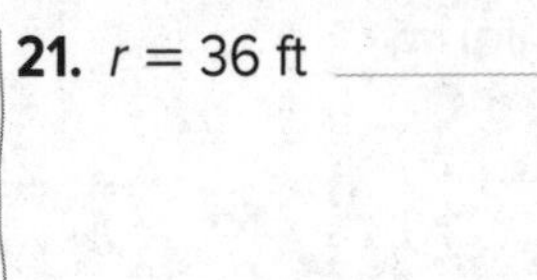

Find the circumference of each circle. Use 3.14 or $\frac{22}{7}$ for π.

22.

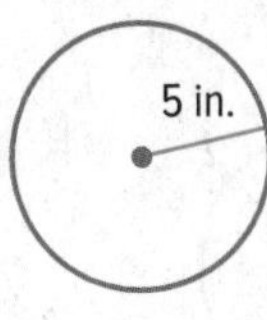

______

23.

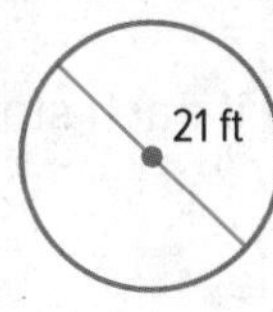

______

24.

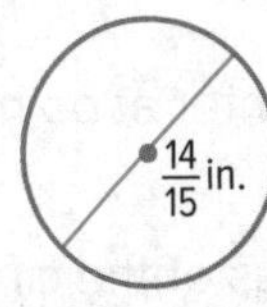

______

25. a button with a radius of 21 millimeters

______

26. a dunk tank with a radius of 36 inches

______

27. The diameter of a music CD is 12 centimeters. Find the circumference of a CD to the nearest tenth. ______

28. At a local park, Sara can choose between two circular paths to walk. One path has a diameter of 120 yards, and the other has a radius of 45 yards. How much farther can Sara walk on the longer path than the shorter path if she walks around the path once? ______

29. MP **Identify Repeated Reasoning** The diagram at the right is made up of circles with the same center. The innermost circle has a diameter of 1 unit. Each circle moving outward has a diameter one more unit than the previous. Without calculating, how much longer is the circumference of each circle? ______

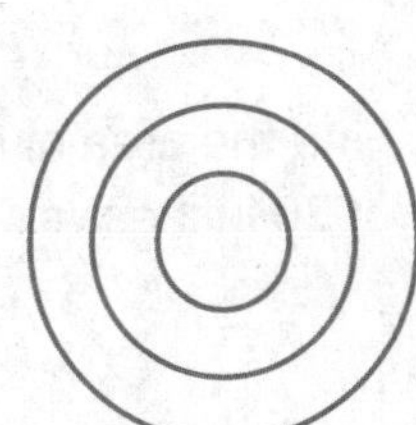

Power Up! Common Core Test Practice

30. A bicycle tire has a radius of 12.5 inches. Select values to complete the equation below to find the circumference of the wheel. Use 3.14 for π.

0.5	4
2	12.5
3.14	25

$C \approx$ ______ $\times$ ______ $\times$ ______

How far does the tire roll in one complete revolution?

______

31. A circle with center at point O is shown. Determine if each statement is true or false.

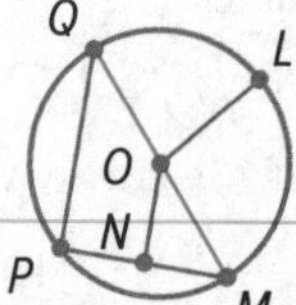

a. $\overline{ON}$ is a radius of the circle. ☐ True ☐ False

b. $\overline{QM}$ is a diameter of the circle. ☐ True ☐ False

c. To find the circumference, multiply the length of $\overline{OL}$ by π. ☐ True ☐ False

Common Core Spiral Review

Find the area of each trapezoid. 6.G.1

32.

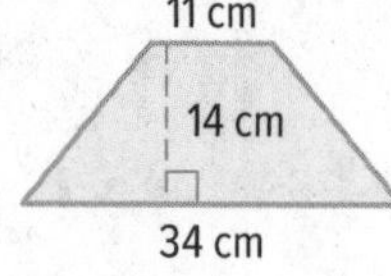

______

33.

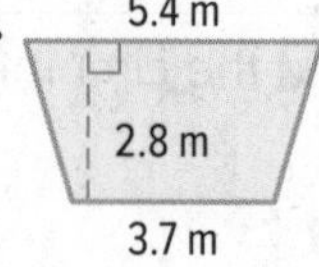

______

34.

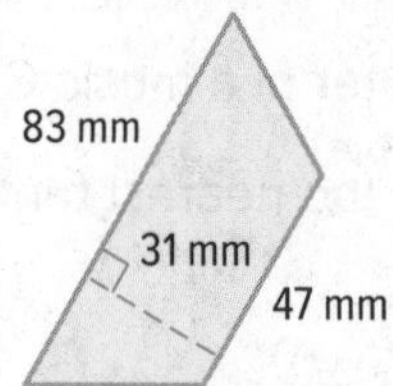

______

35. Find the area of glass used on the side of the parallelogram-shaped building shown. 6.G.1

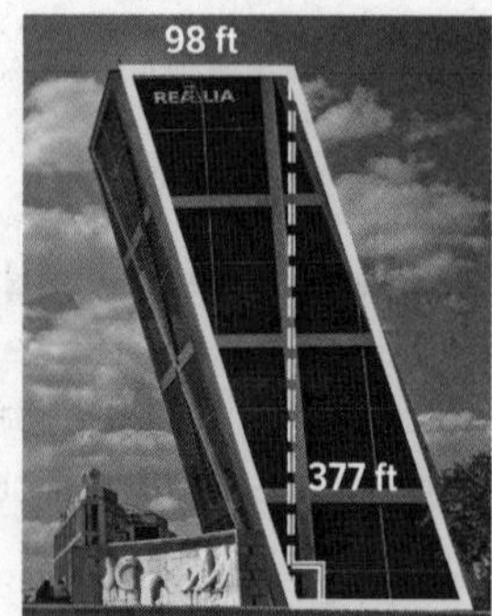

______

36. Find the area of a triangle with a base of 25 inches and a height of 30 inches. 6.G.1

______

Inquiry Lab
Area of Circles

 HOW are the circumference and area of a circle related?

 Content Standards
7.G.4

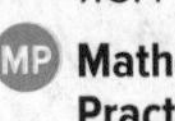 Mathematical Practices
1, 3, 6

Mrs. Allende wants to create a family message center on a wall in her house. There are 4 family members, including Mrs. Allende. She decides to paint 1 circle for each family member using magnetic paint. Each circle will have a 12-inch radius. How do you find the area of a circle?

Hands-On Activity

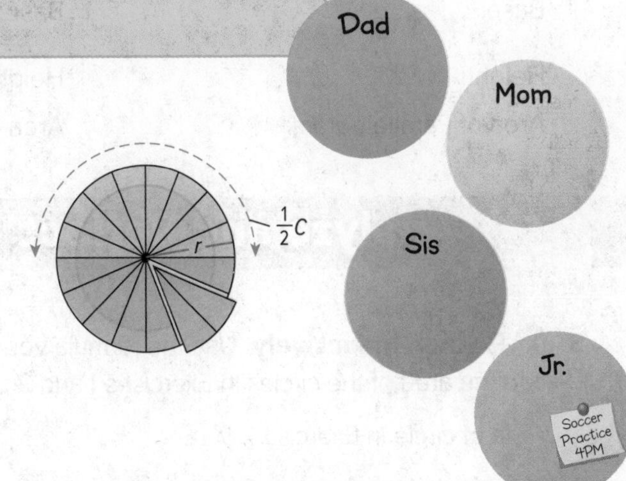

Let's develop a formula for finding the area of a circle.

Step 1 Fold a paper plate in half four times to divide it into 16 equal sections.

Step 2 Label the radius r as shown. Let C represent the circumference of the circle.

Step 3 Cut out each section. Reassemble the sections to form a parallelogram-shaped figure.

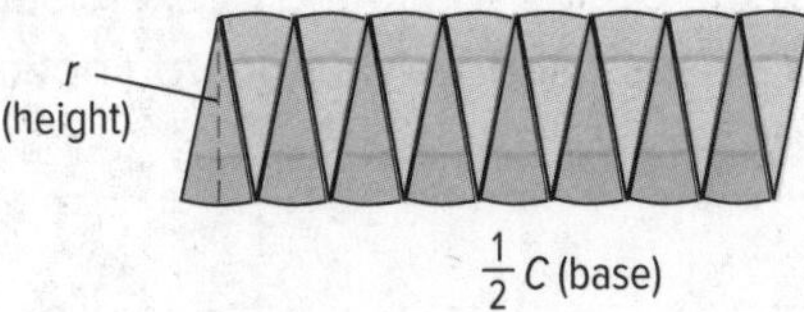

What expressions represent the measurements of the base and the height?

Base: ______________ Height: ______________

Substitute these values into the formula for the area of a parallelogram, $A = b \times h$. Write the new formula. ______________

Replace C with the expression for the circumference of a circle, $2\pi r$. Simplify the equation and describe what it represents.

____

Investigate

Work with a partner. Use the circle to draw and label a parallelogram that would result from cutting and reassembling the circle. Use 3.14 for π.

1.

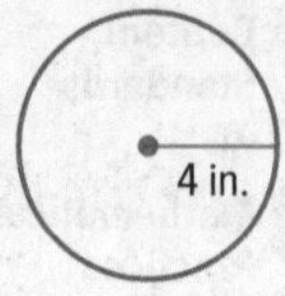

Base: ________

Height: ________

Area of Parallelogram: ________

2.

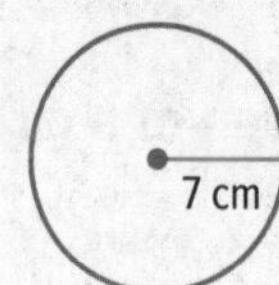

Base: ________

Height: ________

Area of Parallelogram: ________

Analyze and Reflect

3. MP **Reason Inductively** Use the formula you wrote on the previous page to find the area of the circles in Exercises 1 and 2 above. Use 3.14 for π.

Area of circle in Exercise 1: ________

Area of circle in Exercise 2: ________

4. Compare the area of the circles you found in Exercise 3 to the area of the parallelograms in Exercises 1 and 2. What do you notice? Explain.

____

____

Create

5. MP **Model with Mathematics** Find a real-world example of a circle. Measure the radius of the circle. Draw a resulting parallelogram from reassembling the circle. Then calculate the circle's area. ________

6. Inquiry HOW are the circumference and area of a circle related?

____

____

Lesson 2
Area of Circles

Real-World Link

Pets Adrienne bought an 8-foot leash for her dog.

1. Adrienne wants to find the distance the dog runs when it runs one circle with the leash fully extended. Should she calculate the circumference or area? Explain.

2. Suppose she wants to find the amount of running room the dog has with the leash fully extended. Should she calculate the circumference or area? Explain.

3. Describe a real-world situation that would involve finding the area of a circle.

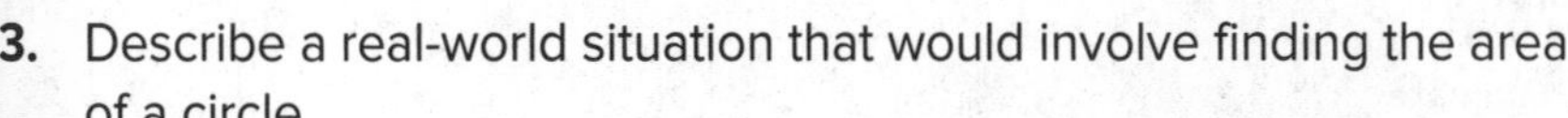

4. Describe a real-world situation that would involve finding the circumference of a circle.

Which MP **Mathematical Practices** did you use? Shade the circle(s) that applies.

① Persevere with Problems	⑤ Use Math Tools
② Reason Abstractly	⑥ Attend to Precision
③ Construct an Argument	⑦ Make Use of Structure
④ Model with Mathematics	⑧ Use Repeated Reasoning

Essential Question

HOW do measurements help you describe real-world objects?

Vocabulary

semicircle

Common Core State Standards

Content Standards
7.G.4

MP Mathematical Practices
1, 3, 4

Key Concept

Find the Area of a Circle

Words The area A of a circle equals the product of π and the square of its radius r.

Model

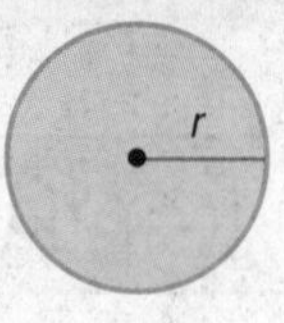

Symbols $A = \pi r^2$

Work Zone

Examples

1. Find the area of the circle. Use 3.14 for π.

2 in.

Estimate $3 \times 2 \times 2 = 12$

$A = \pi r^2$ Area of a circle

$A \approx 3.14 \cdot 2^2$ Replace r with 2.

$A \approx 3.14 \cdot 4$ $2^2 = 2 \cdot 2$ or 4

$A \approx 12.56$ Multiply.

Check for Reasonableness $12.56 \approx 12$ ✓

The area of the circle is approximately 12.56 square inches.

2. Find the area of a circle with a radius of 14 centimeters. Use $\frac{22}{7}$ for π.

Estimate $3 \times 14 \times 14 = 588$

$A = \pi r^2$ Area of a circle

$A \approx \frac{22}{7} \cdot 14^2$ Replace π with $\frac{22}{7}$ and r with 14.

$A \approx \frac{22}{7} \cdot 196$ $14^2 = 14 \cdot 14$ or 196

$A \approx \frac{22}{\cancel{7}_1} \cdot \cancel{196}^{28}$ Divide by the GCF, 7.

$A \approx 616$ Multiply.

Check for Reasonableness $616 \approx 588$ ✓

The area of the circle is approximately 616 square centimeters.

Cross out the formula that is not used for finding the area of a circle.

$A = \pi r^2$ $A = 3.14r^2$

$A = \frac{22}{7}r^2$ $A = \frac{1}{2}bh$

Got it? Do this problem to find out.

a. Find the area of a circle with a radius of 3.2 centimeters. Round to the nearest tenth.

a. ________

Example

3. Find the area of the face of the Virginia quarter with a diameter of 24 millimeters. Use 3.14 for π. Round to the nearest tenth if necessary.

The radius is $\frac{1}{2}(24)$ or 12 millimeters.

$A = \pi r^2$	Area of a circle
$A \approx \mathbf{3.14} \cdot \mathbf{12}^2$	Replace r with 12.
$A \approx 452.16$	Multiply.

The area is approximately 452.2 square millimeters.

Calculating with π

When evaluating expressions involving π, using the π key on a calculator will result in a different approximation.

Got it? Do this problem to find out.

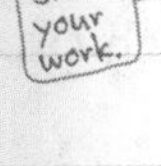

b. The bottom of a circular swimming pool with a diameter of 30 feet is painted blue. How many square feet are blue?

b. ______________

Area of Semicircles

A **semicircle** is half of a circle. The formula for the area of a semicircle is $A = \frac{1}{2}\pi r^2$.

Example

4. Find the area of the semicircle. Use 3.14 for π. Round to the nearest tenth.

16 in.

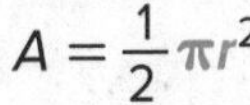

$A = \frac{1}{2}\pi r^2$	Area of a semicircle
$A \approx \frac{1}{2}(\mathbf{3.14})\mathbf{8}^2$	Replace r with 8.
$A \approx 0.5(3.14)(64)$	$8^2 = 8 \cdot 8$ or 64
$A \approx 100.5$	Simplify.

The area of the semicircle is approximately 100.5 square inches.

Got it? Do this problem to find out.

c. Find the approximate area of a semicircle with a radius of 6 centimeters.

c. ______________

Example

5. On a basketball court, there is a semicircle above the free-throw line that has a radius of 6 feet. Find the area of the semicircle. Use 3.14 for π. Round to the nearest tenth.

$A = \frac{1}{2}\pi r^2$ Area of a semicircle

$A \approx 0.5(3.14)(6^2)$ Replace π with 3.14 and r with 6.

$A \approx 0.5(3.14)(36)$ $6^2 = 6 \cdot 6$ or 36

$A \approx 56.5$ Multiply.

So, the area of the semicircle is approximately 56.5 square feet.

Guided Practice

Find the area of each circle. Round to the nearest tenth. Use 3.14 or $\frac{22}{7}$ for π. (Examples 1–3)

1.

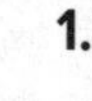

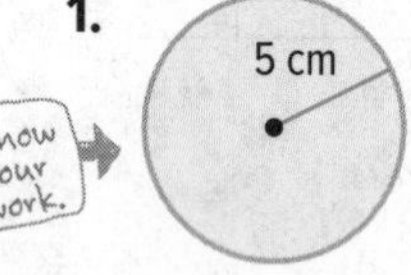

2.

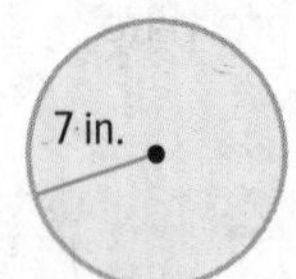

3. diameter = 16 m

4. Rondell draws the semicircle shown at the right. What is the area of the semicircle?

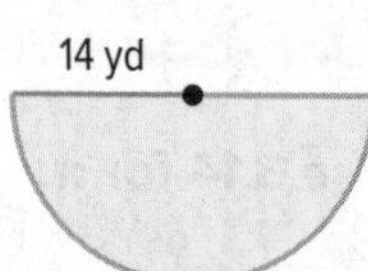

Use 3.14 for π. (Examples 4 and 5)

5. **Building on the Essential Question** Name one way the circumference and area of a circle are the same and one way they are different.

Rate Yourself!

Are you ready to move on? Shade the section that applies.

YES ? NO

For more help, go online to access a Personal Tutor.

Name ______________________ My Homework ______________________

Independent Practice

Go online for Step-by-Step Solutions 

Find the area of each circle. Round to the nearest tenth. Use 3.14 or $\frac{22}{7}$ for π. (Examples 1–3)

1. 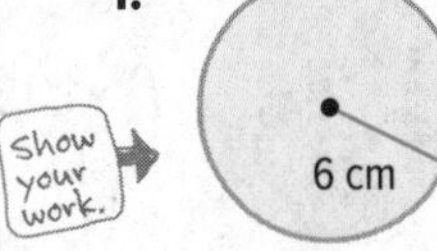

2. 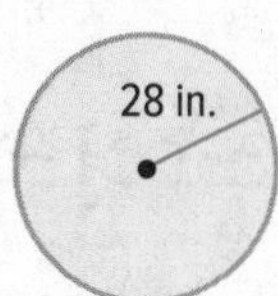

3. 11 ft

4. diameter = 10.5 in.

5. radius = 6.3 mm

6. radius = $3\frac{1}{4}$ yd

7. Refer to the pets problem at the beginning of this lesson. Find the area, to the nearest tenth, of grass that Adrianne's dog may run in if the leash is 9 feet long. (Example 3) ______________________

8. A rotating sprinkler that sprays water at a radius of 11 feet is used to water a lawn. Find the area of the lawn that is watered. Use 3.14 for π. (Example 3) ______________________

Find the area of each semicircle. Round to the nearest tenth. Use 3.14 for π. (Example 4)

9. 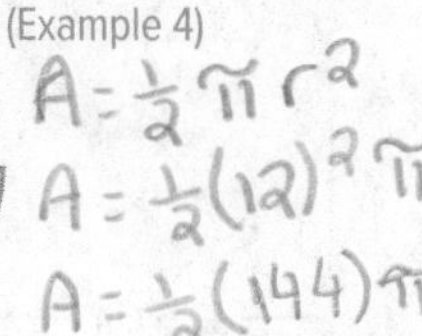

10. 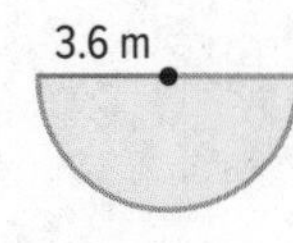

11. 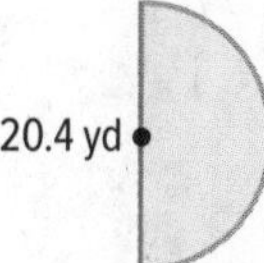

12. The tunnel opening shown is a semicircle. Find the area, to the nearest tenth, of the opening of the tunnel enclosed by the semicircle. (Example 5)

13. **MP Justify Conclusions** Harry's Pizzeria is having a sale on medium and large pizzas. Medium pizzas are 10 inches in diameter and cost $7.99. Large pizzas are 14 inches in diameter and cost $14.99. Which size pizza is the better deal? Explain. (*Hint*: Find the cost per square inch of each pizza.)

The 10inch pizza is the better deal because 1inch is 8¢ and for the 14inch one 1inch is $1.07

The 14inch pizza is the better deal.

10 in
A=78.5sq in
$0.102/sq in

14 in
A=153.9sq
$0.09/sq

H.O.T. Problems Higher Order Thinking

14. **MP Model with Mathematics** Write a real-world problem that involves finding the area of two circles. Then solve your problem.

15. **MP Reason Inductively** If the length of the radius of a circle is doubled, how does that affect the circumference and area? Explain.

MP Persevere with Problems Find the area of the shaded region in each figure. Round to the nearest tenth.

16.

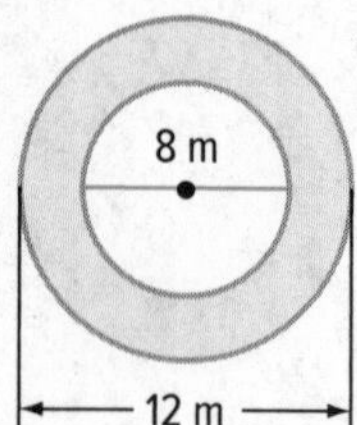

17.

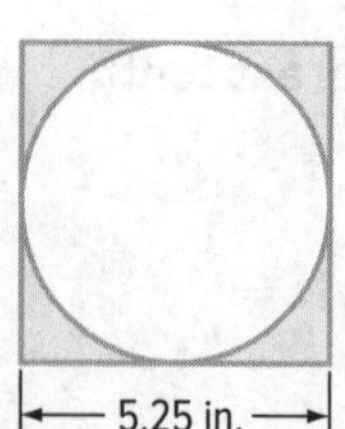

18.

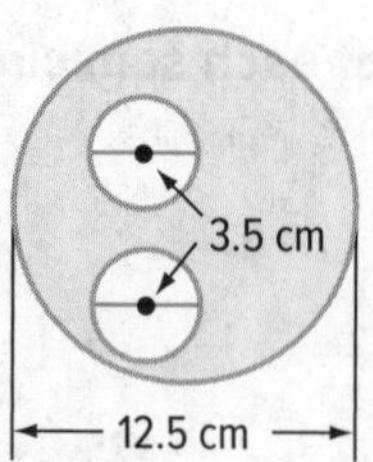

19. **MP Persevere with Problems** Explain how you could find the area of the quarter circle shown at the right. Then write a formula that could be used to find the area of a quarter circle and use the formula to find the area to the nearest tenth.

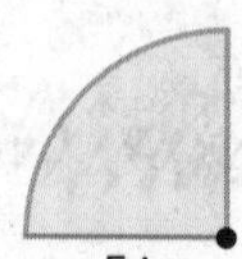

Name ____________ My Homework ____________

Extra Practice

Find the area of each circle. Round to the nearest tenth. Use 3.14 or $\frac{22}{7}$ for π.

20.

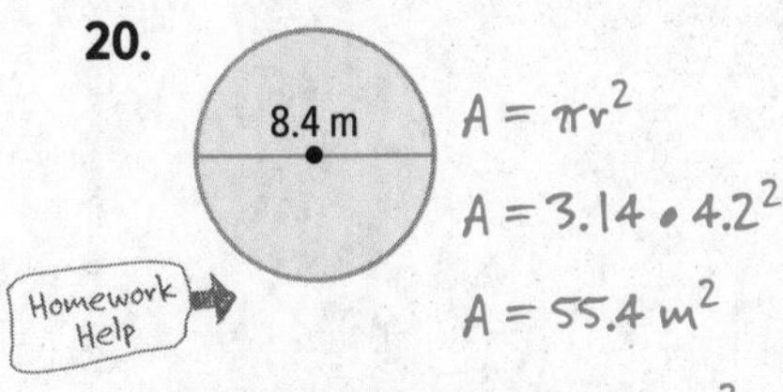

$3.14 \times 4.2 \times 4.2 = 55.4\ m^2$

21.

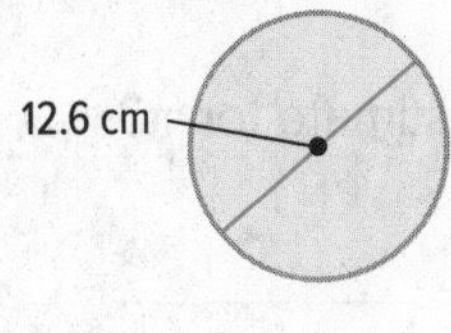

22.

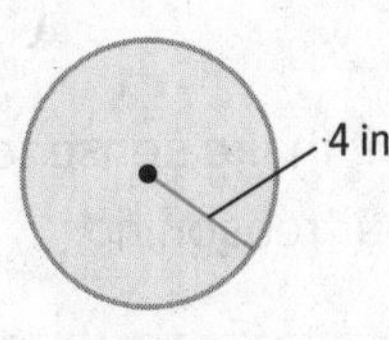

23. diameter = 10.8 yd

24. radius = $3\frac{4}{5}$ ft

25. radius = 9.3 mm

26. Find the area of the Girl Scout patch shown if the diameter is 1.25 inches. Round to the nearest tenth.

Find the area of each semicircle. Round to the nearest tenth. Use 3.14 for π.

27.

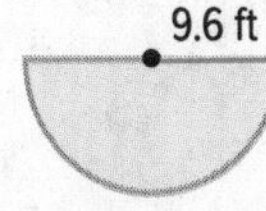

28.

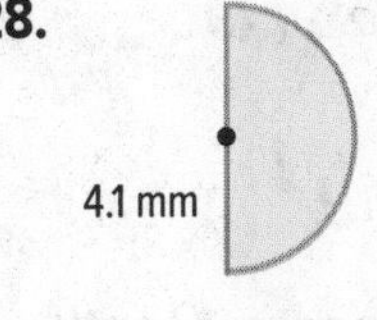

29.

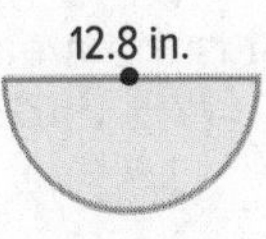

30. A window that is in the shape of a semicircle has a diameter of 28 inches. Find the area of the window. Round to the nearest tenth.

31. MP **Justify Conclusions** Which has a greater area, a triangle with a base of 100 feet and a height of 100 feet or a circle with diameter of 100 feet? Justify your selection.

32. A radio station sends a signal in a circular area with an 80-mile radius. Find the approximate area in square kilometers that receives the signal. (*Hint*: 1 square mile ≈ 2.6 square kilometers)

Power Up! Common Core Test Practice

33. A large pizza at a restaurant has the dimensions shown. Find the area of the pizza. Use $\frac{22}{7}$ for π.

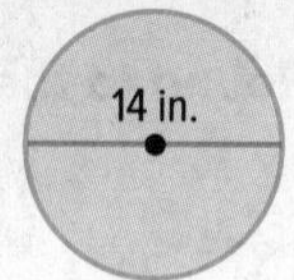

Why does it make sense to use $\frac{22}{7}$ as the estimate for π? Explain your reasoning.

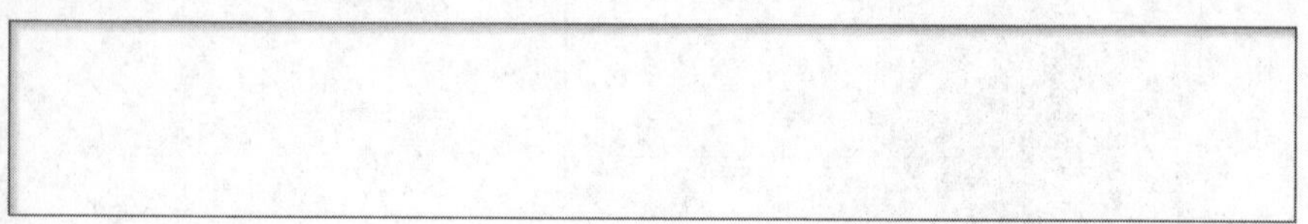

34. Refer to the figures shown below. Which figures have the same area? Select all that apply.

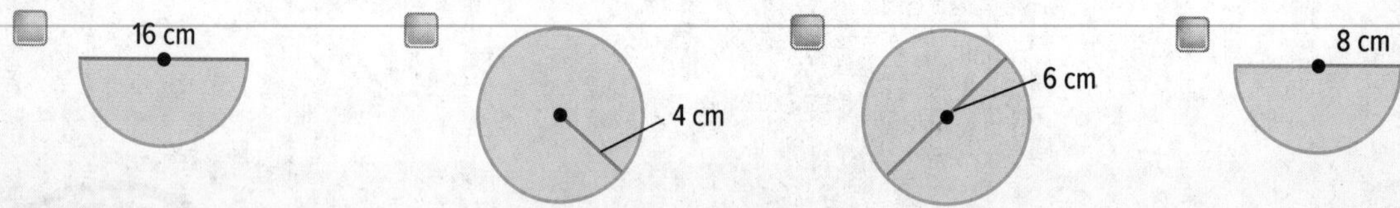

Common Core Spiral Review

35. A frame for a collage of pictures is in the shape of a trapezoid. The two bases are 15 inches and 20 inches. The height of the trapezoid is 12 inches. What is the area enclosed by the frame? **6.G.1** ________

Find the area of each parallelogram. Round to the nearest tenth if necessary. 6.G.1

36.

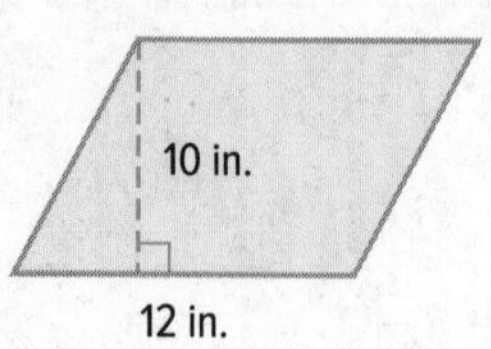

____

37.

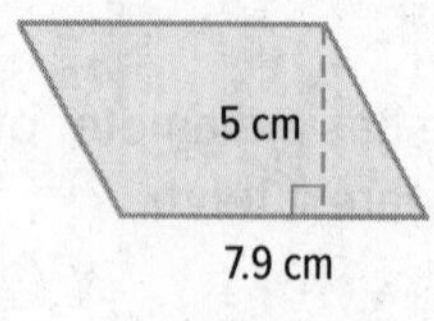

____

38.

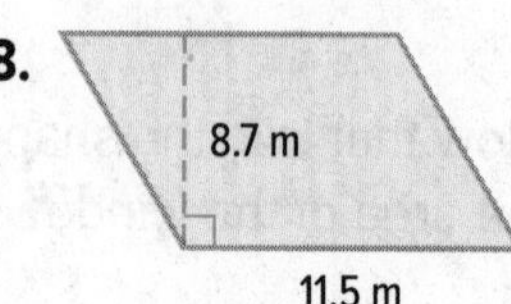

____

Lesson 3

Area of Composite Figures

Real-World Link

Stained Glass Windows An image of a stained glass window is shown below.

1. Identify two of the shapes that make up the window.

2. How could you find the area of the entire window except for the shapes you identified in Exercise 1?

3. Draw a figure that is made up of a triangle and a rectangle on the grid below. Then find the area of your figure by counting square units.

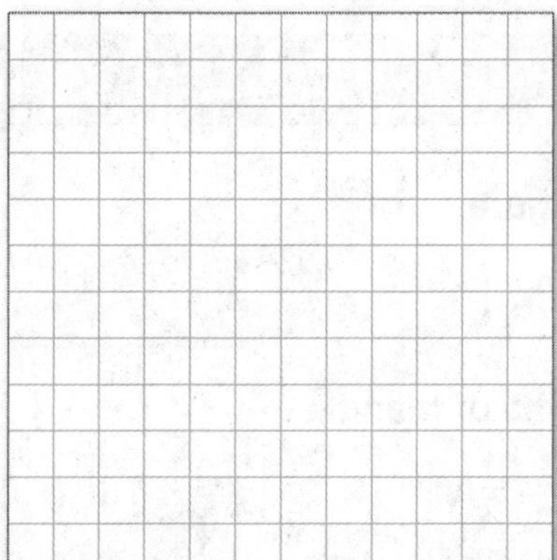

Area: ____________ square units

Essential Question

HOW do measurements help you describe real-world objects?

Vocabulary

composite figure

Common Core State Standards

Content Standards
7.G.4, 7.G.6

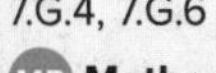

Mathematical Practices
1, 2, 3, 4

Which MP Mathematical Practices did you use? Shade the circle(s) that applies.

① Persevere with Problems
② Reason Abstractly
③ Construct an Argument
④ Model with Mathematics
⑤ Use Math Tools
⑥ Attend to Precision
⑦ Make Use of Structure
⑧ Use Repeated Reasoning

Work Zone

Find the Area of a Composite Figure

A **composite figure** is made up of two or more shapes.

To find the area of a composite figure, decompose the figure into shapes with areas you know. Then find the sum of these areas.

Shape	Words	Formula
Parallelogram	The area A of a parallelogram is the product of any base b and its height h.	$A = bh$
Triangle	The area A of a triangle is half the product of any base b and its height h.	$A = \frac{1}{2}bh$
Trapezoid	The area A of a trapezoid is half the product of the height h and the sum of the bases, b_1 and b_2.	$A = \frac{1}{2}h(b_1 + b_2)$
Circle	The area A of a circle is equal to π times the square of the radius r.	$A = \pi r^2$

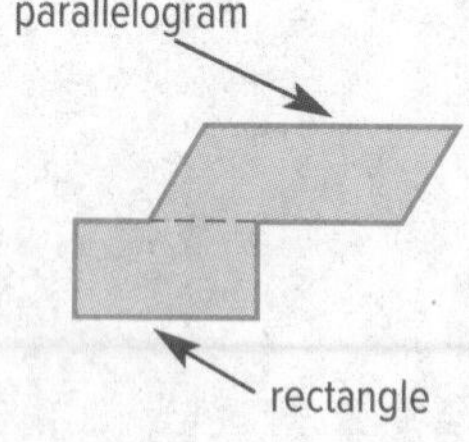

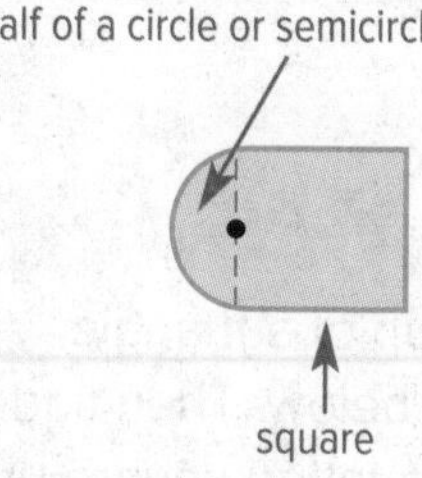

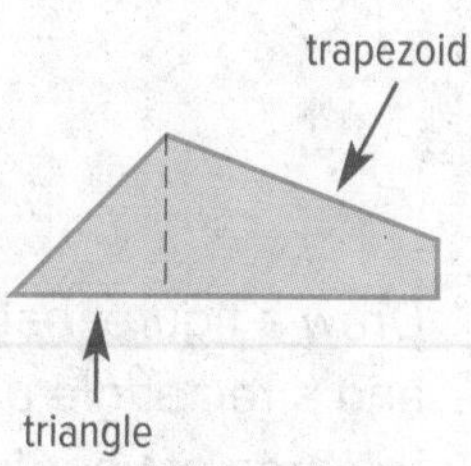

Example

1. Find the area of the composite figure.

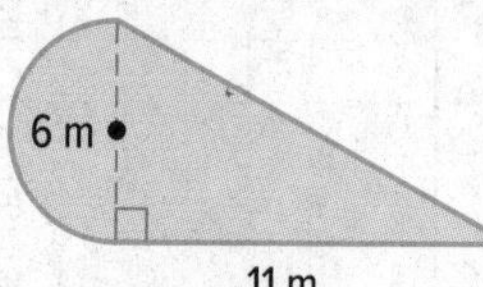

The figure can be separated into a semicircle and a triangle.

Area of semicircle

$A = \frac{1}{2}\pi r^2$

$A \approx \frac{1}{2} \cdot 3.14 \cdot 3^2$

$A \approx 14.1$

Area of triangle

$A = \frac{1}{2}bh$

$A = \frac{1}{2} \cdot 11 \cdot 6$

$A = 33$

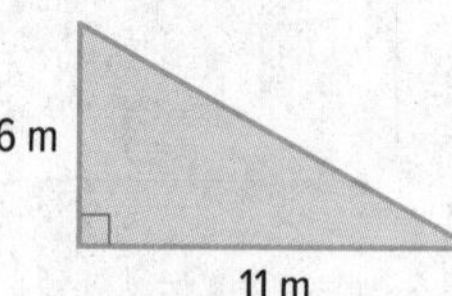

The area of the figure is about 14.1 + 33 or 47.1 square meters.

Got it? **Do this problem to find out.**

a. Find the area of the figure. Round to the nearest tenth if necessary.

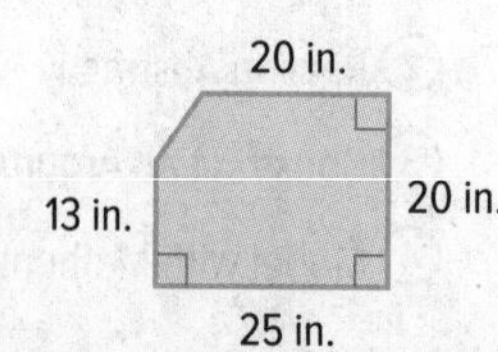

a. ____________

Example

2. A miniature golf hole is composed of a trapezoid and a parallelogram. How many square feet of turf does the hole cover?

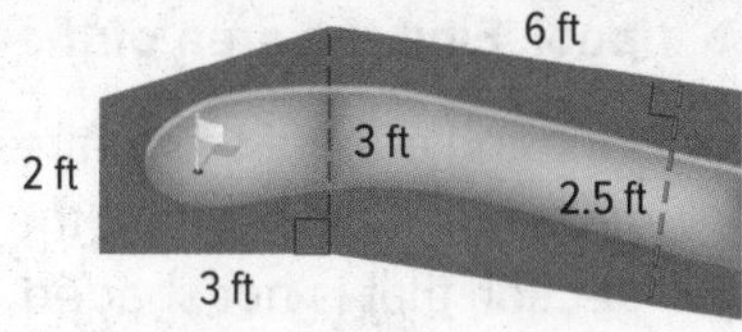

Area of trapezoid

$A = \frac{1}{2}h(b_1 + b_2)$

$A = \frac{1}{2}(3)(2 + 3)$

$A = 7.5$

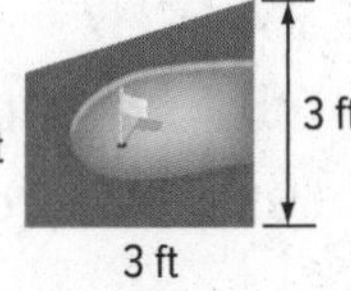

Area of parallelogram

$A = bh$

$A = 6 \cdot 2.5$

$A = 15$

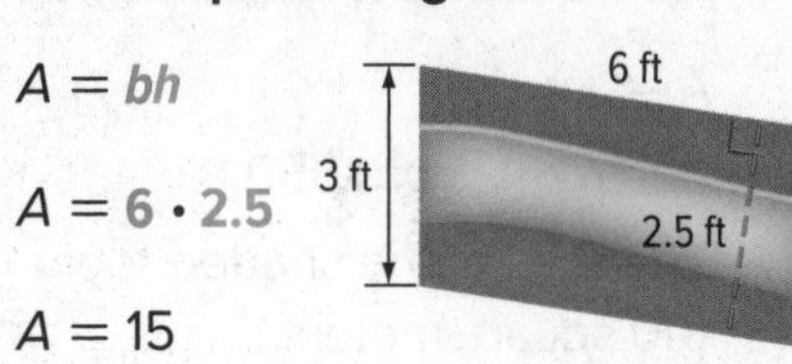

So, 7.5 + 15 or 22.5 square feet of turf will be needed.

Got it? **Do this problem to find out.**

b. Pedro's father is building a shed. How many square feet of wood are needed to build the back of the shed shown at the right?

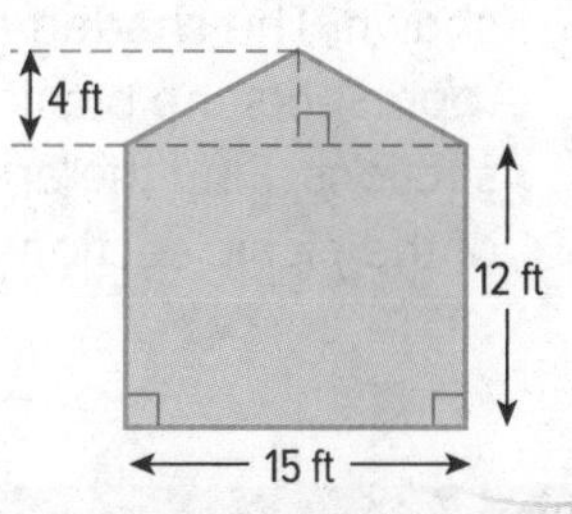

b. ______________

Find the Area of a Shaded Region

Use the areas you know to find the area of a shaded region.

Examples

3. Find the area of the shaded region.

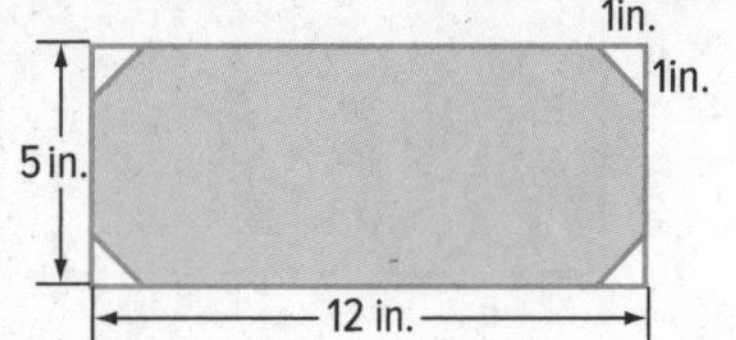

Find the area of the rectangle and subtract the area of the four congruent triangles.

Area of rectangle

$A = \ell w$

$A = 12 \cdot 5$ $\ell = 12, w = 5$

$A = 60$ Simplify.

Area of triangles

$A = 4 \cdot \left(\frac{1}{2}bh\right)$

$A = 4 \cdot \frac{1}{2} \cdot 1 \cdot 1$ $b = 1, h = 1$

$A = 2$ Simplify.

The area of the shaded region is 60 − 2 or 58 square inches.

Congruent Triangles
Congruent triangles have corresponding sides and angles that are congruent.

4. **The blueprint for a hotel swimming area is represented by the figure shown. The shaded area represents the pool. Find the area of the pool.**

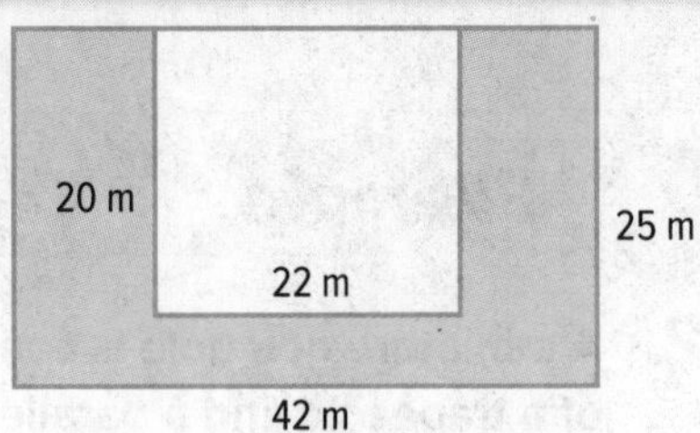

Find the area of the entire rectangle and subtract the section that is not shaded.

Area of the entire rectangle

$A = \ell w$

$A = 42 \cdot 25$ or 1,050

Area not shaded

$A = \ell w$

$A = 22 \cdot 20$ or 440

The area of the shaded region is 1,050 − 440 or 610 square meters.

Got it? **Do this problem to find out.**

c. A diagram for a park is shown. The shaded area represents the picnic sections. Find the area of the picnic sections.

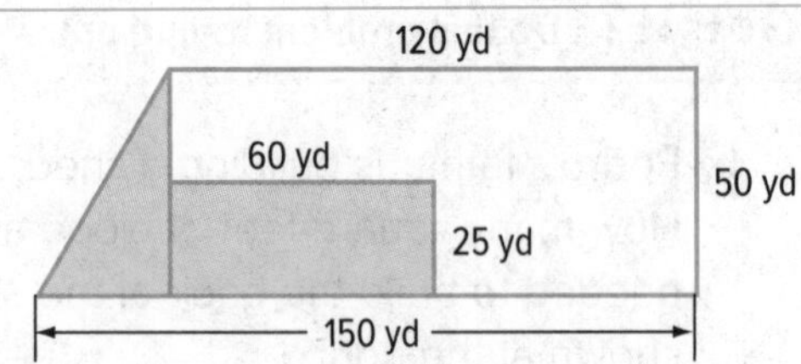

c. ____________

Guided Practice

1. Mike installed the window shown. How many square feet is the window? Round to the nearest tenth. Use 3.14 for π.

(Examples 1 and 2) ____________

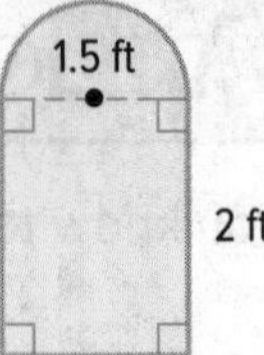

2. A triangle is cut from a rectangle. Find the area of the shaded region.

(Examples 3 and 4) ____________

11 ft

6 ft

4 ft

3. **Building on the Essential Question** Is your answer to Exercise 1 an exact or approximate answer? Explain.

____

____

____

Rate Yourself!

How confident are you about finding the area of composite figures? Check the box that applies.

For more help, go online to access a Personal Tutor.

Name ______________________ My Homework ______________________

Independent Practice

Go online for Step-by-Step Solutions eHelp

Find the area of each figure. Round to the nearest tenth if necessary. (Example 1)

1.

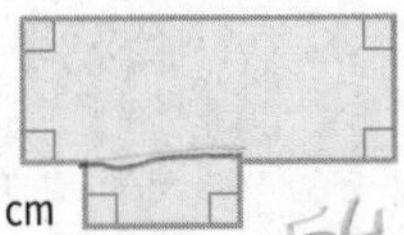

12 cm
4.5 cm
2 cm
5 cm

Show your work.

2.

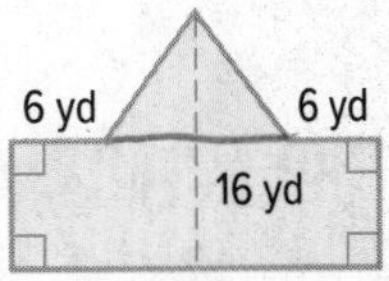

6 yd
6 yd
16 yd
8 yd
24 yd

3.

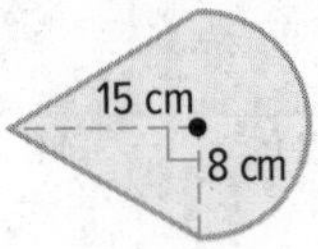

15 cm
8 cm

4.

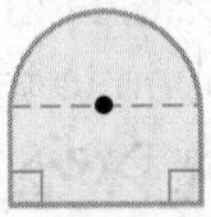

7 m
15 m

5.

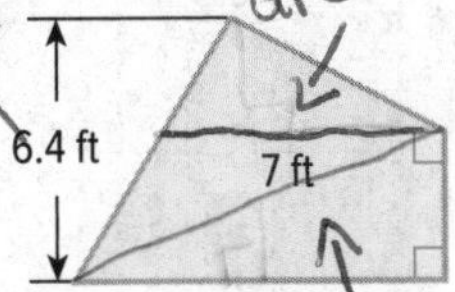

6.4 ft
7 ft
3.6 ft
9 ft

6.

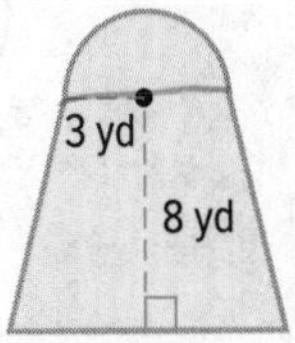

3 yd
8 yd
10 yd

7. Daniel is constructing a deck like the one shown. What is the area of the deck? (Example 2)

5 ft
3.5 ft
12 ft

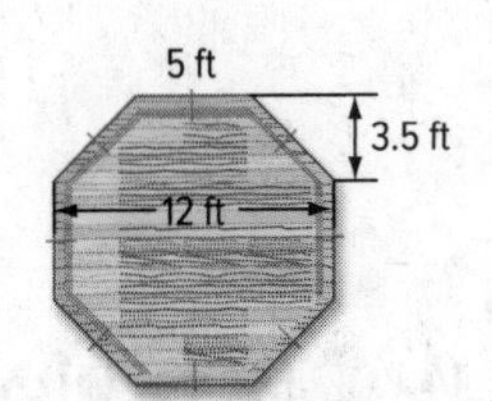

Find the area of the shaded region. Round to the nearest tenth if necessary. (Examples 3 and 4)

8.

10 yd
6 yd
9 yd
15 yd

9.

13 cm
3 cm
7cm
4 cm
1cm
2 cm

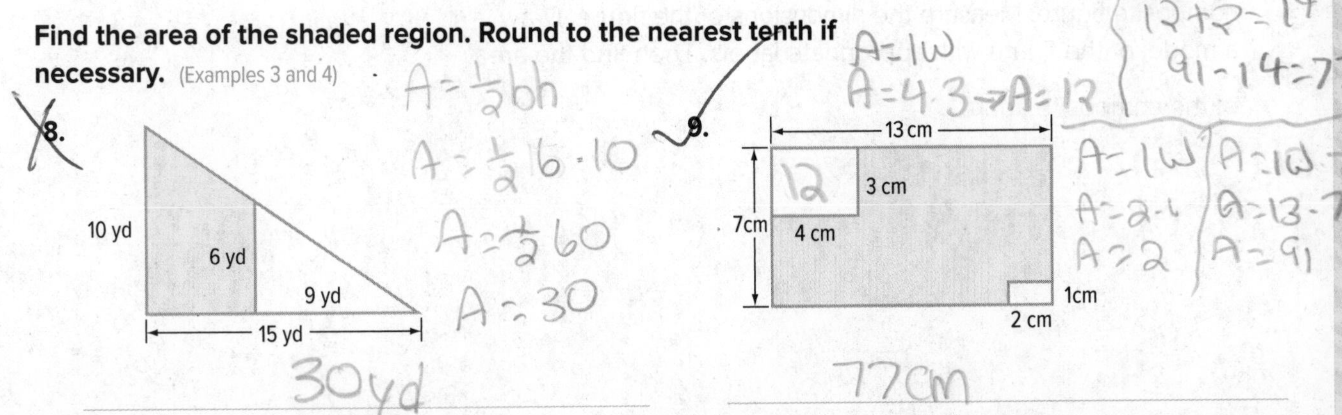

10. **MP Persevere with Problems** Zoe's mom is carpeting her bedroom and needs to know the amount of floor space. How many square feet of carpeting are needed for the room? If she is also installing baseboards on the bottom of all the walls, how many feet of baseboards are needed? ______

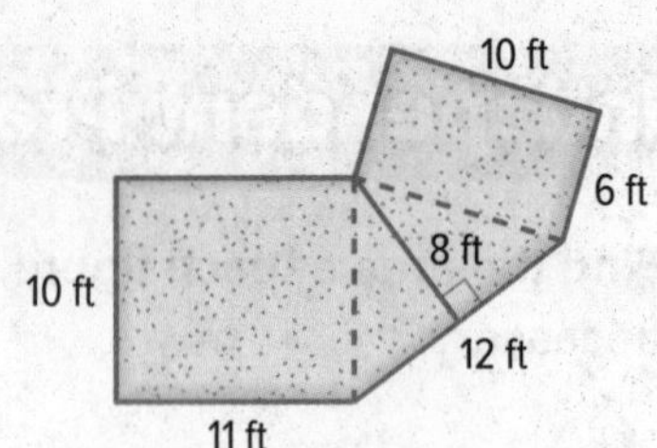

H.O.T. Problems Higher Order Thinking

11. **MP Persevere with Problems** The composite figure shown is made from a rectangle and part of a circle. Find the approximate area and perimeter of the entire figure. Round to the nearest tenth.

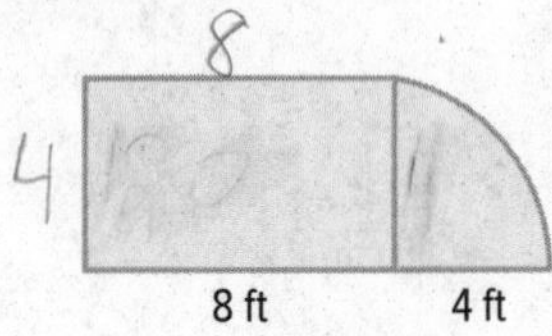

______

12. **MP Reason Abstractly** The side length of the square in the figure at the right is x units. Write expressions that represent the perimeter and area of the figure.

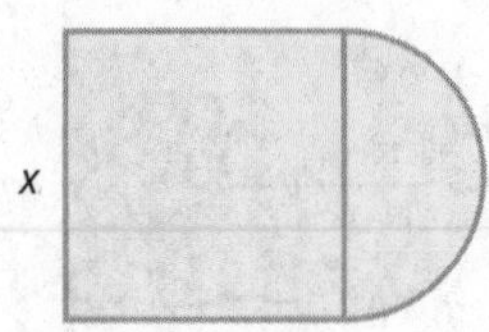

______

13. **MP Persevere with Problems** In the diagram shown at the right, a 2-foot-wide flower border surrounds the heart-shaped pond. What is the area of the border?

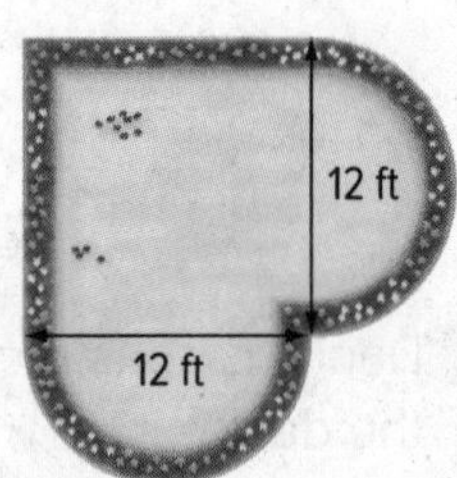

______

14. **MP Model with Mathematics** Find a real-world object that is a composite figure. Measure the dimensions of the figure. Draw a model of the figure with appropriate labels. Then find the area of the composite figure. ______

Name ______________________ My Homework ______________________

Extra Practice

Find the area of each figure. Round to the nearest tenth if necessary.

15. 87.5 m^2

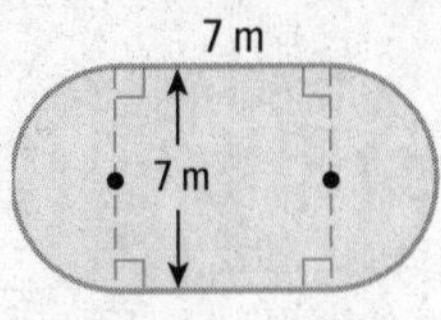
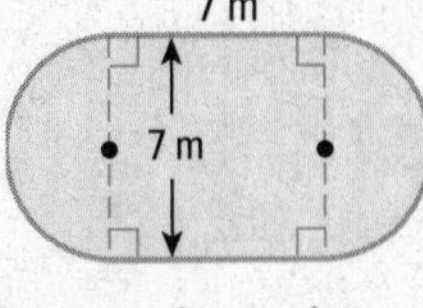

Homework Help

Area of circle

$A = \pi r^2$

$A = 3.14 \cdot 3.5^2$ or 38.5

Area of square

$A = lw$

$A = 7 \cdot 7$ or 49

$38.5 + 49 = 87.5$

16. ______

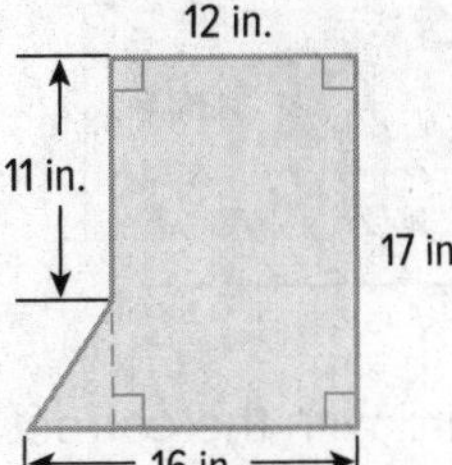

17. ______

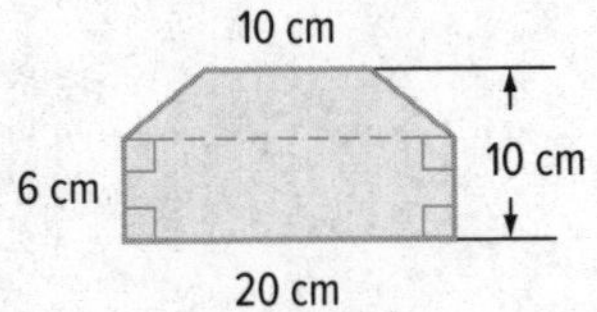

18. ______

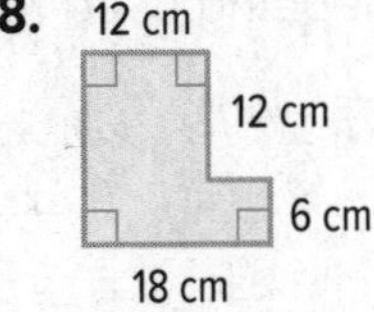

19. A necklace comes with a gold pendant. What is the area of the pendant in square centimeters? ______

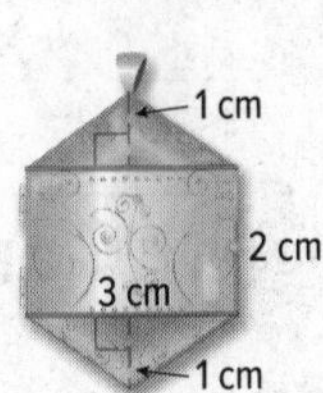

Find the area of the shaded region. Round to the nearest tenth if necessary.

20.

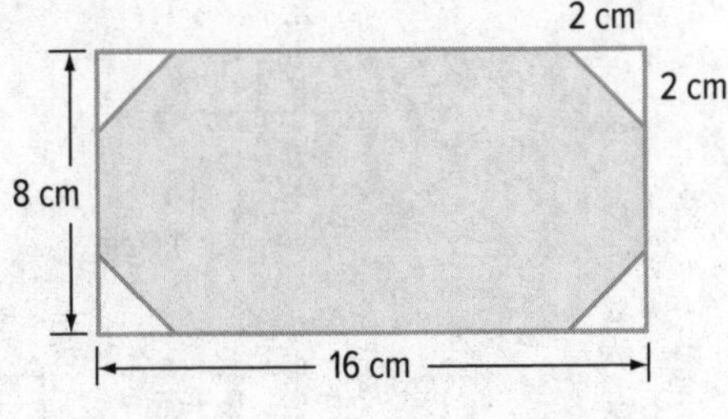

______

21.

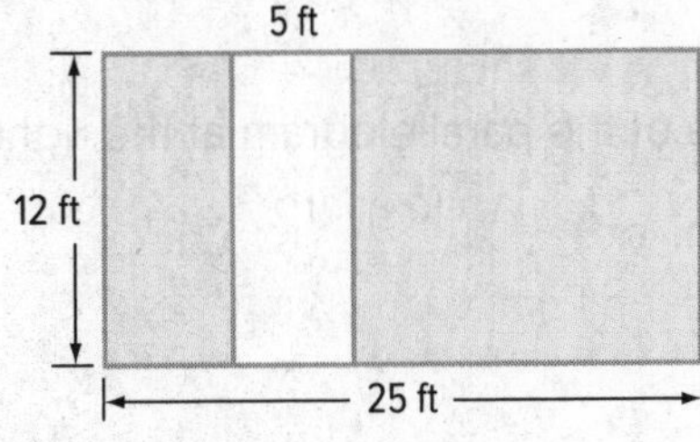

______

Power Up! Common Core Test Practice

22. The Patel's backyard has a rectangular vegetable garden and a triangular pet exercise area.

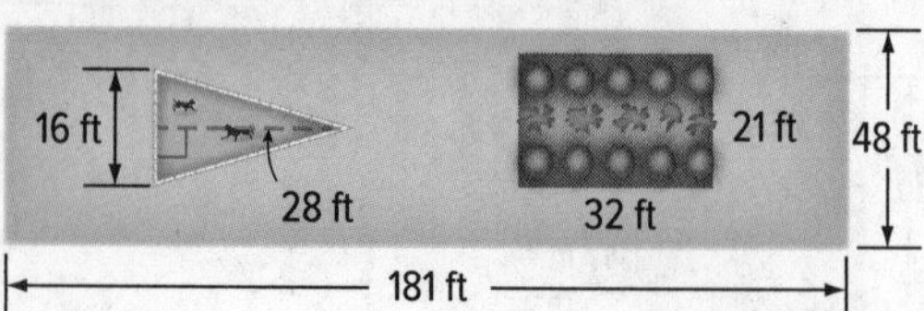

Match each part of the yard with the correct area.

106	672
224	1,092
458	7,520
544	8,688

Pet Exercise Area: ______ ft^2

Vegetable Garden Area: ______ ft^2

Total Backyard Area: ______ ft^2

How much of the backyard is not being used for the vegetable garden or pet exercise area?

23. The figure is made up of a square and four semicircles. Fill in each box to complete each statement. Round to the nearest hundredth.

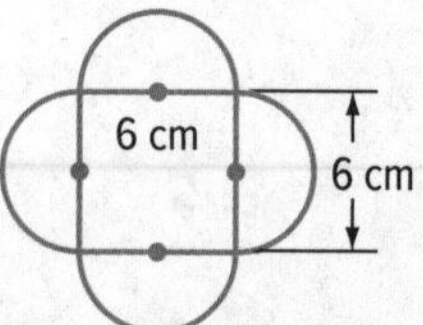

a. The area of the square is ______ cm^2.

b. The area of each semicircle is about ______ cm^2.

c. The total area of the figure is about ______ cm^2.

Common Core Spiral Review

24. What is the area of a triangle with a base of 52 feet and a height of 38 feet? **6.G.1** ______

25. Find the area of the parallelogram at the right. Round to the nearest tenth. **6.G.1** ______

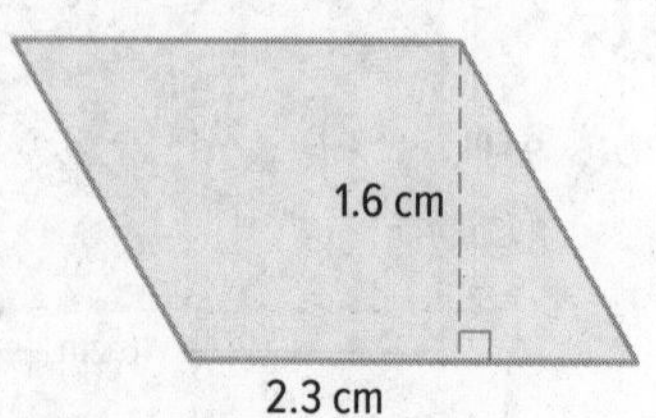

26. Find the height of a parallelogram with an area of 104 square yards and a base of 8 yards. **6.G.1**

______

27. Find the base of a parallelogram with a height of 3.2 meters and an area of 15.04 square meters. **6.G.1**

______

Lesson 4
Volume of Prisms

Vocabulary Start-Up

Recall that a prism is a polyhedron with two parallel, congruent bases. The bases of a *rectangular prism* are rectangles, and the bases of a *triangular prism* are triangles.

Write *rectangular prism* or *triangular prism* on the line below each figure.

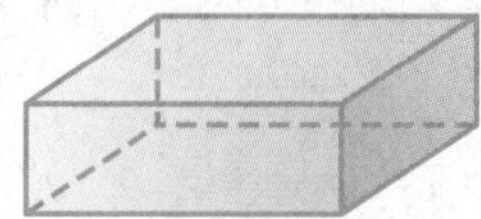

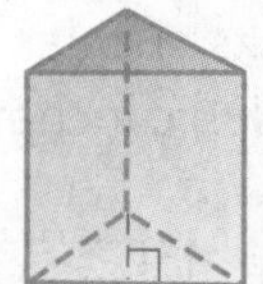

_______________ _______________

Essential Question

HOW do measurements help you describe real-world objects?

Vocabulary

volume

Common Core State Standards

Content Standards
7.G.6

MP Mathematical Practices
1, 2, 3, 4

Real-World Link

1. Suppose you observed the camping tent shown from directly above. What geometric figure would you see?

____

2. What formula would you use to find the area of the figure?

____

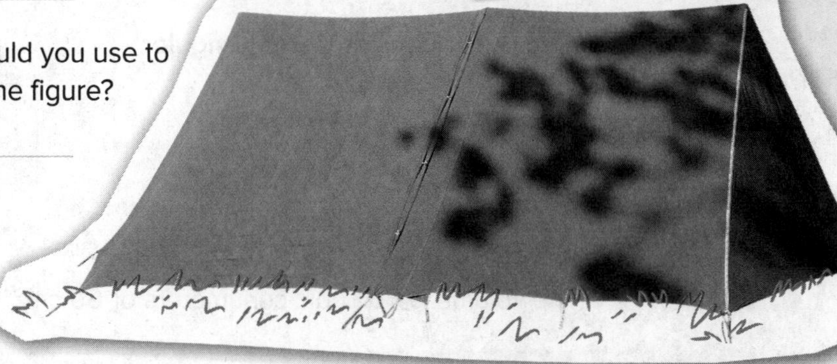

Which MP Mathematical Practices did you use? Shade the circle(s) that applies.

① Persevere with Problems
② Reason Abstractly
③ Construct an Argument
④ Model with Mathematics
⑤ Use Math Tools
⑥ Attend to Precision
⑦ Make Use of Structure
⑧ Use Repeated Reasoning

Key Concept

Volume of a Rectangular Prism

Words The volume V of a rectangular prism is the product of the length ℓ, the width w, and the height h. It is also the area of the base B times the height h.

Model

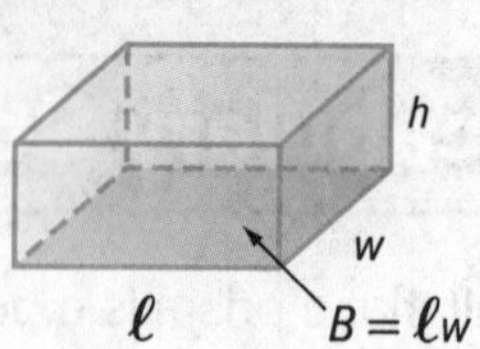

Symbols $V = \ell wh$ or $V = Bh$

Work Zone

The **volume** of a three-dimensional figure is the measure of space it occupies. It is measured in cubic units such as cubic centimeters (cm^3) or cubic inches (in^3).

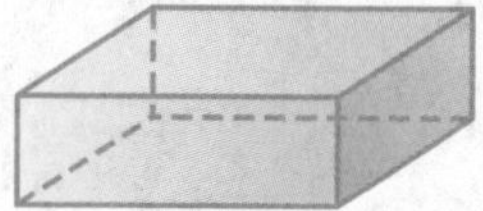

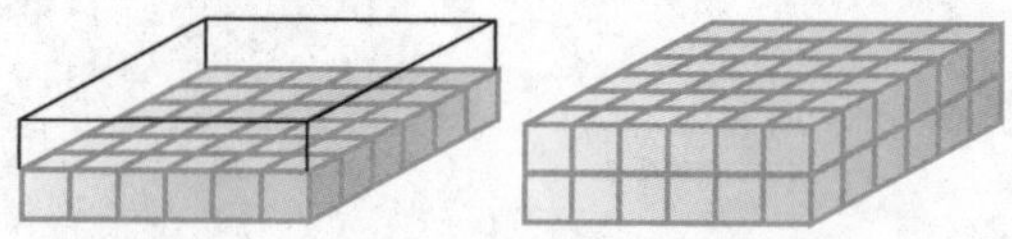

It takes 2 layers of 36 cubes to fill the box. So, the volume of the box is 72 cubic centimeters.

Decomposing Figures

Think of the volume of the prism as consisting of three congruent slices. Each slice contains the base area, 20 square centimeters, and a height of 1 centimeter.

Example

Tutor

1. Find the volume of the rectangular prism.

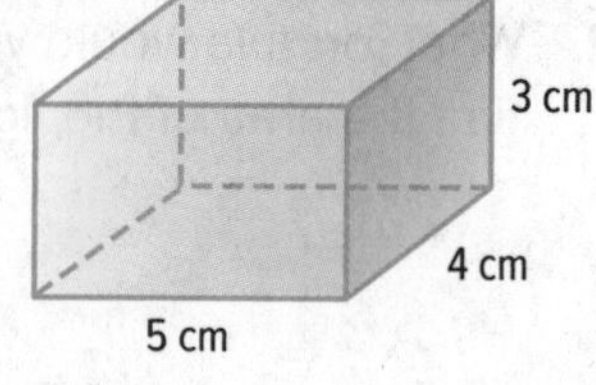

$V = \ell wh$ Volume of a prism

$V = 5 \cdot 4 \cdot 3$ $\ell = 5$, $w = 4$, and $h = 3$

$V = 60$ Multiply.

The volume is 60 cubic centimeters or 60 cm^3.

Got it? **Do this problem to find out.**

Show your work.

a. Find the volume of the rectangular prism shown below.

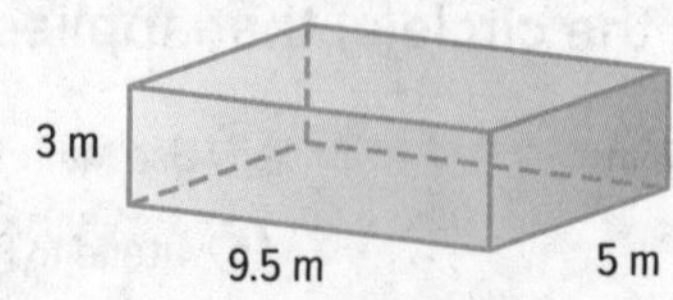

a. ____________

Volume of a Triangular Prism

Key Concept

Words The volume V of a triangular prism is the area of the base B times the height h.

Symbols $V = Bh$, where B is the area of the base.

Model

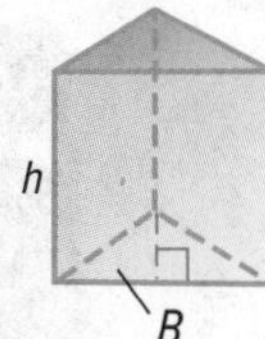

Height

Do not confuse the height of the triangular base with the height of the prism.

The diagram below shows that the volume of a triangular prism is also the product of the area of the base B and the height h of the prism.

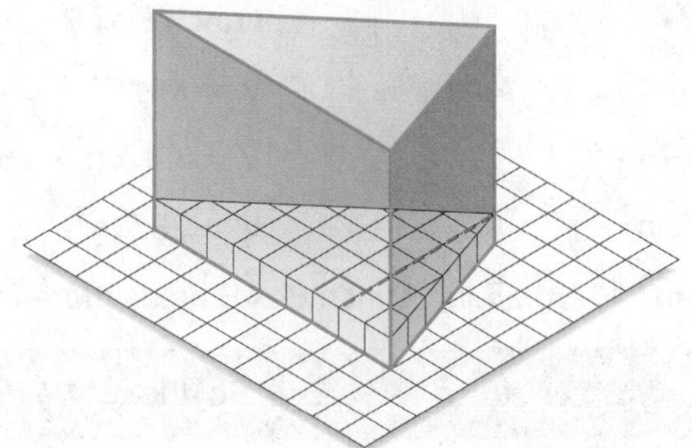

Example

2. Find the volume of the triangular prism shown.

The area of the triangle is $\frac{1}{2} \cdot 6 \cdot 8$, so replace B with $\frac{1}{2} \cdot 6 \cdot 8$.

$V = Bh$ — Volume of a prism

$V = \left(\frac{1}{2} \cdot 6 \cdot 8\right)h$ — Replace B with $\frac{1}{2} \cdot 6 \cdot 8$.

$V = \left(\frac{1}{2} \cdot 6 \cdot 8\right)9$ — The height of the prism is 9.

$V = 216$ — Multiply.

The volume is 216 cubic feet or 216 ft^3.

6 ft
9 ft
8 ft

Before finding the volume of a prism, identify the base. In Example 2, the base is a triangle, so you replace B with $\frac{1}{2}bh$.

Got it? **Do this problem to find out.**

b. Find the volume of the triangular prism.

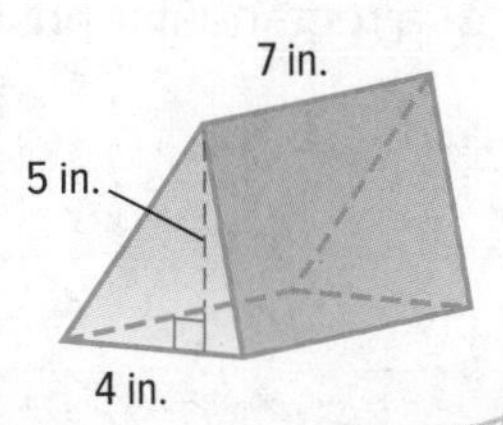

b. ________

Example

3. **Which lunch box holds more food?**

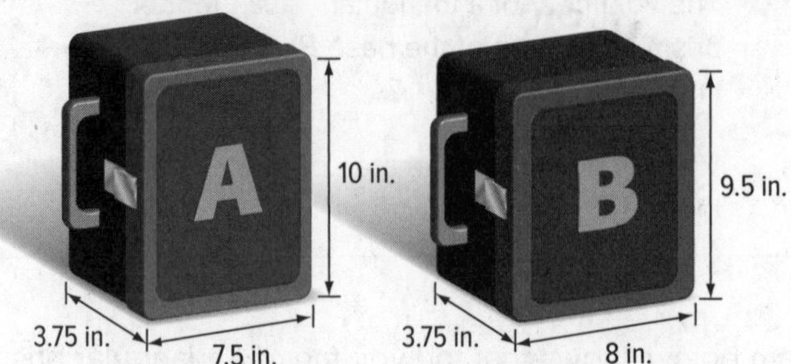

Find the volume of each lunch box. Then compare.

Lunch Box A	**Lunch Box B**
$V = \ell wh$	$V = \ell wh$
$V = 7.5 \cdot 3.75 \cdot 10$	$V = 8 \cdot 3.75 \cdot 9.5$
$V = 281.25 \text{ in}^3$	$V = 285 \text{ in}^3$

Since $285 \text{ in}^3 > 281.25 \text{ in}^3$, Lunch Box B holds more food.

Guided Practice

Find the volume of each prism. Round to the nearest tenth if necessary.
(Examples 1–2)

1. ______

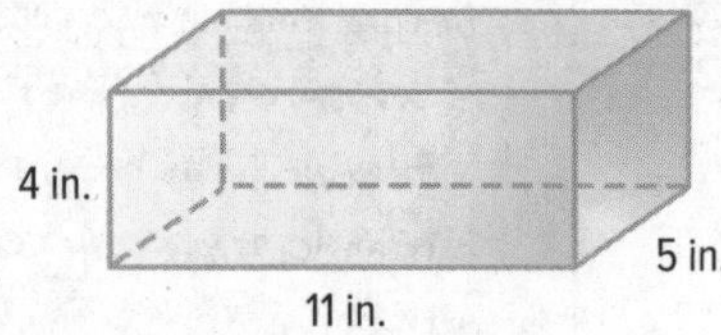

2. ______

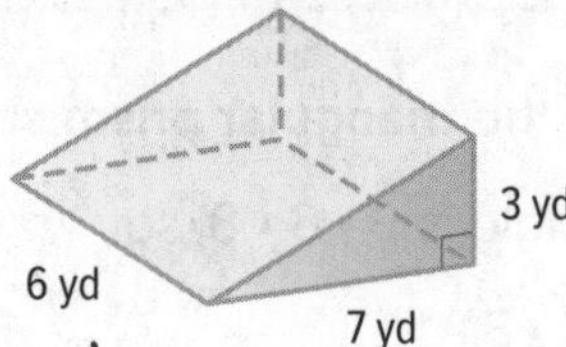

3. One cabinet measures 3 feet by 2.5 feet by 5 feet. A second measures 4 feet by 3.5 feet by 4.5 feet. Which volume is greater? Explain. (Example 3)

______

4. **Building on the Essential Question** Compare and contrast finding the volume of a rectangular prism and a triangular prism. ______

______

______

Rate Yourself!

How confident are you about finding volume for prisms? Check the box that applies.

For more help, go online to access a Personal Tutor.

FOLDABLES Time to update your Foldable!

Name ______________________ My Homework ______________________

Independent Practice

Go online for Step-by-Step Solutions

Find the volume of each prism. Round to the nearest tenth if necessary.
(Examples 1–2)

1.

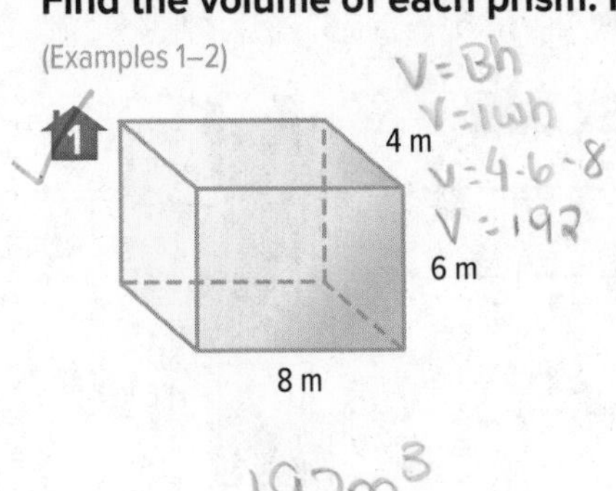

$V = Bh$
$V = lwh$
$V = 4 \cdot 6 \cdot 8$
$V = 192$

192 m^3

2.

9 ft
11 ft
8 ft

$V = Bh$
$V = (\frac{1}{2}bh)h$
$V = (\frac{1}{2} \cdot 8 \cdot 9)11$
$V = 396$

396 ft^3

3.

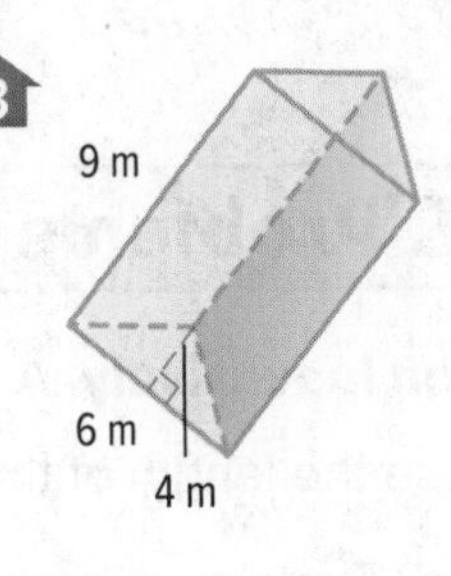

____

Show your work.

4. Which container holds more detergent? Justify your answer. (Example 3)

Soapy Suds holds more detergent because it has a volume of 1248 in^3 while Clean & Bright had a volume of 468 in^3

$V = Bh$
$V = lwh$
$V = 13 \cdot 12 \cdot 8$
$V = 1248$

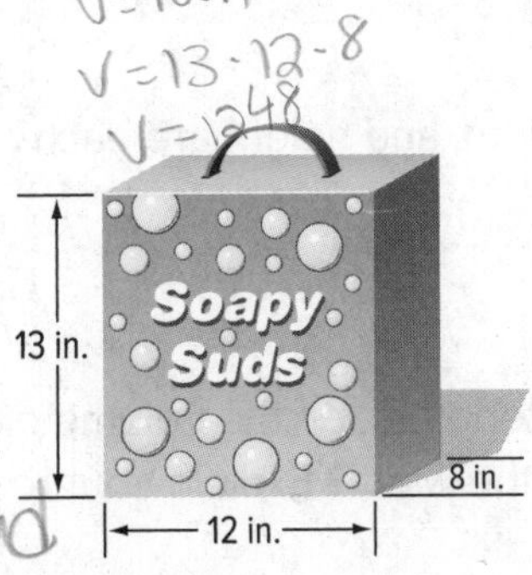

$V = Bh$
$V = (\frac{1}{2}bh)h$
$V = (\frac{1}{2} 13 \cdot 8)9$
$V = 468$

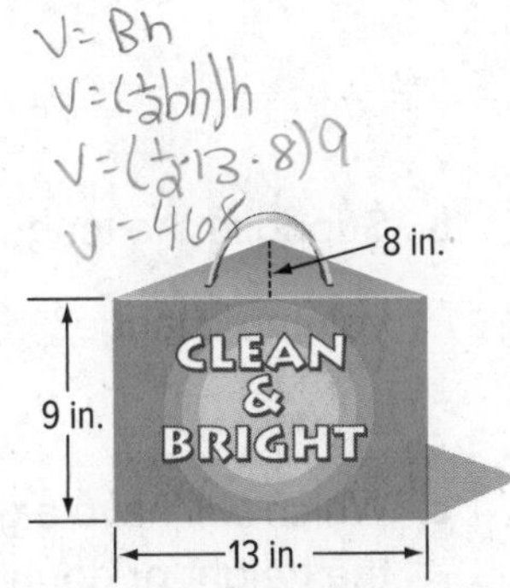

5. **MP Model with Mathematics** Refer to the graphic novel frame below. The table shows possible dimensions for the dunk tank.

Length (ft)	Width (ft)	Height (ft)	Surface Area (ft^2)
2	12	4	136
4	4	8	144
4	7	6	160
8	5	4	144
10	4	3	124

a. Find the volume of each given dunk tank.

____

b. Which dimensions are reasonable for a dunk tank? Explain.

____

____

6. The diagram shows the dimensions of an office. It costs about \$0.11 per year to air condition one cubic foot of space. On average, how much does it cost to air condition the office for one month? ______

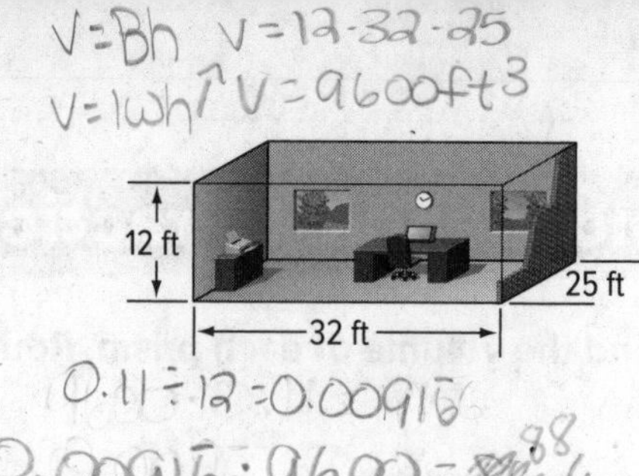

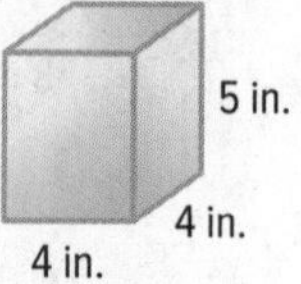

H.O.T. Problems Higher Order Thinking

7. **MP Reason Inductively** A rectangular prism is shown.

5 in.
4 in.
4 in.

a. Suppose the length of the prism is doubled. How does the volume change? Explain your reasoning. ______

b. Suppose the length, width, and height are each doubled. How does the volume change? ______

c. Which will have a greater effect on the volume of the prism: doubling the height or doubling the width? Explain your reasoning.

8. **MP Persevere with Problems** The prism shown has a base that is a trapezoid. Find the volume of the prism. ______

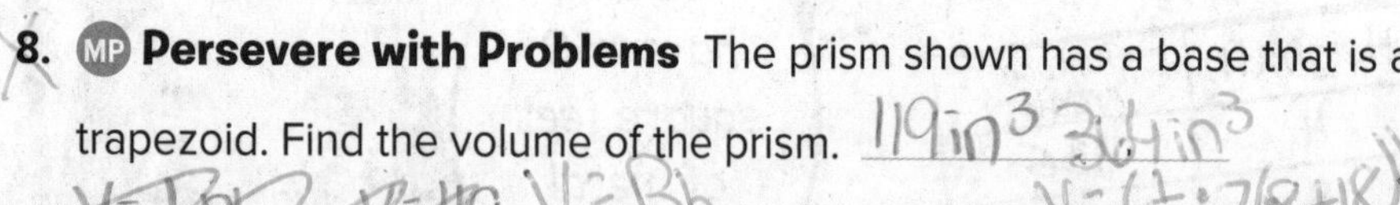

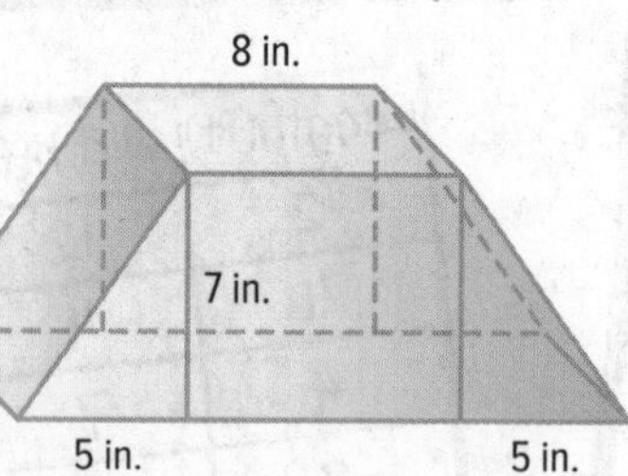

9. **MP Model with Mathematics** Find the volume of a real-world object that is in the shape of a rectangular or triangular prism using appropriate units. Draw a model of the prism including the dimensions. ______

Extra Practice

Find the volume of each prism. Round to the nearest tenth if necessary.

10. 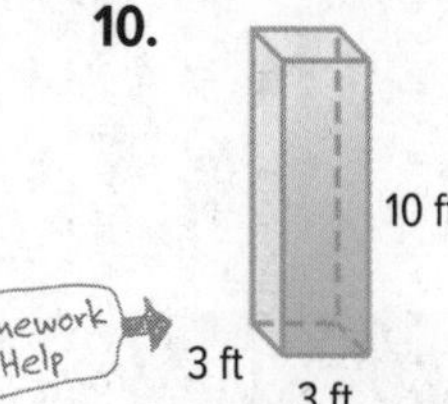

V = lwh
V = 3 · 3 · 10
V = 90

90 ft³

11. 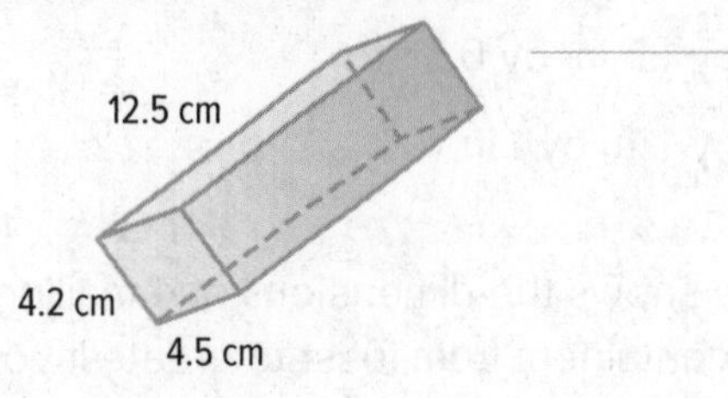

______

12. 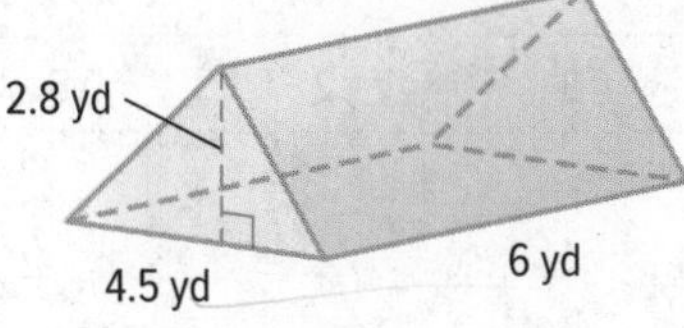

______

13. 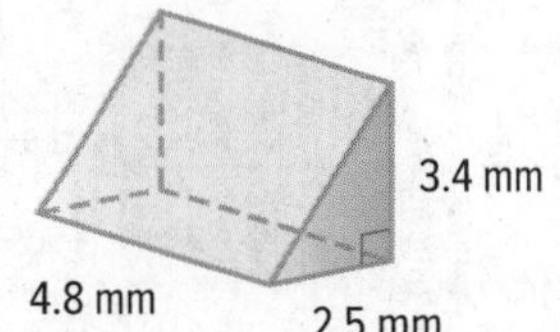

______

14. A toy company makes rectangular sandboxes that measure 6 feet by 5 feet by 1.2 feet. A customer buys a sandbox and 40 cubic feet of sand. Did the customer buy too much or too little sand? Justify your answer.

______

15. The base of a rectangular prism has an area of 19.4 square meters and the prism has a volume of 306.52 cubic meters. Write an equation that can be used to find the height h of the prism. Then find the height of the prism.

______

Find the volume of each prism.

16. 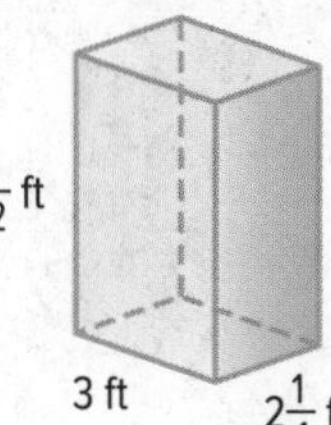

______

17. 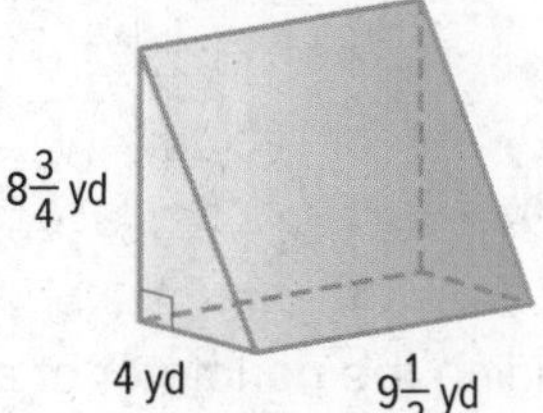

______

18. MP **Reason Abstractly** Write a formula for finding the volume of a cube. Use an exponent and the variable s to represent the side lengths. Then use the formula to find the volume of a cube with side lengths of 7 inches.

______

Power Up! Common Core Test Practice

19. The volume of a paperclip box is 1.5 cubic inches. Which of the following are possible dimensions of the box? Select all that apply.

- ☐ 2 in. by 1.5 in. by 0.5 in.
- ☐ 3 in. by 0.5 in. by 1.5 in.
- ☐ 2 in. by 1 in. by 1 in.
- ☐ 3 in. by 1 in. by 0.5 in.

20. The table shows the dimensions of 4 mailing containers. Sort the containers from least to greatest volume.

Container	ℓ (ft)	w (ft)	h (ft)
A	2	2	2
B	1	3	3
C	3	4	0.5
D	3	2	0.5

	Container	Volume (ft^3)
Least		
Greatest		

Which container has the greatest volume? ☐

Common Core Spiral Review

Find the perimeter of each figure. 4.MD.3

21.

4.3 m
4.3 m 4.3 m
4.3 m 4.3 m
4.3 m

____

22.

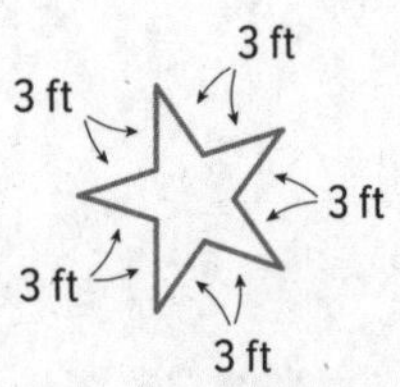

____

23.

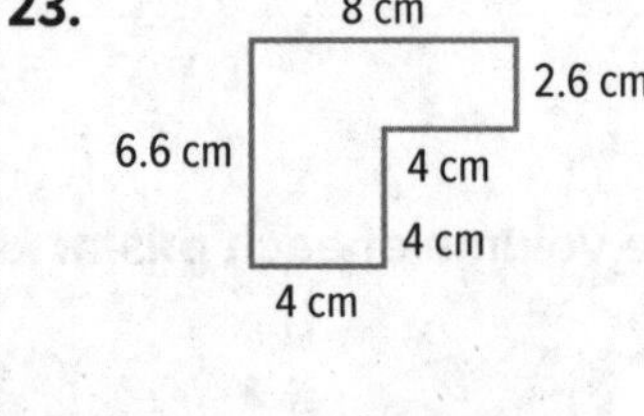

____

24. Write a formula for finding the perimeter of a square. Use your formula to find the perimeter of a square with side length of 0.5 inch. 6.G.3

____

MP Problem-Solving Investigation

Solve a Simpler Problem

Content Standards
7.G.4, 7.G.6

MP Mathematical Practices
1, 3

Case #1 Playgrounds

Liam is helping to mulch the play area at the community center. The diagram shows the dimensions of the play area.

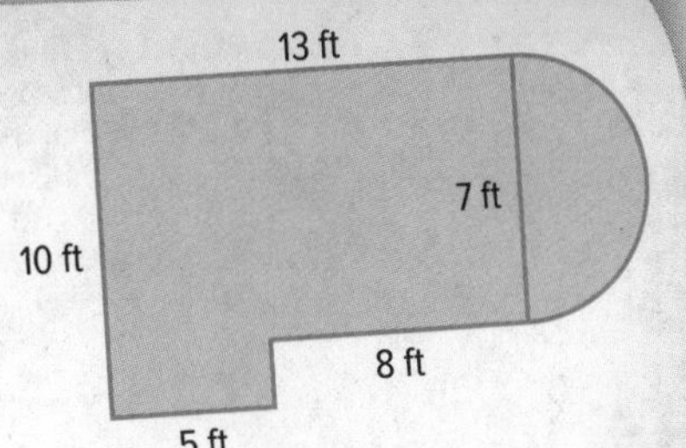

What is the area of the play area to be mulched? Round to the nearest tenth if necessary.

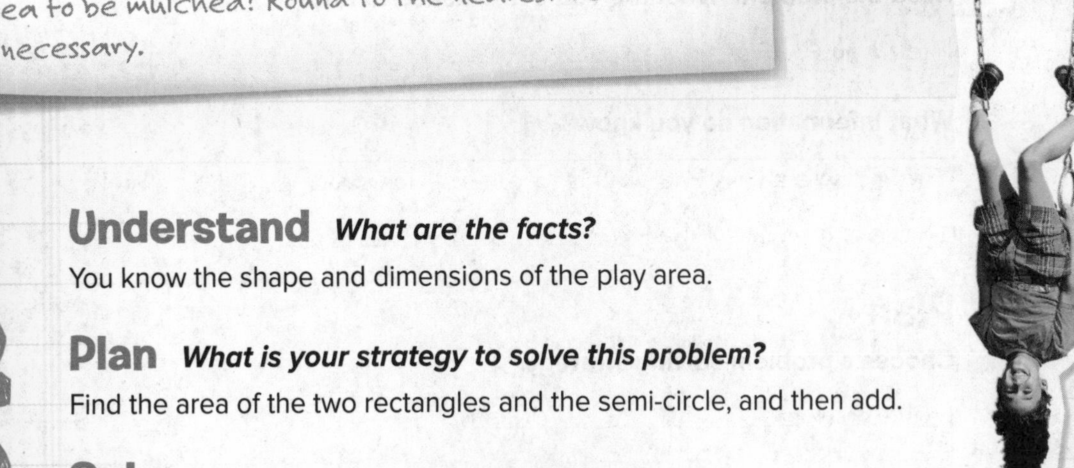

1 Understand ***What are the facts?***

You know the shape and dimensions of the play area.

2 Plan ***What is your strategy to solve this problem?***

Find the area of the two rectangles and the semi-circle, and then add.

3 Solve ***How can you apply the strategy?***

Area of Rectangle 1	Area of Rectangle 2	Area of Semi-Circle
$A = \ell w$	$A = \ell w$	$A = \frac{\pi r^2}{2}$
$A = 5 \cdot 10$	$A = 8 \cdot 7$	$A = \frac{3.14 \cdot (3.5)^2}{2}$
$A =$ ☐	$A =$ ☐	$A =$ ☐

The total area is ☐ + ☐ + ☐ or ☐ square feet.

4 Check ***Does the answer make sense?***

The play area is about 13 • 10 or 130 square feet. So, an answer of ☐ square feet is reasonable.

Analyze the Strategy

Tutor

MP Reason Inductively Why is breaking this problem into simpler parts a good strategy to solve it?

____

____

Case #2 Wallpaper

Dora is painting a wall in her house.

What is the area that will be painted?

1 Understand

Read the problem. What are you being asked to find?

I need to find ______________________________.

What information do you know?

The picture shows the wall is ________ long and ________ high.

There is a window that is ________ by ________.

2 Plan

Choose a problem-solving strategy.

I will use the ______________________________ strategy.

3 Solve

Use your problem-solving strategy to solve the problem.

Find the area of the wall. Then subtract the area of the window.

The dimensions of the wall are ☐ feet by ☐ feet.

So, the area of the wall is ☐ × ☐ = ☐ ft^2.

The dimensions of the window are ☐ feet by ☐ feet.

So, the area of the window is ☐ × ☐ = ☐ ft^2.

☐ − ☐ = ☐

So, ______________________________.

4 Check

Use information from the problem to check your answer.

Use estimation to check the reasonableness of your answer. The area of the wall is approximately 10 × 12 = ☐ ft^2. The answer is reasonable.

Work with a small group to solve the following cases. Show your work on a separate piece of paper.

Case #3 Woodworking

Two workers can make two chairs in two days.

How many chairs can 8 workers working at the same rate make in 20 days?

Case #4 Tips

Ebony wants to leave an 18% tip for a $19.82 restaurant bill. The tax is 6.25%, which is added to the bill before the tip.

How much money does Ebony spend at the restaurant? Explain.

Case #5 Continents

The land area of Earth is 57,505,708 square miles.

To the nearest tenth, how much larger is the land area of Asia than North America? Explain.

Continent	Percent of Earth's Land
Asia	30
Africa	20.2
North America	16.5

Case #6 Fountains

Mr. Flores has a circular fountain with a radius of 5 feet. He plans on installing a brick path around the fountain.

If each brick covers 2 square feet, how many bricks will he need to buy?.

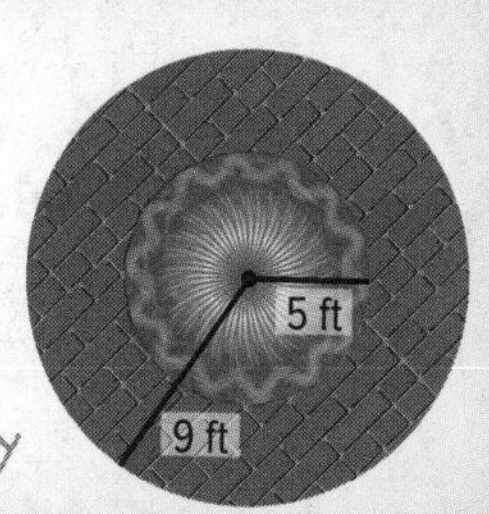

Mid-Chapter Check

Vocabulary Check

1. **Be Precise** Define *circumference*. Explain how to find the circumference of a circle. (Lesson 1)

2. Fill in the blank in the sentence below with the correct term. (Lesson 3)

   A ______________________ is made up of two or more shapes.

Skills Check and Problem Solving

Find the circumference and area of each circle. Use 3.14 for π. Round to the nearest tenth if necessary. (Lessons 1 and 2)

3. circumference = ________

   area = ________

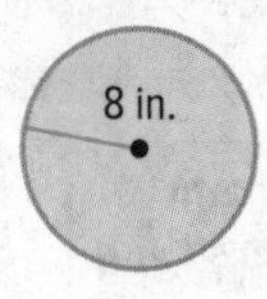

4. circumference = ________

   area = ________

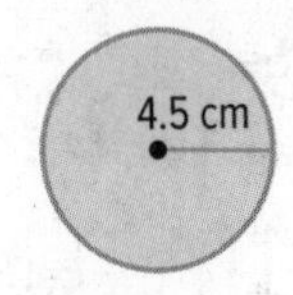

5. circumference = ________

   area = ________

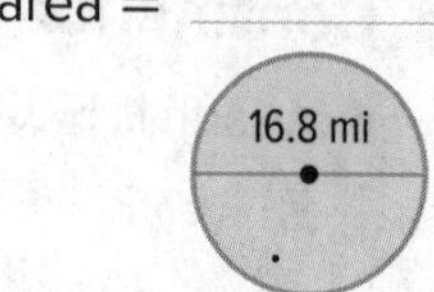

6. The dimensions of a cardboard box are shown in the figure at the right. What is the volume of the box? (Lesson 4)

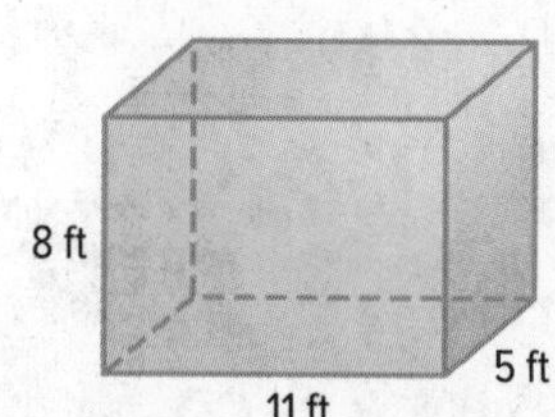

7. **Persevere with Problems** The figure at the right represents the design for a new hole for a miniature golf course. The new turf to cover the hole costs $1.50 per square foot. How much will it cost to cover the entire area? (Lesson 3)

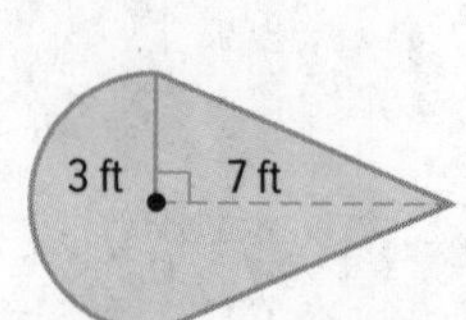

Inquiry Lab

Volume of Pyramids

WHAT is the relationship between the volume of a prism and the volume of a pyramid with the same base area and height?

Content Standards
7.G.6

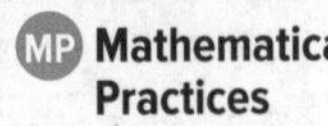

Mathematical Practices
1, 3, 5

A movie theater offers two different containers of popcorn: a square prism and a square pyramid. Both containers are 4 inches tall and have a base area of 16 square inches. Determine the container that holds more popcorn.

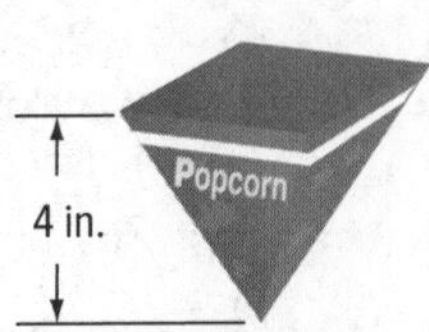

Hands-On Activity

Nets are two-dimensional patterns of three-dimensional figures.

Step 1 Draw the nets of the popcorn containers shown below onto card stock. Cut out and tape each net to form its shape. The prism and pyramid will be open. The pyramid is composed of ☐ congruent isosceles triangles with bases of 4 inches and heights of $4\frac{1}{2}$ inches.

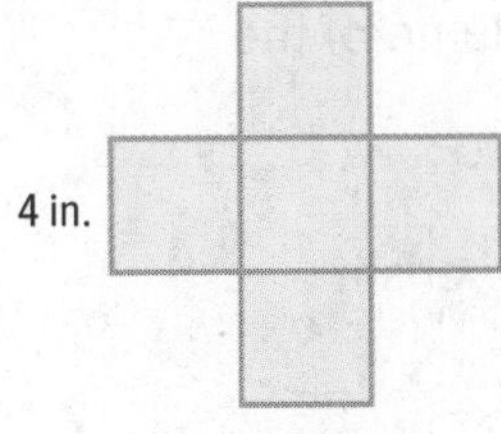

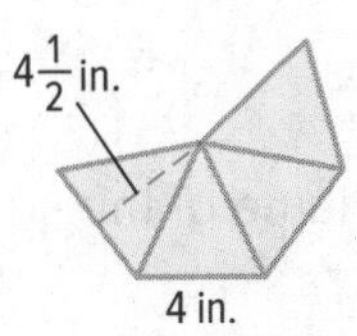

Step 2 Fill the pyramid with rice. Pour the rice from the pyramid into the prism and repeat until the prism is full. Slide a ruler across the top to level the amount.

It took ☐ pyramids of rice to fill the prism.

So, the square ______________ container holds more popcorn than the square ______________ container.

Investigate

Use Math Tools Work with a partner to repeat the Activity with the rectangular prism and the rectangular pyramid shown.

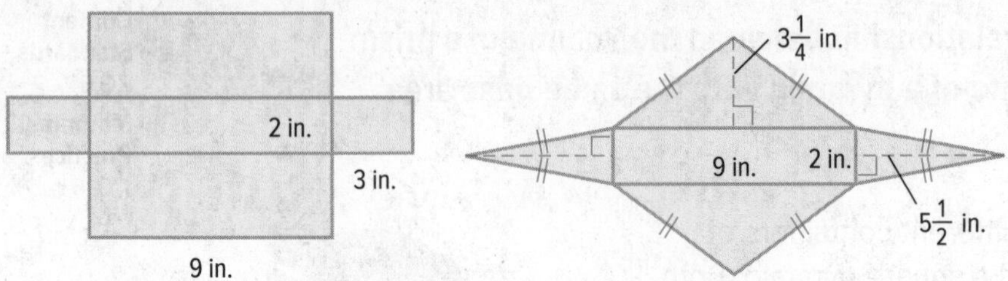

1. How many pyramids of rice did it take to fill the prism?

2. What is true about the bases of your rectangular prism and rectangular pyramid? the heights?

3. Refer to the Activity. What is true about the bases of the square prism and square pyramid? the heights?

Analyze and Reflect

4. What fraction of the volume of the rectangular prism is the volume of the rectangular pyramid? ______

5. Refer to the Activity. What fraction of the volume of the square prism is the volume of the square pyramid? ______

Create

6. **Reason Inductively** How can you find volume of a pyramid given a prism with the same base area and height? Write a formula for the volume of a pyramid based on the formula for the volume of a prism.

7. Inquiry WHAT is the relationship between the volume of a prism and the volume of a pyramid with the same base area and height?

Lesson 5

Volume of Pyramids

Real-World Link

Sand Sculpture Dion is helping his mother build a sand sculpture at the beach in the shape of a pyramid. The square pyramid has a base with a length and width of 12 inches each and a height of 14 inches.

1. Label the dimensions of the sand sculpture on the square pyramid below.

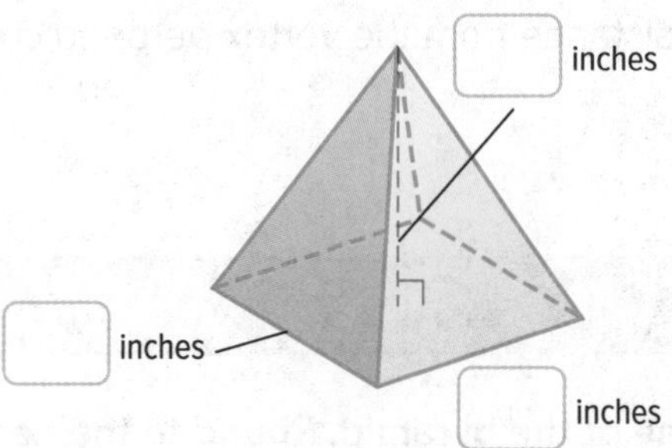

2. What is the area of the base of the pyramid?

3. What is the volume of a square prism with the same dimensions as the pyramid?

Essential Question

HOW do measurements help you describe real-world objects?

Vocabulary

lateral face

Common Core State Standards

Content Standards
7.G.6

MP Mathematical Practices
1, 3, 4, 6

Which MP Mathematical Practices did you use? Shade the circle(s) that applies.

① Persevere with Problems
② Reason Abstractly
③ Construct an Argument
④ Model with Mathematics
⑤ Use Math Tools
⑥ Attend to Precision
⑦ Make Use of Structure
⑧ Use Repeated Reasoning

Key Concept

Volume of a Pyramid

Work Zone

Words The volume V of a pyramid is one third the area of the base B times the height of the pyramid h.

Symbols $V = \frac{1}{3}Bh$

Model

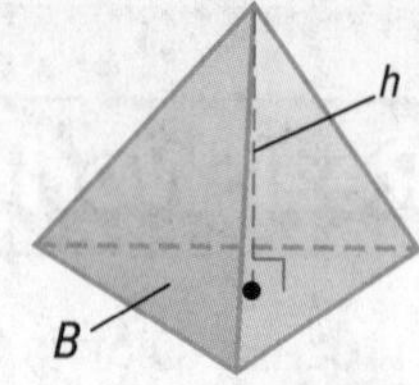

In a polyhedron, any face that is not a base is called a **lateral face**. The lateral faces of a pyramid meet at a common vertex. The height of a pyramid is the distance from the vertex perpendicular to the base.

Examples

Tutor

1. Find the volume of the pyramid. Round to the nearest tenth.

$V = \frac{1}{3}Bh$ Volume of a pyramid

$V = \frac{1}{3}(3.2 \cdot 1.4)2.8$ $B = 3.2 \cdot 1.4, h = 2.8$

$V \approx 4.2$ Simplify.

The volume is about 4.2 cubic inches.

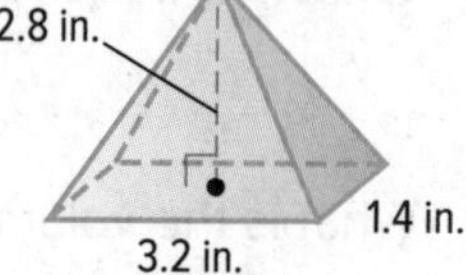

2. Find the volume of the pyramid. Round to the nearest tenth.

$V = \frac{1}{3}Bh$ Volume of a pyramid

$V = \frac{1}{3}\left(\frac{1}{2} \cdot 8.1 \cdot 6.4\right)11$ $B = \frac{1}{2} \cdot 8.1 \cdot 6.4, h = 11$

$V = 95.04$ Simplify.

The volume is about 95.0 cubic meters.

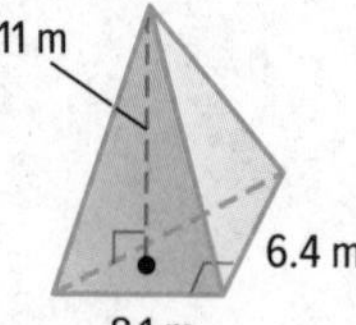

Show your work.

Got it? Do this problem to find out.

a. Find the volume of a pyramid that has a height of 9 centimeters and a rectangular base with a length of 7 centimeters and a width of 3 centimeters.

a. ________

Find the Height of a Pyramid

You can also use the formula for the volume of a pyramid to find a missing height.

Examples

3. The rectangular pyramid shown has a volume of 90 cubic inches. Find the height of the pyramid.

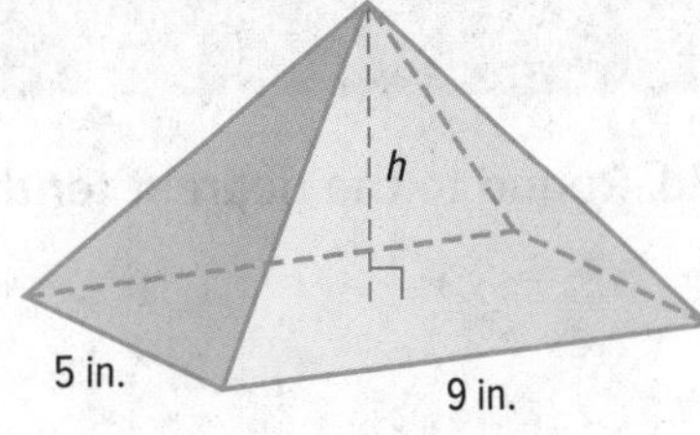

$V = \frac{1}{3}Bh$	Volume of a pyramid
$90 = \frac{1}{3}(9 \cdot 5)h$	$V = 90, B = 9 \cdot 5$
$90 = 15h$	Multiply.
$\frac{90}{15} = \frac{15h}{15}$	Divide by 15.
$6 = h$	Simplify.

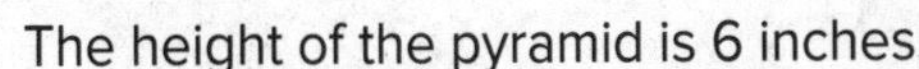

The height of the pyramid is 6 inches.

4. A triangular pyramid has a volume of 44 cubic meters. It has an 8-meter base and a 3-meter height. Find the height of the pyramid.

$V = \frac{1}{3}Bh$	Volume of a pyramid
$44 = \frac{1}{3}\left(\frac{1}{2} \cdot 8 \cdot 3\right)h$	$V = 44, B = \frac{1}{2} \cdot 8 \cdot 3$
$44 = 4h$	Multiply.
$\frac{44}{4} = \frac{4h}{4}$	Divide by 4.
$11 = h$	Simplify.

The height of the pyramid is 11 meters.

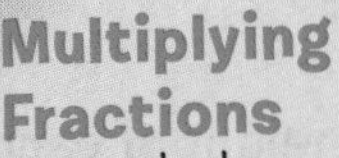

Multiplying Fractions

To find $\frac{1}{3} \cdot \frac{1}{2} \cdot 8 \cdot 3$, multiply $\frac{1}{3} \cdot \frac{1}{2}$ and $8 \cdot 3$ to get $\frac{1}{6}$ and 24, then find $\frac{1}{6}$ of 24.

Got it? Do these problems to find out.

b. A triangular pyramid has a volume of 840 cubic inches. The triangular base has a base length of 20 inches and a height of 21 inches. Find the height of the pyramid.

c. A rectangular pyramid has a volume of 525 cubic feet. It has a base of 25 feet by 18 feet. Find the height of the pyramid.

b. __________

c. __________

Example

5. Kamilah is making a model of the Food Guide Pyramid for a class project. Find the volume of the square pyramid.

12 in.
12 in.
12 in.

$V = \frac{1}{3}Bh$ Volume of a pyramid

$V = \frac{1}{3}(12 \cdot 12)12$ $B = 12 \cdot 12, h = 12$

$V = 576$ Multiply.

The volume is 576 cubic inches.

Guided Practice

Find the volume of each pyramid. Round to the nearest tenth if necessary. (Examples 1 and 2)

1. 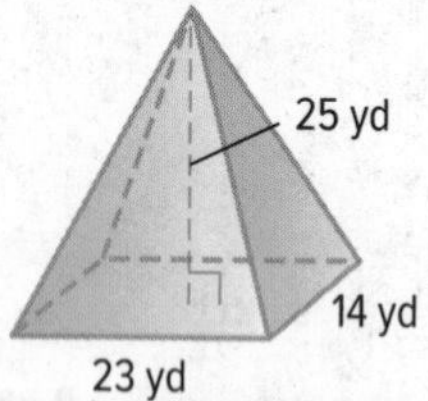

______

2. 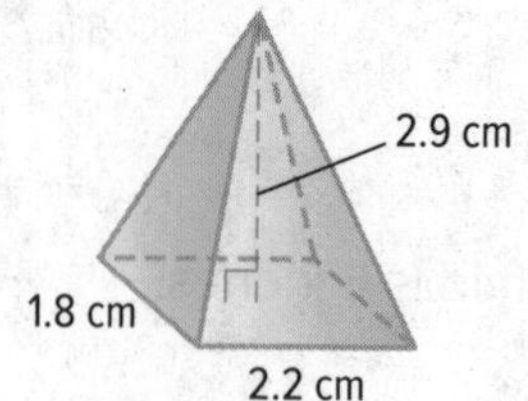

______

Find the height of each pyramid. (Examples 3 and 4)

3. square pyramid: volume 1,024 cm^3; base edge 16 cm ______

4. triangular pyramid: volume 48 in^3; base edge 9 in.; base height 4 in. ______

5. The Transamerica Pyramid is a skyscraper in San Francisco. The rectangular base has a length of 175 feet and a width of 120 feet. The height is 853 feet. Find the volume of the building. (Example 5) ______

6. **Building on the Essential Question** When you are finding the volume of a pyramid, why is it important to know the shape of the base of the pyramid?

______

______

Name ______________________ My Homework ______________________

Independent Practice

Go online for Step-by-Step Solutions

Find the volume of each pyramid. Round to the nearest tenth if necessary. (Examples 1 and 2)

1. 

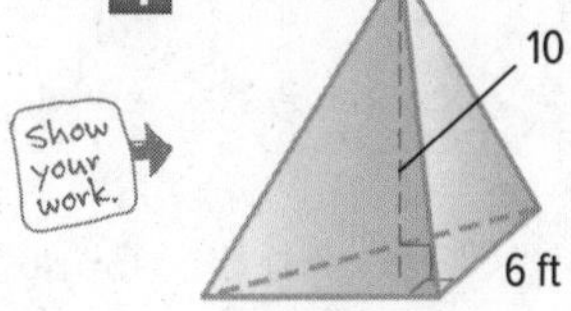

______

2. 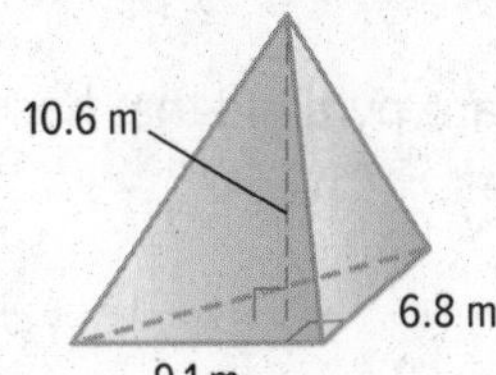

______

3. 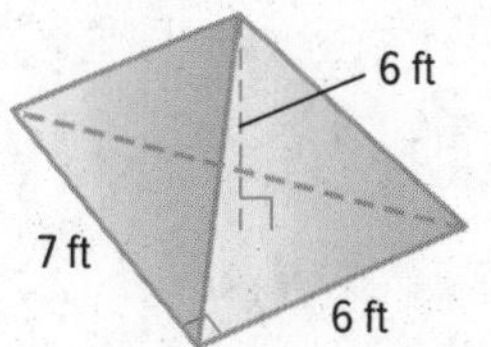

______

4. 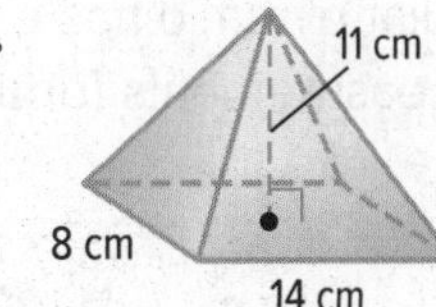

______

Find the height of each pyramid. (Examples 3 and 4)

5. rectangular pyramid: volume 448 in^3; base edge 12 in.; base length 8 in.

______

6. triangular pyramid: volume 270 cm^3; base edge 15 cm; height of base 4 cm

______

7. A glass pyramid has a height of 4 inches. Its rectangular base has a length of 3 inches and a width of 2.5 inches. Find the volume of glass used to create the pyramid. (Example 5)

______

8. The Pyramid Arena in Memphis, Tennessee, is a square pyramid that is 321 feet tall. The base has 600-foot sides. Find the volume of the pyramid. (Example 5)

______

9. **MP Reason Inductively** A rectangular pyramid has a length of 14 centimeters, a width of 9 centimeters, and a height of 10 centimeters. Explain the effect on the volume if each dimension were doubled.

10. Find the height of a square pyramid that has a volume of $25\frac{3}{5}$ meters and a base with 4 meter sides.

H.O.T. Problems Higher Order Thinking

11. **MP Be Precise** A rectangular pyramid has a volume of 160 cubic feet. Find two possible sets of measurements for the base area and height of the pyramid.

12. **MP Persevere with Problems** A square pyramid and a cube have the same bases and volumes. How are their heights related? Explain.

13. **MP Reason Inductively** The two figures shown have congruent bases. How does the volume of the two square pyramids in Figure B compare to the volume of the square pyramid in Figure A?

Figure A

Figure B

14. **MP Reason Inductively** Determine whether the following statement is *true* or *false*. Explain your reasoning.

The volumes of a rectangular-based pyramid and a triangular-based pyramid with congruent heights and equal base areas are equal.

Extra Practice

Find the volume of each pyramid. Round to the nearest tenth if necessary.

15. 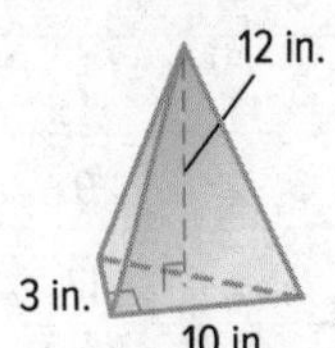

60 in^3

Homework Help

$V = \frac{1}{3}Bh$

$V = \frac{1}{3}\left(\frac{1}{2} \cdot 10 \cdot 3\right)12$

$V = 60$

16. 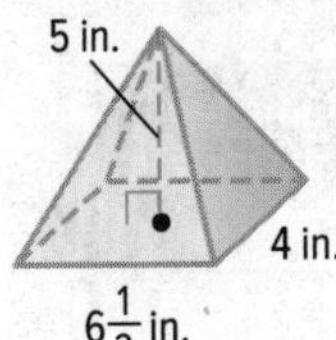

17. 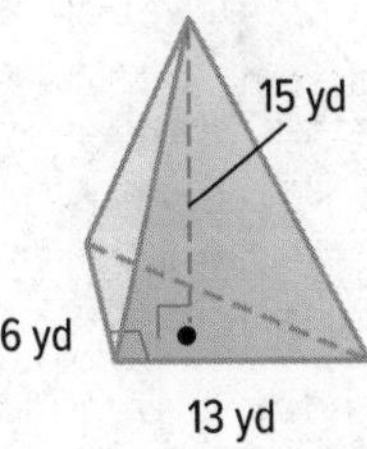

18. 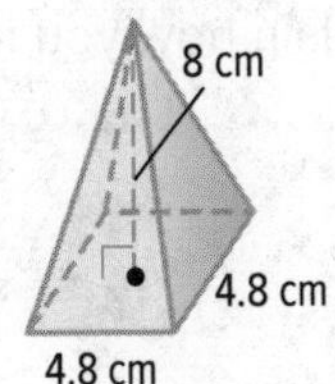

Find the height of each pyramid.

19. square pyramid: volume 297 ft^3; area of the base 81 ft^2

20. hexagonal pyramid: volume 1,320 ft^3; area of the base 120 ft^2

21. square pyramid: volume 550 in^3; area of the base 75 in^2

22. rectangular pyramid: volume 3,800 m^3; area of the base 300 m^2

23. An ancient stone pyramid has a height of 13.6 meters. The edges of the square base are 16.5 meters. Find the volume of the stone pyramid.

Power Up! Common Core Test Practice

24. The table shows the base dimensions and heights of 4 rectangular pyramids. Sort the pyramids from least to greatest volume.

Pyramid	ℓ (yd)	w (yd)	h (yd)
A	4	9	5
B	6	6	7
C	5	5	9
D	3	6	12

	Pyramid	Volume (yd^3)
Least		
Greatest		

Which pyramid has the greatest volume?

25. The rectangular pyramid shown has a volume of 1,560 cubic inches. What is the height of the pyramid? Explain how you found your answer.

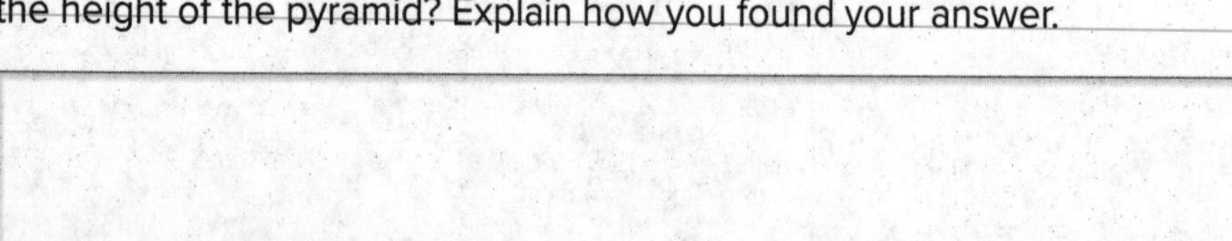

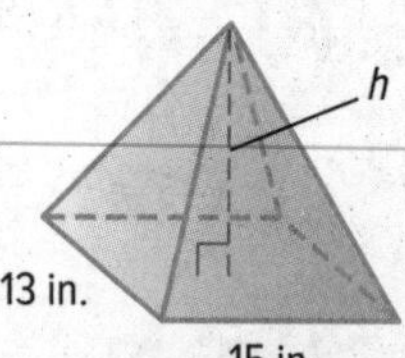

Common Core Spiral Review

Convert each length to feet. Then find the area of each figure in square feet.
5.MD.1

26.

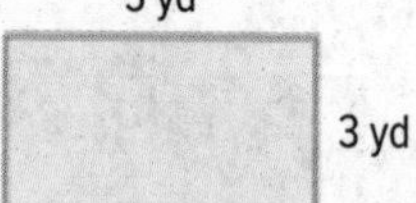

27.

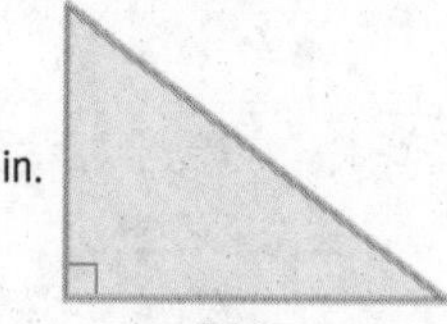

28.

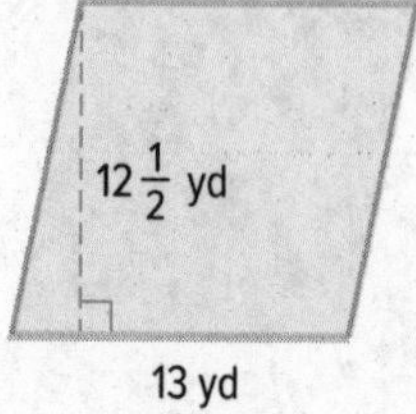

29.

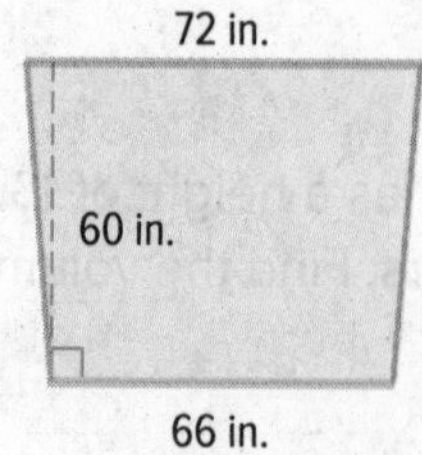

Inquiry Lab

Nets of Three-Dimensional Figures

HOW can models and nets help you find the surface area of prisms?

Content Standards
7.G.6

Mathematical Practices
1, 3, 6

Nets are used to design and manufacture items such as boxes and labels. Find the shapes that make up the net of a cereal box.

Hands-On Activity 1

Make a net from a rectangular prism.

Step 1 Use an empty cereal box. Cut off one of the two top flaps. The remaining top flap is the top face.

Step 2 Label the top and bottom faces using a green marker. Label the front and back faces using a blue marker. Label the left and right faces using a red marker.

Step 3 Carefully cut along the three edges of the top face. Then cut down each vertical edge.

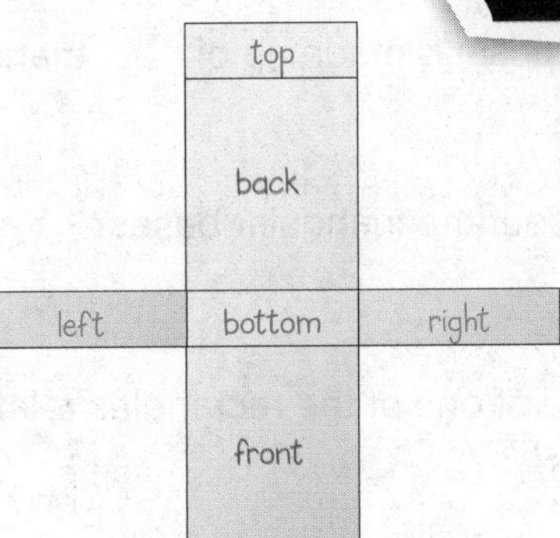

The net of a cereal box is made up of a total of ☐ rectangles.

What do you notice about the top and bottom faces, the left and right faces, and the front and back faces?

____

Hands-On Activity 2

Make a triangular prism from a net.

Step 1 Draw a net on a piece of card stock with the dimensions shown below.

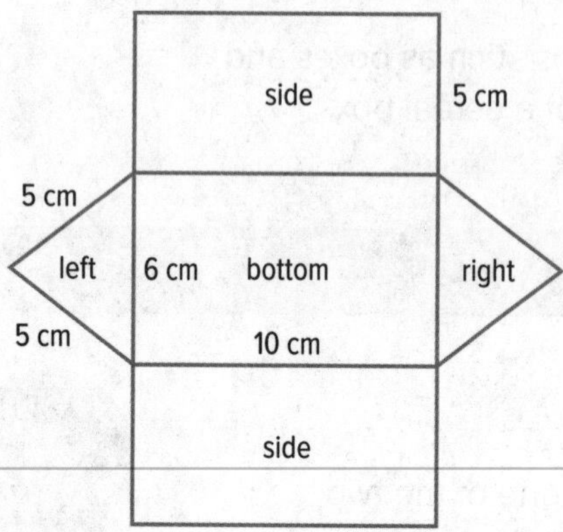

Step 2 Fold the net into a triangular prism. Tape together adjacent edges.

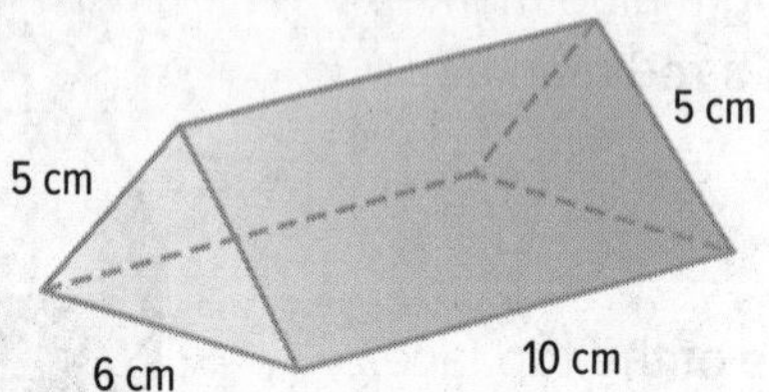

The triangular prism is made up of ☐ triangles and ☐ rectangles.

What is true about the triangular bases?

____

How is the side of one of the rectangles related to the base of one of the triangles?

____

Explain one way to find the total surface area of a triangular prism.

____

____

____

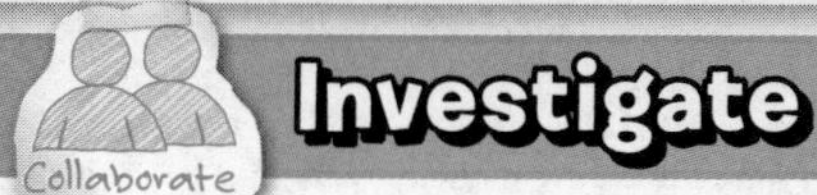

Work with a partner to solve each problem.

1. A net of a rectangular prism that is 24 inches by 18 inches by 4 inches is shown. The net of the prism is labeled with *top*, *bottom*, *side*, and *end*. Fill in the boxes to find the total area of the rectangular prism.

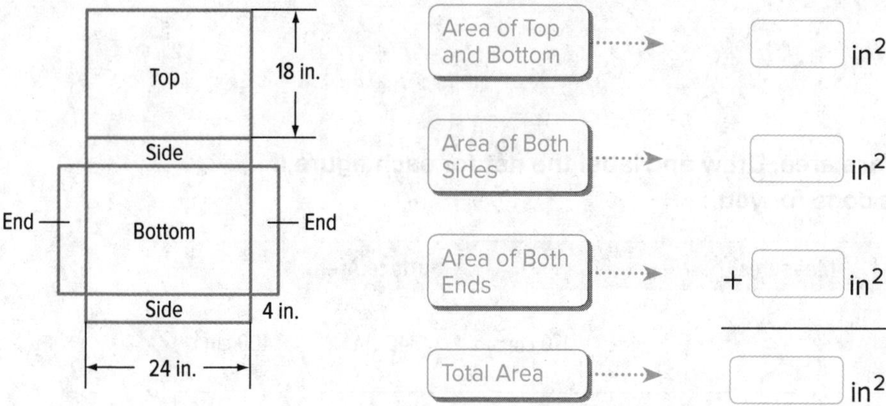

2. Describe in words how you could find the total surface area of a rectangular prism.

3. A net of a triangular prism is shown. Fill in the boxes to find the total area of the triangular prism.

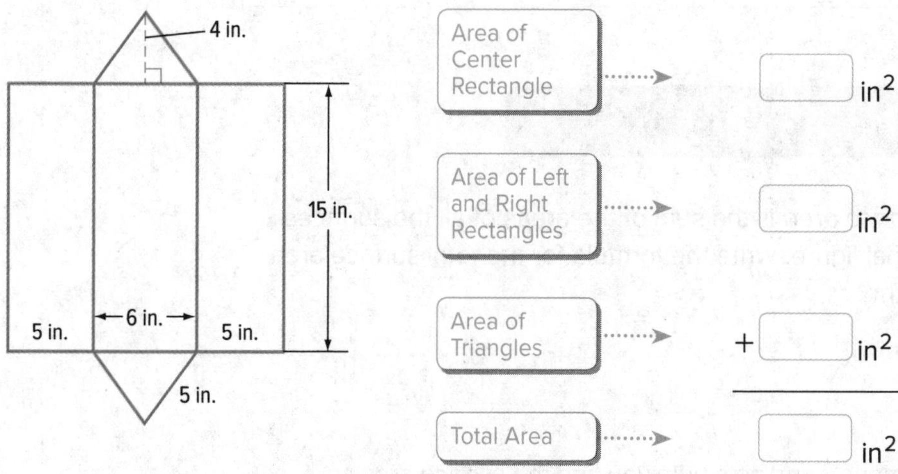

4. Describe in words how you could find the total surface area of a triangular prism.

Analyze and Reflect

Work with a partner.

5. **Reason Inductively** Suppose Ladell wants to wrap a present in a container that is a rectangular prism. How can he determine the amount of wrapping paper that he will need? ______________________________

____

____

Circle each correct surface area. Draw and label the net for each figure if needed. The first one is done for you.

	Prism	Measures	Surface Area		
	Rectangular	Length: 10 cm Width: 8 cm Height: 5 cm	170 cm^2	340 cm^2	400 cm^2
6.	Rectangular	Length: 3 ft Width: 2 ft Height: 5 ft	30 ft^2	31 ft^2	62 ft^2
7.	Rectangular	Length: 2 m Width: 1 m Height: 1.5 m	3 m^2	6.5 m^2	13 m^2
8.	Triangular	Area of Top and Bottom Triangles: 3 mm^2 Area of Center Rectangle: 12 mm^2 Area of Left and Right Rectangles: 10 mm^2	25 mm^2	28 mm^2	38 mm^2
9.	Triangular	Area of Top and Bottom Triangles: 6 in^2 Area of Center Rectangle: 50.4 in^2 Area of Left and Right Rectangles: 56 in^2	174.4 in^2	118.4 in^2	112.4 in^2

Create

10. **Be Precise** *Surface area* is the sum of the areas of all the surfaces of a three-dimensional figure. Write the formula for the total surface area of a rectangular prism.

____

____

11. **Inquiry** HOW can models and nets help you find the surface area of prisms? ______________________________

____

____

Lesson 6

Surface Area of Prisms

Real-World Link

Message Board Members of a local recreation center are permitted to post messages on 8.5-inch by 11-inch paper on the board. Assume the signs are posted vertically and do not overlap, as shown below.

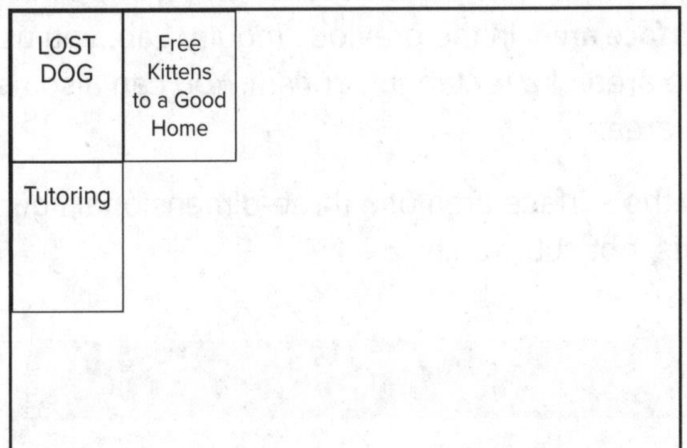

1. Suppose 6 messages fit across the board widthwise.

   What is the width of the board in inches? ☐ inches

2. Suppose 3 messages fit down the board lengthwise.

   What is the length of the board in inches? ☐ inches

3. What is the area in square inches of the message board?

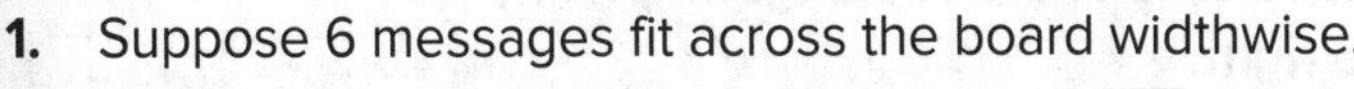

4. Messages can also be posted on the other side of the board. What is the total area of the front and back of the board in square inches?

___

Which MP **Mathematical Practices** did you use? Shade the circle(s) that applies.

① Persevere with Problems
② Reason Abstractly
③ Construct an Argument
④ Model with Mathematics
⑤ Use Math Tools
⑥ Attend to Precision
⑦ Make Use of Structure
⑧ Use Repeated Reasoning

Essential Question

HOW do measurements help you describe real-world objects?

Vocabulary

surface area

Common Core State Standards

Content Standards
7.G.6

MP **Mathematical Practices**
1, 3, 4, 6

Key Concept

Surface Area of a Rectangular Prism

Work Zone

Words The surface area *S.A.* of a rectangular prism with base ℓ, width w, and height h is the sum of the areas of its faces.

Model

Symbols $S.A. = 2\ell h + 2\ell w + 2hw$

The sum of the areas of all the surfaces, or faces, of a three-dimensional figure is the **surface area**. In the previous Inquiry Lab, you used a net to find the surface area of a rectangular prism. You can also use a formula to find surface area.

When you find the surface area of a three-dimensional figure, the units are square units, not cubic units.

Example

Watch Tutor

1. Find the surface area of the rectangular prism shown at the right.

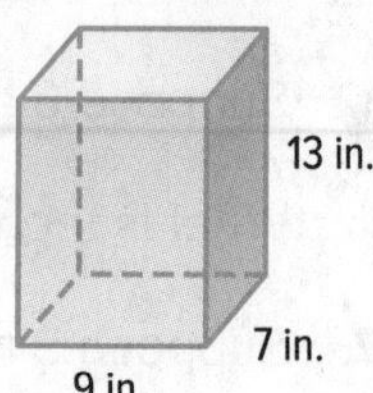

Replace ℓ with 9, w with 7, and h with 13.

$$\begin{aligned} \text{surface area} &= 2\ell h + 2\ell w + 2hw \\ &= 2 \cdot 9 \cdot 13 + 2 \cdot 9 \cdot 7 + 2 \cdot 13 \cdot 7 \\ &= 234 + 126 + 182 \quad \text{Multiply first. Then add.} \\ &= 542 \end{aligned}$$

The surface area of the prism is 542 square inches.

Got it? Do these problems to find out.

Find the surface area of each rectangular prism.

a.

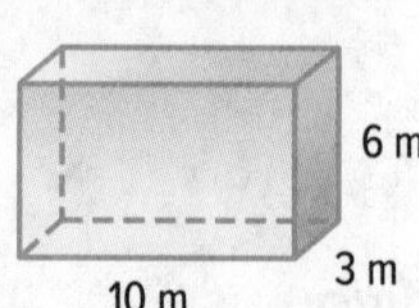

b.

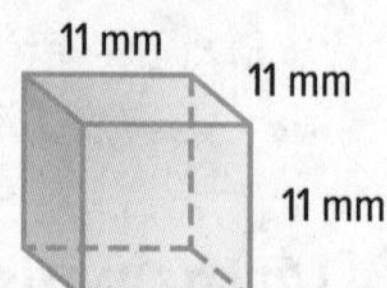

a. ________

b. ________

Example

2. Domingo built a toy box 60 inches long, 24 inches wide, and 36 inches high. He has 1 quart of paint that covers about 87 square feet of surface. Does he have enough to paint the outside of the toy box? Justify your answer.

Step 1 Find the surface area of the toy box.

Replace ℓ with 60, w with 24, and h with 36.

$$\text{surface area} = 2\ell h + 2\ell w + 2hw$$
$$= 2 \cdot 60 \cdot 36 + 2 \cdot 60 \cdot 24 + 2 \cdot 36 \cdot 24$$
$$= 8{,}928 \text{ in}^2$$

Step 2 Find the number of square inches the paint will cover.

$1 \text{ ft}^2 = 1 \text{ ft} \times 1 \text{ ft}$ Replace 1 ft with 12 in.

$= 12 \text{ in.} \times 12 \text{ in.}$ Multiply.

$= 144 \text{ in}^2$

So, 87 square feet is equal to 87 × 144 or 12,528 square inches.

Since 12,528 > 8,928, Domingo has enough paint.

Consistent Units
Since the surface area of the toy box is expressed in inches, convert 87 ft^2 to square inches so that all measurements are expressed using the same units.

Got it? Do this problem to find out.

c. The largest corrugated cardboard box ever constructed measured about 23 feet long, 9 feet high, and 8 feet wide. Would 950 square feet of paper be enough to cover the box? Justify your answer.

c. ____________

Surface Area of Triangular Prisms

To find the surface area of a triangular prism, it is more efficient to find the area of each face and calculate the sum of all of the faces rather than using a formula.

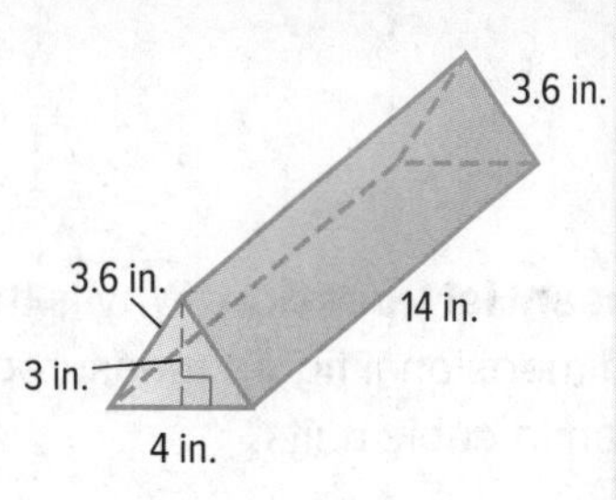

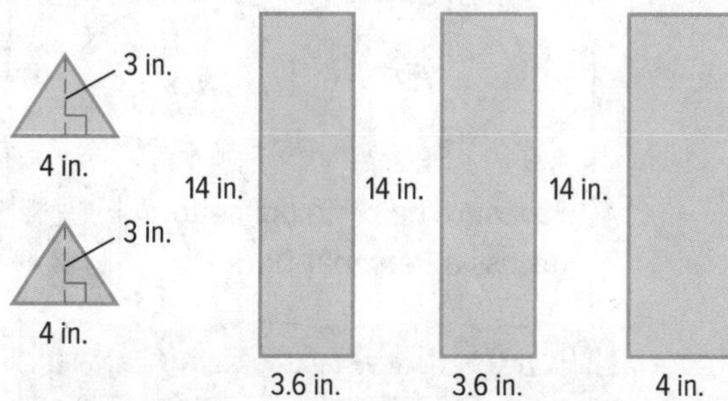

Example

3. Marty is mailing his aunt the package shown. How much cardboard is used to create the shipping container?

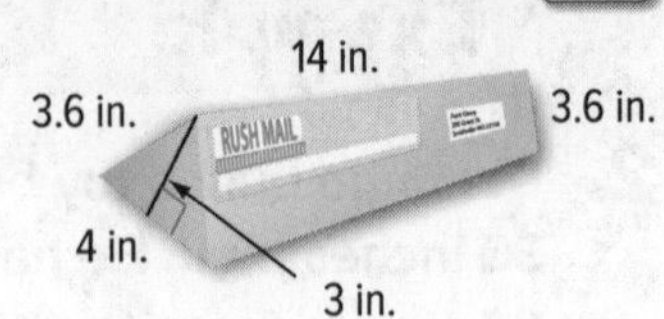

Find the area of each face and add.

The area of each triangle is $\frac{1}{2} \cdot 4 \cdot 3$ or 6.

The area of two of the rectangles is $14 \cdot 3.6$ or 50.4. The area of the third rectangle is $14 \cdot 4$ or 56.

The sum of the areas of the faces is $6 + 6 + 50.4 + 50.4 + 56$ or 168.8 square inches.

Got it? Do this problem to find out.

d. ______

d. Find the surface area of the triangular prism.

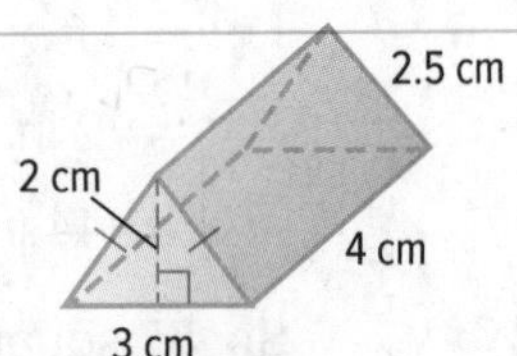

Guided Practice

Find the surface area of each prism. (Examples 1–3)

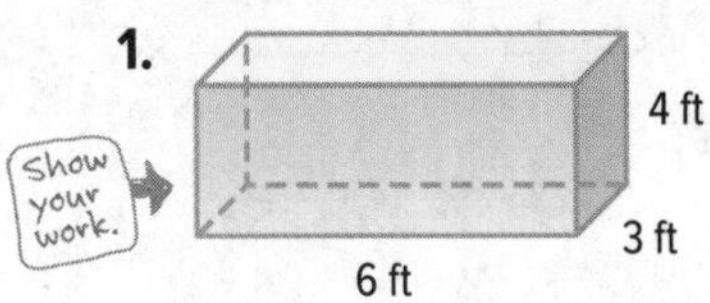

1. ______

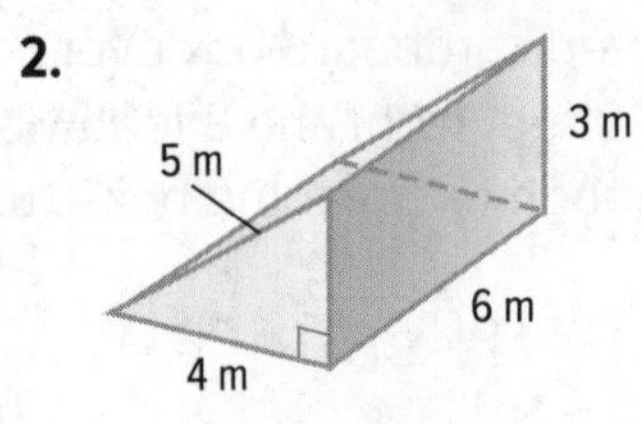

2. ______

3. **Building on the Essential Question** Why is the surface area of a three-dimensional figure measured in square units rather than in cubic units?

______

______

______

Name ______ My Homework ______

Independent Practice

Go online for Step-by-Step Solutions 

Find the surface area of each rectangular prism. Round to the nearest tenth if necessary. (Example 1)

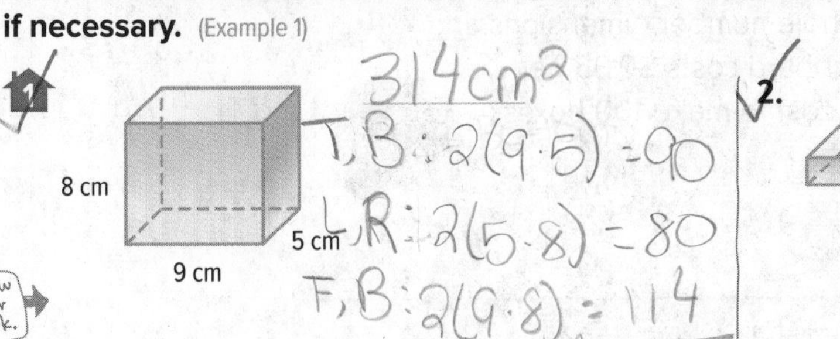

3. When making a book cover, Anwar adds an additional 20 square inches to the surface area to allow for overlap. How many square inches of paper will Anwar use to make a book cover for a book 11 inches long, 8 inches wide, and 1 inch high? (Example 2) ______

Find the surface area of each triangular prism. (Example 3)

4.

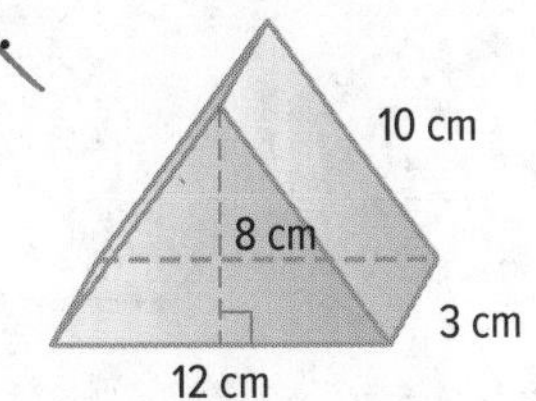

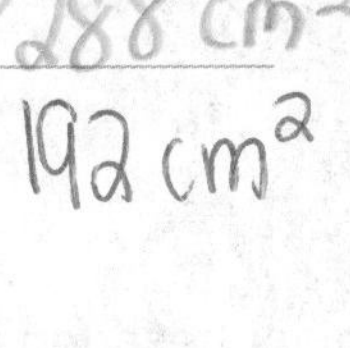

5.

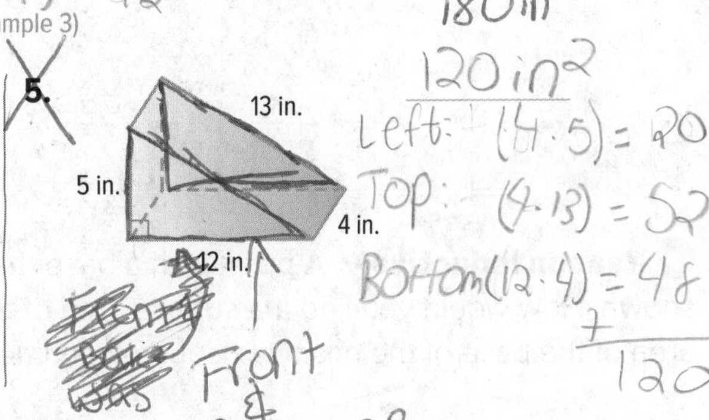

6. **MP Model with Mathematics** Refer to the graphic novel frame below. What whole number dimensions would allow the students to maximize the volume while keeping the surface area at most 160 square feet? Explain. ______

7. Write a formula for the surface area *S.A.* of a cube in which each side measures *x* units.

S.A. = $6x^2$

8. A company will make a cereal box with whole number dimensions and a volume of 100 cubic centimeters. If cardboard costs \$0.05 per 100 square centimeters, what is the least cost to make 100 boxes?

H.O.T. Problems Higher Order Thinking

9. **MP Reason Inductively** Determine if the following statement is *true* or *false*. Explain your reasoning.

If you double one of the dimensions of a rectangular prism, the surface area will double.

False, because when you double the dimensions you are just finding the 2nd side with the same measurments. The Surface Area is not dabled.

10. **MP Reason Inductively** A prism with a base that is a regular hexagon is shown. How would you find the surface area of the hexagonal prism if the area of the base of the prism is *x* square centimeters?

y cm

8 cm

11. **MP Persevere with Problems** The figure at the right is made by placing a cube with 12-centimeter sides on top of another cube with 15-centimeter sides. Find the surface area of the figure.

1,926 cm^2 135 cm^2

$5(12 \cdot 12) = 720$

$15^2 - 12^2$

$225 - 144 = 81$

$5 \cdot 12 = 60$

$5 \cdot 15 = 75$

$1125 + 720 + 81 = 1926$

12 cm

12 cm

15 cm

15 cm

$5(15 \cdot 15) = 1125$

12. **MP Model with Mathematics** Draw and label a rectangular prism that has a total surface area between 100 and 200 square units. Then find the surface area of your prism.

Show your work.

Extra Practice

Find the surface area of each prism. Round to the nearest tenth if necessary.

13. 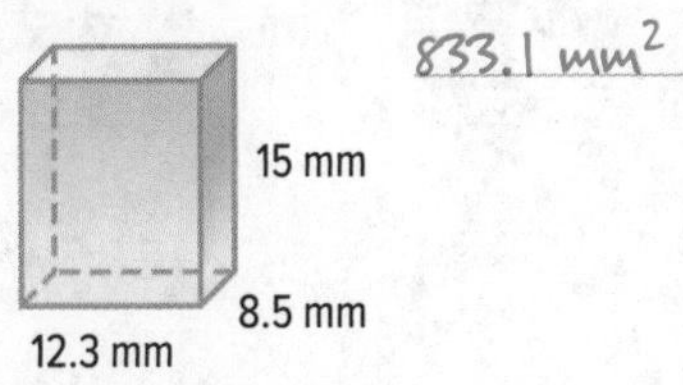

833.1 mm^2

S.A. = 2lh + 2lw + 2hw

Homework Help

= 2 • 12.3 • 15 + 2 • 12.3 • 8.5 + 2 • 15 • 8.5

= 369 + 209.1 + 255

= 833.1

14. 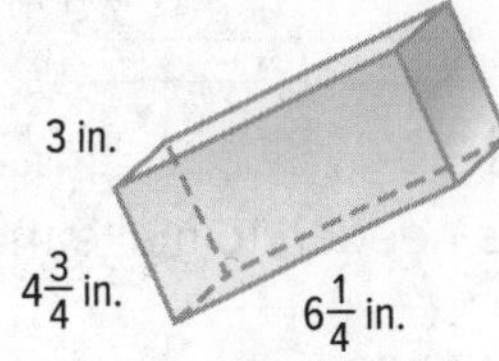

______

15. 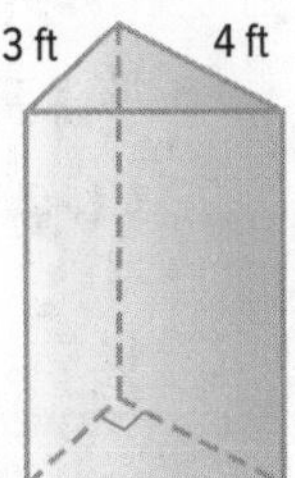

______

16. 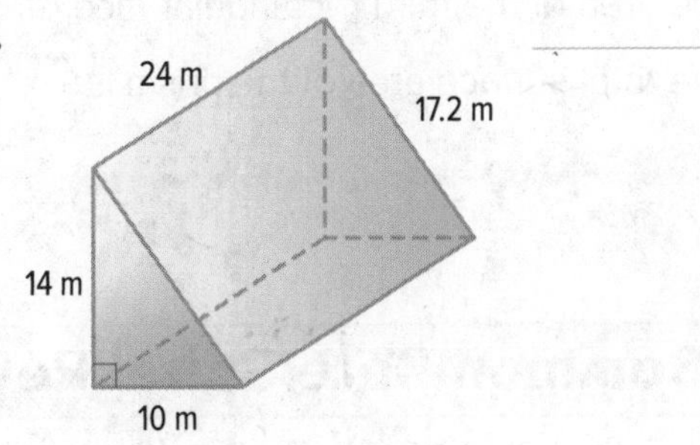

______

17. If one gallon of paint covers 350 square feet, will 8 gallons of paint be enough to paint the inside and outside of the fence shown once? Explain.

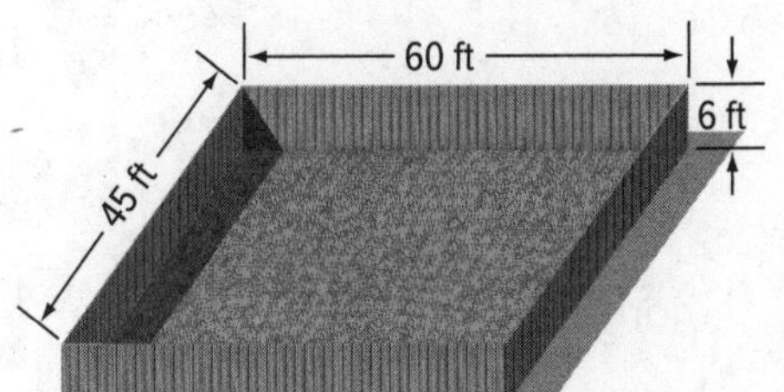

______

18. The attic shown is a triangular prism. Insulation will be placed inside all walls, not including the floor. Find the surface area that will be covered with insulation.

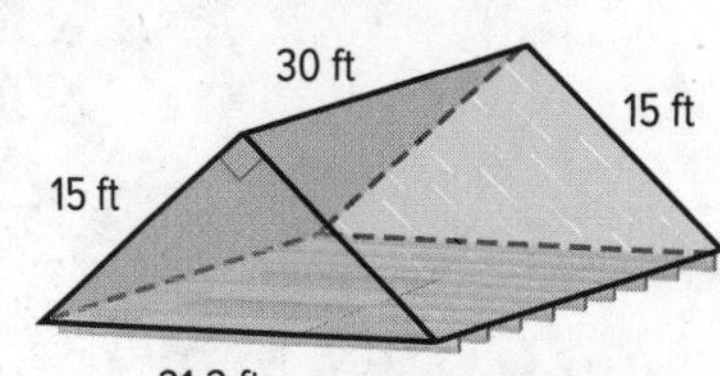

______

19. MP **Be Precise** To the nearest tenth, find the approximate amount of plastic covering the outside of the CD case.

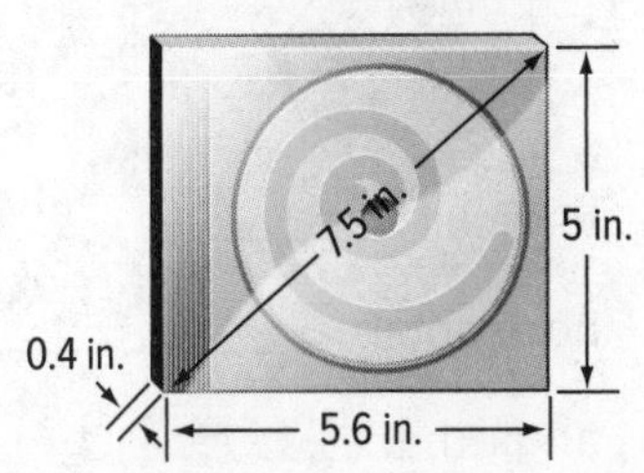

______

Power Up! Common Core Test Practice

20. A cardboard box has the dimensions shown. Select the correct values to complete the formula to find the surface area of the box.

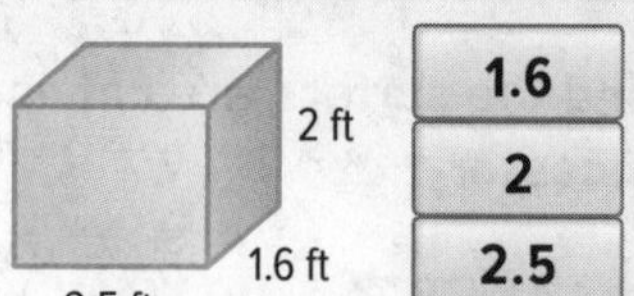

$SA =$ ☐ · ☐ · ☐ + ☐ · ☐ · ☐ + ☐ · ☐ · ☐

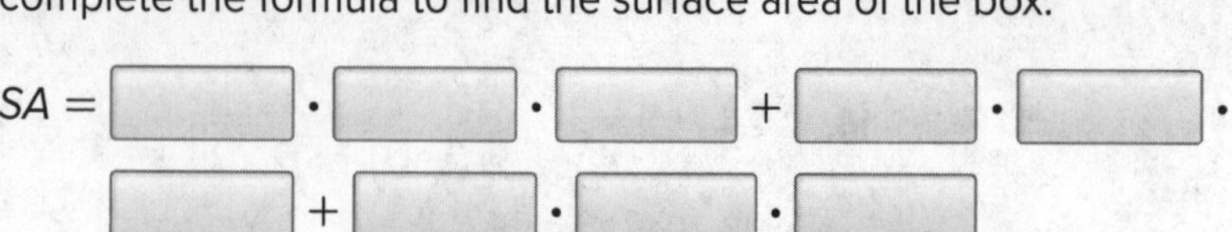

How much cardboard is needed to make the box? ☐

21. A triangular prism has the dimensions shown. Fill in each box to complete each statement.

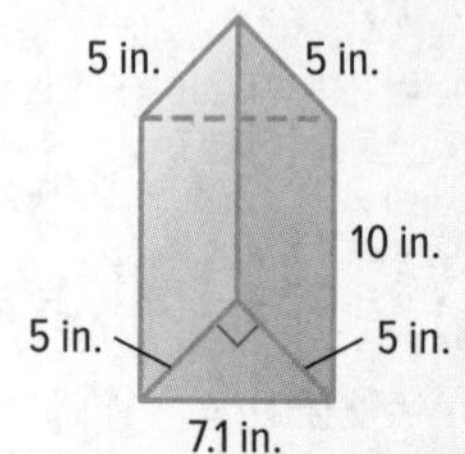

a. The area of each triangular base is ☐ square inches.

b. The area of each of the two congruent rectangular faces is ☐ square inches.

c. The area of the third rectangular face is ☐ square inches.

d. The total surface area of the prism is ☐ square inches.

Common Core Spiral Review

Describe the shape resulting from a vertical, horizontal, and angled cross section for each figure. 7.G.3

22.

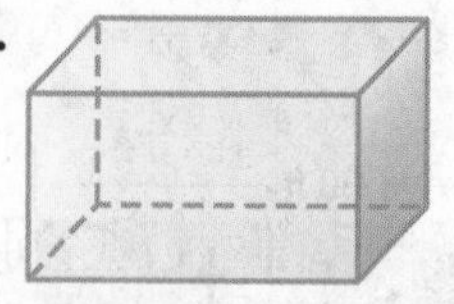

vertical: ______

horizontal: ______

angled: ______

23.

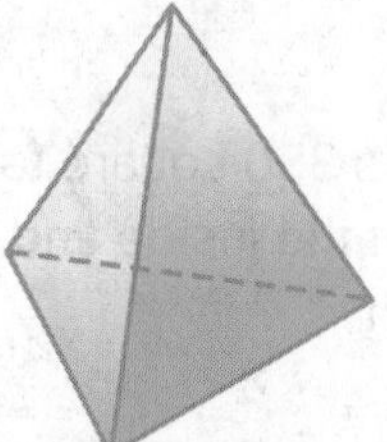

vertical: ______

horizontal: ______

angled: ______

24.

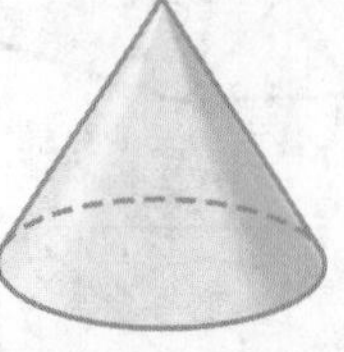

vertical: ______

horizontal: ______

angled: ______

25.

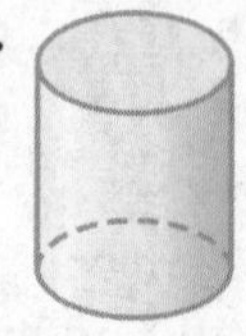

vertical: ______

horizontal: ______

angled: ______

Inquiry Lab

Relate Surface Area and Volume

HOW does the shape of a rectangular prism affect its volume and surface area?

Content Standards
7.G.6

Mathematical Practices
1, 3, 4

You can arrange blocks in many ways. How can you arrange 8 blocks to create the least possible surface area?

What do you know? ______________________________

What do you need to find? ______________________________

Hands-On Activity 1

Tools

Step 1 Create a rectangular prism using 8 centimeter cubes. Record the dimensions in the table below. Find and record the volume and surface area of the prism.

Rectangular Prism	Length (cm)	Width (cm)	Height (cm)	Volume (cm^3)	Surface Area (cm^2)
1	2	2			
2					
3					

Step 2 Repeat Step 1 for as many different rectangular prisms as you can create with 8 cubes.

Does the volume change when the prism changes? Explain.

____

____

The rectangular prism measuring ☐ × ☐ × ☐ has the least surface area.

Hands-On Activity 2

Suppose you make structures in the shape of the ones shown below. What is the volume of each structure? Which structure has the lesser surface area? Draw a net if necessary.

Figure 1

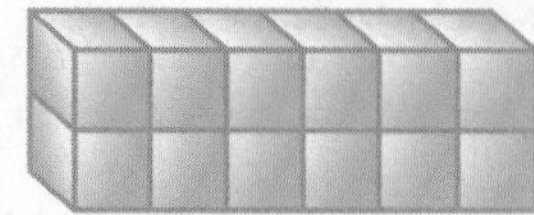
Figure 2

Step 1 Use centimeter cubes to create the rectangular prism shown in Figure 1. Write its dimensions, volume, and surface area in the table below.

Rectangular Prism	Length (cm)	Width (cm)	Height (cm)	Volume (cm^3)	Surface Area (cm^2)
Figure 1	3				
Figure 2					

Step 2 Use centimeter cubes to create the rectangular prism shown in Figure 2. Write its dimensions, volume, and surface area in the table.

Step 3 Compare the volume and surface areas of Figure 1 and Figure 2.

What do you notice about the volume of Figure 1 and Figure 2?

____

The surface area of Figure 1 is ___ square centimeters.

The surface area of Figure 2 is ___ square centimeters.

Compare the surface areas using an inequality.

___ square centimeters < ___ square centimeters

So, Figure ___ has the lesser surface area.

Investigate

Work with a partner. Compare the two figures that have the same volume. Then determine which figure has a greater surface area. Draw a net if necessary.

1.

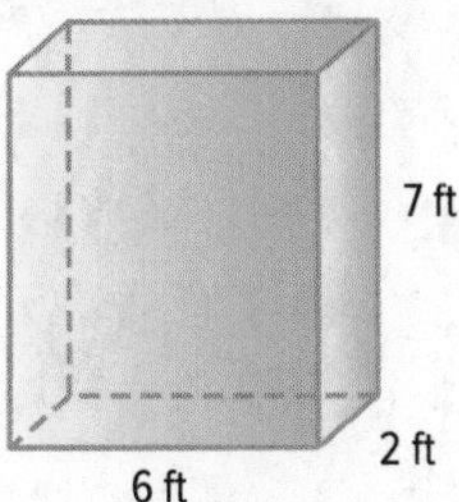

Figure 1

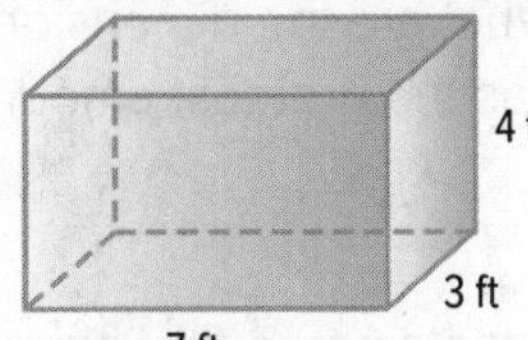

Figure 2

Show your work.

Surface Area: ____________ Surface Area: ____________

2.

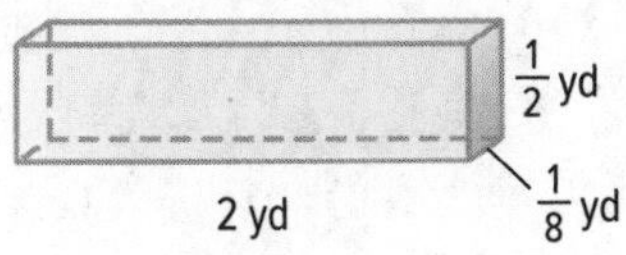

Figure 1

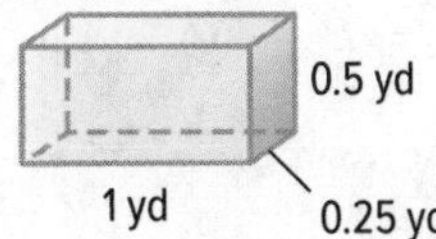

Figure 2

Surface Area: ____________ Surface Area: ____________

3.

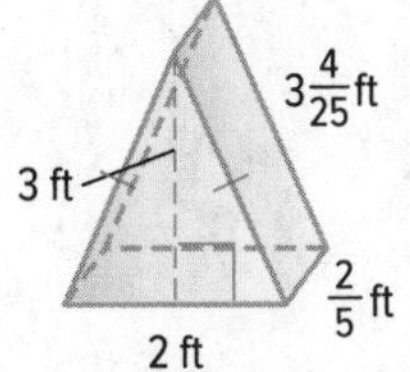

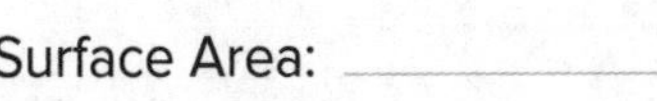

Figure 1

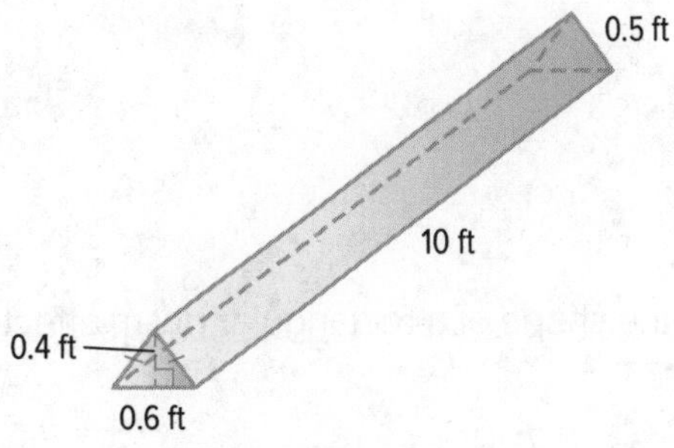

Figure 2

Surface Area: ____________ Surface Area: ____________

Analyze and Reflect

Work with a partner to solve the following problems. Draw a net if necessary.

4. Monique sews together pieces of fabric to make rectangular gift boxes. She only uses whole numbers. What are the dimensions of a box with a volume of 50 cubic inches that has the greatest amount of surface area?

5. Thomas is creating a decorative container to fill with colored sand. He uses only whole numbers. The top of the container is open. What are the dimensions of the rectangular prism that holds 100 cubic inches with the least amount of surface area?

6. MP **Construct an Argument** Zack needs to melt a stick of butter that measures 5 inches by 1 inch by 1 inch. He is going to put the butter in a pan on top of the stove. Explain why cutting the butter into smaller pieces will help the butter melt faster.

Create

7. MP **Model with Mathematics** Draw a sketch of a triangular prism with a volume of 120 cubic units and a surface area of 184 square units.

8. Inquiry HOW does the shape of a rectangular prism affect its volume and surface area?

Lesson 7

Surface Area of Pyramids

Vocabulary Start-Up

Pyramids Ancient Egyptians built pyramids, such as the one shown in the photo below. A right square pyramid has a square base and four isosceles triangles that make up the lateral faces. The **lateral surface area** is the sum of the areas of all its lateral faces. The height of each lateral face is called **slant height**.

Essential Question

HOW do measurements help you describe real-world objects?

Vocabulary

lateral surface area
slant height
regular pyramid

Common Core State Standards

Content Standards
7.G.6

MP Mathematical Practices
1, 3, 4, 5

1. Fill in the blanks on the diagram below with the terms *slant height* and *lateral face*.

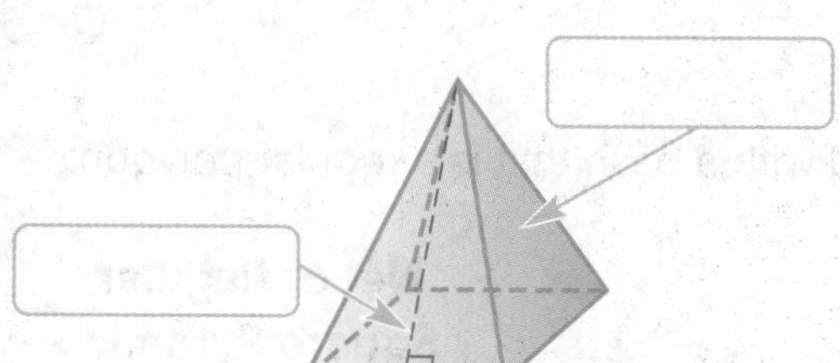

2. Draw a net of a square pyramid.

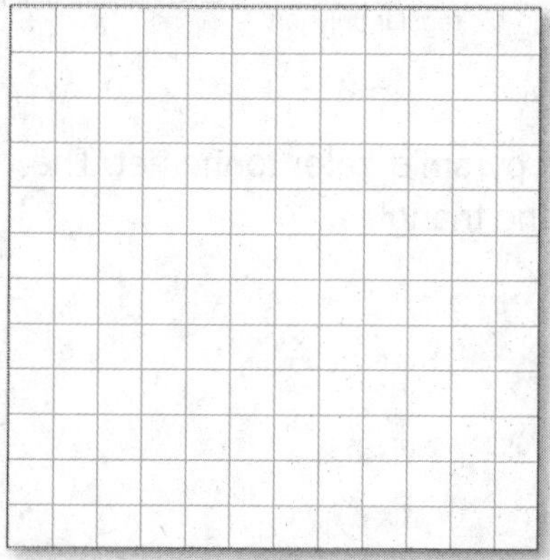

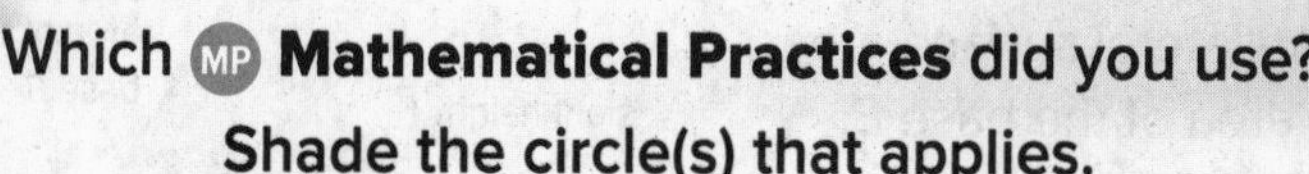

Which MP Mathematical Practices did you use? Shade the circle(s) that applies.

① Persevere with Problems
② Reason Abstractly
③ Construct an Argument
④ Model with Mathematics
⑤ Use Math Tools
⑥ Attend to Precision
⑦ Make Use of Structure
⑧ Use Repeated Reasoning

Key Concept

Surface Area of a Pyramid

Work Zone

Lateral Area

Words The lateral surface area *L.A.* of a regular pyramid is half the perimeter *P* of the base times the slant height ℓ.

Model slant height ℓ

Symbols $L.A. = \frac{1}{2}P\ell$

Total Surface Area

Words The total surface area *S.A.* of a regular pyramid is the lateral area *L.A.* plus the area of the base *B*.

Symbols $S.A. = B + L.A.$ or $S.A. = B + \frac{1}{2}P\ell$

A **regular pyramid** is a pyramid with a base that is a regular polygon.

Model of Regular Square Pyramid **Net of Regular Square Pyramid**

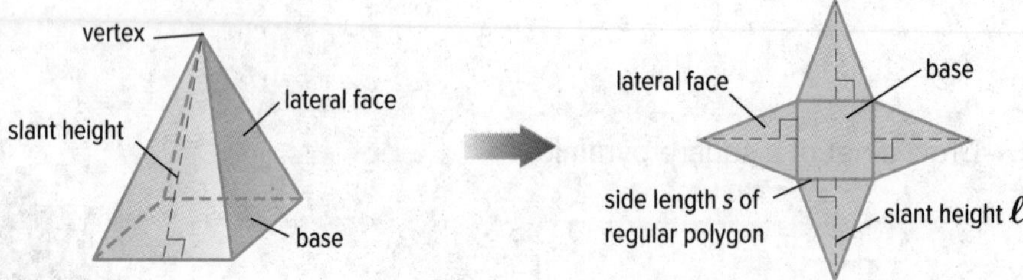

To find the lateral area *L.A.* of a regular pyramid, refer to the net. The lateral area is the sum of the areas of the triangles.

$L.A. = 4\left(\frac{1}{2}s\ell\right)$ Area of the lateral faces

$L.A. = \frac{1}{2}(4s)\ell$ Commutative Property of Multiplication

$L.A. = \frac{1}{2}P\ell$ The perimeter of the base *P* is 4*s*.

The total surface area of a regular pyramid is the lateral surface area *L.A.* plus the area of the base *B*.

$$S.A. = B + \frac{1}{2}P\ell$$

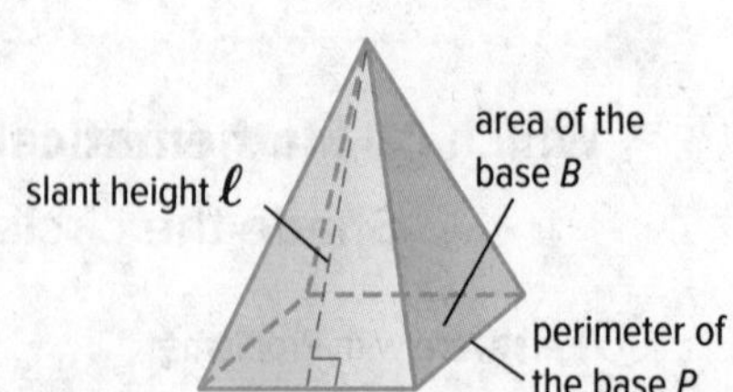

Examples

1. Find the total surface area of the pyramid. Round to the nearest tenth.

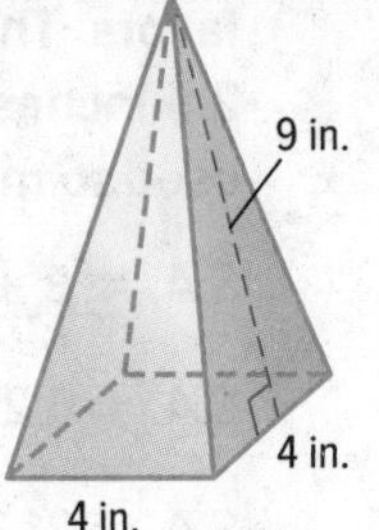

$S.A. = B + \frac{1}{2}P\ell$ — Surface area of a pyramid

$S.A. = 16 + \frac{1}{2}(16 \cdot 9)$ — $B = 4 \cdot 4$, $P = 4 \cdot 4$ or 16, $\ell = 9$

$S.A. = 88$ — Simplify.

The surface area is 88 square inches.

2. Find the total surface area of the pyramid with a base area of 111 square meters.

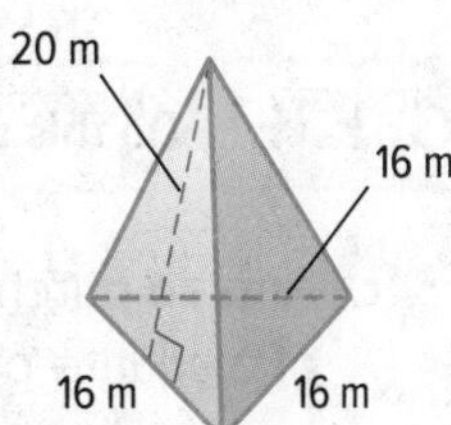

$S.A. = B + \frac{1}{2}P\ell$ — Surface area of a pyramid

$S.A. = 111 + \frac{1}{2}(48 \cdot 20)$ — $B = 111$, $P = 16 + 16 + 16$ or 48, $\ell = 20$

$S.A. = 591$ — Simplify.

The surface area of the pyramid is 591 square meters.

3. Find the total surface area of the pyramid.

$S.A. = B + \frac{1}{2}P\ell$ — Surface area of a pyramid

$S.A. = 43.5 + \frac{1}{2}P\ell$ — $B = \frac{1}{2} \cdot 10 \cdot 8.7$ or 43.5

$S.A. = 43.5 + \frac{1}{2}(30 \cdot 12)$ — $P = 10 + 10 + 10$ or 30, $\ell = 12$

$S.A. = 223.5$ — Simplify.

The surface area is 223.5 square feet.

Perimeter of a Square

The formula for the perimeter of a square is $P = 4s$.

Got it? Do these problems to find out.

a. Find the surface area of a square pyramid that has a slant height of 8 centimeters and a base length of 5 centimeters.

b. Find the total surface area of the pyramid shown.

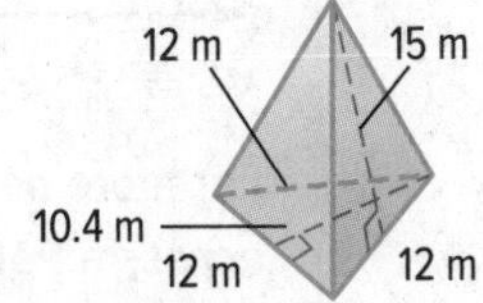

a. ____________

b. ____________

Example

4. Sal is wrapping gift boxes that are square pyramids for party favors. They have a slant height of 3 inches and base edges 2.5 inches long. How many square inches of card stock are used to make one gift box?

$S.A. = B + \frac{1}{2}P\ell$ — Surface area of a pyramid

$S.A. = 6.25 + \frac{1}{2}(10 \cdot 3)$ — $B = 2.5^2$ or 6.25, $P = 4(2.5)$ or 10, $\ell = 3$

$S.A. = 21.25$ — Simplify.

So, 21.25 square inches of card stock are used to make one gift box.

Got it? **Do this problem to find out.**

Show your work.

c. Amado purchased a bottle of perfume that is in the shape of a square pyramid. The slant height of the bottle is 4.5 inches and the base is 2 inches. Find the surface area.

c. _______________

Guided Practice

Find the total surface area of each pyramid. Round to the nearest tenth. (Examples 1–3)

1. 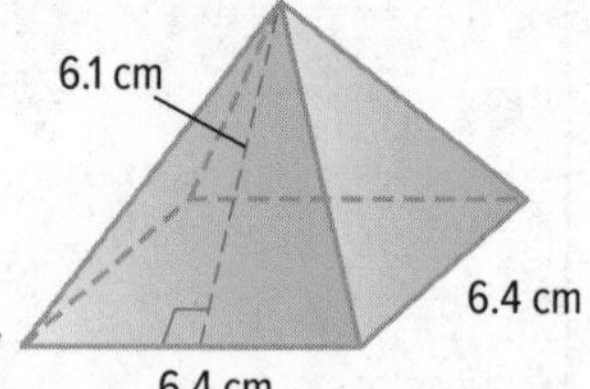

____

Show your work.

2. 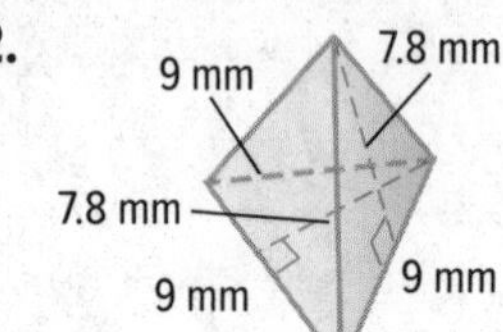

____

3. The Washington Monument is an obelisk with a square pyramid top. The slant height of the pyramid is 55.5 feet, and the square base has sides of 34.5 feet. Find the lateral area of the pyramid. (Example 4) _______________

4. **Building on the Essential Question** Justify the formula for the surface area of a pyramid.

Rate Yourself!

ow confident are you about finding the surface area of pyramids? Check the box that applies.

For more help, go online to access a Personal Tutor.

FOLDABLES Time to update your Foldable!

Name ________________ My Homework ________________

Independent Practice

Go online for Step-by-Step Solutions

Find the total surface area of each pyramid. Round to the nearest tenth. (Examples 1–3)

1. Show your work.

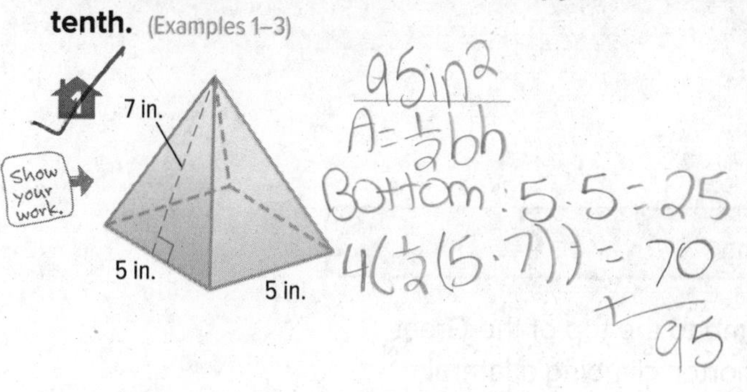

2.

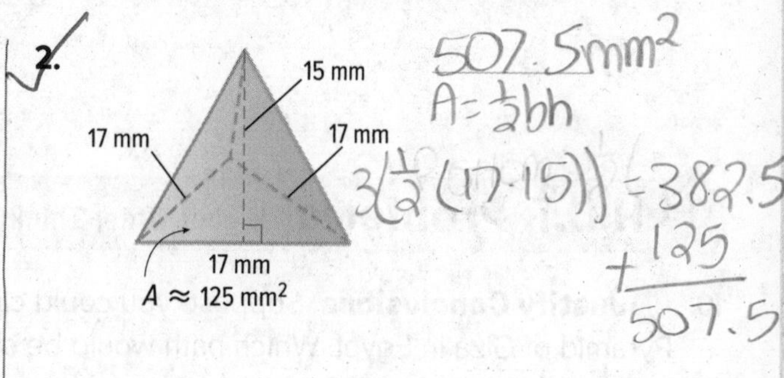

3. 4.

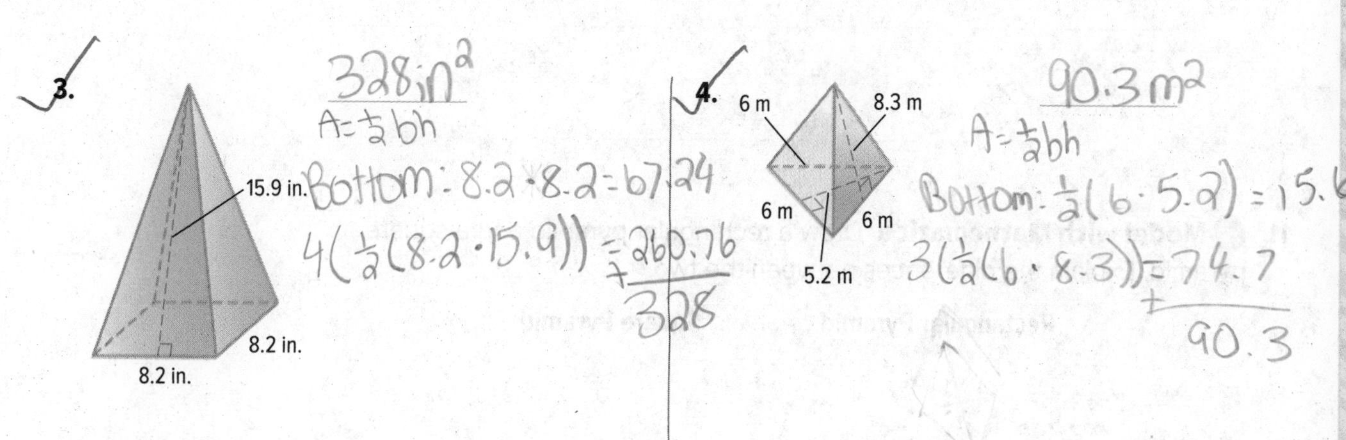

5. A triangular pyramid has a slant height of 0.75 foot. The equilateral triangular base has a perimeter of 1.2 feet and an area of about 0.07 square foot. Find the approximate surface area. (Example 4)

____

6. The gemstone shown is a square pyramid that has a base with sides 3.4 inches long. The slant height of the pyramid is 3.8 inches. Find the surface area of the gemstone. (Example 4)

____

7. Isaac is building a birdhouse for a class project. The birdhouse is a regular hexagonal pyramid. The base has side lengths of 3 inches and an area of about 24 square inches. The slant height is 6 inches. Find the approximate surface area of the birdhouse. (Example 4)

____

8. **Persevere with Problems** A square pyramid has a surface area of 175 square inches. The square base has side lengths of 5 inches. Find the slant height of the pyramid.

9. A square pyramid has a lateral area of 107.25 square centimeters and a slant height of 8.25 centimeters. Find the length of each side of its base.

H.O.T. Problems Higher Order Thinking

10. **Justify Conclusions** Suppose you could climb to the top of the Great Pyramid of Giza in Egypt. Which path would be shorter, climbing a lateral edge or the slant height? Justify your response.

11. **Model with Mathematics** Draw a rectangular pyramid and a square pyramid. Explain the differences between the two.

Rectangular Pyramid **Square Pyramid**

12. **Persevere with Problems** The total height of the figure shown is 20 inches and the slant height is 17 inches. Which has a greater surface area, the prism or the pyramid? Explain your reasoning.

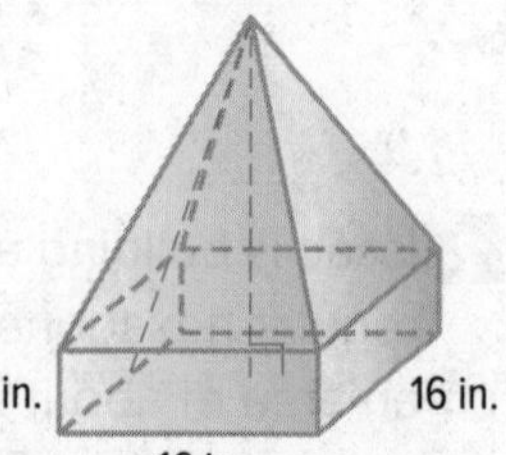

Name ______________________ My Homework ______________________

Extra Practice

Find the total surface area of each pyramid. Round to the nearest tenth.

13. 197.1 m^2

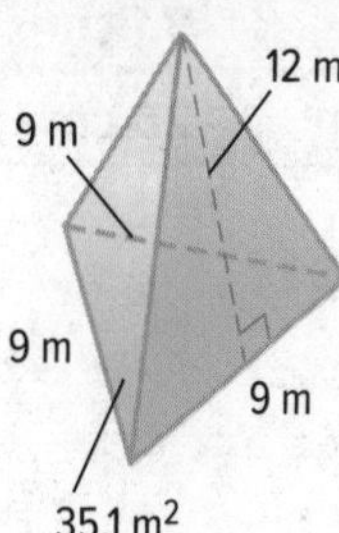

S.A. = $B + \frac{1}{2}Pl$

S.A. = $35.1 + \frac{1}{2}(27 \cdot 12)$

S.A. = 197.1

Homework Help

14. ______

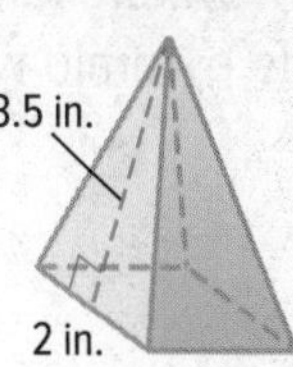

15. ______

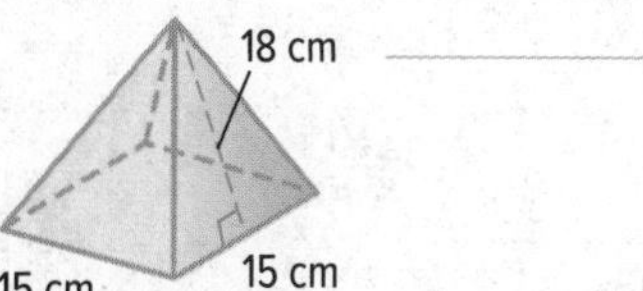

16. ______

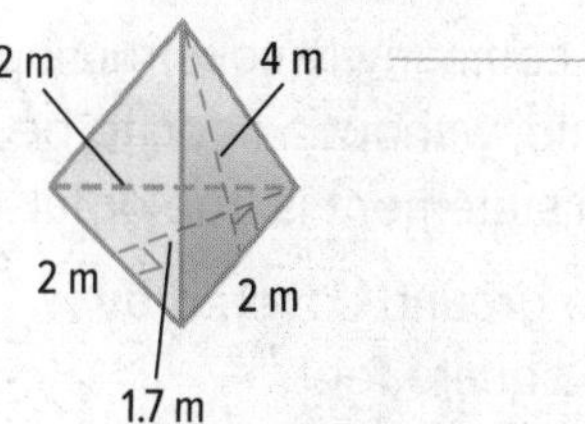

17. A square pyramid has a slant height of $4\frac{2}{3}$ feet. The base has side lengths of $2\frac{1}{4}$ feet. Find the surface area. ______

18. A building in San Francisco is shaped like a square pyramid. It has a slant height of 856.1 feet and each side of its base is 145 feet long. Find the lateral area of the building. ______

19. MP **Use Math Tools** Complete the organizer below to help you remember what each part of the formula for the surface area of a pyramid represents.

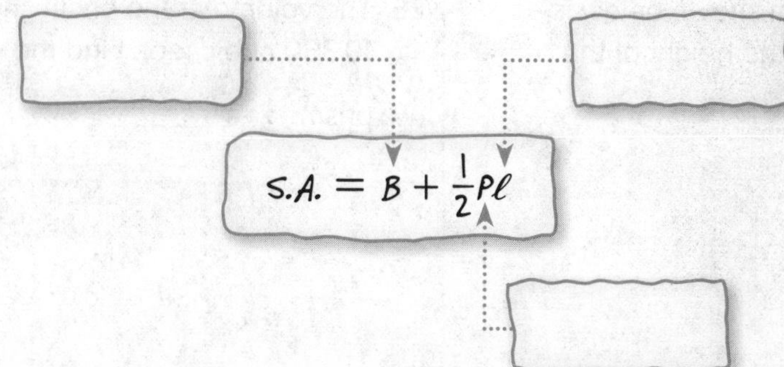

Power Up! Common Core Test Practice

20. The base of a square pyramid has a perimeter of 28 centimeters. The height of the pyramid is 2.1 centimeters longer than the side length of the base. Label the net of the pyramid with the correct dimensions.

2.1 cm	7 cm
4 cm	9.1 cm
4.9 cm	28 cm

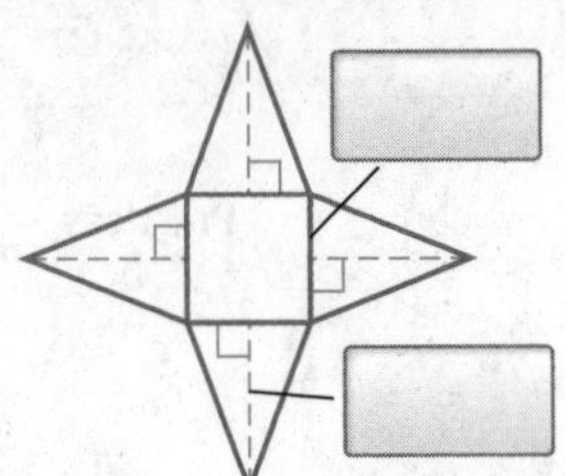

What is the total surface area of the pyramid?

21. An entertainment company is constructing a tent in the shape of a square pyramid, without a floor, to be used at a party. Determine if each statement is true or false.

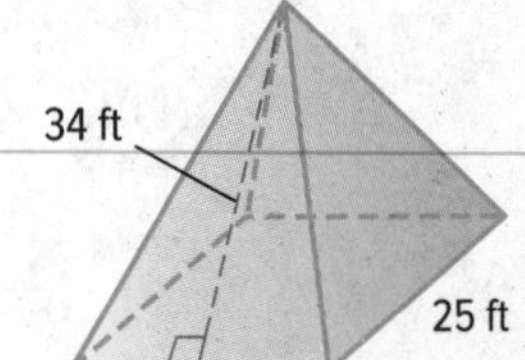

a. The area of the ground covered by the tent is 625 square feet. ☐ True ☐ False

b. The area of each triangular face of the tent is 425 square feet. ☐ True ☐ False

c. The amount of material needed to make the tent is 2,325 square feet. ☐ True ☐ False

Common Core Spiral Review

Find the surface area of each prism. 7.G.6

22. ____________

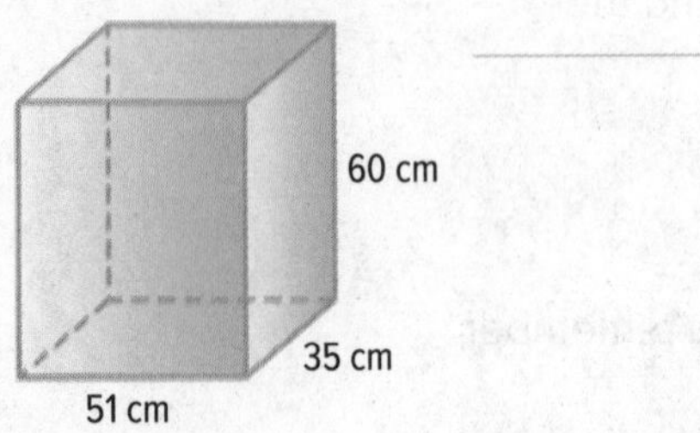

23. ____________

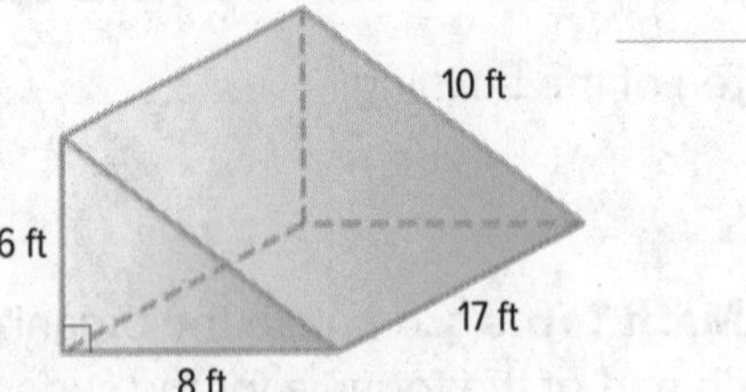

24. The volume of the prism shown below is 140 cubic meters. Find the height of the prism. 7.G.6 ____________

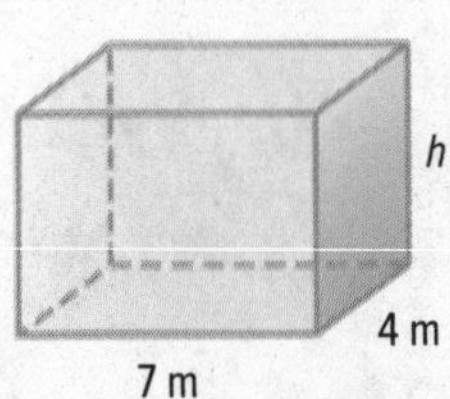

25. The volume of the prism shown below is 10,360 cubic feet. Find the width of the prism. 7.G.6 ____________

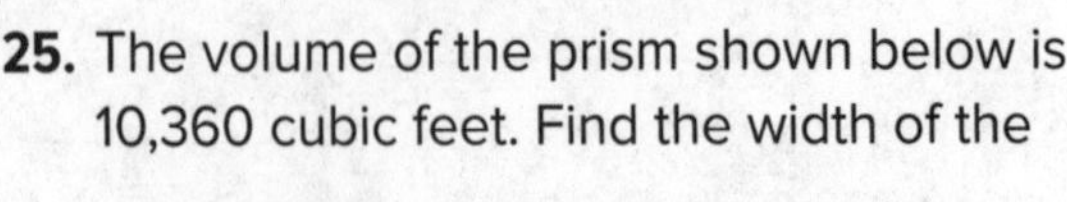

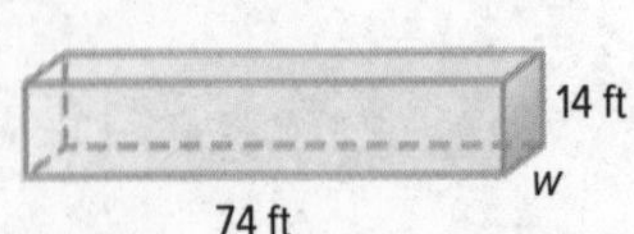

Inquiry Lab
Composite Figures

HOW can you find the volume and surface area of a composite figure?

Content Standards 7.G.6
Mathematical Practices 1, 3, 4

A company made a model of a new office building. The building is composed of rectangular prisms. You can use centimeter cubes to find the volume of the building model.

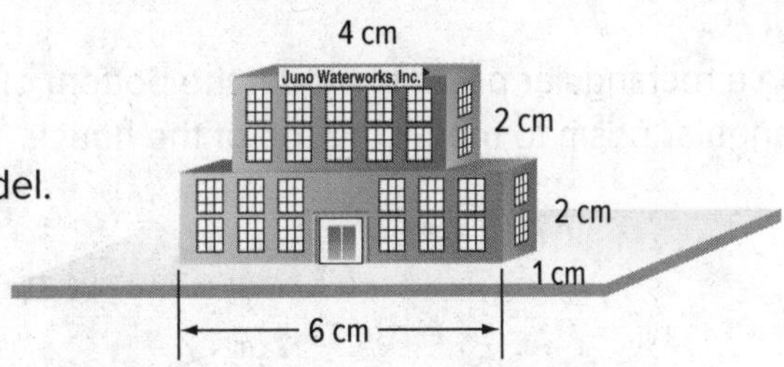

Hands-On Activity 1

The model is a *composite figure* because it is made from two rectangular prisms.

Step 1 Model the top and bottom rectangular prisms using cubes.

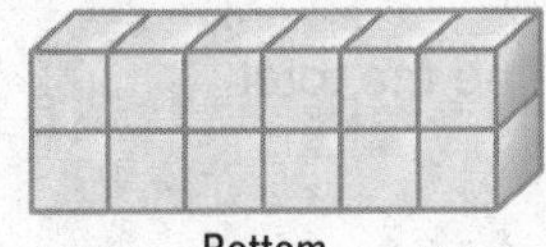
Bottom

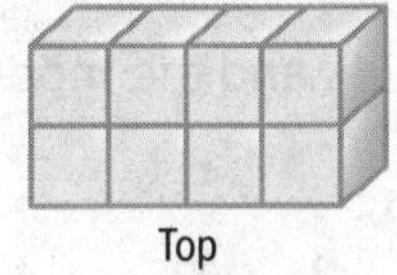
Top

Step 2 Count the cubes to find the dimensions. Write the dimensions in the table below. Then use the cube models to find the volume of both prisms. Write these measures in the table below step 3.

Step 3 Use the table to find the volume of the entire building model. Write these measures in the composite row of the table.

Model	Length (cm)	Width (cm)	Height (cm)	Volume (cm^3)
Bottom	6	1		
Top				
Composite				

Mr. Wendell's class made a model of a house. The model was composed of a rectangular prism and a triangular prism. Determine the volume and surface area of the model house.

Hands-On Activity 2

Step 1 Use a rectangular prism to model the bottom of the house. Use a triangular prism to model the top of the house.

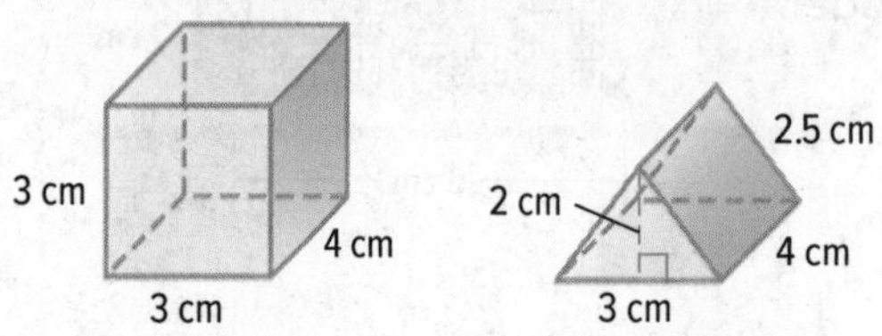

Step 2 Complete the tables below using the models from Step 1.

Prism	Length (cm)	Width (cm)	Height (cm)
Rectangular	4	3	

Prism	Length (cm)	Base (cm)	Height (cm)
Triangular	4	3	

Step 3 Use the information from the tables and the models to find the total volume of the model house.

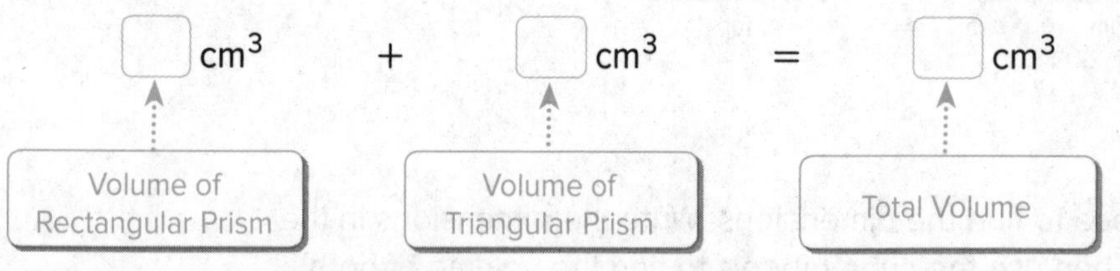

Step 4 Use the information from the tables and the models to find the total surface area of the model house.

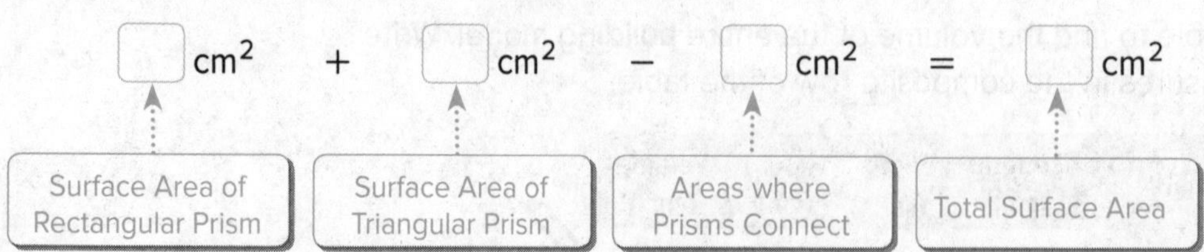

The total volume of the model house is ☐ cubic centimeters. The total surface area is ☐ square centimeters.

Investigate

Work with a partner.

1. **MP Model with Mathematics** Use the top, side, and front views to build a figure using centimeter cubes.

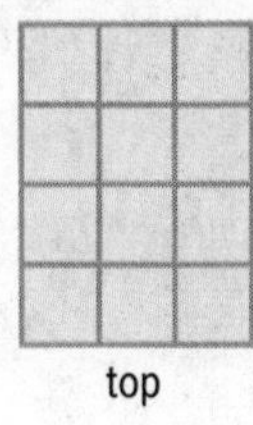

top

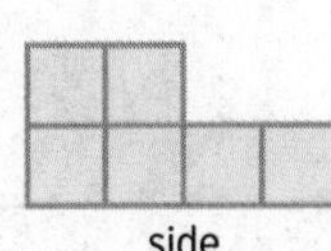

side

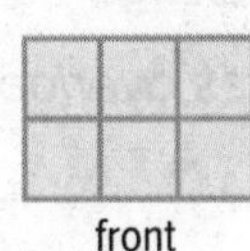

front

a. Make a sketch of the figure you built.

b. Find the volume and surface area of the figure.

Volume: ________ Surface Area: ________

Refer to the figure at the right for Exercises 2–4.

2. The figure is comprised of a ________ and a square ________.

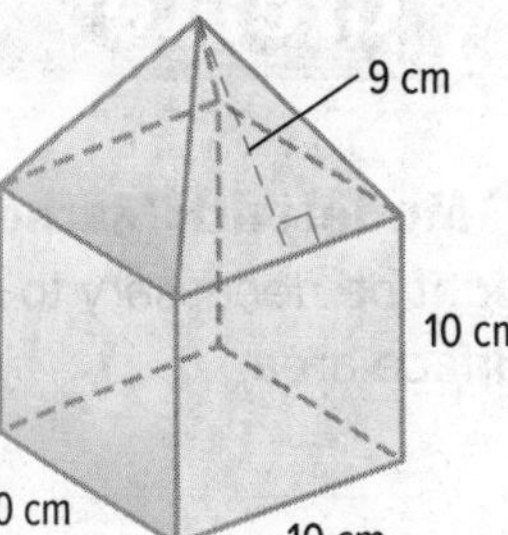

3. Complete the following to find the volume of the figure.

a. The volume of the cube is ☐ cubic centimeters.

b. The volume of the square pyramid is 250 cubic centimeters.

c. So, the volume of the composite figure is ☐ cubic centimeters.

4. Complete the following to find the surface area of the figure.

a. The surface area of the cube is ☐ square centimeters.

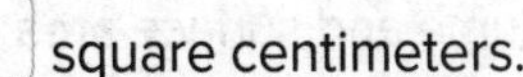

b. The surface area of the square pyramid is ☐ square centimeters.

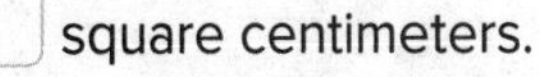

c. The area where the figures overlap is ☐ square centimeters.

d. The surface area of the composite figure is ☐ square centimeters.

Analyze and Reflect

MP Reason Inductively **Work with a partner. Write each of the following statements in the correct location. One statement is done for you.**

5. *measured in square units*
6. *measured in cubic units*
7. *involves adding measures of each figure*
8. *involves subtracting where figures overlap*

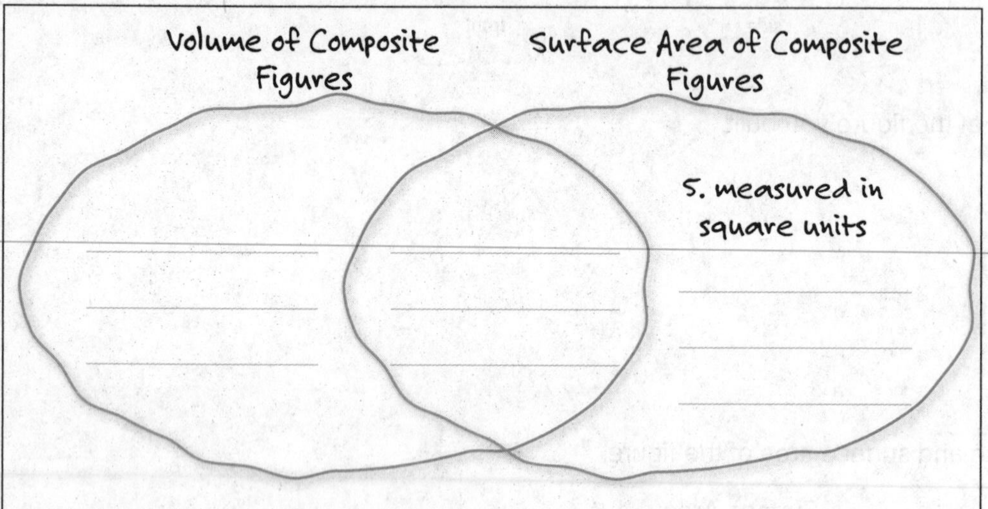

Create

9. **MP Model with Mathematics** Describe a real-world situation where it might be necessary to use a model or a drawing to find the volume or surface area.

10. **Inquiry** HOW can you find the volume and surface area of a composite figure?

Volume and Surface Area of Composite Figures

Real-World Link

Kaylee and Miles are making a bat house for their backyard like the one shown. They need to determine the surface area to find how much wood they will need.

1. Look at the largest bat house. What three-dimensional figures make up the bat house?

2. What method could you use to find the surface area of the bat house?

3. Suppose you wanted to find the volume of the bat house. What method could you use?

Essential Question

HOW do measurements help you describe real-world objects?

Common Core State Standards

Content Standards
7.G.6

Mathematical Practices
1, 3, 4

Which MP **Mathematical Practices** did you use?
Shade the circle(s) that applies.

① Persevere with Problems
② Reason Abstractly
③ Construct an Argument
④ Model with Mathematics
⑤ Use Math Tools
⑥ Attend to Precision
⑦ Make Use of Structure
⑧ Use Repeated Reasoning

Work Zone

Volume of a Composite Figure

The volume of a composite figure can be found by decomposing the figure into solids whose volumes you know how to find.

Examples

1. Find the volume of the composite figure.

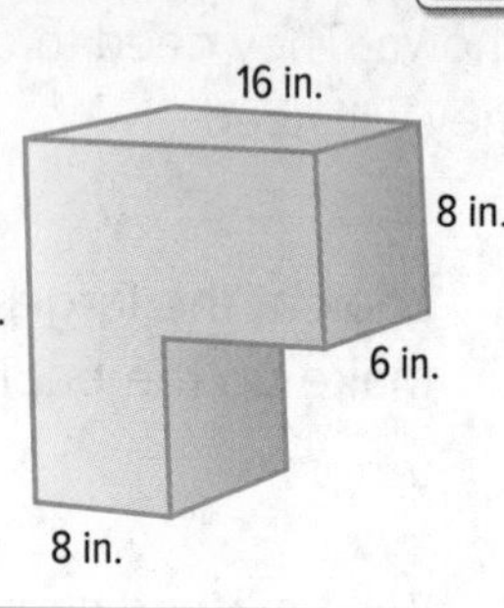

Find the volume of each prism.

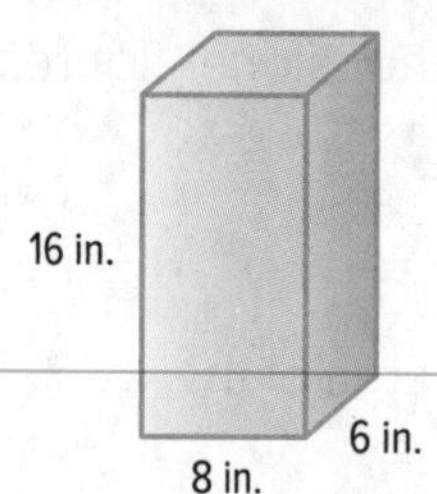

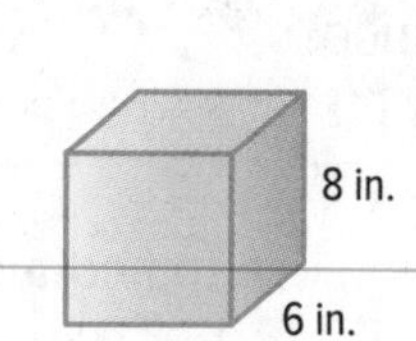

$V = \ell wh$ $\quad$ $V = \ell wh$

$V = 8 \cdot 6 \cdot 16$ or 768 $\quad$ $V = 8 \cdot 6 \cdot 8$ or 384

The volume is $768 + 384$ or 1,152 cubic inches.

2. Find the volume of the composite figure.

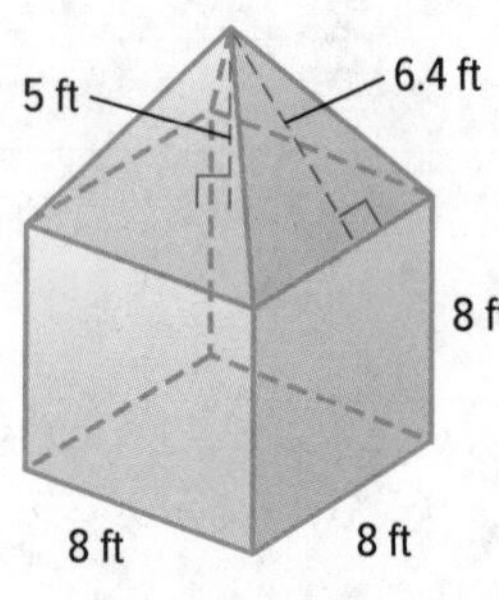

Find the volume of the cube and the pyramid. Round to the nearest tenth.

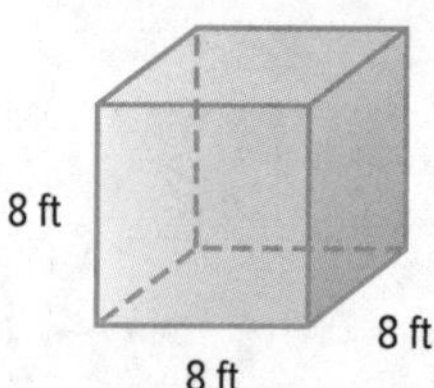

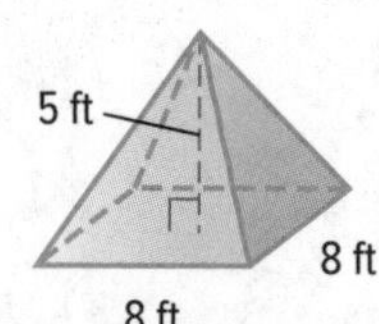

$V = \ell wh$ $\quad$ $V = \frac{1}{3}Bh$

$V = 8 \cdot 8 \cdot 8$ or 512 $\quad$ $V = \frac{1}{3}(8 \cdot 8)5$ or 106.7

The volume is $512 + 106.7$ or 618.7 cubic feet.

Got it? Do this problem to find out.

a. Find the volume of the composite figure.

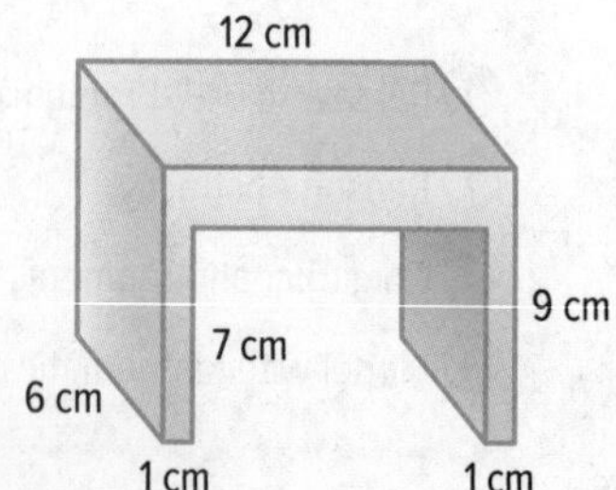

a. ________

Surface Area of a Composite Figure

You can also find the surface area of composite figures by finding the areas of the faces that make up the composite figure.

Examples

3. Find the surface area of the figure in Example 1.

The surface is made up of three different polygons.

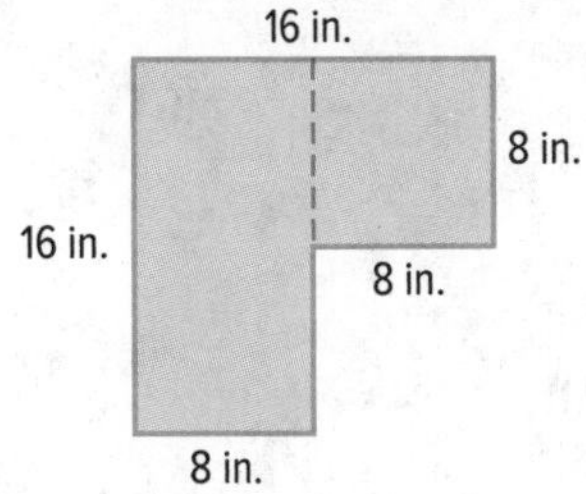

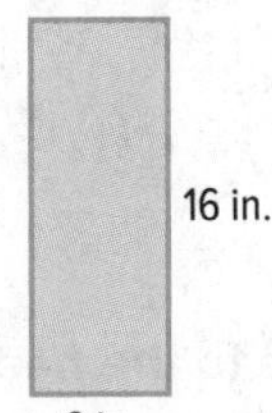

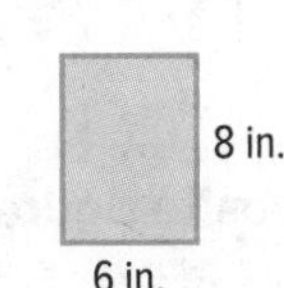

$A = \ell w + \ell w$
$A = (8 \cdot 16) + (8 \cdot 8)$
$A = 128 + 64$ or 192

$A = \ell w$
$A = 6 \cdot 16$
$A = 96$

$A = \ell w$
$A = 6 \cdot 8$
$A = 48$

The total surface area is $2(192) + 2(96) + 4(48)$ or 768 square inches.

4. Find the surface area of the composite figure in Example 2.

The figure is made up of two different polygons.

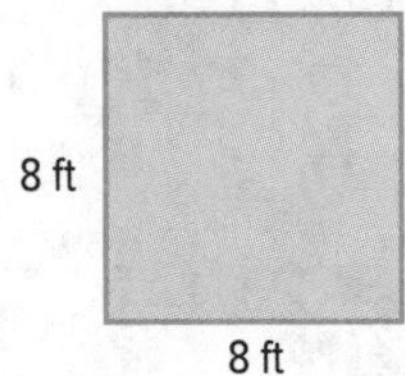

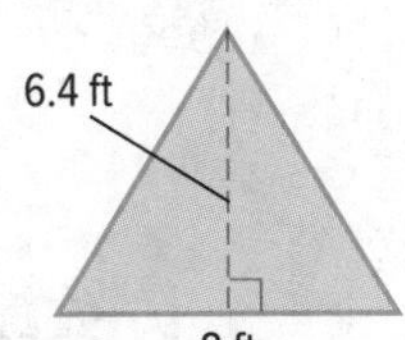

$A = \ell w$
$A = 8 \cdot 8$ or 64

$A = \frac{1}{2}bh$
$A = \frac{1}{2} \cdot 8 \cdot 6.4$ or 25.6

The total surface area is $5(64) + 4(25.6)$ or 422.4 square feet.

Surface Area
To make it easier to see each face, sketch the faces and label the dimensions of each.

Got it? Do this problem to find out.

b. Find the surface area of the steps that are represented by the composite figure shown.

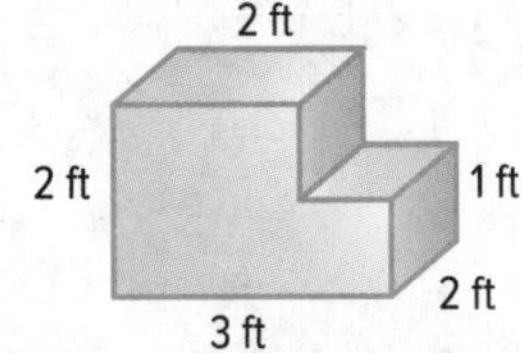

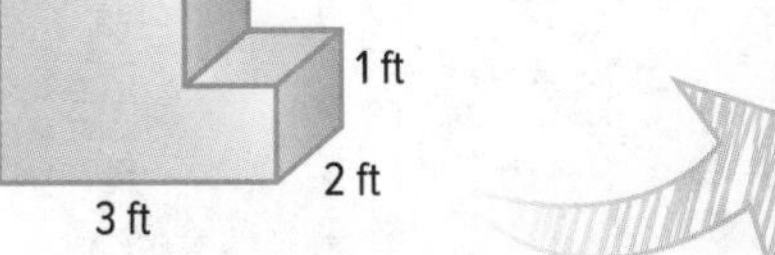

b. __________

Guided Practice

Find the volume of each composite figure. Round to the nearest tenth if necessary. (Examples 1 and 2)

1. ______

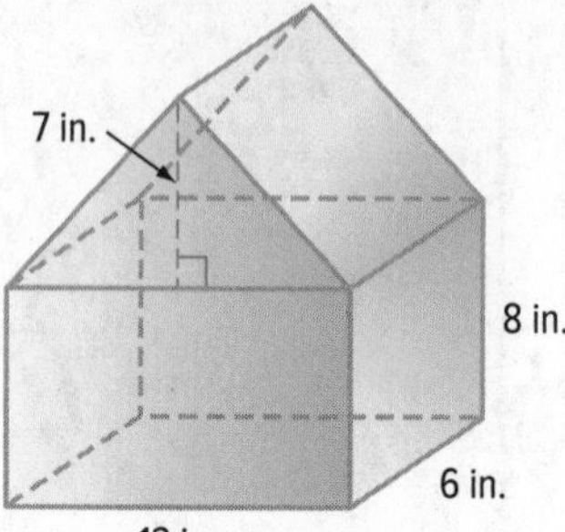

Show your work.

2. ______

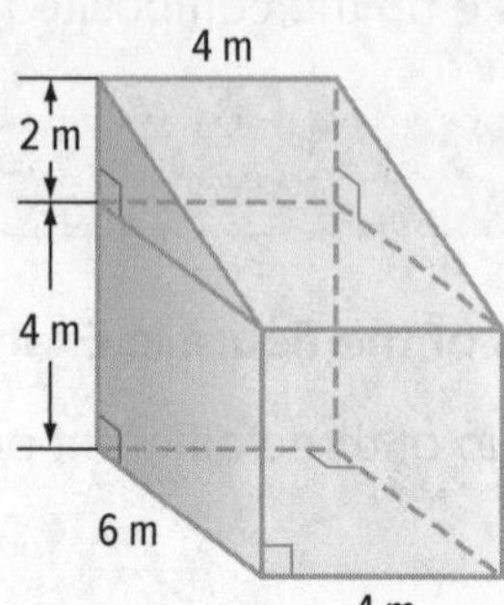

Find the surface area of each composite figure. Round to the nearest tenth if necessary. (Examples 3 and 4)

3. ______

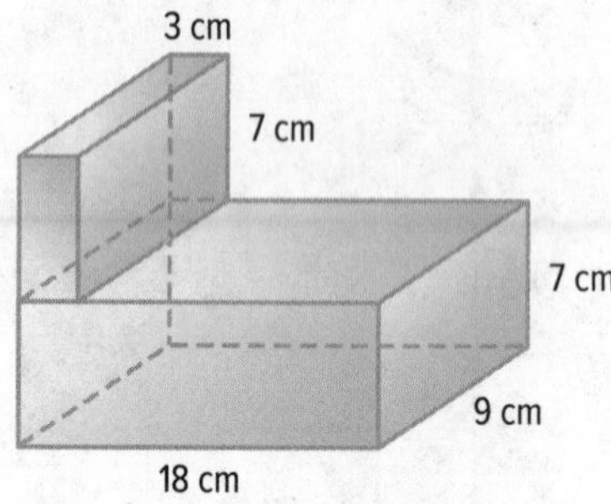

4. ______

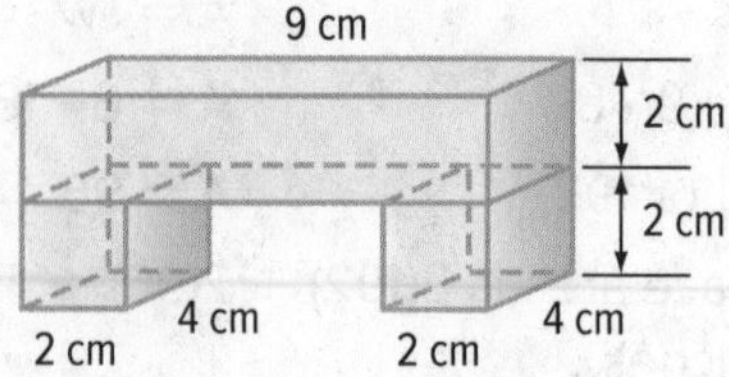

5. **Building on the Essential Question** How do the previous lessons in this chapter help you find the surface area and volume of a composite figure?

______

______

______

______

Rate Yourself!

Are you ready to move on? Shade the section that applies.

For more help, go online to access a Personal Tutor.

Name ______________________________ My Homework ______________________________

Independent Practice

Go online for Step-by-Step Solutions

Find the volume of each composite figure. Round to the nearest tenth if necessary. (Examples 1 and 2)

1. ______________

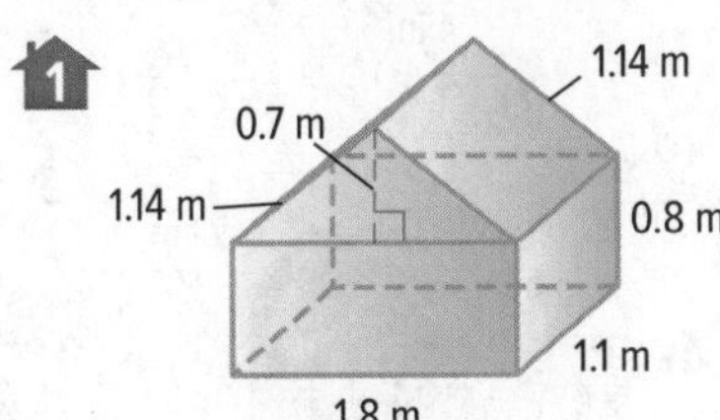

2. ______________

2.5 m
3 m
2 m
4 m

Find the surface area of each composite figure. Round to the nearest tenth if necessary. (Examples 3 and 4)

3. ______________

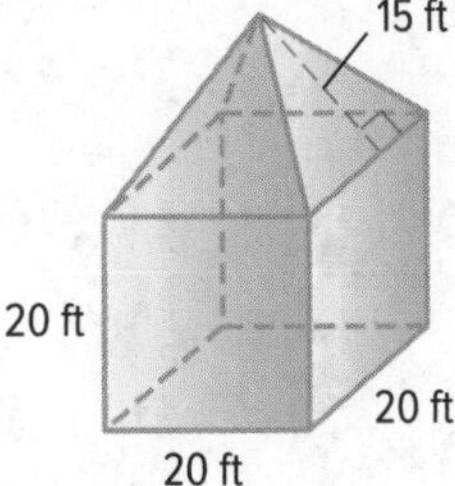

4. ______________

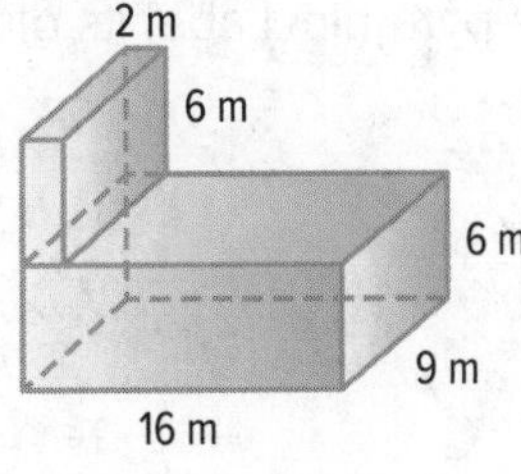

5. Find the volume of the figure at the right in cubic feet. Round to the nearest tenth. (Examples 1 and 2)

____

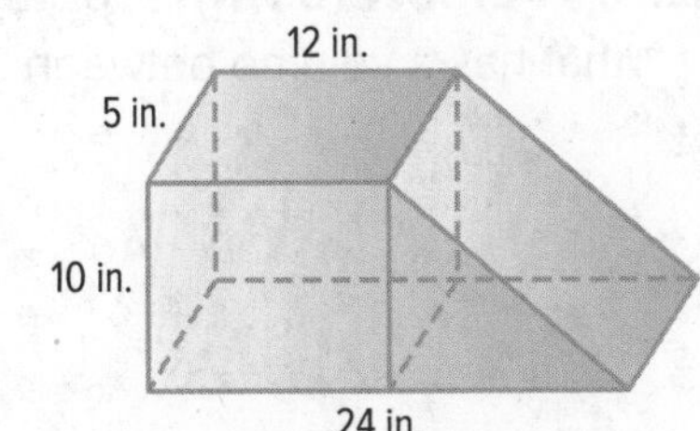

6. **Reason Inductively** The swimming pool at the right is being filled with water. Find the number of cubic feet that it will take to fill the swimming pool. (*Hint:* The area of a trapezoid is $A = \frac{1}{2}h(b_1 + b_2)$.) (Examples 1 and 2)

____

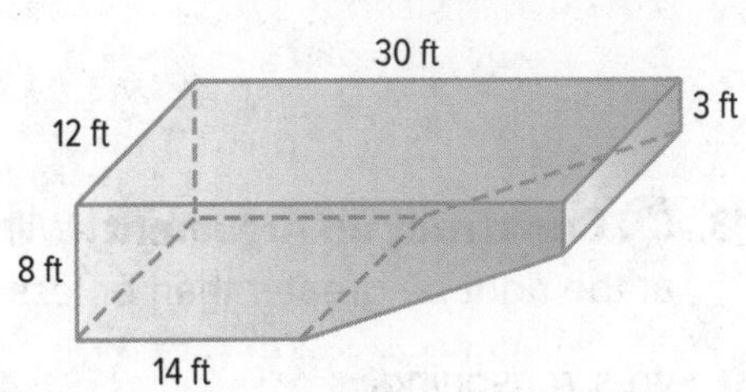

Copy and Solve For Exercises 7–8, show your work on a separate piece of paper. Round to the nearest tenth. (Examples 1–4)

7. Find the surface area of the figure in Exercise 1.

8. Find the volume of the figure in Exercise 4.

9. A carryout container is shown. The bottom base is a 4-inch square and the top base is a 4-inch by 6-inch rectangle. The height of the container is 5 inches. Find the volume of food that it holds. ____________________

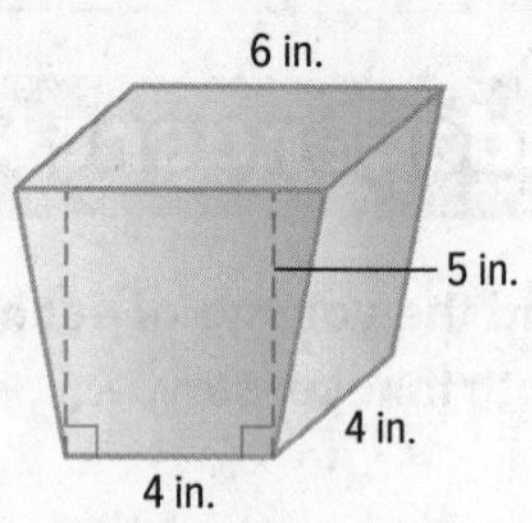

10. Refer to the house shown. Find the surface area and volume of the house. Do not include the bottom of the house when calculating the surface area. ____________________

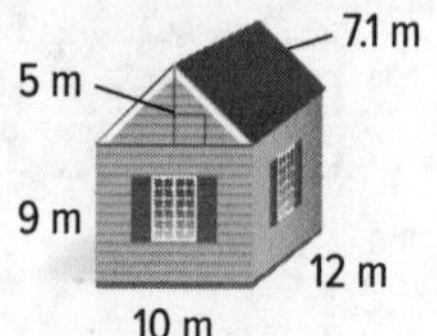

H.O.T. Problems Higher Order Thinking

11. MP **Model with Mathematics** Draw a composite figure that is made up of a cube and a square pyramid. Label its dimensions and find the volume of the figure. ____________________

12. MP **Persevere with Problems** Draw an example of a composite figure that has a volume between 250 and 300 cubic units.

13. MP **Construct an Argument** Will the surface area of the figure at the right be greater than or less than 180 square inches? Explain your reasoning. ____________________

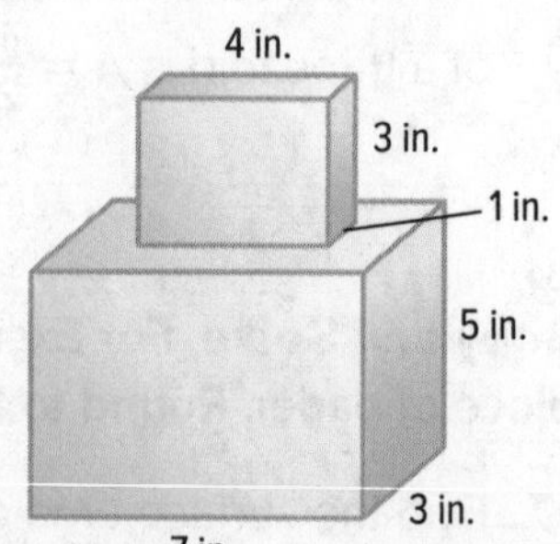

Name ______________________ My Homework ______________

Extra Practice

Find the volume of each composite figure. Round to the nearest tenth if necessary.

14. 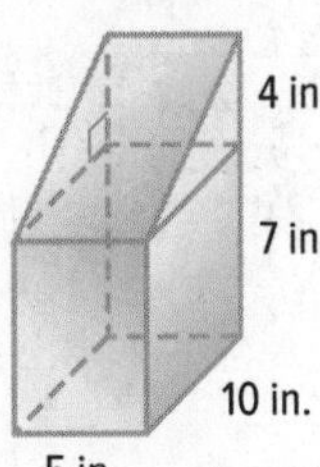

450 in^3

Homework Help

Rectangular Prism

$V = lwh$

$V = 5 \cdot 10 \cdot 7$

$V = 350$

Triangular Prism

$V = Bh$

$V = \frac{1}{2} \cdot 10 \cdot 4 \cdot 5$

$V = 100$

Total Volume = 350 + 100 or 450 in^3

15. 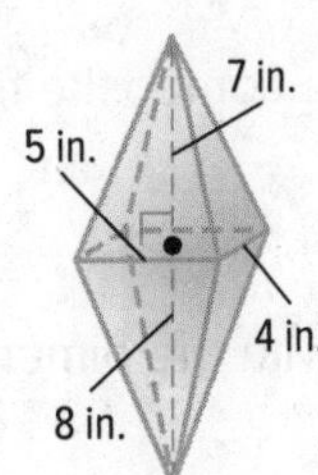

____

Find the surface area of each composite figure. Round to the nearest tenth if necessary.

16. 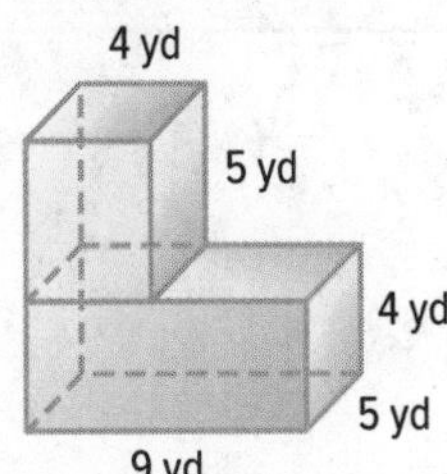

____

17. 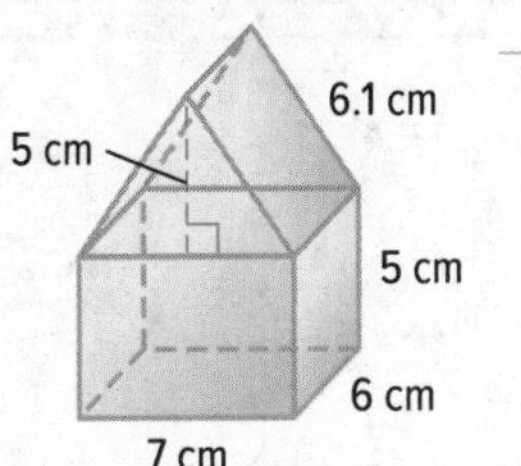

____

18. MP **Find the Error** Seth is finding the surface area of the composite figure shown. Find his mistake and correct it.

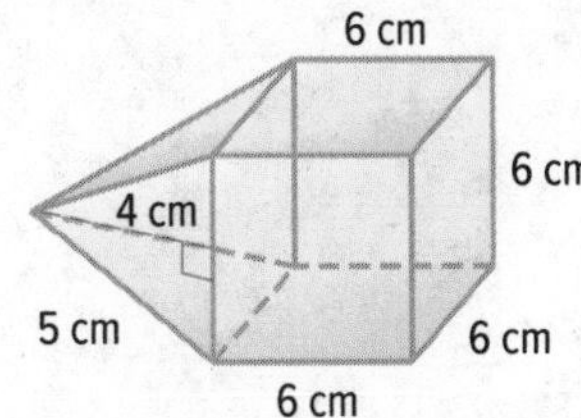

$V = \frac{1}{3}Bh + s^3$

$V = \frac{1}{3} \cdot 36 \cdot 4 + 6^3$

$V = 264\ cm^3$

____

____

____

Power Up! Common Core Test Practice

19. Refer to the composite figure with the dimensions shown. Fill in the boxes to complete each statement.

 a. The volume of the composite figure is ☐.

 b. The total surface area of the composite figure is ☐.

2 ft
2 ft
2.2 ft
0.2 ft
2 ft
2.2 ft

20. Refer to the composite figure with the dimensions shown.

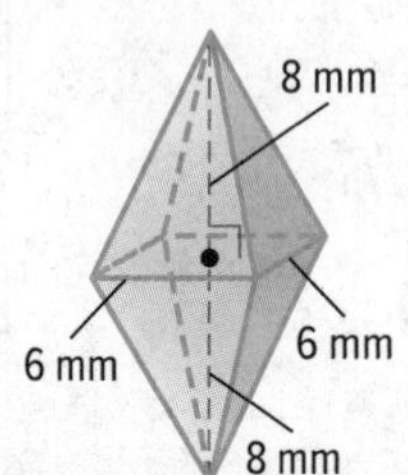

2
3
4
6
8

Select the correct values to complete the expression to find the volume of the figure.

$V = \square \cdot \frac{1}{3} \cdot \square \cdot \square \cdot \square$

What is the volume of the composite figure?

☐

Common Core Spiral Review

Draw a net for each figure. 6.G.4

21.

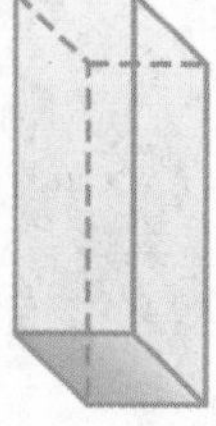

Show your work.

22.

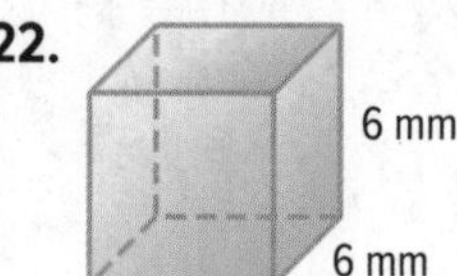

23.

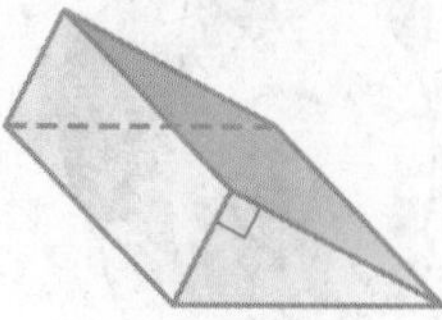

24.

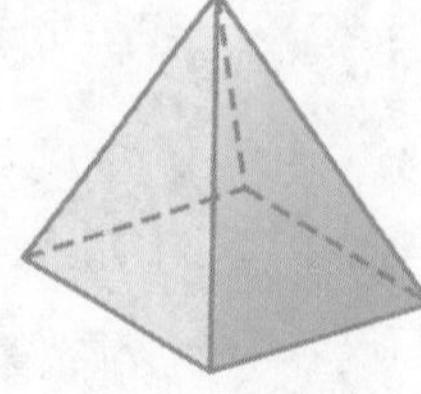

21ST CENTURY CAREER in Landscape Architecture

Landscape Architect

Do you have an artistic side, and do you enjoy being outdoors? If so, a career in landscape design might be a perfect fit for you. Landscape architects design outside areas such as yards, parks, playgrounds, campuses, shopping centers, and golf courses. Their designed areas are not only meant to be beautiful, but also functional and compatible with the natural environment. A landscape architect must be proficient in mathematics, science, and the use of computer-aided design.

Is This the Career for You?

Are you interested in a career as a landscape architect? Take some of the following courses in high school.

- Algebra
- Architectural Design
- Botany
- Drafting/Illustrative Design Technology
- Geometry

Find out how math relates to a career in Landscape Architecture.

Planting in Circles

For each problem, use the information in the designs.

1. In Design 2, what is the radius of the larger grassy area? ____________

2. The small circular fountain in Design 1 is surrounded by a stone wall. Find the circumference of the wall. Use $\frac{22}{7}$ for π.

____

3. Find the circumference of the smaller grassy area in Design 2. Use 3.14 for π.

____

4. In Design 2, how much greater is the lawn area in the larger circle than in the smaller circle? Use 3.14 for π. ____________

5. In Design 2, the smaller circle is surrounded by a path 1 meter wide. What is the outer circumference of the path? Use the π key on a calculator and round to the nearest tenth.

____

6. In Design 1, the area of the large circular patio is about 201.1 square feet. What is the radius of the patio? Round to the nearest foot. ____________

Design 1

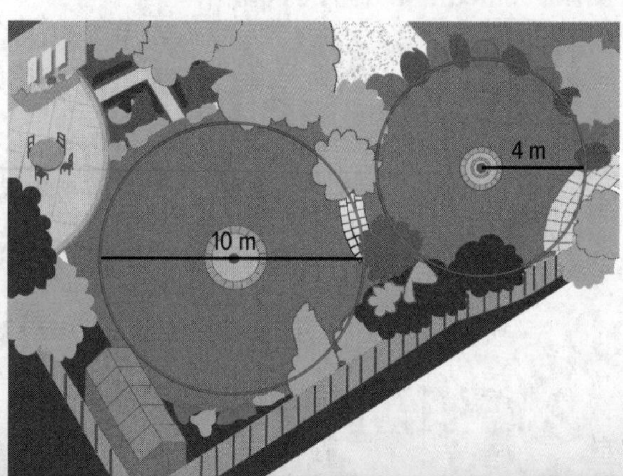

Design 2

Career Project

It's time to update your career portfolio! Download free landscaping software from the Internet and use it to create your own landscape design. Include a list of all the plants, materials, and hard elements used in your design. Also, provide an estimate of the total cost of the landscaping project.

____

____

____

____

What is something you really want to do in the next ten years?

- ____________
- ____________
- ____________
- ____________
- ____________

Chapter Review

Vocabulary Check

Complete each sentence using the vocabulary list at the beginning of the chapter. Then circle the word that completes the sentence in the word search.

1. The distance across a circle through its center is called the ______________.
2. The ______________ is the distance from the center to any point on the circle.
3. A ______________ is the set of all points in a plane that are the same distance from a point.
4. The point in a circle from which all other points are equidistant is called the ______________.
5. The distance around a circle is the ______________.
6. The ratio of circumference to diameter is called ______________.
7. A ______________ is half of a circle.
8. A ______________ figure is made up of two or more shapes.
9. The ______________ of a three-dimensional figure is the measure of the space it occupies.
10. The sum of the areas of all the faces of a three-dimensional figure is the ______________ area.
11. The triangular faces of a pyramid that are not bases are ______________ faces.
12. The height of each lateral face of a pyramid is called the ______________ height.

G	U	H	M	I	R	H	X	P	E	S	F	S	S	Q	W	C	P	A	C	O	Q
P	I	B	A	E	V	O	L	U	M	E	E	V	W	Z	A	X	S	D	O	O	C
P	G	J	T	V	C	T	Y	A	G	E	R	M	R	Q	A	Q	I	E	M	I	O
C	N	N	C	D	E	D	Z	M	M	L	R	L	I	I	N	A	D	P	P	I	D
M	E	Z	C	K	G	H	N	L	O	C	P	I	V	C	M	C	W	Z	O	C	Z
C	H	C	I	R	C	U	M	F	E	R	E	N	C	E	I	P	M	C	S	N	P
X	E	O	V	G	B	L	H	V	Z	I	B	B	T	P	T	R	S	O	I	C	N
S	Y	V	T	D	K	L	A	B	W	C	O	E	E	T	J	L	C	V	T	D	L
S	P	M	T	Y	S	U	Z	R	I	T	R	J	Q	O	R	Y	B	L	E	Y	Q
Y	P	V	S	Z	B	E	Z	O	E	H	N	W	G	J	C	T	T	Q	E	H	S
P	R	D	B	T	T	L	D	W	T	T	N	W	Y	X	Z	S	V	J	X	L	J
E	K	A	J	N	J	G	W	C	M	P	A	C	D	M	K	L	M	O	S	L	P
A	I	Y	D	A	Z	X	F	Q	F	C	X	L	Y	M	Y	A	W	X	F	Z	F
V	X	Y	M	I	N	H	P	I	W	D	U	Z	L	A	T	N	Q	L	O	Q	J
Z	U	O	Y	M	U	U	K	K	K	L	I	C	E	W	H	T	A	J	M	U	H
E	C	A	F	R	U	S	G	Z	E	X	K	W	S	T	R	X	J	X	K	I	K

Key Concept Check

Use Your FOLDABLES

Use your Foldable to help review the chapter.

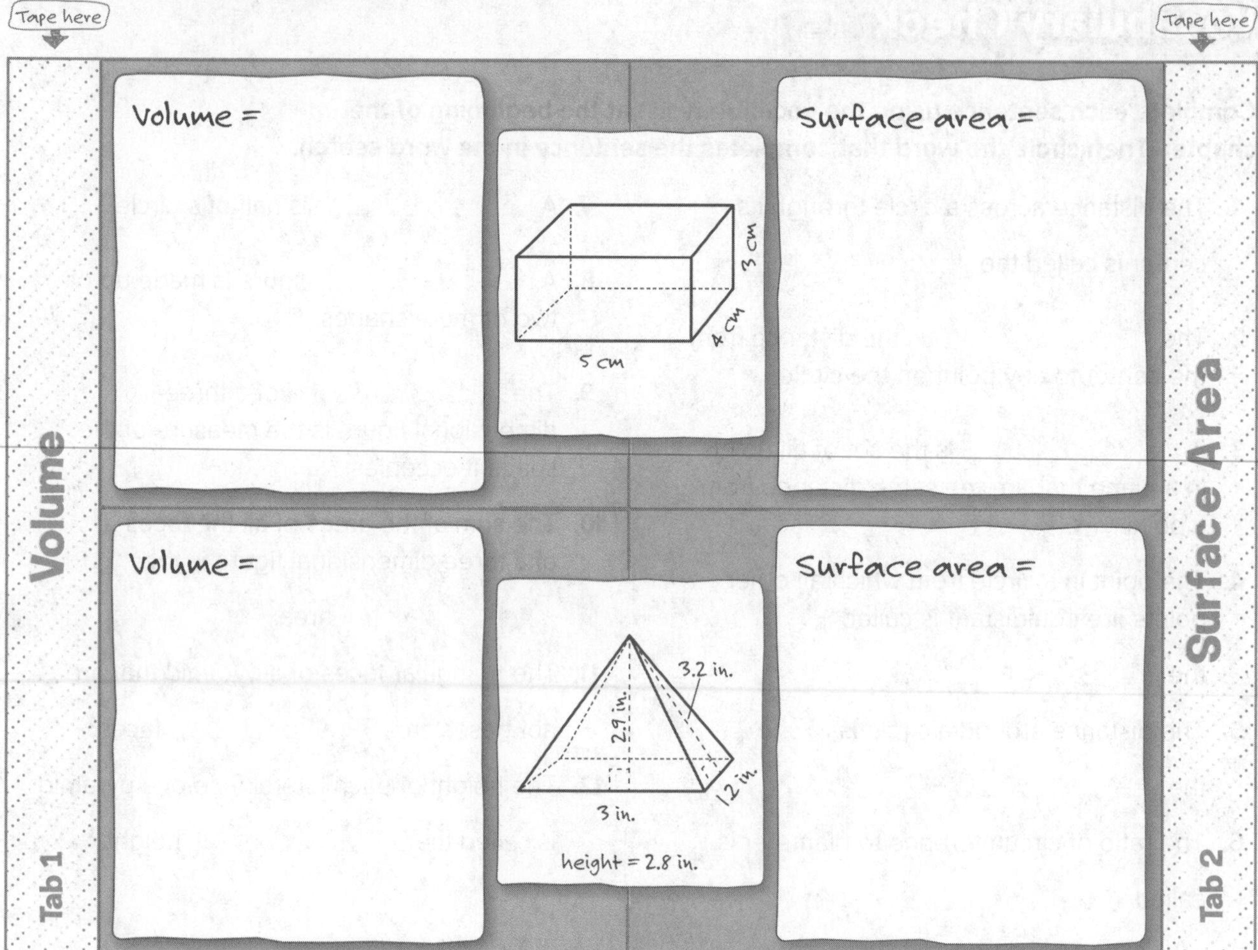

Got it?

Circle the correct term or number to complete each sentence.

1. The diameter of a circle is (twice, three times) its radius.

2. The area of a circle equals the product of pi and the square of its (radius, diameter).

3. The volume of a rectangular prism can be found by multiplying the area of the base times the (length, height).

4. To find the surface area of a triangular prism, find the area of each face and calculate the (sum, product) of all the faces.

Juice Box Packaging

Supreme Packaging Company manufactures juice boxes for juice companies. They are examining different ways to make the juice boxes using various lengths, widths, and heights. The measurements of one juice box are shown.

Write your answers on another piece of paper. Show all of your to receive full credit.

Part A

What is the volume of the juice box shown? The company received an order to make a jumbo juice box that has twice the volume as the one shown. Could you double the current dimensions to make the jumbo juice box at the suggested volume? Explain.

Part B

Draw and label a net to find the surface area of the original juice box. It costs Supreme Packaging $0.02 per square inch to create one juice box. The company groups eight juice boxes together as one package. How much does it cost to create one package?

Part C

An artist created the picture of the citrus fruit on the label. The picture is a circle and has an area of 12.56 square inches. Will the artist's picture fit on the juice box label? Explain. Use 3.14 for π.

Reflect

Answering the Essential Question

Use what you learned about measuring figures to complete the graphic organizer.

Circumference

Area

Essential Question

HOW do measurements help you describe real-world objects?

Volume

Surface Area

Answer the Essential Question. HOW do measurements help you describe real-world objects?

UNIT PROJECT

Watch

Turn Over a New Leaf The flatness of leaves serves an important purpose. In this project you will:

- **Collaborate** with your classmates as you research the primary function of leaves.
- **Share** the results of your research in a creative way.
- **Reflect** on how you use different measurements to solve real-life problems.

Collaborate

Go Online Work with your group to research and complete each activity. You will use your results in the Share section on the following page.

1. Suppose you have a cube that is 10 centimeters on each side. Find the volume, surface area, and surface area to volume ratio.

2. Disassemble the cube from Exercise 1 into centimeter cubes. Arrange the cubes in a 50-by-20-by-1 prism. Find the volume, surface area, and surface area to volume ratio.

3. Compare and contrast the volume, surface area, and surface area to volume ratio from Exercises 1 and 2.

4. Trace the outline of a leaf onto centimeter grid paper. Estimate the volume of the leaf. (Assume the height of your leaf is 0.1 centimeter.) Estimate the surface area. (You can ignore the edge of the leaf.) Find the surface area to volume ratio.

5. Do research to find the primary function of a leaf. Explain how the surface area to volume ratio of a leaf aids in its function.

6. Find examples from nature or man-made objects that have a small surface area to volume ratio. Explain the benefits.

With your group, decide on a way to share what you have learned about the surface area to volume ratio of leaves. Some suggestions are listed below, but you could also think of other creative ways to present your information. Remember to show how you used mathematics to complete each of the activities in this project!

- Create a digital presentation that compares two types of leaves. Use what you learned about surface area to volume ratios in your presentation.
- Imagine you discovered a new type of leaf. Create an annotated diagram of your leaf. The annotations should include the type of information you learned in this project.

Check out the note on the right to connect this project with other subjects.

with Science

Environmental Literacy Write a paragraph detailing facts about the leaves you researched. Some questions to consider are:

- What are the names of the trees that dropped these leaves?
- Are these types of trees common in your state?

6. **Answer the Essential Question** How can you use different measurements to solve real-life problems?

 a. How did what you learned about geometric figures help you use different measurements to solve real-life problems in this project?

 b. How did what you learned about measuring figures to help you use different measurements to solve real-life problems in this project?

UNIT 5

CCSS

Statistics and Probability

Essential Question

WHY is learning mathematics important?

Chapter 9
Probability

Probability describes the likelihood of an event occurring. In this chapter, you will develop probability models and find probabilities of simple and compound events.

Chapter 10
Statistics

Statistics can be used to draw conclusions about a population. In this chapter, you will use random samples to make predictions and compare populations.

Unit Project Preview

Math Genes A *Punnett Square* is a diagram that is used to predict the genetic traits of offspring.

A pea plant can be tall (described by TT, TS, or ST) or short (described by SS). Complete the Punnett Square below. What percent of the outcomes indicate that the offspring will be short? ________

At the end of Chapter 10, you'll complete a project in which you use pets' traits to make predictions about their offspring. Put on your lab coat and grab your math tool kit to begin this adventure!

Pea Plants

		Parent 1	
		T	S
Parent 2	T	TT	
	S		

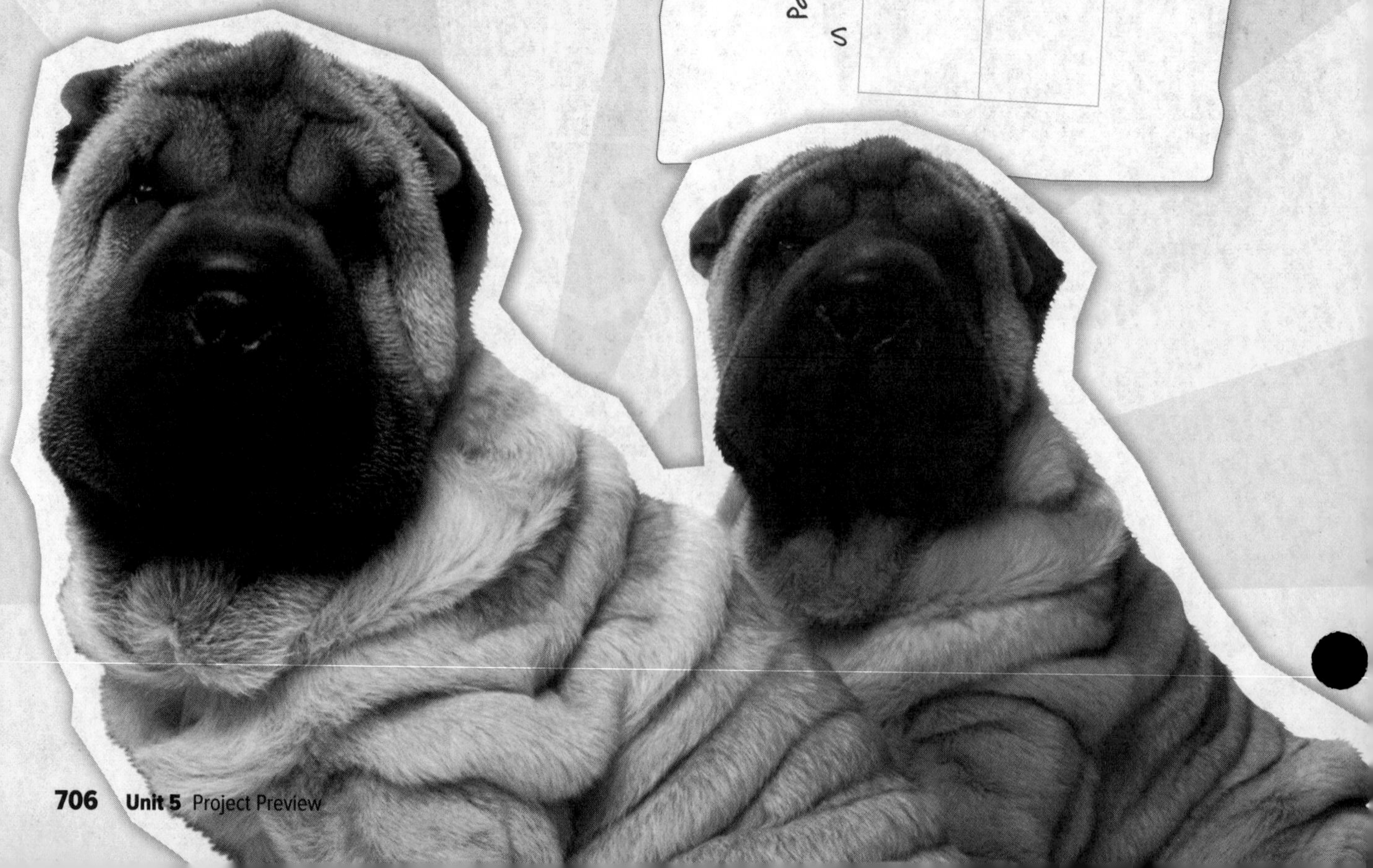

Chapter 9
Probability

Essential Question

HOW can you predict the outcome of future events?

Common Core State Standards

Content Standards
7.SP.5, 7.SP.6, 7.SP.7, 7.SP.7a, 7.SP.7b, 7.SP.8, 7.SP.8a, 7.SP.8b, 7.SP.8c

MP Mathematical Practices
1, 3, 4, 5

Math in the Real World

Probability is the likelihood or chance of an event occurring.

At the beginning of a football game, a coin is tossed to determine which team receives the ball first. Fill in the table below to indicate the number of times a team would expect to win the coin toss based on the number of games played.

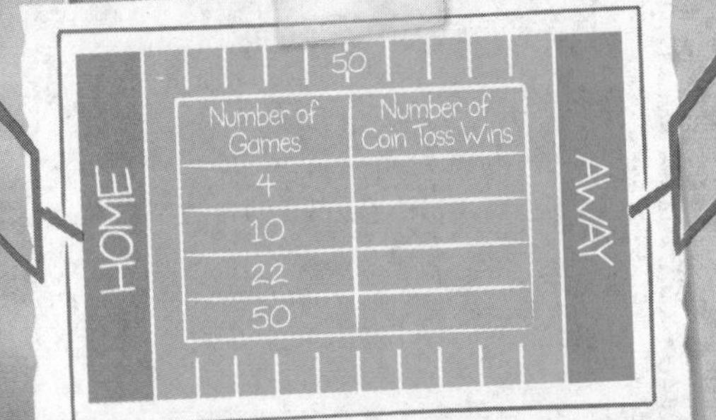

Number of Games	Number of Coin Toss Wins
4	
10	
22	
50	

FOLDABLES® Study Organizer

Cut out the Foldable on page FL11 of this book.

Place your Foldable on page 786.

Use the Foldable throughout this chapter to help you learn about probability.

What Tools Do You Need?

Vocabulary

complementary events
compound event
dependent events
experimental probability
fair
Fundamental Counting Principle
independent events
outcome
permutation
probability
random
relative frequency
sample space
simple event
simulation
theoretical probability
tree diagram
uniform probability model
unfair

Review Vocabulary

Fractions, Decimals, and Percents Equivalent rational numbers are numbers that have the same value. For example, three-fourths is equivalent to 0.75 or 75%.

A probability can be expressed as a fraction, decimal, or percent. For each rational number, write the missing equivalent values. Write fractions in simplest form.

What Do You Already Know?

List three things you already know about probability in the first section. Then list three things you would like to learn about probability in the second section.

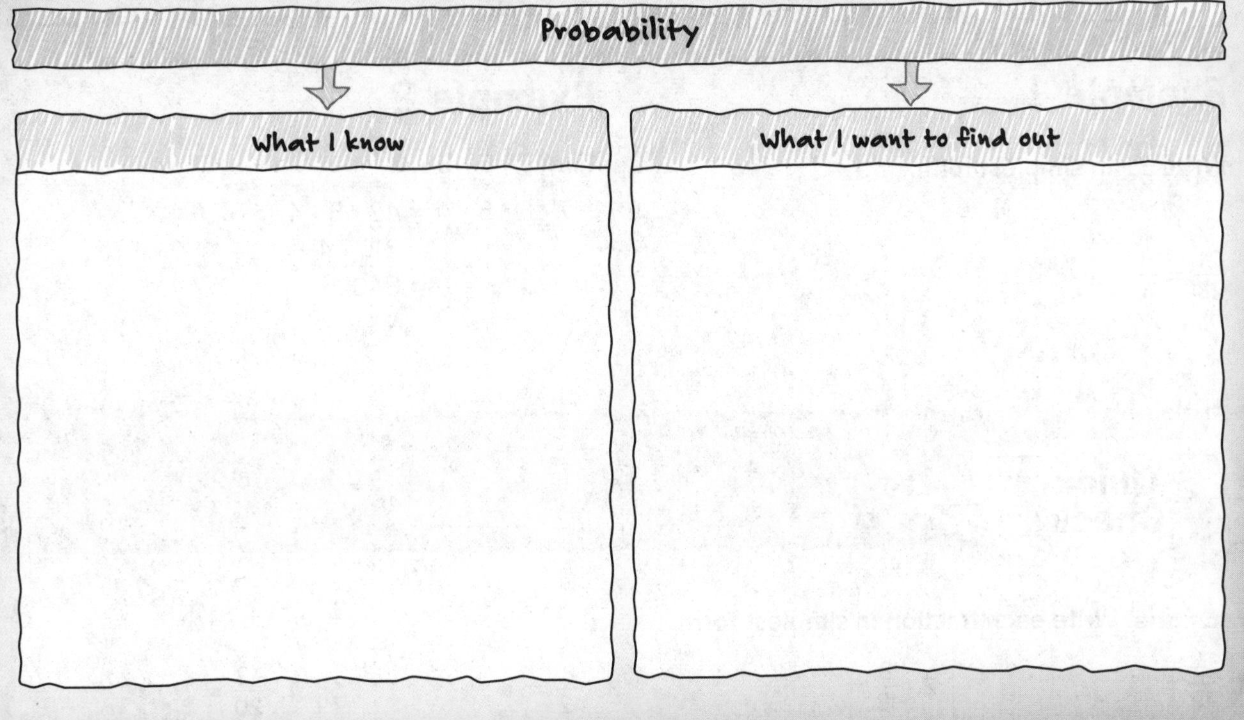

When Will You Use This?

Here are a few examples of how probability is used in the real world.

Activity 1 Have you ever read something like "The chances of winning are 75%." or "30% of the people surveyed said that they prefer vanilla ice cream."? Use the Internet to find an example like the ones given. Describe your example and what it means to you.

Jamar and Theresa in

Radio Surveys

Are you ready for our first task as radio station interns?

I'm ready! It's gonna be fun! What should we do first?

Activity 2 Go online at **connectED.mcgraw-hill.com** to read the graphic novel ***Radio Surveys***. What type of information do Jamar and Theresa want to learn from their survey?

Try the Quick Check below.
Or, take the Online Readiness Quiz.

Quick Review

Common Core Review 5.NBT.5, 6.NS.4

Example 1

Write $\frac{21}{28}$ in simplest form.

$$\frac{21}{28} \overset{\div 7}{=} \frac{3}{4}$$

Divide the numerator and denominator by the GCF, 7.

Example 2

Find 7 • 6 • 5 • 4.

$$7 \cdot 6 \cdot 5 \cdot 4 = 42 \cdot 5 \cdot 4$$
$$= 210 \cdot 4$$
$$= 840$$

Multiply from left to right.

Quick Check

Fractions Write each fraction in simplest form.

1. $\frac{5}{15}$ = ______

2. $\frac{3}{18}$ = ______

3. $\frac{8}{12}$ = ______

4. $\frac{12}{20}$ = ______

Products Find each product.

5. 6 • 5 = ______

6. 10 • 9 • 8 = ______

7. 4 • 3 • 2 • 1 = ______

8. Suppose you listen to 9 songs each hour for 5 hours every day this week. How many songs will you have listened to this week?

______

Which problems did you answer correctly in the Quick Check? Shade those exercise numbers below.

① ② ③ ④ ⑤ ⑥ ⑦ ⑧

Lesson 1

Probability of Simple Events

Vocabulary Start-Up

Probability is the chance that some event will occur. A **simple event** is one outcome or a collection of outcomes. What is an **outcome**?

Math Definition	Real-World Definition
A possible result in a probability experiment.	

Outcome

Essential Question

HOW can you predict the outcome of future events?

Vocabulary

probability
outcome
simple event
random
complementary events

Common Core State Standards

Content Standards
7.SP.5, 7.SP.7, 7.SP.7a

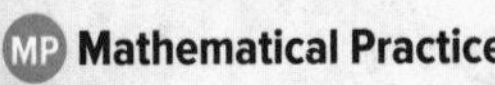

1, 3, 4

Real-World Link

For a sledding trip, you randomly select one of the four hats shown. Complete the table to show the possible outcomes.

Hat Selection Outcomes			
Outcome 1	green hat	**Outcome 3**	
Outcome 2		**Outcome 4**	

1. Write a ratio that compares the number of blue hats to the total number of hats. ________

2. Describe a hat display in which you would have a better chance of selecting a red hat.

____

____

Which MP Mathematical Practices did you use?
Shade the circle(s) that applies.

① Persevere with Problems
② Reason Abstractly
③ Construct an Argument
④ Model with Mathematics
⑤ Use Math Tools
⑥ Attend to Precision
⑦ Make Use of Structure
⑧ Use Repeated Reasoning

Key Concept

Probability

Words The probability of an event is a ratio that compares the number of favorable outcomes to the number of possible outcomes.

Symbols $P(event) = \frac{\text{number of favorable outcomes}}{\text{number of possible outcomes}}$

Work Zone

STOP and Reflect

In the space below, describe an example of a simple event that is certain to occur.

The probability of a chance event is a number between 0 and 1 that expresses the likelihood of the event occurring. Greater numbers indicate greater likelihood. A probability near 0 indicates an unlikely event, a probability around $\frac{1}{2}$ indicates an event that is neither unlikely nor likely, and a probability near 1 indicates a likely event.

Probability can be written as a fraction, decimal, or percent.

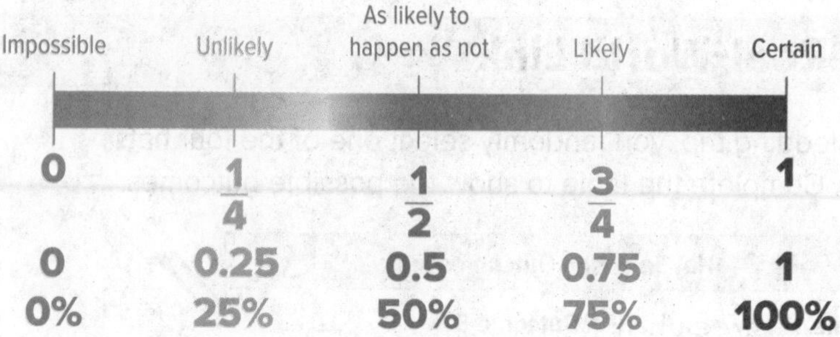

0	$\frac{1}{4}$	$\frac{1}{2}$	$\frac{3}{4}$	1
0	0.25	0.5	0.75	1
0%	25%	50%	75%	100%

Outcomes occur at **random** if each outcome is equally likely to occur.

Example

There are six equally likely outcomes if a number cube with sides labeled 1 through 6 is rolled.

1. Find *P*(6) or the probability of rolling a 6.

There is only one 6 on the number cube.

$P(6) = \frac{\text{number of favorable outcomes}}{\text{number of possible outcomes}}$

$= \frac{1}{6}$

The probability of rolling a 6 is $\frac{1}{6}$, or about 17%, or about 0.17.

Got it? **Do this problem to find out.**

a. A coin is tossed. Find the probability of the coin landing on heads. Write your answer as a fraction, percent, and decimal.

a. ______________

Example

2. Find the probability of rolling a 2, 3, or 4 on the number cube.

The word *or* indicates that the number of favorable outcomes needs to include the numbers 2, 3, and 4.

$P(2, 3, \text{or } 4) = \dfrac{\text{number of favorable outcomes}}{\text{number of possible outcomes}}$

$= \dfrac{3}{6} \text{ or } \dfrac{1}{2}$ Simplify.

The probability of rolling a 2, 3, or 4 is $\frac{1}{2}$, 50%, or 0.5.

Got it? **Do these problems to find out.**

Show your work.

The spinner at the right is spun once. Find the probability of each event. Write each answer as a fraction, percent, and decimal.

b. $P(\text{F})$ **c.** $P(\text{D or G})$ **d.** $P(\text{vowel})$

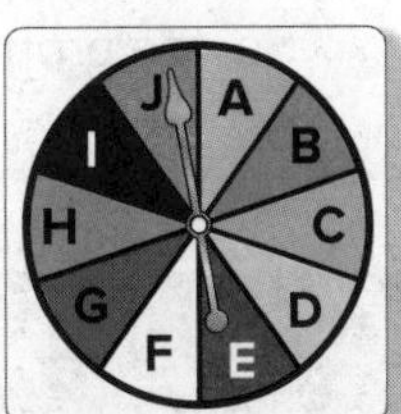

b. ____________

c. ____________

d. ____________

Find Probability of the Complement

Complementary events are two events in which either one or the other must happen, but they cannot happen at the same time. For example, a coin can either land on heads or *not* land on heads. The sum of the probability of an event and its complement is 1 or 100%.

Complement
In everyday language complement means the quantity required to make something complete. This is similar to the math meaning.

Example

3. Find the probability of *not* rolling a 6 in Example 1.

The probability of *not* rolling a 6 and the probability of rolling a 6 are complementary. So, the sum of the probabilities is 1.

$P(6) + P(\text{not } 6) = 1$ $P(6)$ and $P(\text{not } 6)$ are complements.

$\dfrac{1}{6} + P(\text{not } 6) = 1$ Replace $P(6)$ with $\frac{1}{6}$.

$\dfrac{1}{6} + \dfrac{5}{6} = 1$ **THINK** $\frac{1}{6}$ plus what number equals 1?

The probability of *not* rolling a 6 is $\frac{5}{6}$, or about 83% or 0.83.

Got it? **Do this problem to find out.**

e. A bag contains 5 blue, 8 red, and 7 green marbles. A marble is selected at random. Find the probability the marble is *not* red.

e. ____________

Example

4. Mr. Harada surveyed his class and discovered that 30% of his students have blue eyes. Identify the complement of this event. Then find its probability.

The complement of having blue eyes is *not* having blue eyes. The sum of the probabilities is 100%.

$P(\text{blue eyes}) + P(\textit{not}\text{ blue eyes}) = 100\%$	P(blue eyes) and P(*not* blue eyes) are complements.
$30\% + P(\textit{not}\text{ blue eyes}) = 100\%$	Replace P(blue eyes) with 30%.
$30\% + 70\% = 100\%$	**THINK** 30% plus what number equals 100%?

So, the probability that a student does *not* have blue eyes is 70%, 0.7, or $\frac{7}{10}$.

Guided Practice

A letter tile is chosen randomly. Find the probability of each event. Write each answer as a fraction, percent, and decimal. (Examples 1–3)

1. P(D) ______

2. P(S, V, or L) ______

3. P(*not* D) ______

4. The probability of choosing a "Go Back 1 Space" card in a board game is 25%. Describe the complement of this event and find its probability. (Example 4) ______

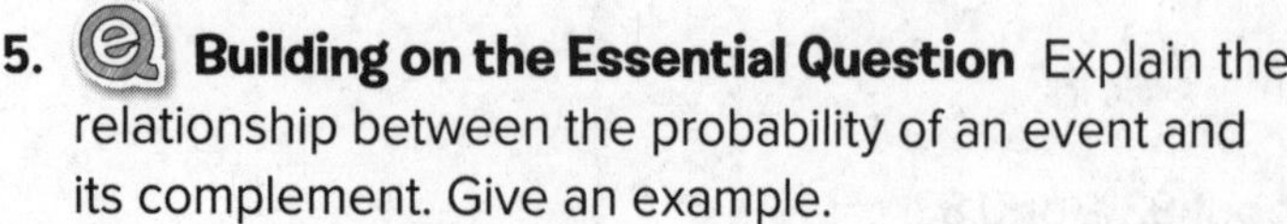

5. **Building on the Essential Question** Explain the relationship between the probability of an event and its complement. Give an example.

Rate Yourself!

How confident are you about finding the probability of simple events? Shade the ring on the target.

For more help, go online to access a Personal Tutor.

FOLDABLES Time to update your Foldable!

Name ______________________ My Homework ______________________

Independent Practice

Go online for Step-by-Step Solutions eHelp

The spinner shown is spun once. Find the probability of each event. Write each answer as a fraction, percent, and decimal. (Examples 1–3)

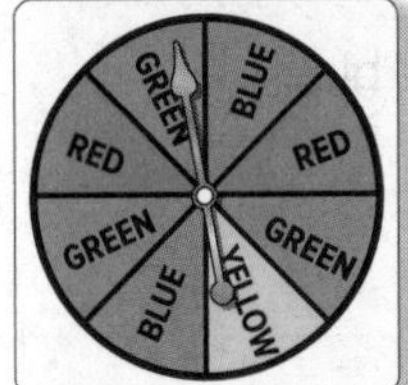

1. *P*(blue)

2. *P*(red or yellow)

3. *P*(*not* brown)

4. *P*(*not* green)

5. Refer to the table on air travel at selected airports. Suppose a flight that arrived at El Centro is selected at random. What is the probability that the flight did *not* arrive on time? Write the answer as a fraction, decimal, and percent. Explain your reasoning. (Example 4)

Air Travel	
Airport	Arrivals (Percent on-time)
El Centro (CA)	80
Baltimore (MD)	82

6. **MP Model with Mathematics** Refer to the graphic novel frame below. Jamar and Theresa decide to create a music mix and include an equal number of songs from each genre. What is the probability that any given song would be from the hip-hop genre? ______________

One jelly bean is picked, without looking, from the dish. Write a sentence that explains how likely it is for each event to happen.

7. black

8. purple, red, or yellow

H.O.T. Problems Higher Order Thinking

9. MP **Persevere with Problems** The probability of landing in a certain section on a spinner can be found by considering the size of the angle formed by that section. On spinner shown, the angle formed by the yellow section is one-fourth of the angle formed by the entire circle. So, $P(\text{yellow}) = \frac{1}{4}$, 0.25, or 25%.

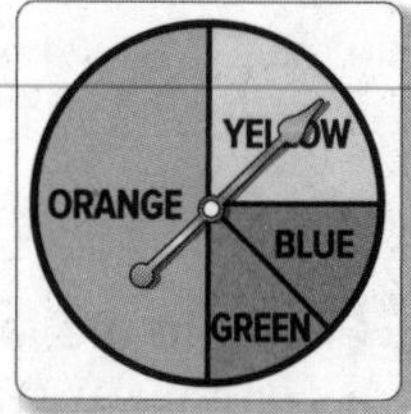

   a. Determine *P*(green) and *P*(orange) for the spinner. Write the probabilities as fractions, decimals, and percents.

   b. Determine *P*(*not* yellow).

10. MP **Persevere with Problems** A bag contains 6 red, 4 blue, and 8 green marbles. How many marbles of each color should be added so that the total number of marbles is 27, but the probability of randomly selecting one marble of each color remains unchanged?

11. MP **Which One Doesn't Belong?** Circle the pair of probabilities that does not belong with the other three. Explain your reasoning.

$0.625, \frac{3}{8}$	0.38, 62%	$\frac{7}{8}$, 0.125	70%, $\frac{1}{3}$

Name ____________________ My Homework ____________________

Extra Practice

Ten cards numbered 1 through 10 are mixed together and then one card is drawn. Find the probability of each event. Write each answer as a fraction, percent, and decimal.

12. $P(8)$

$\frac{1}{10}$, 10%, or 0.1

Only 1 card has an 8. So, $P(8)$ is $\frac{1}{10}$, 10%, or 0.1.

Homework Help

13. $P(7 \text{ or } 9)$

$\frac{1}{5}$, 20%, or 0.2

There is 1 card with a 7 and 1 card with a 9. So, $P(7 \text{ or } 9)$ is $\frac{1}{5}$, or 20%, or 0.2.

14. P(less than 5)

15. P(greater than 3)

16. P(odd)

17. P(even)

18. P(*not* a multiple of 4)

19. P(*not* 5, 6, 7, or 8)

20. P(divisible by 3)

21. Of the students at Grant Middle School, 63% are girls. The school newspaper is randomly selecting a student to be interviewed. Describe the complement of selecting a girl and find the probability of the complement. Write the answer as a fraction, decimal, and percent.

22. The table shows the number of dogs and cats at a groomer. If a pet is selected at random to be groomed, find the probability that Patches the cat will be selected. Then find the probability that a cat will be selected.

Pets at the Groomer	
Cats	**Dogs**
12	16

23. MP **Persevere with Problems** For a certain game, the probability of choosing a card with the number 13 is $\frac{8}{1{,}000}$. Find the probability of *not* choosing a card with the number 13. Then describe the likelihood of the event occurring.

CCSS Power Up! Common Core Test Practice

24. The types of songs on Max's MP3 player are shown on the graph. Max will play one of the songs at random. Complete the model below to find *P*(country or R&B).

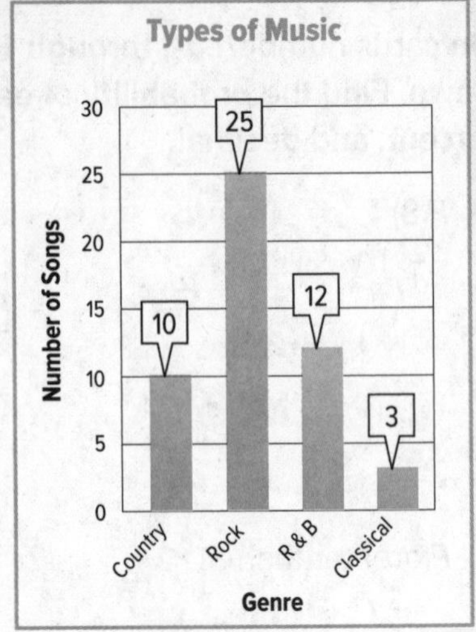

P(country or R&B) =

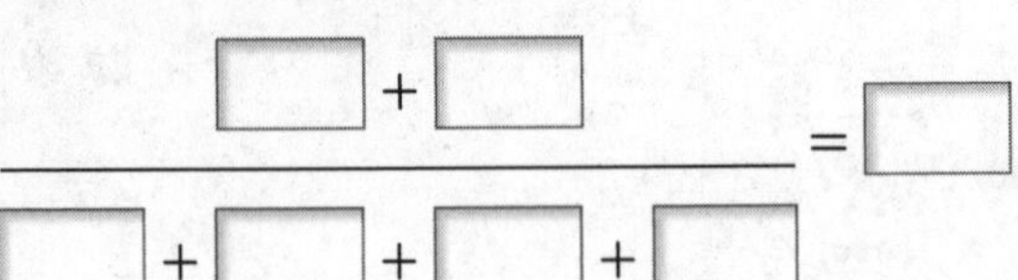

25. Joel has a bowl containing the numbers of colored candies shown in the table. Which of the following probabilities are correct? Select all that apply.

Color	Number
Red	5
Orange	3
Yellow	1
Green	6

- ☐ $P(\text{red}) = \frac{1}{4}$
- ☐ $P(\text{orange}) = \frac{1}{5}$
- ☐ $P(\text{yellow}) = \frac{1}{10}$
- ☐ $P(\text{green}) = \frac{2}{5}$

CCSS Common Core Spiral Review

Compare each decimal using <, >, or =. 5.NBT.3b

26. 0.2 ◯ 0.3

27. 0.75 ◯ 0.7

28. 5.89 ◯ 5.899

29. Dwayne misses 12% of his foul shots and Bryan misses 0.2 of his foul shots. Write 12% and 0.2 as fractions in simplest form. Then compare the fractions to determine who misses more foul shots. 6.NS.7b

Inquiry Lab

Relative Frequency

HOW is probability related to relative frequency?

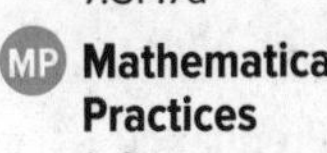

Content Standards
7.SP.6, 7.SP.7, 7.SP.7a

Mathematical Practices
1, 3

In a board game, you get an extra turn if you roll doubles or two of the same number.

You can conduct an experiment to find the relative frequency of rolling doubles using two number cubes. **Relative frequency** is the ratio of the number of experimental successes to the number of experimental attempts.

Hands-On Activity

Step 1 Complete the table to show all of the possible outcomes for rolling two number cubes. Shade all of the possible outcomes that are doubles.

(1, 1)	(2, 1)				
(1, 2)	(2, 2)				
(1, 3)	(2, 3)				
(1, 4)					
(1, 5)					
(1, 6)					

The probability of rolling doubles is ______.

How many times would you expect doubles to be rolled if you roll the number cubes 50 times? Explain. ______

______

Step 2 Roll two number cubes and record the number of doubles in the table. Repeat the experiment 50 times.

Number of Rolls	50
Number of Doubles	

Step 3 Find the relative frequency of rolling doubles. Use the ratio $\frac{\text{number of times doubles were rolled}}{\text{number of rolls}}$. ______

Compare the ratios in Steps 1 and 3. What do you notice? Explain.

______

Suppose the number cubes are rolled 100 times. Would you expect the results to be the same? Explain why or why not.

______

Collaborate

Investigate

Work with a partner.

1. Place a paperclip around the tip of a pencil. Then place the tip on the center of the spinner. Spin the paperclip 40 times. Record the results in the table below.

Section	A	B	C	D
Frequency				
Relative Frequency				

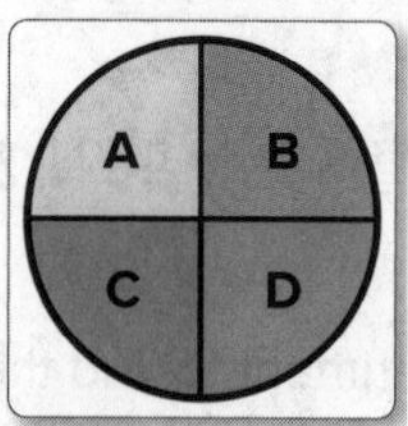

The spinner above is spun once. Find the probability of each event.

2. $P(A)$ ____

3. $P(B)$ ____

4. $P(C)$ ____

5. $P(D)$ ____

Analyze and Reflect

6. Based on your results from the spinner experiment, are the outcomes of A, B, C, or D equally likely? ____________________

7. MP **Reason Inductively** What would you expect to happen to the long-run relative frequency of spinning an A as you increase the number of spins from 40 to 1,000? ____________________

____

Create

8. MP **Justify Conclusions** If you rolled a number cube 600 times, approximate the relative frequency of rolling a 3 or 6. Explain your reasoning to a classmate. ____________________

9. Inquiry HOW is probability related to relative frequency?

____

____

Lesson 2

Theoretical and Experimental Probability

Real-World Link

Watch

Carnival Games The prize wheels for a carnival game are shown. You receive a less expensive prize if you spin and win on wheel A. You receive a more expensive prize if you spin and win on wheel B.

Wheel A

Wheel B

In a **uniform probability model**, each outcome has an equal probability of happening.

1. Which wheel has uniform probability? ________

2. Use a paperclip and the tip of your pencil to spin each wheel 4 times. Record your results.

Spin	Wheel A	Wheel B
1		
2		
3		
4		

3. Why do you think winners on wheel A receive a less expensive prize than winners on wheel B?

____

____

Which MP **Mathematical Practices** did you use? Shade the circle(s) that applies.

① Persevere with Problems
② Reason Abstractly
③ Construct an Argument
④ Model with Mathematics
⑤ Use Math Tools
⑥ Attend to Precision
⑦ Make Use of Structure
⑧ Use Repeated Reasoning

Essential Question

HOW can you predict the outcome of future events?

Vocabulary

uniform probability model
theoretical probability
experimental probability

Common Core State Standards

Content Standards
7.SP.7, 7.SP.7a, 7.SP.7b

MP **Mathematical Practices**
1, 3, 4

Work Zone

Experimental and Theoretical Probability

Theoretical probability is based on uniform probability — what *should* happen when conducting a probability experiment. **Experimental probability** is based on relative frequency — what *actually* occurs during such an experiment.

The theoretical probability and the experimental probability of an event may or may not be the same. As the number of attempts increases, the theoretical probability and the experimental probability should become closer in value.

Trials
A trial is one experiment in a series of successive experiments.

Examples

1. The graph shows the results of an experiment in which a spinner with 3 equal sections is spun sixty times. Find the experimental probability of spinning red for this experiment.

The graph indicates that the spinner landed on red 24 times, blue 15 times, and green 21 times.

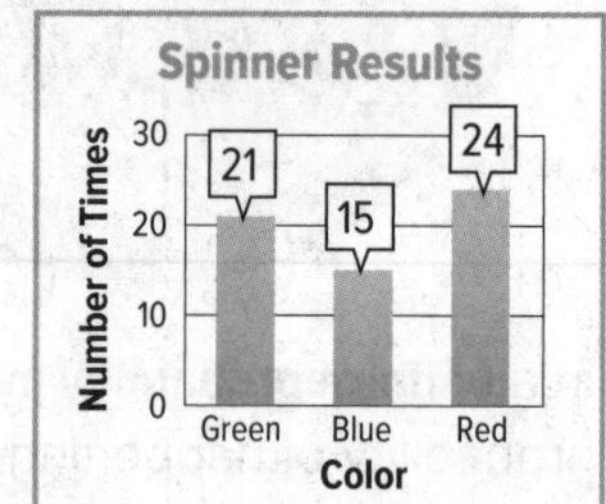

$$P(\text{red}) = \frac{\text{number of times red occurs}}{\text{total number of spins}}$$

$$= \frac{24}{60} \text{ or } \frac{2}{5}$$

The experimental probability of spinning red is $\frac{2}{5}$.

2. Compare the experimental probability you found in Example 1 to its theoretical probability.

The spinner has three equal sections: red, blue, and green. So, the theoretical probability of spinning red is $\frac{1}{3}$. Since $\frac{2}{5} \approx \frac{1}{3}$, the experimental probability is close to the theoretical probability.

Show your work.

a. ____________

b. ____________

Got it? Do these problems to find out.

a. Refer to Example 1. If the spinner was spun 3 more times and landed on green each time, find the experimental probability of spinning green for this experiment.

b. Compare the experimental probability you found in Exercise **a** to its theoretical probability.

Examples

3. Two number cubes are rolled together 20 times. A sum of 9 is rolled 8 times. What is the experimental probability of rolling a sum of 9?

$$P(9) = \frac{\text{number of times a sum of 9 occurs}}{\text{total number of rolls}}$$

$$= \frac{8}{20} \text{ or } \frac{2}{5}$$

The experimental probability of rolling a sum of 9 is $\frac{2}{5}$.

4. Compare the experimental probability you found in Example 3 to its theoretical probability. If the probabilities are not close, explain a possible reason for the discrepancy.

When rolling two number cubes, there are 36 possible outcomes. The theoretical probability of rolling a sum of 9 is $\frac{4}{36}$ or $\frac{1}{9}$.

Rolls with Sum of 9	
First Cube	**Second Cube**
3	6
4	5
5	4
6	3

Since $\frac{1}{9}$ is not close to $\frac{2}{5}$, the experimental probability is *not* close to the theoretical probability. One possible explanation is that there were not enough trials.

Got it? **Do these problems to find out.**

c. In Example 3, what is the experimental probability of rolling a sum that is *not* 9?

d. Two coins are tossed 10 times. Both coins land on heads 6 times. Compare the experimental probability to the theoretical probability. If the probabilities are not close, explain a possible reason for the discrepancy.

e. Suppose three coins are tossed 10 times. All three coins land on heads 1 time. Compare the experimental probability to the theoretical probability. If the probabilities are not close, explain a possible reason for the discrepancy.

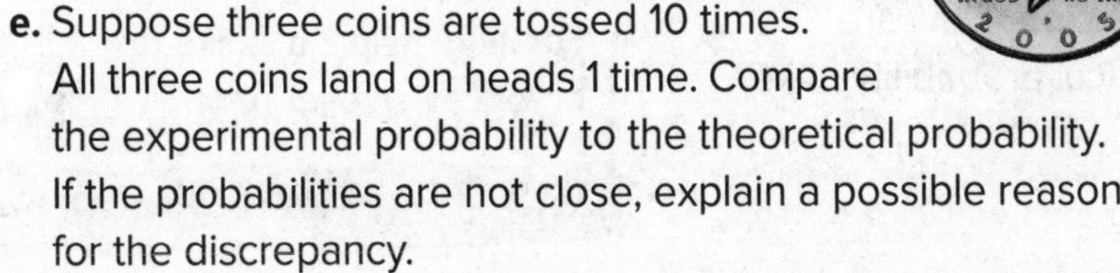

c. ____________

d. ____________

e. ____________

Predict Future Events

Theoretical and experimental probability can be used to make predictions about future events.

Example

Solving Proportions

The cross products of any proportion are equal.

$\frac{29}{100} = \frac{x}{5{,}000}$

5. Last year, a DVD store sold 670 action DVDs, 580 comedy DVDs, 450 drama DVDs, and 300 horror DVDs. A media buyer expects to sell 5,000 DVDs this year. Based on these results, how many comedy DVDs should she buy? Explain.

2,000 DVDs were sold and 580 were comedy. So, the probability is $\frac{580}{2{,}000}$ or $\frac{29}{100}$.

$\frac{29}{100} = \frac{x}{5{,}000}$ Write a proportion.

$29 \cdot 5{,}000 = 100 \cdot x$ Find the cross products.

$145{,}000 = 100x$ Multiply.

$1{,}450 = x$ Divide each side by 100.

She should buy about 1,450 comedy DVDs.

Guided Practice

1. A coin is tossed 50 times, and it lands on heads 28 times. Find the experimental probability and the theoretical probability of the coin landing on heads. Then, compare the experimental and theoretical probabilities. (Examples 1–4)

2. Yesterday, 50 bakery customers bought muffins and 11 of those customers bought banana muffins. If 100 customers buy muffins tomorrow, how many would you expect to buy a banana muffin? (Example 5)

3. **Building on the Essential Question** How are experimental probability and theoretical probability alike?

Rate Yourself!

Are you ready to move on? Shade the section that applies.

For more help, go online to access a Personal Tutor.

FOLDABLES Time to update your Foldable!

Independent Practice

Go online for Step-by-Step Solutions

1. A number cube is rolled 20 times and lands on 1 two times and on 5 four times. Find each experimental probability. Then compare the experimental probability to the theoretical probability. (Examples 1–4)

   a. landing on 5

   ____

   ____

   b. *not* landing on 1

   ____

   ____

2. The spinner at the right is spun 12 times. It lands on blue 1 time. (Examples 1–4)

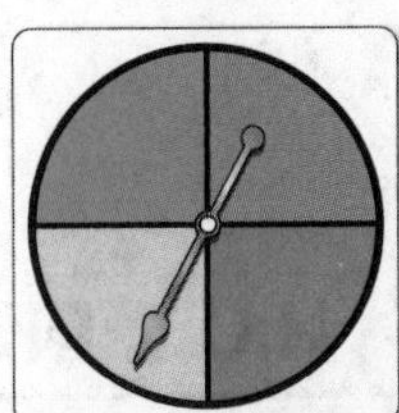

   a. What is the experimental probability of the spinner landing on blue?

   ____

   b. Compare the experimental and theoretical probabilities of the spinner landing on blue. If the probabilities are not close, explain a possible reason for the discrepancy.

   ____

   ____

3. The frequency table shows the results of a survey of 70 zoo visitors who were asked to name their favorite animal exhibit. (Example 5)

What is your Favorite Animal Exhibit?		
Exhibit	**Tally**	**Frequency**
Bears	~~||||~~		6
Elephants	~~||||~~ ~~||||~~ ~~||||~~ ||	17
Monkeys	~~||||~~ ~~||||~~ ~~||||~~ ~~||||~~		21
Penguins	~~||||~~ ~~||||~~ |||	13
Snakes	~~||||~~ ~~||||~~ |||	13

   a. Suppose 540 people visit the zoo. Predict how many people will choose the monkey exhibit as their favorite. ____________

   b. Suppose 720 people visit the zoo. Predict how many people will choose the penguin exhibit as their favorite. ____________

4. MP **Make a Conjecture** Cross out the part of the concept circle that does *not* belong. Then describe the relationship among the remaining parts.

   ____

   ____

   ____

a coin landing on tails 8 out of 10 times	results based on an experiment
outcomes that should happen	rolling a sum of 9 twice in 5 trials

5 **MP Multiple Representations** A spinner with three equal-sized sections marked A, B, and C is spun 100 times.

a. **Numbers** What is the theoretical probability of landing on A?

b. **Numbers** The results of the experiment are shown in the table. What is the experimental probability of landing on A? on C?

Section	Frequency
A	24
B	50
C	26

c. **Models** Make a drawing of what the spinner might look like based on its experimental probabilities. Explain.

H.O.T. Problems Higher Order Thinking

6. **MP Persevere with Problems** The experimental probability of a coin landing on heads is $\frac{7}{12}$. If the coin landed on tails 30 times, find the number of tosses.

7. **MP Reason Inductively** Twenty sharpened pencils are placed in a box containing an unknown number of unsharpened pencils. Suppose 15 pencils are removed at random and five of the removed pencils are sharpened. Based on this, is it reasonable to assume that the number of unsharpened pencils was 40? Explain your reasoning.

8. **MP Reason Inductively** The results of spinning a spinner with six equal sections are shown. Determine the minimum number of additional spins needed and their frequency of landing on each color so that the experimental probabilities will be equal to the theoretical probabilities.

Explain your reasoning.

Color	Frequency
Blue	8
Green	6
Orange	12
Purple	10
Red	8
Yellow	4

Extra Practice

For Exercises 9 and 10, find each experimental probability. Then compare the experimental probability to its theoretical probability. If the probabilities are not close, explain a possible reason for the discrepancy.

9. A coin is tossed 20 times. It lands on heads 9 times.

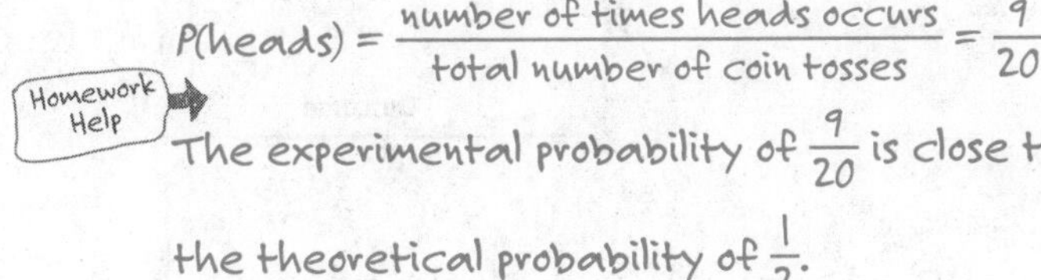

the theoretical probability of $\frac{1}{2}$.

10. A heart is randomly chosen 7 out of 12 times from the cards shown.

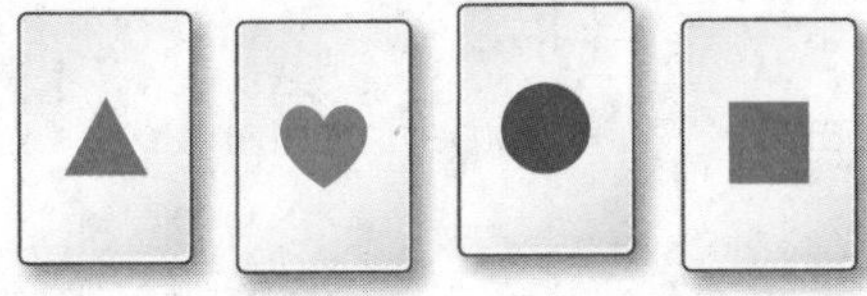

____

____

____

____

Solve.

11. Last month, customers at a gift shop bought 40 birthday cards, 19 congratulations cards, 20 holiday cards, and 21 thank you cards. Suppose 125 customers buy greeting cards next month. How many would you expect to buy a birthday card?

____

12. Use the graph at the right.

a. What is the probability that a mother received a gift of flowers or plants? Write the probability as a fraction in simplest form.

____

b. Suppose 400 mothers will receive a gift. Predict how many will receive flowers or plants.

____

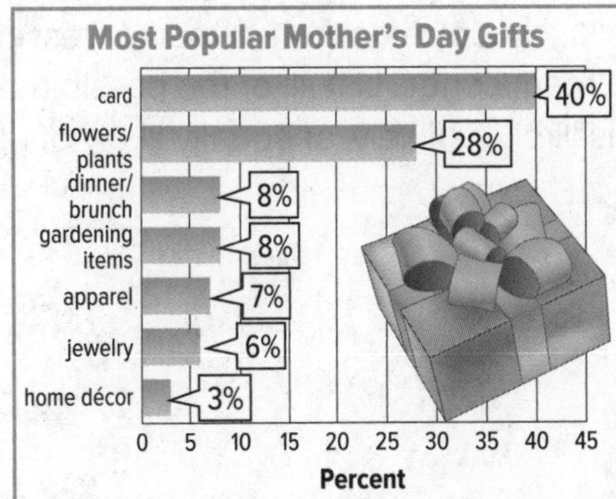

CCSS Power Up! Common Core Test Practice

13. J.R. tossed a coin 100 times. Fill in the boxes to complete each statement.

Based on J.R.'s results, the [ ] probability of tossing heads is [ ] %. This is [ ] than the theoretical probability of tossing heads with a coin.

Tossing a Coin

Number of Tosses: 0, 20, 40, 60, 80

Outcome: Heads, Tails

14. Determine if each situation represents experimental or theoretical probability.

a. Saul flips a coin 20 times and determines that the probability of flipping heads is 0.55. ☐ experimental ☐ theoretical

b. Kelly has made 16 out of 25 free throws. The probability that she will make her next free throw is 64%. ☐ experimental ☐ theoretical

c. There are 4 pennies, 2 nickels, 5 dimes, and 5 quarters in a jar. The probability that a randomly selected coin is a penny is $\frac{1}{4}$. ☐ experimental ☐ theoretical

CCSS Common Core Spiral Review

For Exercises 15 and 16, circle the greater probability. 7.SP.5

15. The spinner at the right is spun.

P(red) *P*(not red)

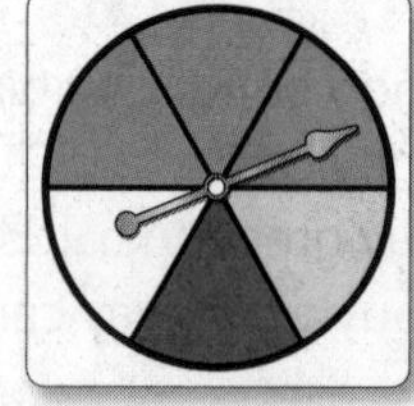

16. A number cube is rolled.

P(multiple of 3) *P*(prime number)

17. A restaurant offers three flavors of ice cream on its dessert menu: vanilla, chocolate, and strawberry. Dessert options are sundaes or ice cream cones. List all of the possible desserts. Then determine if it is likely, unlikely, or equally likely of randomly choosing a sundae.

7.SP.5 ______________________________

____

Inquiry Lab

Fair and Unfair Games

HOW can you determine if a game is fair?

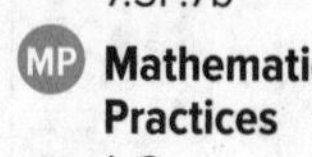

Content Standards
7.SP.7, 7.SP.7a, 7.SP.7b

MP Mathematical Practices
1, 3

In a counter-toss game, players toss three two-color counters. The winner of each game is determined by how many counters land with either the red or yellow side facing up. Find out if this game is fair or unfair.

Mathematically speaking, a two-player game is **fair** if each player has an equal chance of winning. A game is **unfair** if there is not such a chance.

Tools

Work in pairs to play the game described above.

Step 1 Player 1 tosses the counters. If 2 or 3 counters land red-side up, Player 1 wins. If 2 or 3 counters land yellow-side up, Player 2 wins. Record the results in the table below. Place a check in the winner's column for each game.

Game	Player 1	Player 2	Game	Player 1	Player 2
1			6		
2			7		
3			8		
4			9		
5			10		

Step 2 Player 2 then tosses the counters and the results are recorded.

Step 3 Continue alternating turns until the counters have been tossed 10 times.

Based on your results, do you think the game is fair or unfair? Circle your response below.

Fair Unfair

Investigate

Work with a partner.

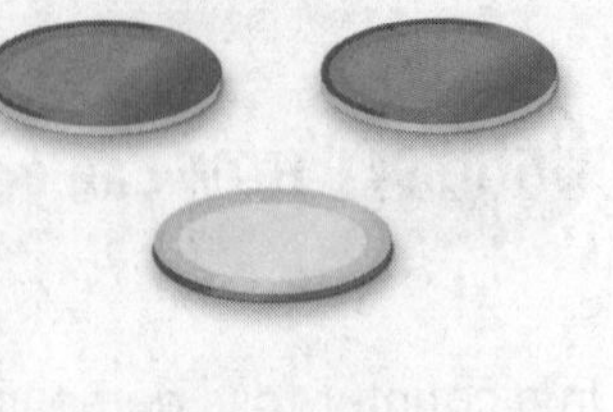

1. Complete the organized list of all the possible outcomes resulting from one toss of the three counters described in Activity 1.

Counter 1	Counter 2	Counter 3	Outcome
red	red	red	red, red, red

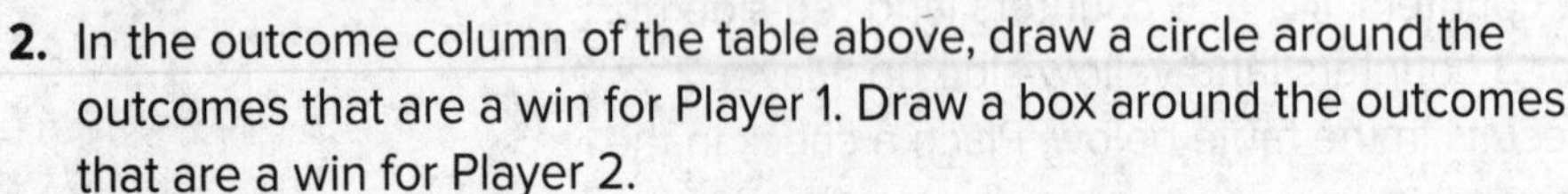

2. In the outcome column of the table above, draw a circle around the outcomes that are a win for Player 1. Draw a box around the outcomes that are a win for Player 2.

3. Calculate the theoretical probability of each player winning. Write each probability as a fraction and as a percent. Is the game fair or unfair?

4. Use your results from Activity 1 to calculate the experimental probability of each player winning.

Analyze and Reflect

5. MP **Justify Conclusions** Compare the probabilities you found in Exercises 3 and 4. Explain any discrepancies.

6. MP **Reason Inductively** Predict the number of times Player 1 would win if the game were played 100 times. Explain your reasoning.

David and Lyn made up a game using a plastic cup. A cup is tossed. If it lands right-side up or open-end down, David wins. If it lands on its side, Lyn wins. Is this game fair?

Hands-On Activity 2

Work in pairs to play the game and determine if David and Lyn created a fair game.

Step 1 Player 1 tosses the cup. If it lands right-side up or open-end down, Player 1 gets a point. If the cup lands on its side, Player 2 gets a point. Record your results in the table below.

Toss	Player 1	Player 2	Toss	Player 1	Player 2
1			6		
2			7		
3			8		
4			9		
5			10		

Step 2 Player 2 then tosses the cup and the results are recorded.

Step 3 Continue alternating turns until there is a total of 10 tosses.

Based on your results, do you think the game David and Lyn created is fair or unfair? Circle your response below.

Fair Unfair

There are three possible outcomes when tossing the cup and David wins if two of those outcomes happen. It may appear that David has a better chance of winning, however this is not necessarily true.

Explain why Lyn actually has a better chance at winning the game.

What was the experimental probability for the cup landing right-side up or open-end down?

Investigate

Work with a partner.

7. A game involves rolling two number cubes. Player 1 wins the game if the total of the numbers rolled is 5 or if a 5 is shown on one or both number cubes. Otherwise, Player 2 wins. Fill in the table for all of the possible outcomes of rolling two number cubes.

	1	2	3	4	5	6
1	1 + 1 = 2	1 + 2 = 3	1 + 3 = 4	1 + 4 = 5	1 + 5 = 6	1 + 6 = 7
2	2 + 1 = 3					
3						
4						
5						
6						

8. Shade in the cells of the table in which Player 1 is a winner.

Analyze and Reflect

9. For the number cube game, calculate the theoretical probability of each player winning. Write each probability as a fraction and as a percent.

10. MP **Justify Conclusions** Is the number cube game fair? Explain.

Create

11. MP **Model with Mathematics** Design and describe a game in which the outcome is not fair. Then explain how you could change the game to make it fair.

12. Inquiry HOW can you determine if a game is fair?

Lesson 3

Probability of Compound Events

Real-World Link

Travel Aimee wants to pack enough items to create 6 different outfits. She packs 1 jacket, 3 shirts, and 2 pairs of jeans. Can Aimee create 6 different outfits from her clothing items?

1. Complete the table below.

Outfit	Clothing Items
1	jacket, shirt 1, jeans 1
2	jacket, shirt 1, jeans 2
3	jacket, shirt 2, jeans 1
4	jacket, shirt 2,
5	jacket, shirt 3,
6	jacket,

2. The table is an example of an organized list. What is another way to show the different outfits that Aimee can create?

3. Describe another situation for which you might want to list all of the possible outcomes.

Essential Question

HOW can you predict the outcome of future events?

Vocabulary

sample space
tree diagram
compound event

Common Core State Standards

Content Standards
7.SP.8, 7.SP.8a, 7.SP.8b

MP Mathematical Practices
1, 3, 4, 5

Which **MP Mathematical Practices** did you use?
Shade the circle(s) that applies.

① Persevere with Problems
② Reason Abstractly
③ Construct an Argument
④ Model with Mathematics
⑤ Use Math Tools
⑥ Attend to Precision
⑦ Make Use of Structure
⑧ Use Repeated Reasoning

Work Zone

Find a Sample Space

The set of all of the possible outcomes in a probability experiment is called the **sample space**. Organized lists, tables, and **tree diagrams** can be used to represent the sample space.

Examples

1. The three students chosen to represent Mr. Balderick's class in a school assembly are shown. All three of them need to sit in a row on the stage. Use a list to find the sample space for the different ways they can sit in a row.

Students
Adrienne
Carlos
Greg

Use A for Adrienne, C for Carlos, and G for Greg. Use each letter exactly once.

ACG AGC CAG CGA GAC GCA

So, the sample space consists of 6 outcomes.

2. A car can be purchased in blue, silver, red, or purple. It also comes as a convertible or hardtop. Use a table or a tree diagram to find the sample space for the different styles in which the car can be purchased.

Color	Top
blue	convertible
blue	hardtop
silver	convertible
silver	hardtop
red	convertible
red	hardtop
purple	convertible
purple	hardtop

Color	Top	Sample Space
Blue	Convertible	BC
	Hardtop	BH
Silver	Convertible	SC
	Hardtop	SH
Red	Convertible	RC
	Hardtop	RH
Purple	Convertible	PC
	Hardtop	PH

Using either method, the sample space consists of 8 outcomes.

Show your work.

Got it? Do this problem to find out.

a. The table shows the sandwich choices for a picnic. Find the sample space using a list, table, or tree diagram for a sandwich consisting one type of meat and one type of bread.

Meat	Bread
ham turkey	rye sourdough white

a. ______

Find Probability

A **compound event** consists of two or more simple events. The probability of a compound event, just as with simple events, is the fraction of outcomes in the sample space for which the compound event occurs.

Example

3. Suppose you toss a quarter, a dime, and a nickel. Find the sample space. What is the probability of getting three tails?

Make a tree diagram to show the sample space.

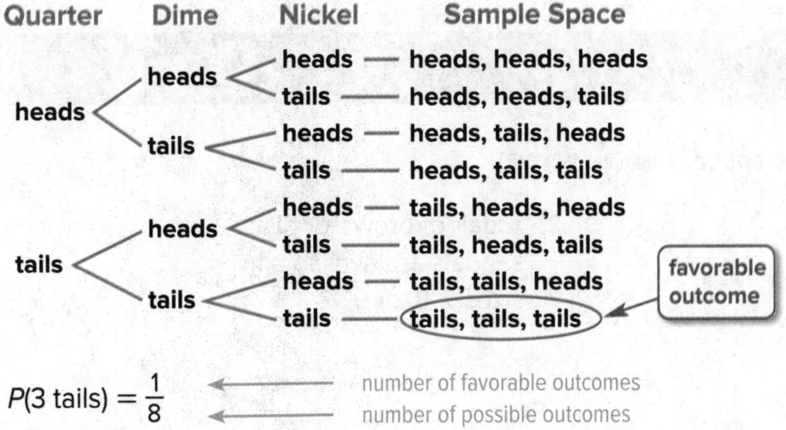

$P(\text{3 tails}) = \frac{1}{8}$ ← number of favorable outcomes / number of possible outcomes

So, the probability of getting three tails is $\frac{1}{8}$.

Got it? Do this problem to find out.

b. The animal shelter has both male and female Labrador Retrievers in yellow, brown, or black. There is an equal number of each kind. What is the probability of choosing a female yellow Labrador Retriever? Show your work in the space below.

Random

When choosing an outcome, assume that each outcome is chosen randomly.

Example

4. To win a carnival prize, you need to choose one of 3 doors labeled 1 through 3. Then you need to choose a red, yellow, or blue box behind each door. What is the probability that the prize is in the blue or yellow box behind door 2?

The table shows that there are 9 total outcomes. Two of the outcomes are favorable.

So, the probability that the prize is in a blue or yellow box behind door 2 is $\frac{2}{9}$.

Outcomes	
door 1	red box
door 1	yellow box
door 1	blue box
door 2	red box
door 2	yellow box
door 2	blue box
door 3	red box
door 3	yellow box
door 3	blue box

Guided Practice

For each situation, find the sample space. (Examples 1–2)

1. A coin is tossed twice.

2. A pair of brown or black sandals are available in sizes 7, 8, or 9.

3. Gerardo spins a spinner with four equal sections, labeled A, B, C, and D, twice. If letter A is spun at least once, Gerardo wins. Otherwise, Odell wins. Use a list to find the sample space. Then find the probability that Odell wins. (Examples 3–4)

4. **Bulding on the Essential Question** How do tree diagrams, tables, and lists help you find the probability of a compound event?

Rate Yourself!

☐ I understand how to show a sample space.

Great! You're ready to move on!

☐ I still have questions about showing a sample space.

No Problem! Go online to access a Personal Tutor.

Tutor

Name ______________________________ My Homework ______________________________

Independent Practice

Go online for Step-by-Step Solutions

For each situation, find the sample space. (Examples 1–2)

1. tossing a coin and spinning the spinner at the right

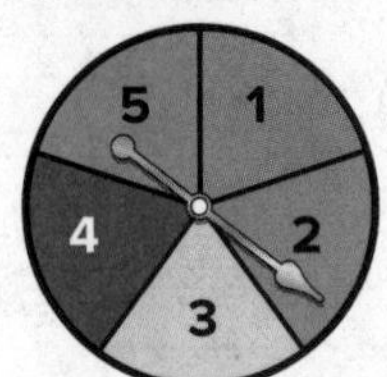

2. picking a number from 1 to 5 and choosing the color red, white, or blue

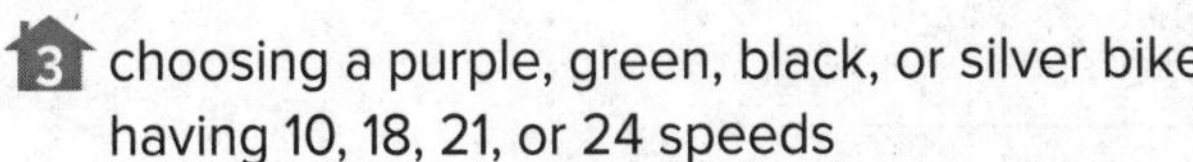

3. choosing a purple, green, black, or silver bike having 10, 18, 21, or 24 speeds

4. choosing a letter from the word SPACE and choosing a consonant from the word MATH

For each game, find the sample space. Then find the indicated probability. (Examples 3–4)

5. Alana tosses 2 number cubes. She wins if she rolls double sixes.

   Find *P*(Alana wins). ______________________________

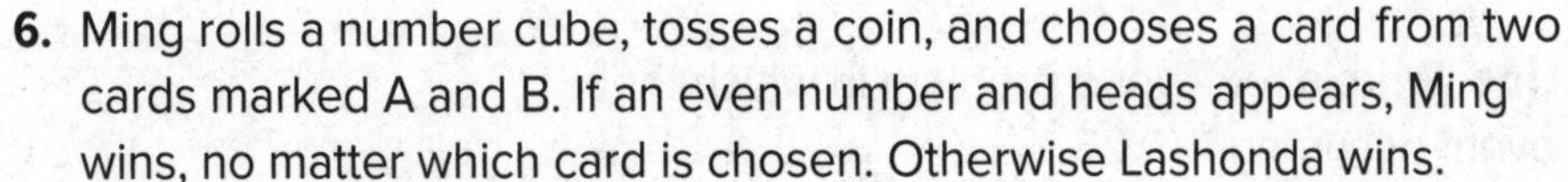

6. Ming rolls a number cube, tosses a coin, and chooses a card from two cards marked A and B. If an even number and heads appears, Ming wins, no matter which card is chosen. Otherwise Lashonda wins.

   Find *P*(Ming wins). ______________________________

7 MP **Persevere with Problems** The following is a game for two players.

- Three counters are labeled according to the table at the right.
- Toss the three counters.
- If exactly 2 counters match, Player 1 scores a point. Otherwise, Player 2 scores a point.

Counters	Side 1	Side 2
Counter 1	red	blue
Counter 2	red	yellow
Counter 3	blue	yellow

Find the probability that each player scores a point.

H.O.T. Problems Higher Order Thinking

8. MP **Persevere with Problems** Refer to Exercise 7. Do the two players both have an equal chance of winning? Explain.

9. MP **Find the Error** Caitlyn wants to determine the probability of guessing correctly on two true-false questions on her history test. She draws the tree diagram below using C for correct and I for incorrect. Find her mistake and correct it.

Question 1	Question 2	Sample Space
C	C	CC
	I	CI
I	C	CC
	I	II

10. MP **Model with Mathematics** Write a real-world problem in which the probability of a compound event occurring is 0.25.

Name ______________________ My Homework ______________________

Extra Practice

11. Three-course dinners can be made from the menu shown. Find the sample space for a dinner consisting of an appetizer, entrée, and dessert.

Appetizers	Entrees	Desserts
Soup	Steak	Carrot cake
Salad	Chicken	Apple pie

Homework Help

Appetizer	Entree	Dessert	Sample Space
S	S	C	SSC
		A	SSA
	C	C	SCC
		A	SCA
Sa	S	C	SaSC
		A	SaSA
	C	C	SaCC
		A	SaCA

12. Mr. and Mrs. Romero are expecting triplets. Suppose the chance of each child being a boy is 50% and of being a girl is 50%. Find the probability of each event.

a. *P*(all three children will be boys) ______

b. *P*(at least one boy and one girl) ______

c. *P*(two boys and one girl) ______

d. *P*(at least two girls) ______

Copy and Solve **For Exercises 13 and 14, show your work on a separate piece of paper.**

13. The University of Oregon's football team has many different uniforms. The coach can choose from four colors of jerseys and pants: green, yellow, white, and black. There are three helmet options: green, white, and yellow. Also, there are the same four colors of socks and two colors of shoes, black and yellow.

a. How many jersey/pant combinations are there?

b. If the coach picks a jersey/pant combination at random, what is the probability he will pick a yellow jersey with green pants?

c. Use a tree diagram to find all of the possible shoe and sock combinations.

14. MP **Use Math Tools** Use the Internet or another source to find the top five best-selling animated movies. Then create a list of the possibilities for choosing a movie and choosing a wide-screen or full-screen version.

CCSS Power Up! Common Core Test Practice

15. Mr. Skeels will choose one student from each of the two groups to present their history report to the class. Which of the following represent possible outcomes? Select all that apply.

Group 1
Ava
Antoine
Greg

Group 2
Mario
Brooke

- ☐ (Ava, Brooke)
- ☐ (Greg, Brooke)
- ☐ (Antoine, Greg)
- ☐ (Antoine, Mario)

16. Campers choose one activity from each of the morning, afternoon, and evening activities shown below.

Morning	Afternoon	Evening
Hiking (H)	Archery (A)	Horseback Riding (R)
Canoeing (C)	Bird Watching (B)	Campfire Building (F)
		Navigating (N)

Make a list to show the sample space for the possible morning, afternoon, and evening activities.

What is the probability that a randomly selected camper will be horseback riding in the evening?

CCSS Common Core Spiral Review

Eight cards numbered 1–8 are shuffled together. A card is drawn at random. Find the probability of each event. 7.SP.5

17. *P*(8) ________

18. *P*(greater than 5) ________

19. *P*(even) ________

20. *P*(3 or 7) ________

21. What is the probability of rolling a number greater than 4 on a number cube? Explain. 7.SP.5

Lesson 4

Simulations

Real-World Link

Music Downloads A new electronics store is opening at the mall. One out of six new customers will receive a free music download. The winners are chosen at random. On Monday, the store had 50 customers. You can act out or *simulate* 50 random customers by using the random number generator on a graphing calculator.

Type in the following keystrokes to set 1 as the lower bound and 6 as the upper bound for 50 trials.

Keystrokes: MATH 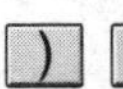 5 1 , 6 , 50) ENTER

The screen should look similar to the screen shown below.

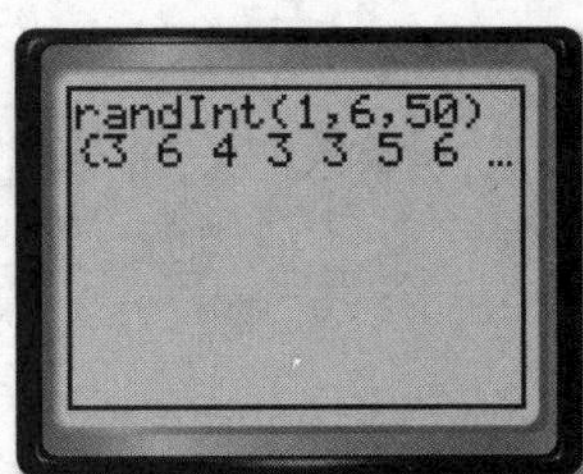

A set of 50 numbers ranging from 1 to 6 appears. Use the right arrow key to see the next number in the set.

1. Let the number 3 represent a customer who wins a free download. Write the experimental probability of winning a download.

2. Compare the experimental probabilities found in Exercise 1 to the theoretical probability of winning a download.

Which MP **Mathematical Practices** did you use? Shade the circle(s) that applies.

① Persevere with Problems
② Reason Abstractly
③ Construct an Argument
④ Model with Mathematics
⑤ Use Math Tools
⑥ Attend to Precision
⑦ Make Use of Structure
⑧ Use Repeated Reasoning

Essential Question

HOW can you predict the outcome of future events?

Vocabulary

simulation

Common Core State Standards

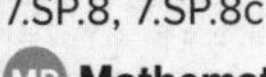

Content Standards

7.SP.8, 7.SP.8c

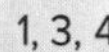

Mathematical Practices

1, 3, 4

Work Zone

Model Equally Likely Outcomes

A **simulation** is an experiment that is designed to model the action in a given situation. For example, you used a random number generator to simulate rolling a number cube. Simulations often use models to act out an event that would be impractical to perform.

Example

1. A cereal company is placing one of eight different trading cards in its boxes of cereal. If each card is equally likely to appear in a box of cereal, describe a model that could be used to simulate the cards you would find in 15 boxes of cereal.

Choose a method that has 8 possible outcomes, such as tossing 3 coins. Let each outcome represent a different card.

For example, the outcome of all three coins landing heads up could simulate finding card 1.

Toss 3 coins to simulate the cards that might be in 15 boxes of cereal. Repeat 15 times.

Coin Toss Simulation

Outcome	Card	Outcome	Card
HHH	1	TTT	5
HHT	2	TTH	6
HTH	3	THT	7
HTT	4	THH	8

Got it? Do this problem to find out.

a. A restaurant is giving away 1 of 5 different toys with its children's meals. If the toys are given out randomly, describe a model that could be used to simulate which toys would be given with 6 children's meals.

a. ________

Example

2. Every student who volunteers at the concession stand during basketball games will receive a free school T-shirt. The T-shirts come in 3 different designs.

Design a simulation that could be used to model this situation. Use your simulation to find how many times a student must volunteer in order to get all 3 T-shirts.

Use a spinner divided into 3 equal sections. Assign each section one of the T-shirts. Spin the spinner until you land on each section.

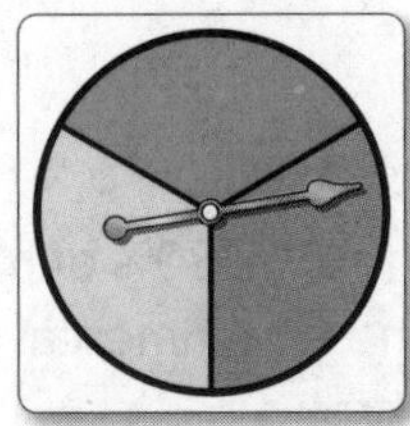
first spin

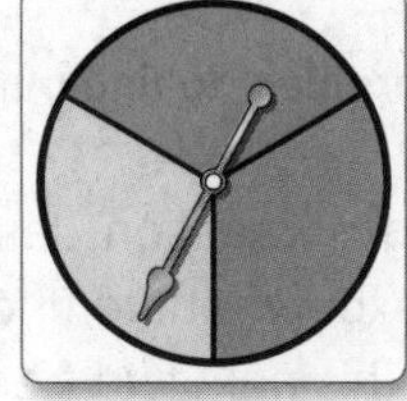
second spin

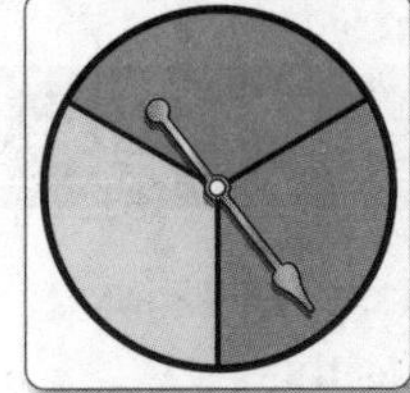
third spin

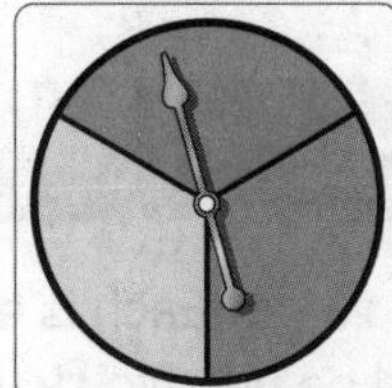
fourth spin

Based on this simulation, a student should volunteer 4 times in order to get all 3 T-shirts.

Got it? **Do this problem to find out.**

b. Mr. Chen must wear a dress shirt and a tie to work. Each day he picks one of his 6 ties at random. Design a simulation that could be used to model this situation. Use your simulation to find how many days Mr. Chen must work in order to wear all of his ties.

b. _______________

Model Unequally Likely Outcomes

Simulations can also be used to model events in which the outcomes are not equally likely.

STOP and Reflect

How could you simulate a 20% chance? Write your answer below.

Example

3. There is a 60% chance of rain for each of the next two days. Describe a method you could use to find the experimental probability of having rain on both of the next two days.

Place 3 red and 2 blue marbles in a bag. Let 60% or $\frac{3}{5}$ of them represent rain. Let 40% or $\frac{2}{5}$ of them represent no rain.

Randomly pick one marble to simulate the first day. Replace the marble and pick again to simulate the second day. Find the probability of rain on both days.

Got it? **Do this problem to find out.**

c. During the regular season, Jason made 80% of his free throws. Describe an experiment to find the experimental probability of Jason making his next two free throws.

c. ______________

Guided Practice

1. An ice cream store offers waffle cones or sugar cones. Each is equally likely to be chosen. Describe a model that could be used to simulate this situation. Based on your simulation, how many people must order an ice cream cone in order to sell all possible combinations? (Examples 1 and 2)

2. An electronics store has determined that 45% of its customers buy a wide-screen television. Describe a model that you could use to find the experimental probability that the next three television-buying customers will buy a wide-screen television. (Example 3)

3. **Building on the Essential Question** Explain how using a simulation is related to experimental probability.

Rate Yourself!

How well do you understand simulations? Circle the image that applies.

Clear | Somewhat Clear | Not So Clear

For more help, go online to access a Personal Tutor.

Independent Practice

1. The questions on a multiple-choice test each have 4 answer choices. Describe a model that you could use to simulate the outcome of guessing the correct answers to a 50-question test. (Example 1)

2. A game requires drawing balls numbered 0 through 9 for each of four digits to determine the winning number. Describe a model that could be used to simulate the selection of the number. (Example 1)

MP **Model with Mathematics Describe a model you could use to simulate each event.**

3. A jar of cookies contains 18 different types of cookies. Each type is equally likely to be chosen. Based on your simulation, how many times must a cookie be chosen in order to get each type? (Example 2)

4. A cooler contains 5 bottles of lemonade, 4 bottles of water, and 3 bottles of juice. Each type is equally likely to be chosen. Based on your simulation, how many times must a drink be chosen in order to get each type? (Example 3)

5. Players at a carnival game win about 30% of the time. Based on your simulation, what is the experimental probability that the next four players will win. (Example 3)

6. **Model with Mathematics** Suppose a mouse is placed in the maze at the right. If each decision about direction is made at random, create a simulation to determine the probability that the mouse will find its way out before coming to a dead end or going out the In opening.

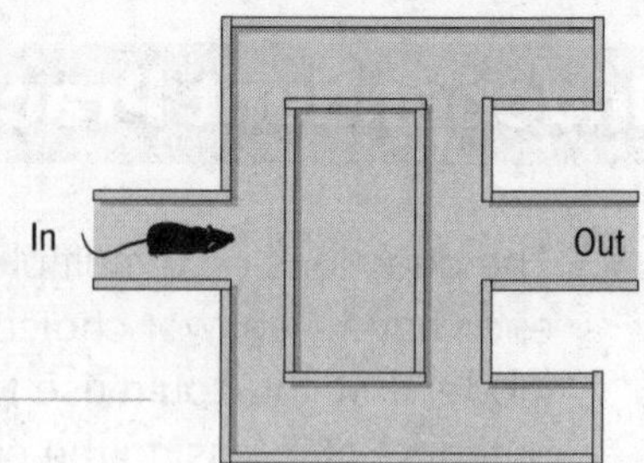

H.O.T. Problems Higher Order Thinking

7. **Model with Mathematics** Describe a situation that could be represented by a simulation. What objects could be used in your simulation?

8. **Persevere with Problems** A simulation uses cards numbered 0 through 9 to generate five 2-digit numbers. A card is selected for the tens digit and not replaced. Then a card for the ones digit is drawn and not replaced. The process is repeated until all the cards are used. If the simulation is performed 10 times, about how many times could you expect a 2-digit number to begin with a 5? Explain.

9. **Justify Conclusions** Determine whether the following statement is *sometimes*, *always*, or *never* true. Justify your answer.

A spinner can be used to model equally likely outcomes.

10. **Justify Conclusions** Barton believes that the coin his teacher uses for an experiment gives an advantage to one team of students. His teacher has students toss the coin 50 times each and record their results. Based on the results in the table, do you think the coin is fair? Explain.

Student	Heads	Tails
1	17	33
2	22	28
3	28	22
4	21	29
5	13	37
6	20	30

Name ______________________ My Homework ______________________

Extra Practice

11. A store employee randomly gives scratch-off discount cards to the first 50 customers. The cards offer discounts of 10%, 20%, 25%, 30%, or 40%. There is an equal chance of receiving any of the 5 cards. Describe a model that could be used to simulate the discount received by 4 customers.

Use a spinner with 5 equal sections to represent the 5 different discounts. Spin 4 times to simulate 4 customers receiving cards.

12. On average, 75% of the days in Henderson county are sunny, with little or no cloud cover. Describe a model that you could use to find the experimental probability of sunny days each day for a week in Henderson county.

MP Model with Mathematics **Describe a model you could use to simulate each event.**

13. Every student who participated in field day activities received a water bottle. The water bottles came in 2 different colors. Based on your simulation, how many students had to receive a water bottle in order to distribute water bottles in both colors?

14. A field hockey team wins 80% of its games. Based on your simulation, what is the experimental probability of the team winning its next 3 games?

15. There are 4 different magazines on Hannah's nightstand. Each evening, she randomly selects one magazine to read. Based on your simulation, how many days must she select a magazine in order to read all 4 magazines?

CCSS Power Up! Common Core Test Practice

16. The table shows the chance of rain this weekend. Select values to fill in the boxes in the model below to describe a method you could use to find the experimental probability of having rain on both days.

Day	Saturday	Sunday
Chance of Rain	30%	30%

2	4	6	7	red
10	30	60	70	blue

Place 3 red and ☐ blue marbles in a bag. Let the ☐ marbles represent rain since ☐ % of the marbles are this color. Let the ☐ marbles represent no rain since ☐ % of the marbles are this color.

Randomly pick one marble to simulate the first day. Replace the marble and pick again to simulate the second day. Find the experimental probability of rain on both days. Do you think it matters how many trials of the simulation you conduct? Will conducting more trials result in a better prediction? Explain your reasoning.

17. At a restaurant, 1 out of every 6 kids' meals wins a prize. Determine which probability models could be used to simulate winning a prize. Select all that apply.

- ☐ Toss a coin. Let tossing heads represent winning a prize and let tossing tails represent not winning a prize.
- ☐ Spin a spinner with equal size spaces labeled A, B, C, D, E, and F. Let spinning A represent winning a prize and let spinning other letters represent not winning a prize.
- ☐ Roll a number cube. Let rolling a 1 represent winning a prize and let rolling a 2, 3, 4, 5, or 6 represent not winning a prize.

CCSS Common Core Spiral Review

18. A local video store has advertised that one out of every four customers will receive a free box of popcorn with their video rental. So far, 15 out of 75 customers have received popcorn. Compare the experimental and theoretical probabilities of receiving popcorn. **7.SP.5, 7.SP.7**

Inquiry Lab

Simulate Compound Events

HOW do simulations help you understand the probability of events happening?

Content Standards
7.SP.8, 7.SP.8c

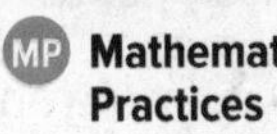

Mathematical Practices
1, 3

A local shop randomly gives coupons to 3 out of every 8 customers. Use a spinner to determine the probability that a customer will receive a coupon two days in a row.

Hands-On Activity 1

Step 1 A spinner with eight equal sections can be used to simulate the situation. Label three of the sections with the letter C to represent the people that receive a coupon. Label five of the sections with the letter D to represent the people that do not receive a coupon.

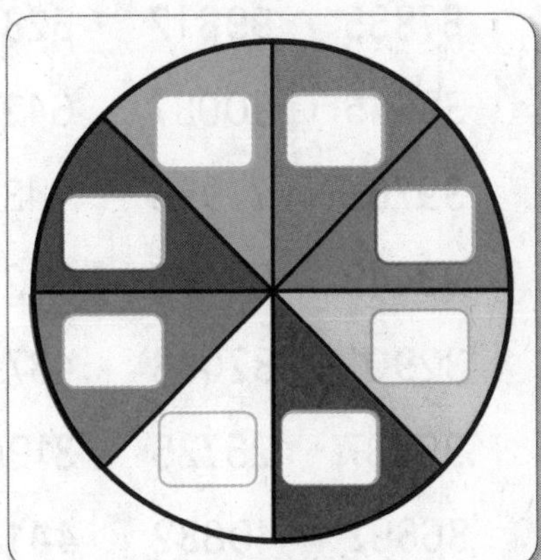

Step 2 Every two spins of the spinner represents one trial. Use a paperclip and the tip of your pencil to spin the spinner twice and record the results in the table. Perform a total of 15 trials.

Trial	Spin 1	Spin 2	Trial	Spin 1	Spin 2	Trial	Spin 1	Spin 2
1			6			11		
2			7			12		
3			8			13		
4			9			14		
5			10			15		

Based on your results, what is the experimental probability that a customer will receive a coupon two days in a row?

____

You can also use a random number table to simulate a compound event.

There is a 10% chance of rain for a city on Sunday and a 20% chance of rain on Monday. Use a random number table to find the probability that it will rain on both days.

Hands-On Activity 2

Step 1 A random number table has random digits in rows that can be grouped in different combinations as needed. These digits are arranged in groups of 5, but the grouping often does not matter. Since the situation we want to represent involves two days, continue drawing lines to separate the numbers into two-digit numbers.

48587	49460	89640	30270
19507	87835	99812	52353
11364	35645	90087	64254
87045	39769	77995	14316
69913	93449	68497	31270
81827	32901	82033	43714
33386	99637	25725	31900
41575	86692	40882	44123
77351	12790	62795	77307

Step 2 Using the digits 0 through 9, assign one digit in the tens place for rain on Sunday and assign two different digits in the ones place for rain on Monday. For example, the digit 1 in the tens place can represent rain occurring on Sunday and the digits 1 and 2 in the ones place can represent rain occurring on Monday.

Step 3 Find the numbers in the table that have a 1 in the tens place and either a 1 or 2 in the ones place. Those numbers are 11 and 12. Circle those numbers in the table.

Step 4 Find the probability using the numbers found in Step 3.

There were ☐ instances of the random numbers 11 and 12 occurring out of 90 random numbers.

So, the probability that it will rain on both days is $\frac{\square}{90}$ or $3\frac{1}{3}\%$.

Investigate

Work with a partner.

1. Luke plays goalie on his soccer team. He usually stops 2 out of every 6 penalty kicks. Label the sections of the spinner at the right. Then use the spinner to determine the experimental probability that Luke stops 2 penalty kicks in a row.

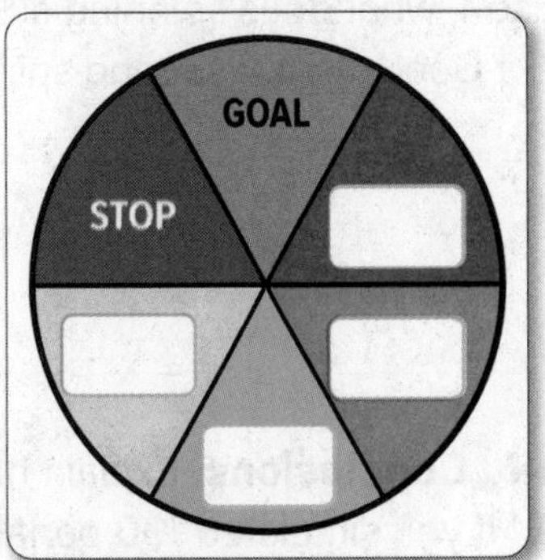

Trial	Spin 1	Spin 2	Trial	Spin 1	Spin 2	Trial	Spin 1	Spin 2
1			6			11		
2			7			12		
3			8			13		
4			9			14		
5			10			15		

The experimental probability is ______________.

2. Suppose 40% of customers who enter a pet store own a cat. What is the probability that it will take at least 4 customers before a cat owner enters the store? Use a random number table to simulate this compound event.

In the table below, separate the numbers into groups of 4. Then use the digits 0, 1, 2, and 3 to represent people who own cats. You are looking for groups of 4 numbers that do *not* contain a 0, 1, 2, or 3. Circle those groups.

18771	47374	36541	83454
97907	40978	34947	78482
26071	12644	94567	35467
02459	78467	06161	85897
44480	71716	13166	44096
72769	18974	24186	50866
35842	78478	45468	15441
58438	37487	16187	89892
83711	54631	19846	08483

In this case, the probability is $\frac{\square}{45}$ or 15.6%.

So, the experimental probability that it takes at least 4 customers before a cat owner enters the store is 15.6%.

Analyze and Reflect

3. In Exercise 1, what does spinning a Stop on your first spin, and spinning a Goal on your second spin represent in this situation?

4. MP **Justify Conclusions** Explain how your results might change for Exercise 1 if you simulated 100 penalty kicks.

5. In Exercise 2, why were the numbers from the random number table separated into groups of four?

6. In Exercise 2, you could have used any 4 numbers to represent cat owners. Complete the simulation four more times using the numbers in the table to represent the cat owners.

Numbers that Represent Cat Owners	Experimental Probability
4, 5, 6, 7	$\frac{\square}{45}$
0, 1, 8, 9	$\frac{\square}{45}$
3, 4, 5, 6	$\frac{\square}{45}$

Create

7. MP **Model with Mathematics** Design a simulation that could be used to predict the probability of taking a four question multiple-choice test with four answer choices and getting all four questions correct by guessing. Conduct 50 trials of the experiment. Then calculate the experimental probability of getting all four questions correct by guessing.

8. Inquiry HOW do simulations help you understand the probability of events happening?

 **Problem-Solving Investigation**

Act It Out

Content Standards
7.SP.8, 7.SP.8c

Mathematical Practices
1, 3, 4

Case #1 Winning Serves

Edie has been practicing her volleyball serve every day after school. She hits a good serve an average of 3 out of 4 times.

What is the probability that Edie will hit two good serves in a row?

Understand *What are the facts?*

You know that Edie hits a good serve an average of 3 out of 4 times. Act it out with a spinner.

Plan *What is your strategy?*

Spin a spinner, numbered 1 to 4, two times. If the spinner lands on 1, 2, or 3, she hits a good serve. If the spinner lands on 4, she doesn't. Repeat the experiment 10 times.

Solve *How can you apply the strategy?*

Here are some possible results. Circle the columns that show two good serves. The first two are done for you.

Trials	1	2	3	4	5	6	7	8	9	10
First Spin	4	1	4	3	1	2	2	1	3	2
Second Spin	2	3	3	2	1	4	1	4	3	3

The circled columns show that six out of 10 trials resulted in two good serves in a row. So, the probability is %.

Check *Does the answer make sense?*

Repeat the experiment several times to see whether the results agree.

Analyze the Strategy

Reason Inductively Describe an advantage of using the *act it out* strategy?

Case #2 Tests

Tools

James uses a spinner with four equal sections to answer a five-question multiple-choice quiz. Each question has choices A, B, C, and D.

Is this a good way to answer the quiz questions?

Understand

- **Read the problem. What are you being asked to find?**

I need to find ____

____.

- **What information do you know?**

The spinner has 4 equal parts. There are 5 multiple-choice questions.

The answer choices are A, B, C, and D.

2 Plan

- **Choose a problem-solving strategy.**

I will use the ______________________ strategy.

3 Solve

Use your problem-solving strategy to solve the problem.

Spin a spinner with four equal parts labeled A, B, C, and D five times. Repeat the experiment two times. Make a table of the results.

Question	1	2	3	4	5
Trial 1					
Trial 2					

With each spin there is an equal chance of landing on any section. Since the probability of an answer being A, B, C, or D is __________ likely, any answer choice is possible.

Is using a spinner to answer a multiple-choice question a good idea? ________

Check

Use information from the problem to check your answer.

Repeat the experiment several times to see if the results agree.

Work with a small group to solve the following cases. Show your work on a separate piece of paper.

Case #3 Chess

A chess tournament will be held and 32 students will participate. If a player loses one match, he or she will be eliminated.

How many total games will be played in the tournament?

Case #4 Running

Six runners are entered in a race. Assume there are no ties.

In how many ways can first and second places be awarded?

Case #5 Fair Games

Karla and Jason are playing a game with number cubes. Each number cube is numbered 1 to 6. They roll both number cubes. If the product is a multiple of 3, Jason wins. If the product is a multiple of 4, Karla wins.

Is this game fair or unfair? Justify your response.

Case #6 Algebra

The figure shown at the right is known as Pascal's Triangle.

Make a conjecture for the numbers in the 6th and 7th rows.

Mid-Chapter Check

Vocabulary Check

1. Define *probability*. Give an example of the probability of a simple event. (Lesson 1)

2. Fill in the blank in the sentence below with the correct term. (Lesson 4)

   A(n) ____________ is an experiment that is designed to act out a given situation.

Skills Check and Problem Solving

The table shows the number of science fiction, action, and comedy movies Jason has in his collection. Suppose one movie is selected at random. Find each probability. Write as a fraction in simplest form. (Lesson 1)

Type of Movie	
Science Fiction	10
Action	7
Comedy	3

3. *P*(science fiction) ________

4. *P*(*not* action) ________

5. A coin is tossed 20 times. It lands heads 4 times. Compare the experimental probability to its theoretical probability. If the probabilities are not close, explain a possible reason for the discrepancy. (Lesson 2)

6. A weather forecaster predicts a 30% chance of rain for each of the next three days. Describe a way to simulate the chance that it will rain the next three days. (Lesson 4)

7. MP **Persevere with Problems** Without looking, Santiago took a handful of multi-colored candies from a bag and found that 20% of the candies were yellow and 15% were green. Suppose there were 480 candies in the bag. Based on Santiago's results, how many more yellow candies would you expect there to be than green candies? (Lesson 1)

Fundamental Counting Principle

Real-World Link

Classes Tyler wants to take a class at the community center. The table shows the class options he is considering. All of the classes are offered only on Monday and Tuesday.

Class	Day
Drawing Martial Arts Dance	Monday Tuesday

1. According to the table, how many classes is he considering? ______

2. How many days are the classes offered?

______

3. Complete the tree diagram to find the number of different class and day outcomes.

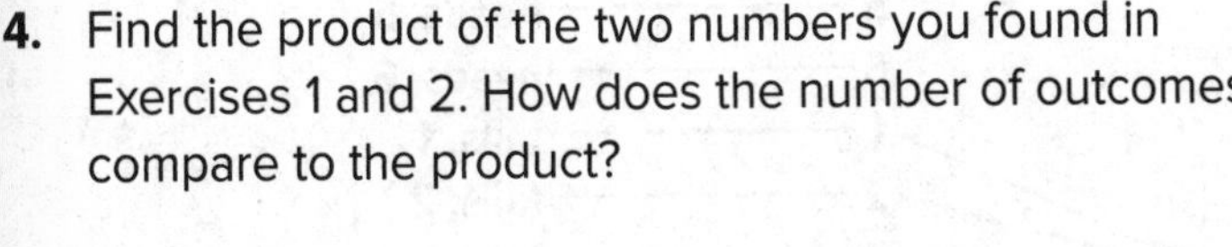

4. Find the product of the two numbers you found in Exercises 1 and 2. How does the number of outcomes compare to the product?

______

Which **MP Mathematical Practices** did you use? Shade the circle(s) that applies.

① Persevere with Problems
② Reason Abstractly
③ Construct an Argument
④ Model with Mathematics
⑤ Use Math Tools
⑥ Attend to Precision
⑦ Make Use of Structure
⑧ Use Repeated Reasoning

Essential Question

HOW can you predict the outcome of future events?

Vocabulary

Fundamental Counting Principle

Common Core State Standards

Content Standards
7.SP.5, 7.SP.8, 7.SP.8a, 7.SP.8b

MP Mathematical Practices
1, 3, 4

Key Concept

Fundamental Counting Principle

If event M has m possible outcomes and event N has n possible outcomes, then event M followed by event N has $m \times n$ possible outcomes.

Work Zone

You can use multiplication instead of making a tree diagram to find the number of possible outcomes in a sample space. This is called the **Fundamental Counting Principle**.

Example

1. Find the total number of outcomes when a coin is tossed and a number cube is rolled.

A coin has 2 possible outcomes. A number cube has 6 possible outcomes. Multiply the possible outcomes of each event.

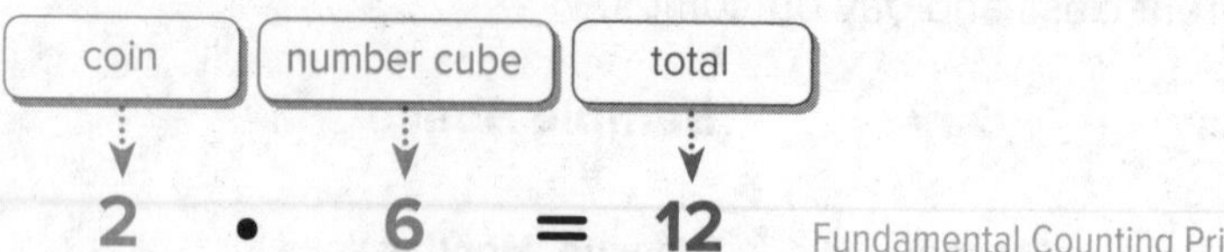

Fundamental Counting Principle

There are 12 different outcomes.

Check Draw a tree diagram to show the sample space.

Coin	Number Cube	Sample Space
heads	1	heads, 1
	2	heads, 2
	3	heads, 3
	4	heads, 4
	5	heads, 5
	6	heads, 6
tails	1	tails, 1
	2	tails, 2
	3	tails, 3
	4	tails, 4
	5	tails, 5
	6	tails, 6

The tree diagram also shows that there are 12 outcomes. ✓

Got it? Do this problem to find out.

a. Find the total number of outcomes when choosing from bike helmets that come in three colors and two styles.

a. ________

Find Probability

You can use the Fundamental Counting Principle to help find the probability of events.

Examples

2. Find the total number of outcomes from rolling a number cube with sides labeled 1–6 and choosing a letter from the word NUMBERS. Then find the probability of rolling a 6 and choosing an M.

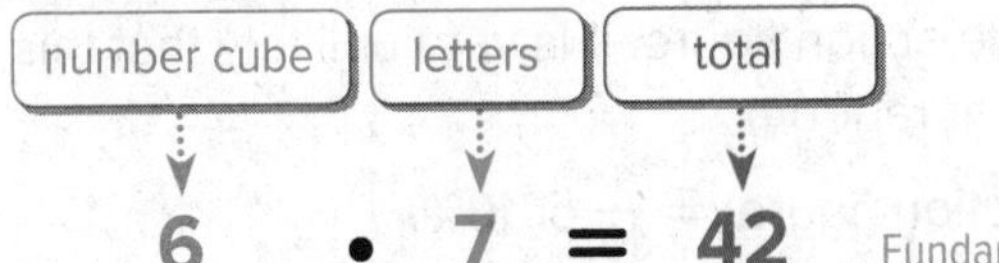

$6 \cdot 7 = 42$ Fundamental Counting Principle

There are 42 different outcomes.

There is only one favorable outcome. So, the probability of rolling a 6 and choosing an M is $\frac{1}{42}$ or about 2%.

3. Find the number of different jeans available at The Jeans Shop. Then find the probability of randomly selecting a size 32 × 34 slim fit. Is it likely or unlikely that the jeans would be chosen?

The Jeans Shop

Waist Size	Length (in.)	Style
30	30	slim fit
32	32	bootcut
34	34	loose fit
36		
38		

Jean Size

In men's jeans, the size is labeled waist × length. So, a 32 × 34 is a 32-inch waist with a 34-inch length.

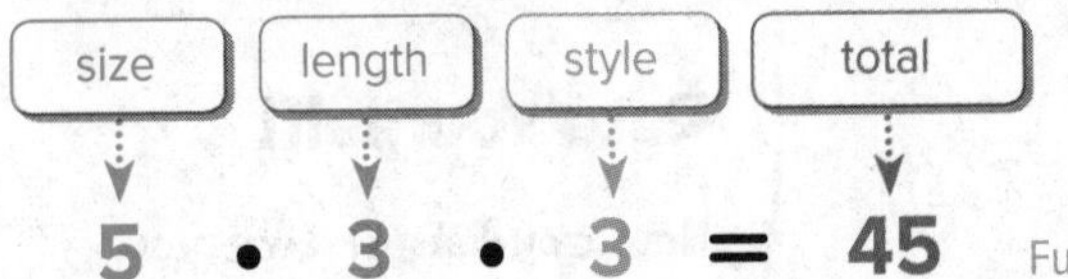

$5 \cdot 3 \cdot 3 = 45$ Fundamental Counting Principle

There are 45 different types of jeans to choose. Out of the 45 possible outcomes, only one is favorable. So, the probability of randomly selecting a 32 × 34 slim fit is $\frac{1}{45}$ or about 2%.

It is very unlikely that the size would be chosen at random.

Got it? Do this problem to find out.

b. Two number cubes are rolled. What is the probability that the sum of the numbers on the cubes is 12? How likely is it that the sum would be 12?

b. ________

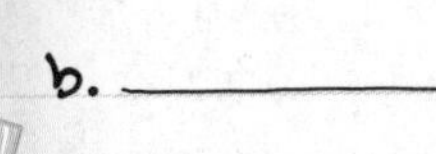

Example

4. A box of toy cars contains blue, orange, yellow, red, and black cars. A separate box contains a male and a female action figure. What is the probability of randomly choosing an orange car and a female action figure? Is it likely or unlikely that this combination is chosen?

First, find the number of possible outcomes.

There are 5 choices for the car and 2 choices for the action figure.

$5 \cdot 2 = 10$ Fundamental Counting Principal

There are 10 possible outcomes. There is one way to choose an orange car and a female action figure. It is very unlikely that this combination is chosen at random.

$P(\text{orange car, female action figure}) = \frac{1}{10}$ or 10%.

Guided Practice

1. Use the Fundamental Counting Principle to find the number of outcomes from tossing a quarter, a dime, and a nickel. (Example 1)

2. How many outcomes are possible when rolling a number cube and picking a cube from 4 different colored cubes? (Example 1)

3. Find the number of different outfits that can be made from 3 sweaters, 4 blouses, and 6 skirts. Then find the probability of randomly selecting a particular sweater-blouse-skirt outfit. Is the probability of this event likely or unlikely? (Examples 2–4)

4. **Building on the Essential Question** Compare and contrast tree diagrams and the Fundamental Counting Principle.

Rate Yourself!

How confident are you about using the Fundamental Counting Principle? Shade the ring on the target.

For more help, go online to access a Personal Tutor.

Name ______________________ My Homework ______________________

Independent Practice

Go online for Step-by-Step Solutions

Use the Fundamental Counting Principle to find the total number of outcomes for each situation. (Example 1)

1. choosing a bagel with one type of cream cheese from the list shown in the table

______

Bagels	Cream Cheese
Plain Blueberry Cinnamon raisin Garlic	Plain Chive Sun-dried tomato

2. choosing a sandwich and a side from the list shown in the table

______

Sandwiches	Sides
Ham Turkey Roast Beef Tuna Salad Vegetarian	Pasta Salad Fruit Cup Potato Chips Side Salad

3. picking a month of the year and a day of the week ______

4. choosing from a comedy, horror, or action movie each shown in four different theaters

______

5. Find the number of possible routes from Eastland to Johnstown that pass through Harping. Then find the probability that State and Fairview will be used if a route is selected at random. State the probability as a fraction and percent. (Examples 2–3)

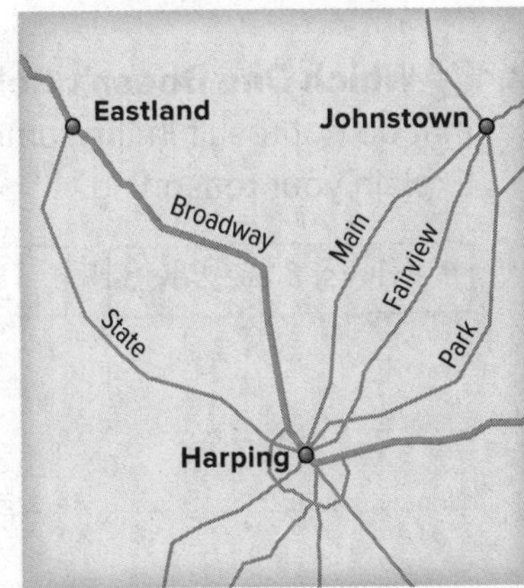

____

6. Find the number of possible choices for a 2-digit number that is greater than 19. Then find the number of possible choices for a 4-digit Personal Identification Number (PIN) if the digits cannot be repeated. (Example 1)

____

7. An electronics company makes educational apps for 5 subjects, including math. The app has 10 versions, with a different avatar in each version. One version has an avatar that looks similar to a lion. The company is randomly giving free apps to its customers. Find the probability of randomly receiving a math app with a lion avatar. How likely is the probability of receiving this app at random? (Examples 2–4)

____

8. A sandwich shop offers 4 different meats and 2 different cheeses. Suppose the sandwich shop offers 24 different meat-cheese sandwiches. How many different breads does the sandwich shop use?

9. **Justify Conclusions** A store offers 32 different T-shirt designs and 11 choices of color. Is the store's advertisement true? Explain.

H.O.T. Problems Higher Order Thinking

10. **Persevere with Problems** Determine the number of possible outcomes when tossing one coin, two coins, and three coins. Then determine the number of possible outcomes for tossing *n* coins. Describe the strategy you used.

11. **Which One Doesn't Belong?** Identify the choices for events *M* and *N* that do not result in the same number of outcomes as the other two. Explain your reasoning.

9 drinks, 8 desserts	18 shirts, 4 pants	10 groups, 8 activities

12. **Justify Conclusions** Marcus has a choice of a white, gray, or black shirt to wear with a choice of tan, black, brown, or denim pants. Without calculating the number of possible outcomes, how many more outfits can he make if he buys a green shirt? Explain your reasoning to a classmate.

13. **Persevere with Problems** Write an algebraic expression to find the number of outcomes if a number cube is rolled *x* times.

Name ______ My Homework ______

Extra Practice

Use the Fundamental Counting Principle to find the total number of outcomes for each situation.

14. rolling a number cube and spinning a spinner with eight equal sections 48

6 • 8 = 48

15. tossing a coin and selecting one letter from the word MATH ______

16. selecting one sweatshirt from a choice of five sweatshirts and one pair of pants from a choice of four pairs of pants ______

17. selecting one entrée from a choice of nine entrées and one dessert from a choice of three desserts ______

18. rolling a number cube and tossing two coins ______

19. choosing tea in regular, raspberry, lemon, or peach; sweetened or unsweetened; and in a glass or bottle ______

20. A cafeteria offers oranges, apples, or bananas as its fruit option. It offers peas, green beans, or carrots as the vegetable option. Find the number of fruit and vegetable options. If the fruit and the vegetable are chosen at random, what is the probability of getting an orange and carrots? Is it likely or unlikely that a customer would get an orange and carrots?

______

21. MP **Justify Conclusions** The table shows cell phone options offered by a wireless phone company. If a phone with one payment plan and one accessory is given away at random, predict the probability that it will be Brand B and have a headset. Explain your reasoning.

Phone Brands	Payment Plans	Accessories
Brand A Brand B Brand C	Individual Family Business Government	Leather case Car mount Headset Travel charger

______

Power Up! Common Core Test Practice

22. A restaurant has 24 different lunch combinations. Which of the following could describe the lunch options? Select all that apply.

- ☐ 3 drink sizes, 4 main dishes, 2 side dishes
- ☐ 2 appetizers, 6 main dishes, 3 desserts
- ☐ 3 kinds of bread, 8 kinds of sandwiches
- ☐ 2 drink sizes, 7 appetizers, 2 main dishes

23. Hat Shack sells 9 different styles of hats in several different colors for 2 different sports teams. The company makes 108 kinds of hats in all. Select the correct values to complete the formula below to find the number of different color hats the Hat Shack makes.

Hat Shack		
Styles	Colors	Teams
9	?	2

☐ = ☐ × ☐ × ☐

2
9
108
c

How many different colors does the company use for hats?

☐

Common Core Spiral Review

Find each probability. 7.SP.8

24. A coin is tossed and a spinner with 4 equal sections labeled W–Z is spun. Find *P*(heads and Z).

______

25. A pizza shop offers a single item pizza with choice of pepperoni, green peppers, pineapple, sausage, or mushroom toppings. The pizza can be thick crust or thin crust. Find *P*(thick crust).

______

Describe a model that could be used to simulate each situation. 7.SP.8c

26. There is a fifty percent chance of rain on Monday.

______

27. A restaurant randomly gives away 1 of 6 toys. Determine the number of times a child needs to visit the restaurant to receive all 6 toys.

______

Lesson 6

Permutations

Real-World Link

Scheduling Colt is planning his Saturday. He wants to mow the grass, go swimming, and do his homework. How many different ways are there to arrange what he wants to do?

Fill in the blanks of the organized list below to find all of the possible arrangements of the activities.

1: Mowing	2: Swimming	3: Homework
1: Mowing	2: Homework	3: ______
1: Swimming	2: Mowing	3: Homework
1: Swimming	2: Homework	3: ______
1: Homework	2: ______	3: ______
1: ______	2: ______	3: ______

1. How many choices does Colt have for his first activity?

____

2. Once the first activity is selected, how many choices does Colt have for the second activity?

____

3. Once the first and second activities are selected, how many choices does Colt have for the third activity?

____

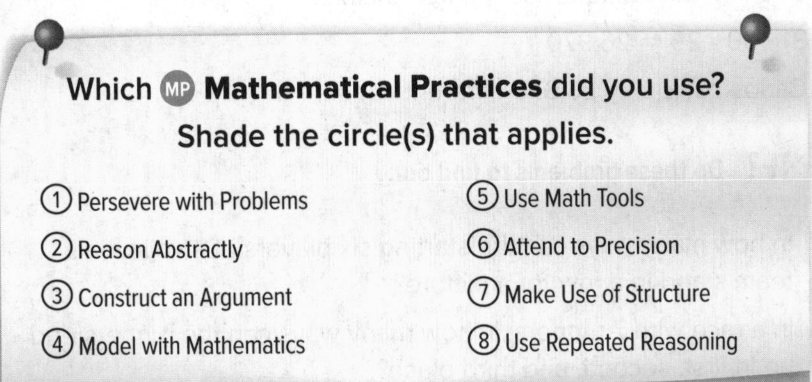

Which MP Mathematical Practices did you use? Shade the circle(s) that applies.

① Persevere with Problems
② Reason Abstractly
③ Construct an Argument
④ Model with Mathematics
⑤ Use Math Tools
⑥ Attend to Precision
⑦ Make Use of Structure
⑧ Use Repeated Reasoning

Essential Question

HOW can you predict the outcome of future events?

Vocabulary

permutation

Common Core State Standards

Content Standards
7.SP.8, 7.SP.8a

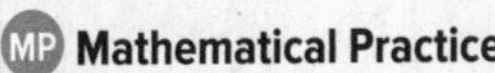

Mathematical Practices
1, 3, 4

Work Zone

Find a Permutation

A **permutation** is an arrangement, or listing, of objects in which order is important.

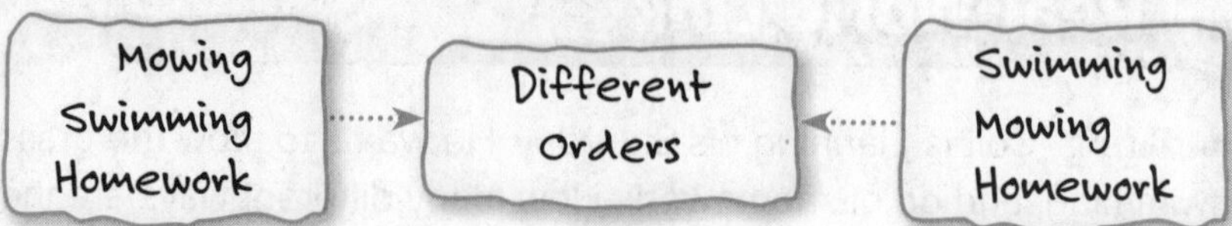

You can use the Fundamental Counting Principle to find the number of permutations.

Examples

1. Julia is scheduling her first three classes. Her choices are math, science, and language arts. Use the Fundamental Counting Principle to find the number of different ways Julia can schedule her first three classes.

There are **3** choices for the first class.

There are **2** choices that remain for the second class.

There is **1** choice that remains for the third class.

$$3 \cdot 2 \cdot 1 = 6$$ ← the number of permutations of 3 classes

There are 6 possible arrangements, or permutations, of the 3 classes.

2. An ice cream shop has 31 flavors. Carlos wants to buy a three-scoop cone with three different flavors. How many cones could he buy if the order of the flavors is important?

There are 31 choices for the first scoop, 30 choices for the second scoop, and 29 choices for the third scoop.

Use the Fundamental Counting Principle.

$31 \cdot 30 \cdot 29 = 26{,}970$

Carlos could buy 26,970 different cones.

Got it? **Do these problems to find out.**

a. ______

b. ______

a. In how many ways can the starting six players of a volleyball team stand in a row for a picture?

b. In a race with 7 runners, in how many ways can the runners end up in first, second, and third place?

The symbol $P(31, 3)$ represents the number of permutations of 31 things taken 3 at a time.

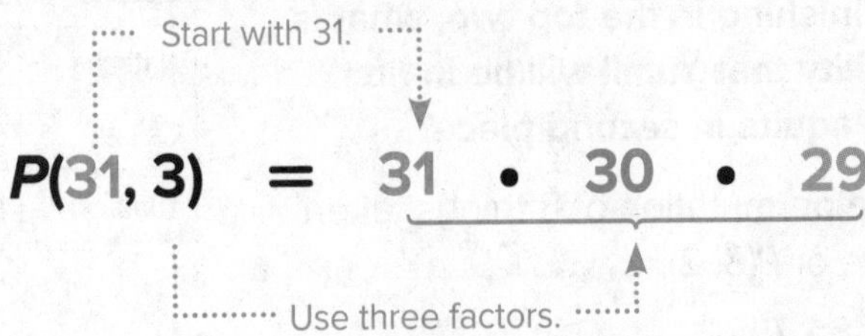

Example

3. Find $P(8, 3)$.

$P(8, 3) = 8 \cdot 7 \cdot 6$ or 336 8 things taken 3 at a time

Got it? Do these problems to find out.

c. $P(12, 2)$ d. $P(4, 4)$ e. $P(10, 5)$

c. ______

d. ______

Show your work

e. ______

Find Probabililty

Permutations can be used when finding probabilities of real-world situations.

Examples

4. Ashley's MP3 player has a setting that allows the songs to play in a random order. She has a playlist that contains 10 songs. What is the probability that the MP3 player will randomly play the first three songs in order?

First find the permutation of ten things taken three at a time or $P(10, 3)$.

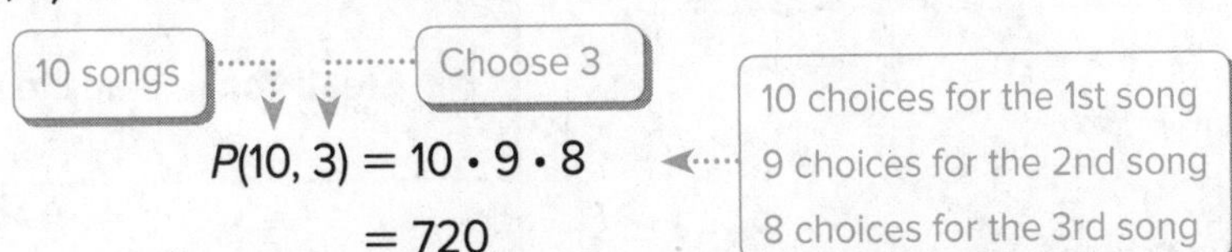

$$P(10, 3) = 10 \cdot 9 \cdot 8$$
$$= 720$$

So, there are 720 different ways to play the first 3 songs. Since you want the first three songs in order, there is only 1 out of the 720 ways to do this.

So, the probability that the first 3 songs will play in order is $\frac{1}{720}$.

Notation

In Example 4, the notation P(10, 3) indicates a permutation while the notation P(playing the first three songs in order) indicates probability.

5. **A swimming event features 8 swimmers. If each swimmer has an equally likely chance of finishing in the top two, what is the probability that Yumii will be in first place and Paquita in second place?**

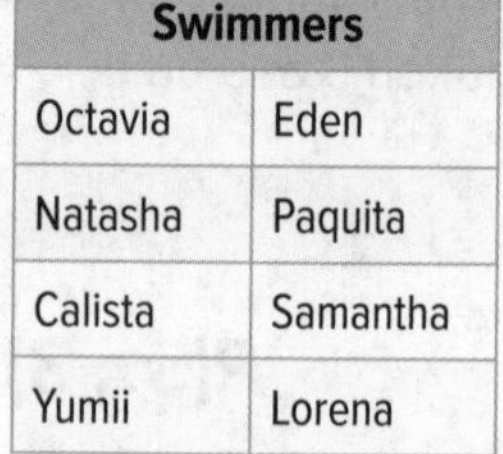

Swimmers	
Octavia	Eden
Natasha	Paquita
Calista	Samantha
Yumii	Lorena

First find the permutation of 8 things taken two at a time or $P(8, 2)$.

$$P(8, 2) = 8 \cdot 7$$
$$= 56$$

There are 56 possible arrangements, or permutations, of the two places. Since there is only one way of having Yumii come in first and Paquita second, the probability of this event is $\frac{1}{56}$.

Reasonable Answers

A possible probability of $\frac{1}{56}$ indicates that it is very unlikely that Yumii will finish first and Paquita will finish second.

Got it? **Do this problem to find out.**

Show your work.

f. Two different letters are randomly selected from the letters in the word *math*. What is the probability that the first letter selected is *m* and the second letter is *h*?

f. ______________

Guided Practice

Check

1. In how many ways can a president, vice president, and secretary be randomly selected from a class of 25 students? (Examples 1 and 2) ______________

2. Find the value of $P(5, 3)$. (Example 3) ______________

3. Adrianne, Julián, and two of their friends will sit in a row at a baseball game. If each friend is equally likely to sit in any seat, what is the probability that Adrianne will sit in the first seat and Julián will sit in the second seat? (Examples 4 and 5)

____

4. **Building on the Essential Question** HOW can you find the number of permutations of a set of objects?

____

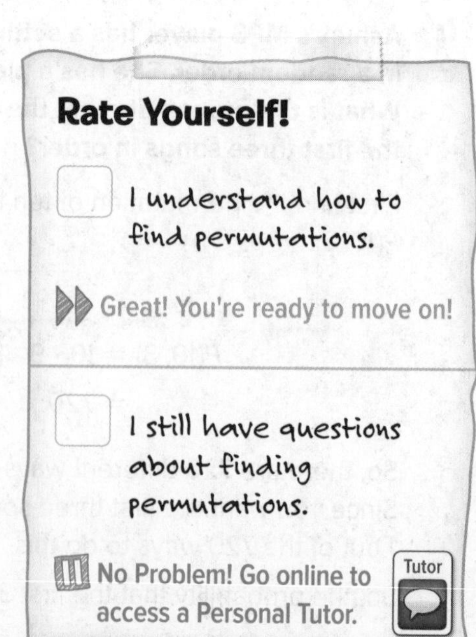

Independent Practice

Go online for Step-by-Step Solutions

1. In the Battle of the Bands contest, in how many ways can the four participating bands perform? (Examples 1 and 2)

____

2. A garage door code has 5 digits. If no digit is repeated, how many codes are possible?

____

Find each value. Use a calculator if needed. (Example 3)

3. $P(7, 4)$ _______________

4. $P(12, 5)$ _______________

5. $P(8, 8)$ _______________

6. You have five seasons of your favorite TV show on DVD. If you randomly select two of them from a shelf, what is the probability that you will select season one first and season two second? (Examples 4 and 5)

____

7. **MP Model with Mathematics** The graphic novel frame below explains how the survey has students rank their favorite kinds of music. In how many ways can the survey be answered? _______________

8. A certain number of friends are waiting in line to board a new roller coaster. They can board the ride in 5,040 different ways. How many friends are in line?

9. The Coughlin family discovered they can stand in a row for their family portrait in 720 different ways. How many members are in the Coughlin family?

10. Howland Middle School assigns a four-digit identification number to each student. The number is made from the digits 1, 2, 3, and 4, and no digit is repeated. If assigned randomly, what is the probability that an ID number will end with a 3?

H.O.T. Problems Higher Order Thinking

11. **MP Model with Mathematics** Describe a real-world situation that has 6 permutations.

12. **MP Persevere with Problems** There are 1,320 ways for three students to win first, second, and third place during a debate match. How many students are there on the debate team? Explain your reasoning.

13. **MP Persevere with Problems** A *combination* is an arrangement where order is *not* important. You can find the number of combinations of items by dividing the number of permutations by the number of ways the smaller set can be arranged. The combination at the right shows the number of combinations if you choose 2 flavors of ice cream out of 5 flavors. Use this method to find each value.

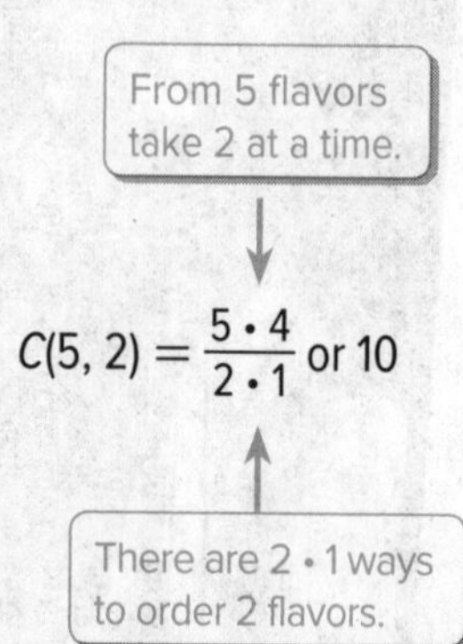

a. $C(6, 4)$ ______ b. $C(10, 3)$ ______

c. $C(5, 3)$ ______ d. $C(8, 6)$ ______

Extra Practice

14. How many permutations are possible of the letters in the word FRIEND? 720

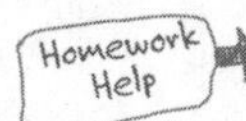

6 • 5 • 4 • 3 • 2 • 1 = 720

15. How many different 3-digit numbers can be formed using the digits 9, 3, 4, 7, and 6? Assume no number can be used more than once. ________

Find each value. Use a calculator if needed.

16. $P(9, 2)$ ________

17. $P(5, 5)$ ________

18. $P(7, 7)$ ________

19. The members of the Evergreen Junior High Quiz Bowl team are listed in the table. If a captain and an assistant captain are chosen at random, what is the probability that Walter is selected as captain and Mi-Ling as co-captain? ________

Evergreen Junior High Quiz Bowl Team	
Jamil	Luanda
Savannah	Mi-Ling
Tucker	Booker
Ferdinand	Nina
Walter	Meghan

20. Alex, Aiden, Dexter, and Dion are playing a video game. If they each have an equally likely chance of getting the highest score, what is the probability that Dion will get the highest score and Alex the second highest? ________

21. A child has wooden blocks with the letters shown. Find the probability that the child randomly arranges the letters in the order TIGER. ________

Power Up! Common Core Test Practice

22. The schools listed in the table are finalists in a science competition. First through third places will win a prize. Each school is equally likely to win the competition. Select values to complete the model below to find the probability that Lincoln wins first place, River Valley wins second place, and Glenwood wins third place.

Finalists
Chester Middle School
Glenwood Middle School
Lincoln Middle School
River Valley Middle School
South Middle School

Find the number of ways the schools can finish in first, second, and third place:

$P($ ☐ , ☐ $) =$ ☐

1	2	3	4	5
10	20	30	60	90

The number of ways that Lincoln can finish first, River Valley second, and Glenwood third is equal to ☐.

P(Lincoln first, River Valley second, Glenwood third) $= \frac{☐}{☐}$

23. The five finalists in a writing contest are Cesar, Teresa, Sean, Nikita, and Alfonso. There will be a first place award and a second place award. Each finalist is equally likely to win an award. Determine if each statement is true or false.

a. There are 10 permutations of 5 finalists taken 2 at a time. ☐ True ☐ False

b. There is only 1 way that Teresa can earn first place and Sean can earn second place. ☐ True ☐ False

c. The probability that Teresa earns first place and Sean earns second place is 0.05. ☐ True ☐ False

Common Core Spiral Review

A card is pulled from a stack of 30 cards labeled 1–30. Find each probability. Write as a fraction in simplest form. 7.SP.5

24. P(greater than 5) ______

25. P(*not* 1) ______

26. P(an even number) ______

27. A cross country athlete has a white, a red, and a gray sweatshirt. She has black and gray running pants. Make a list to show the possible combinations of training outfits. 7.SP.8b ______

Inquiry Lab

Independent and Dependent Events

HOW can one event impact a second event in a probability experiment?

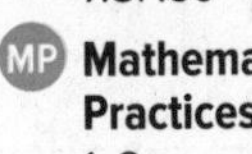

Content Standards
7.SP.8, 7.SP.8b, 7.SP.8c

Mathematical Practices
1, 3

Jeanie wants to go to the movies and Kate wants to go skating. They decide by doing a simulation. They place two red counters in a bag to represent going to the movies and two white counters to represent going skating. If they draw or remove two red counters, they will go to the movies. If they draw two white counters they will go skating. If they draw a red and a white counter, they will stay home.

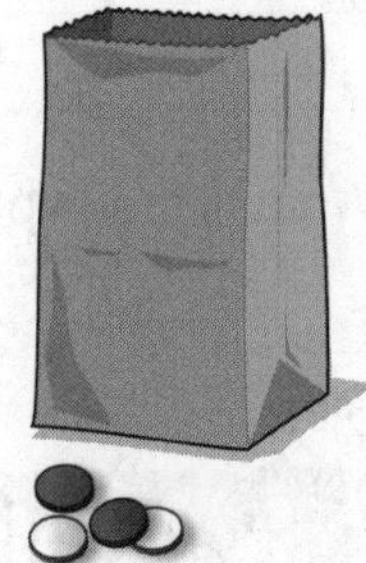

You can simulate this activity using counters.

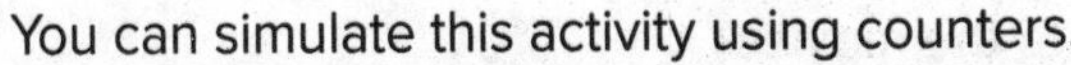

Hands-On Activity

Step 1 Place two red counters and two white counters in a paper bag.

Step 2 Without looking, draw a counter from the bag and record its color in the table below. Place the counter back in the bag.

Step 3 Without looking, draw a second counter and record its color in the table. The two colors are one trial. Place the counter back in the bag.

Step 4 Repeat until you have 18 trials.

Trial	1st Color	2nd Color	Trial	1st Color	2nd Color	Trial	1st Color	2nd Color
1			7			13		
2			8			14		
3			9			15		
4			10			16		
5			11			17		
6			12			18		

What is the experimental probability that the girls will go to the movies?

____

Investigate

Work with a partner.

1. Complete the same experiment from the Activity. Except do not replace the counter after the first draw for each trial. Record your results.

Trial	1st Color	2nd Color	Trial	1st Color	2nd Color	Trial	1st Color	2nd Color
1			7			13		
2			8			14		
3			9			15		
4			10			16		
5			11			17		
6			12			18		

What is the experimental probability that the girls will go to the movies?

Analyze and Reflect

The tree diagrams below represent the possible outcomes for the Activity and for Exercise 1. Use the diagrams to answer Exercises 2–3.

Investigation

1st Draw	2nd Draw	Outcome
R	R	RR
	R	RR
	W	RW
	W	RW
W	R	WR
	R	WR
	W	WW
	W	WW

Exercise 1

1st Draw	2nd Draw	Outcome
R	R	RR
	W	RW
	W	RW
W	R	WR
	R	WR
	W	WW

2. What is the theoretical probability of drawing two reds in the Investigation? In Exercise 1?

3. **MP Reason Inductively** Is there a better chance that the girls will go to the movies if the counters are replaced after the first draw? Explain.

Create

4. Inquiry HOW can one event impact a second event in a probability experiment?

Independent and Dependent Events

Vocabulary Start-Up

When one event does not affect the outcome of the other event, the events are **independent events**. For example, if you toss a coin twice, the first toss has no affect on the second toss. Complete the graphic organizer below.

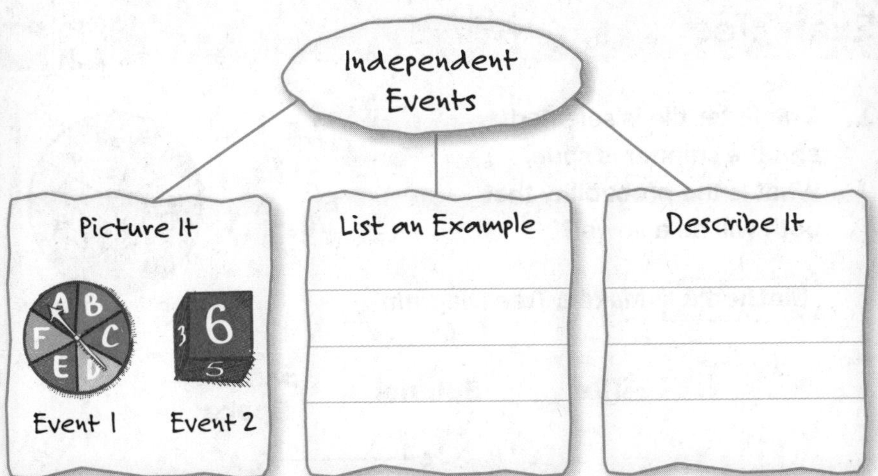

Essential Question

HOW can you predict the outcome of future events?

Vocabulary

independent events
dependent events

Common Core State Standards

Content Standards
7.SP.8, 7.SP.8a, 7.SP.8b

MP Mathematical Practices
1, 3, 4

Real-World Link

Independent is a common word in the English language. Use a dictionary to look up its definition. Explain how the dictionary definition can help you remember the mathematical definition of *independent*. ______________________________

____

____

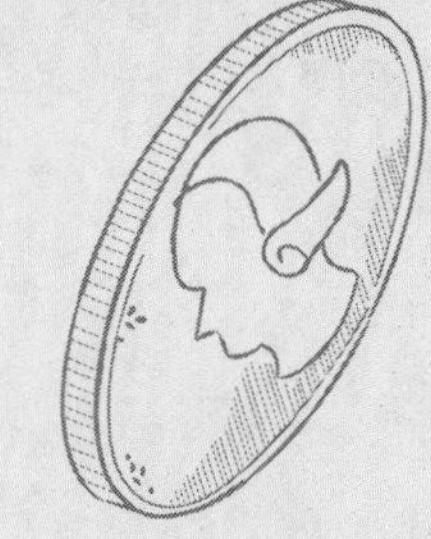

Which MP Mathematical Practices did you use? Shade the circle(s) that applies.

① Persevere with Problems
② Reason Abstractly
③ Construct an Argument
④ Model with Mathematics
⑤ Use Math Tools
⑥ Attend to Precision
⑦ Make Use of Structure
⑧ Use Repeated Reasoning

Key Concept

Probability of Independent Events

Work Zone

Words	The probability of two independent events can be found by multiplying the probability of the first event by the probability of the second event.
Symbols	$P(A \text{ and } B) = P(A) \cdot P(B)$

You can use organized lists, tables, tree diagrams, or multiplication to find the probability of compound events.

Examples

1. One letter tile is selected and the spinner is spun. What is the probability that both will be a vowel?

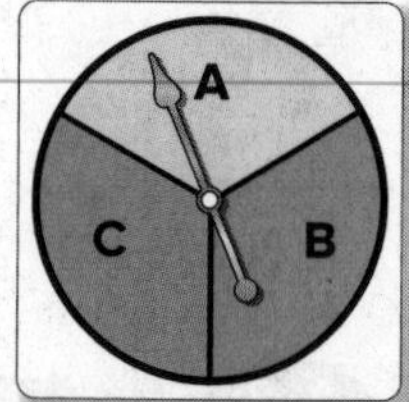

Method 1 **Make a Tree Diagram**

Tile	Spinner	Sample Space
G	A	G, A
	B	G, B
	C	G, C
B	A	B, A
	B	B, B
	C	B, C
E	A	E, A
	B	E, B
	C	E, C
A	A	A, A
	B	A, B
	C	A, C

There are 12 outcomes. Two outcomes contain only vowels. The probability that both will be a vowel is $\frac{2}{12}$ or $\frac{1}{6}$.

Method 2 **Use Multiplication**

$P(\text{selecting a vowel}) = \frac{2}{4}$ or $\frac{1}{2}$. $P(\text{spinning a vowel}) = \frac{1}{3}$.

$P(\text{both vowels}) = \frac{1}{2} \cdot \frac{1}{3}$ or $\frac{1}{6}$. Multiply the probabilities.

So, using either method the probability is $\frac{1}{6}$.

2. **The spinner and number cube shown are used in a game. What is the probability of a player *not* spinning blue and then rolling a 3 or 4?**

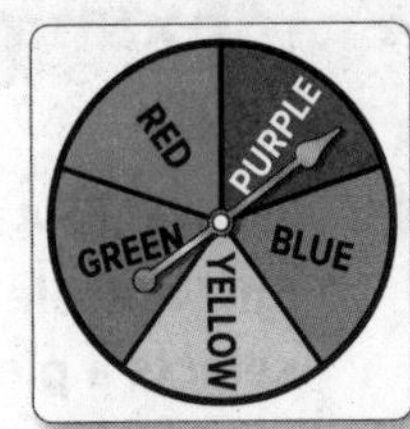

You are asked to find the probability of the spinner *not* landing on blue and rolling a 3 or 4 on a number cube. The events are independent because spinning the spinner does not affect the outcome of rolling a number cube.

First, find the probability of each event.

$P(not\ \text{blue}) = \frac{4}{5}$ ← number of ways not to spin blue / number of possible outcomes

$P(3 \text{ or } 4) = \frac{2}{6} \text{ or } \frac{1}{3}$ ← number of ways to roll 3 or 4 / number of possible outcomes

Then, find the probability of both events occurring.

$P(not\ \text{blue and 3 or 4}) = \frac{4}{5} \cdot \frac{1}{3}$ $P(A \text{ and } B) = P(A) \cdot P(B)$

$= \frac{4}{15}$ Multiply.

The probability is $\frac{4}{15}$.

Check Make an organized list, table, or a tree diagram to show the sample space.

Got it? **Do this problem to find out.**

a. A game requires players to roll two number cubes to move the game pieces. The faces of the cubes are labeled 1 through 6. What is the probability of rolling a 2 or 4 on the first number cube and then rolling a 5 on the second?

a. ________

Probability of Dependent Events

Key Concept

Words If two events *A* and *B* are dependent, then the probability of both events occurring is the product of the probability of *A* and the probability of *B* after *A* occurs.

Symbols $P(A \text{ and } B) = P(A) \cdot P(B \text{ following } A)$

If the outcome of one event affects the outcome of another event, the events are called **dependent events**. For example, you have a bag with blue and green marbles. You pick one marble, do not replace it, and pick another one.

Example

3. There are 4 oranges, 7 bananas, and 5 apples in a fruit basket. Ignacio selects a piece of fruit at random and then Terrance selects a piece of fruit at random. Find the probability that two apples are chosen.

Since the first piece of fruit is not replaced, the first event affects the second event. These are dependent events.

$P(\text{first piece is an apple}) = \frac{5}{16}$ ← number of apples / ← total pieces of fruit

$P(\text{second piece is an apple}) = \frac{4}{15}$ ← number of apples left / ← total pieces of fruit left

$P(\text{two apples}) = \frac{\cancel{5}^{1}}{\cancel{16}_{4}} \cdot \frac{\cancel{4}^{1}}{\cancel{15}_{3}} \text{ or } \frac{1}{12}.$

The probability that two apples are chosen is $\frac{1}{12}$.

b. ____________

c. ____________

Got it? Do these problems to find out.

Refer to the situation above. Find each probability.

b. *P*(two bananas)

c. *P*(orange then apple)

Guided Practice

A penny is tossed and a number cube is rolled. Find each probability. (Examples 1–2)

1. *P*(tails and 3) ____________

2. *P*(heads and odd) ____________

3. Cards labeled 5, 6, 7, 8, and 9 are in a stack. A card is drawn and not replaced. Then, a second card is drawn at random. Find the probability of drawing two even numbers. (Example 3) ____________

4. **Building on the Essential Question** Explain the difference between independent events and dependent events.

____

____

____

____

Rate Yourself!

Are you ready to move on? Shade the section that applies.

I have a few questions. | I'm ready to move on. | I have a lot of questions.

For more help, go online to access a Personal Tutor.

Name ______ My Homework ______

Independent Practice

Go online for Step-by-Step Solutions

A number cube is rolled and a marble is selected at random from the bag at the right. Find each probability. Show your work. (Example 1)

1. P(1 and red) ______

2. P(3 and purple) ______

3. P(even and yellow) ______

4. P(odd and *not* green) ______

5. A carnival game wheel has 12 equal sections. One of the sections contains a star. To win a prize, players must land on the section with the star on two consecutive spins. What is the probability of a player winning? (Example 2) ______

6. A standard set of dominoes contains 28 tiles, with each tile having two sides of dots from 0 to 6. Of these tiles, 7 have the same number of dots on each side. If four players each randomly choose a tile, without replacement, what is the probability that each chooses a tile with the same number of dots on each side? (Example 3) ______

Mrs. Ameldo's class has 5 students with blue eyes, 7 with brown eyes, 4 with hazel eyes, and 4 with green eyes. Two students are selected at random. Find each probability. (Example 3)

7. P(green then brown) ______

8. P(two blue) ______

9. P(hazel then blue) ______

10. P(brown then blue) ______

11. **MP Reason Inductively** You and a friend plan to see 2 movies over the weekend. You can choose from 6 comedy, 2 drama, 4 romance, 1 science fiction, or 3 action movies. You write the movie titles on pieces of paper, place them in a bag, and each randomly select a movie. What is the probability that neither of you selects a comedy? Is this a dependent or independent event? Explain.

H.O.T. Problems Higher Order Thinking

12. **MP Model with Mathematics** There are 9 marbles representing 3 different colors. Write a problem where 2 marbles are selected at random without replacement and the probability is $\frac{1}{6}$.

13. **MP Find the Error** A spinner with equal sections numbered from 1 to 5 is spun twice. Raul is finding the probability that both spins will result in an even number. Find his mistake and correct it.

14. **MP Justify Conclusions** Determine whether the following statement is *true* or *false*. If false, provide a counterexample.
If two events are independent, then the probability of both events is less than 1.

15. **MP Persevere with Problems** A company has determined that 2% of the pudding cups it produces are defective in some way. The pudding cups are sold in packages of two.

a. What is the probability that both pudding cups in a package are defective?

b. The company produces 1,000,000 packages each year. Predict the number of packages in which both cups are defective.

Extra Practice

A number cube is rolled and a letter is selected from the word AMERICA. Find each probability. Show your work.

16. P(less than 4 and vowel) $\frac{1}{7}$

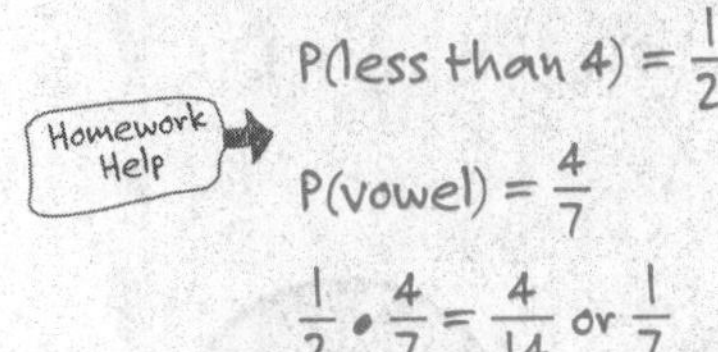

$P(\text{less than } 4) = \frac{1}{2}$

$P(\text{vowel}) = \frac{4}{7}$

$\frac{1}{2} \cdot \frac{4}{7} = \frac{4}{14}$ or $\frac{1}{7}$

17. P(greater than 1 and a consonant) ______

18. A number cube is rolled and a coin is tossed. What is the probability of the cube landing on 5 or 6 and the coin landing on heads? ______

19. A laundry basket contains 18 blue socks and 24 black socks. What is the probability of randomly picking 2 black socks, without replacement, from the basket? ______

20. **MP Persevere with Problems** Corbin is playing a board game that requires rolling two number cubes to move a game piece. He needs to roll a sum of 6 on his first turn and then a sum of 10 on his second turn to land on the next two bonus spaces. What is the probability that Corbin will roll a sum of 6 and then a sum of 10 on his next two turns? ______

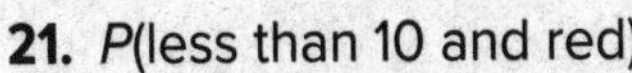

Copy and Solve **Solve Exercises 21–28 on a separate sheet of paper.**

A card is pulled from a stack of 15 cards labeled 1–15 and the spinner shown is spun. Find each probability.

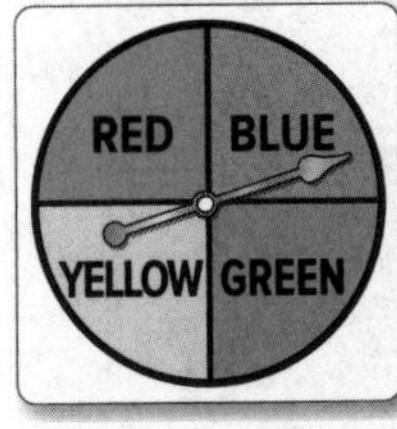

21. P(less than 10 and red)

22. P(odd and red or blue)

23. P(even and blue)

24. P(prime number and blue)

Maddie is packing for a trip. In her closet, there are 3 red, 4 black, 2 green, and 2 yellow blouses. She randomly selects 2 blouses. Find each probability.

25. P(red and red)

26. P(black and yellow)

27. P(red and black)

28. P(green and green)

Power Up! Common Core Test Practice

29. A bag contains letter tiles. There are 6 vowels in the bag and 14 consonants. On Michael's next turn, he will select a letter tile at random from the bag. Without replacing the first tile, he will then select a second letter tile. Determine if each of the following probabilities is true or false.

a. $P(\text{vowel then vowel}) = \frac{3}{38}$ ☐ True ☐ False

b. $P(\text{vowel then consonant}) = \frac{21}{95}$ ☐ True ☐ False

c. $P(\text{consonant then consonant}) = \frac{49}{100}$ ☐ True ☐ False

30. The spinners are each spun once.

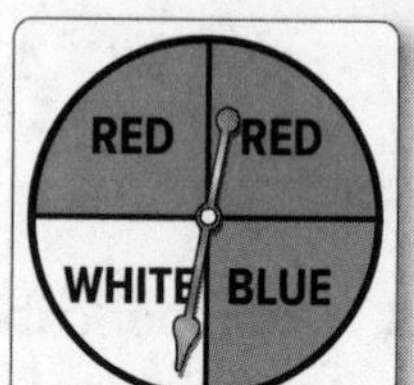

Do the two spins represent independent or dependent events? ______

Select the correct values to complete the model below to find $P(2 \text{ and white})$.

$P(2 \text{ and white}) = \frac{\square}{\square} \times \frac{\square}{\square} = \frac{\square}{\square}$

1	6
2	8
3	12
4	16

Common Core Spiral Review

Solve each proportion. 6.RP.1, 6.RP.3

31. $\frac{1}{4} = \frac{x}{72}$ $x =$ ______

32. $\frac{8}{n} = \frac{0.5}{0.9}$ $n =$ ______

33. $\frac{1}{3} = \frac{m}{153}$ $m =$ ______

34. $\frac{0.2}{a} = \frac{1.8}{18}$ $a =$ ______

35. 9 is 15% of what number? Write an equation. Then solve. 6.RP.3

______

36. A school librarian surveyed students about their favorite type of novel. The results are shown in the table at the right. What percent of students chose science fiction as their favorite type of novel? Round to the nearest whole percent. 6.RP.3c

______

Type of Novel	Number of Students
mystery	18
romance	10
science fiction	26
other	4

21ST CENTURY CAREER in Medicine

Pediatricians

Do you have compassion, a sense of humor, and the ability to analyze data? You might want to consider a career in medicine. Pediatricians care for the health of infants, children, and teenagers. They diagnose illnesses, interpret diagnostic tests, and prescribe and administer treatment.

College & Career READINESS

Is This the Career for You?

Are you interested in a career as a pediatrician? Take some of the following courses in high school.

- Algebra
- Biology
- Calculus
- Chemistry
- Psychology

Find out how math relates to a career in Medicine.

On Call for Kids

Use the information in the table below to solve each problem. Write each answer as a percent rounded to the nearest whole number.

1. What is the probability that one of the patients tested has strep throat? ________
2. If a patient has strep throat, what is the probability that they have a positive test? ________
3. What is the probability that a patient with the disease has a negative test? ________
4. If a patient does not have the disease, what is the probability that they have a positive test? ________
5. What is the probability that a patient that does *not* have strep throat tested negative for the disease? ________
6. The *positive predictive value*, or *PPV*, is the probability that a patient with a positive test result will have the disease. What is the PPV? ________
7. The *negative predictive value*, or *NPV*, is the probability that a patient with a negative test result will not have the disease. What is the NPV? ________

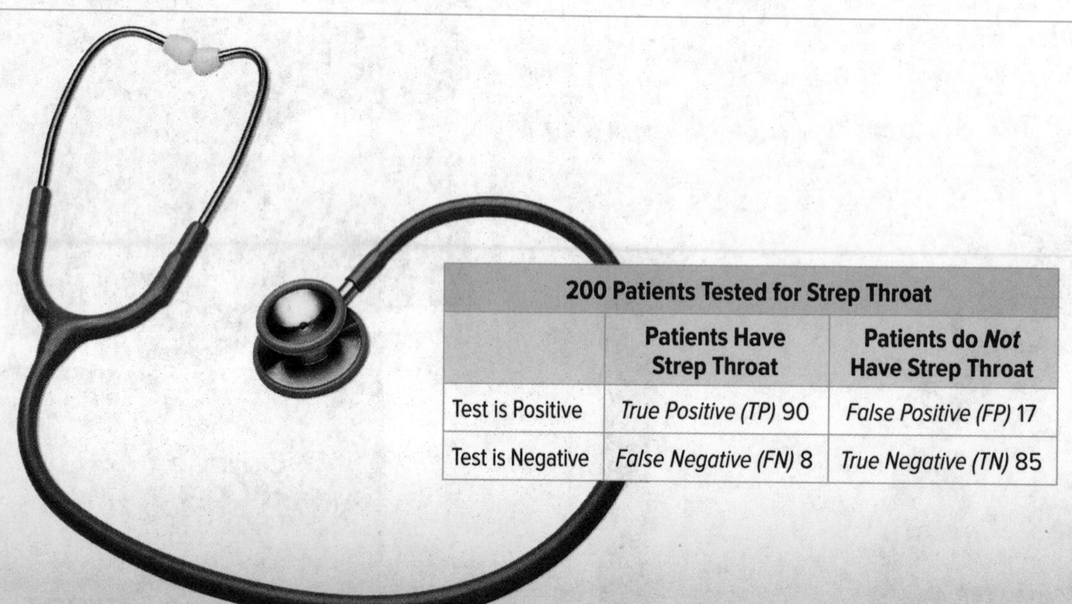

200 Patients Tested for Strep Throat		
	Patients Have Strep Throat	Patients do *Not* Have Strep Throat
Test is Positive	*True Positive (TP)* 90	*False Positive (FP)* 17
Test is Negative	*False Negative (FN)* 8	*True Negative (TN)* 85

Career Project

It's time to update your career portfolio! Interview your pediatrician. Be sure to ask what he or she enjoys most about being a pediatrician and what is most challenging. Include all the interview questions and answers in your portfolio.

What are some short term goals you need to achieve to become a pediatrician?

-
-
-
-
-

Chapter Review

Check

Vocabulary Check

Unscramble each of the clue words. After unscrambling all of the terms, use the numbered letters to find a sentence assoicated with probability.

HELATCORTEI

☐☐☐☐☐☐☐☐☐☐☐ (8th letter: 8; 10th letter: 5)

PORTUNMETAI

☐☐☐☐☐☐☐☐☐☐☐ (10th letter: 2; 11th letter: 9)

LEAPMS ECPAS

☐☐☐☐☐☐ ☐☐☐☐☐ (1st letter: 4; 7th letter: 3; 10th letter: 6)

COAPELMERTMYN

☐☐☐☐☐☐☐☐☐☐☐☐☐ (2nd letter: 7; 10th letter: 1)

☐☐☐☐ ☐ ☐☐☐☐

1 2 3 4 5 6 7 8 9

Complete each sentence using one of the unscrambled words above.

1. The ________________ is the set of all of the possible outcomes of a probability experiment.

2. A ________________ is an arrangement, or listing, of objects in which order is important.

3. The ________________ probability is based on what should happen when conducting a probability experiment.

4. Two events in which one or the other must happen, but they cannot happen at the same time are ________________.

Key Concept Check

Use Your FOLDABLES®

Use your Foldable to help review the chapter.

Tape here

Probability	
Example	Example
Picture	Picture

Got it?

Match each term or phrase on the left with the words on the right.

1. Based on what actually occurred in a probability experiment
2. The outcome of one event affects the outcome of a separate event
3. Consists of two or more simple events
4. Can be used to find the sample space

a. compound event

b. experimental probability

c. Fundamental Counting Principle

d. dependent event

e. tree diagrams

f. organized lists

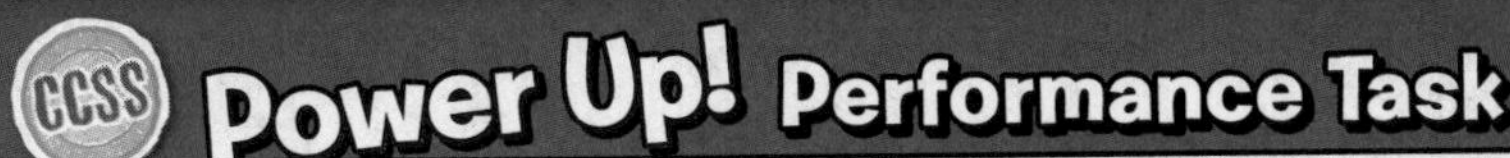

Carnival Prizes

Kelli is in charge of a game booth at the school carnival. The game has two simple rules.

- Randomly pick one blue card and one red card.
- If the product of the two numbers is greater than or equal to 45, you win a prize.

Write your answers on another piece of paper. Show all of your work to receive full credit.

Part A
Create a sample space and find the product of each combination. What is the probability that a person wins the game? Express your answer as a fraction in lowest terms and as a percent rounded to the nearest whole number.

Part B
The sponsor of the booth determines that they are giving away too many prizes. Recommend a minimum winning number that lowers the chance of winning to 25%. Explain your reasoning.

Part C
Participants achieving a winning score of 70 or higher in four consecutive attempts will receive a large stuffed animal. What is the probability of this occurring?

Part D
After changing the game rules, patrons and onlookers are disappointed when the first five games yield products of 12, 21, 32, 35, and 12. Recommend a statement that the sponsor can use to reassure customers that the game is fair.

Reflect

Answering the Essential Question

Use what you learned about probability to complete the graphic organizer.

Theoretical Probability

Experimental Probability

Essential Question

HOW can you predict the outcome of future events?

Sample Space

Simulation

Answer the Essential Question. HOW can you predict the outcome of future events?

Chapter 10
Statistics

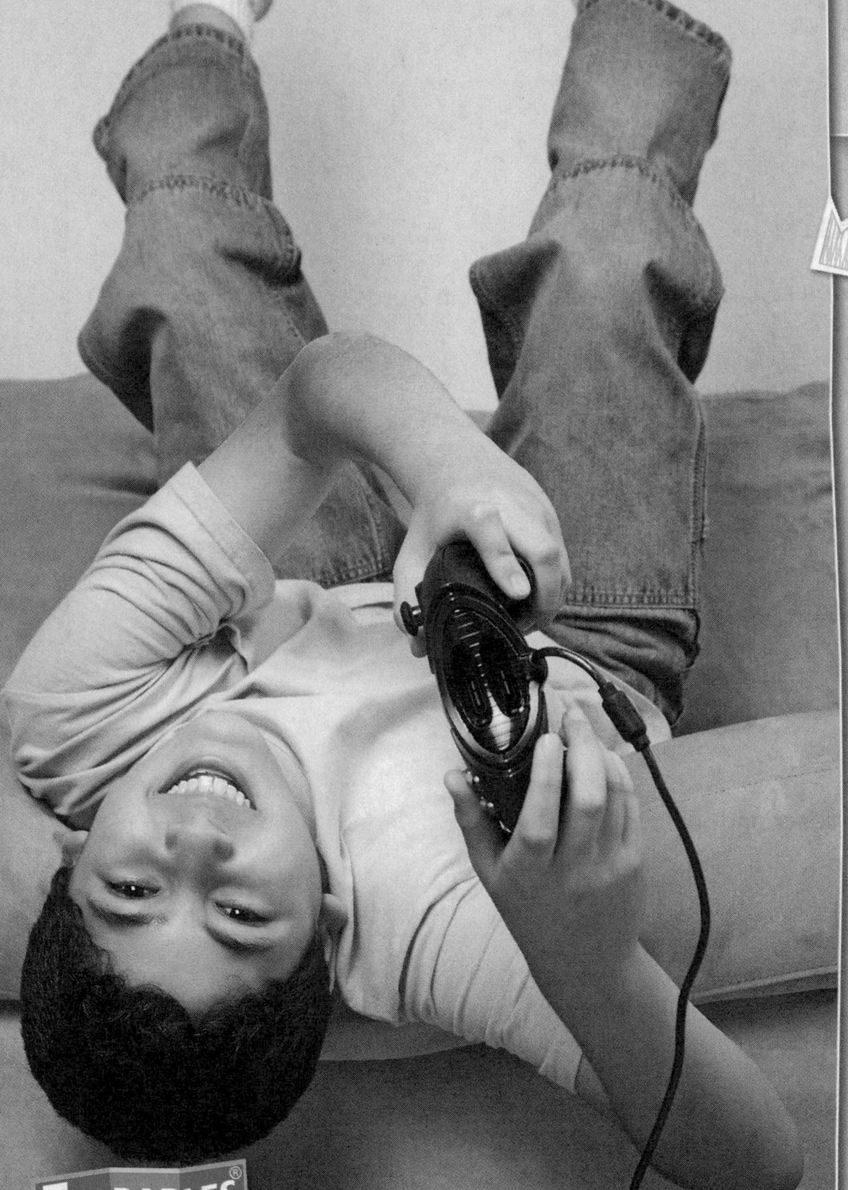

Essential Question

HOW do you know which type of graph to use when displaying data?

Common Core State Standards

Content Standards
7.SP.1, 7.SP.2, 7.SP.3, 7.SP.4

MP **Mathematical Practices**
1, 3, 4, 5, 6,

Math in the Real World

Surveys are used to collect information. Survey results can be shown in graphs.

The results of a survey of 50 middle school students are shown in the table. On the circle graph, write the percent of students who preferred each activity.

Activity	Number of Students
Gaming	22
Social Networking	18
Viewing Movies	6
Other	4

Favorite Online Activity

Viewing Movies ____ %
Other
Social Networking ____ %
Gaming ____ %

FOLDABLES® Study Organizer

1. Cut out the Foldable on page FL13 of this book.
2. Place your Foldable on page 850.
3. Use the Foldable throughout this chapter to help you learn about statistics.

What Tools Do You Need?

Vocabulary

biased sample
convenience sample
double box plot
double dot plot
population
sample
simple random sample
statistics
survey
systematic random sample
unbiased sample
voluntary response sample

Study Skill: Writing Math

Describe Data When you *describe* something, you represent it in words.

The table shows the prices for takeout orders at Lombardo's Restaurant.

Takeout	Price ($)
Main Dish	8.00
Side Dish	2.50
Dessert	4.00

Use the table to complete the following statements.

1. The price of a dessert is ______________________.
2. The price of a main dish is twice the price of ______________________.
3. A ______________________ is the least expensive item.

Write two other statements that describe the data.

4. ______________________
5. ______________________

What Do You Already Know?

Read each statement. Decide whether you agree (A) or disagree (D). Place a checkmark in the appropriate column and then justify your reasoning.

Statistics			
Statement	A	D	Why?
Statistics deal with collecting, organizing, and interpreting data.			
A sample is the same thing as the entire population.			
A biased sample accurately represents the entire population.			
Graphs are sometimes made to influence conclusions by misrepresenting the data.			
A double box plot consists of two box plots that are drawn on the same number line.			
Any type of display can be used to represent data.			

When Will You Use This?

Here are a few examples of how statistics are used in the real world.

Activity 1 Find the average monthly high and low temperatures for the city where you live. Then find the average monthly high and low temperatures for another city. How do these temperatures compare to the temperatures for your city?

____

____

Activity 2 Go online at **connectED.mcgraw-hill.com** to read the graphic novel ***Record Highs***. Blake, Hannah, and Jamar need to compare the record high temperatures for two different cities. What cities do they use? ______________________

Try the Quick Check below.
Or, take the Online Readiness Quiz.

Quick Review

Common Core Review 6.SP.5c, 6.NS.4

Example 1

Which players averaged more than 10 points per game?

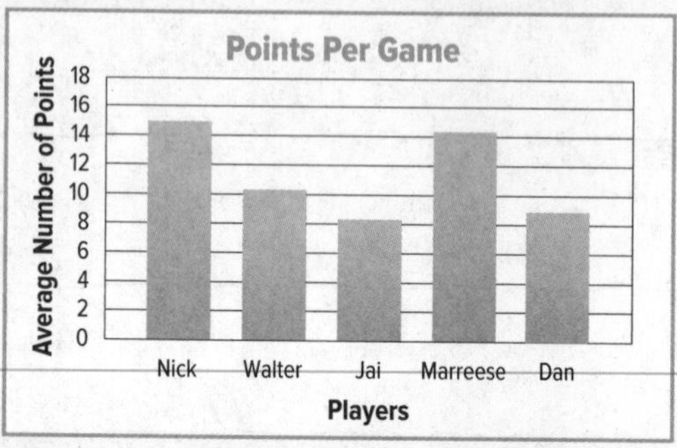

Nick, Walter, and Marreese averaged more than 10 points per game.

Example 2

Use the circle graph. Suppose 300 people were surveyed. How many people have two accounts?

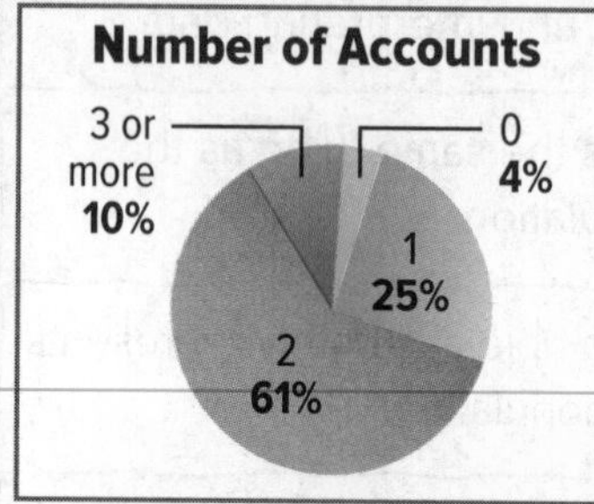

Find 61% of 300.

$$61\% \text{ of } 300 = 61\% \times 300$$
$$= 0.61 \times 300 \text{ or } 183$$

So, 183 people have two accounts.

Quick Check

Graphs **The bar graph at the right shows the number of items each student obtained during a scavenger hunt.**

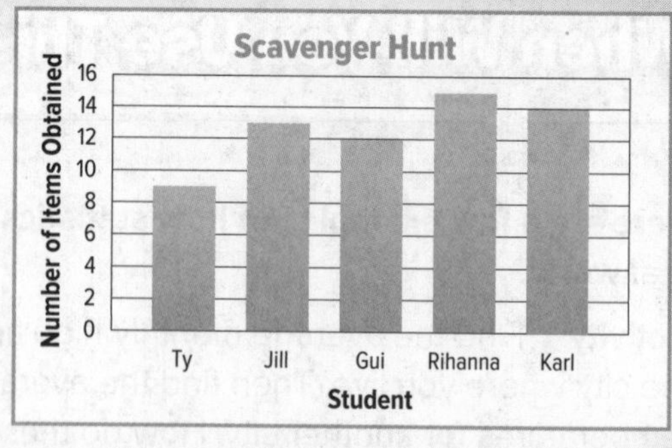

1. Who obtained the most items?

____

2. Who obtained the least items?

____

3. Refer to the circle graph in Example 2. Suppose 300 people were surveyed. How many people have 1 account?

____

How Did You Do?

Which problems did you answer correctly in the Quick Check? Shade those exercise numbers below.

① ② ③

Lesson 1

Make Predictions

Vocabulary Start-Up

Statistics deal with collecting, organizing, and interpreting data. A **survey** is a method of collecting information. The group being studied is the **population**. Sometimes the population is very large. To save time and money, part of the group, called a **sample**, is surveyed.

For each survey topic, determine which set represents the population and which represents a sample of the population. Write *population* or *sample*.

	Survey Topic	Set A	Set B
1.	dress code changes	the students in a middle school	the seventh graders in the middle school
2.	favorite flavors of ice cream	the customers at an ice cream shop in the town	the residents of a town

Essential Question

HOW do you know which type of graph to use when displaying data?

Vocabulary

statistics
survey
population
sample

Common Core State Standards

Content Standards
7.SP.1, 7.SP.2

Mathematical Practices
1, 3, 4

Real-World Link

Logan wants to survey students in his school about their favorite and least favorite zoo exhibit. Describe a possible sample Logan could survey instead of surveying the entire school.

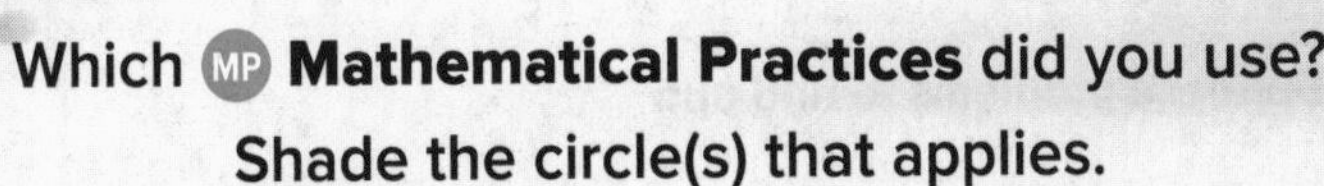

Which **Mathematical Practices** did you use? Shade the circle(s) that applies.

1. Persevere with Problems
2. Reason Abstractly
3. Construct an Argument
4. Model with Mathematics
5. Use Math Tools
6. Attend to Precision
7. Make Use of Structure
8. Use Repeated Reasoning

Work Zone

Make Predictions Using Ratios

You can use the results of a survey or past actions to predict the actions of a larger group. Since the ratios of the responses of a good sample are often the same as the ratios of the responses of the population.

Examples

The students in Mr. Blackwell's class brought photos from their summer break. The table shows how many students brought each type of photo.

Summer Break Photos	
Location	**Students**
beach	6
campground	4
home	7
theme park	11

1. What is the probability that a student brought a photo taken at a theme park?

$$P(\text{theme park}) = \frac{\text{number of theme park photos}}{\text{number of students with a photo}} = \frac{11}{28}$$

So, the probability of a theme park photo is $\frac{11}{28}$.

2. There are 560 students at the school where Mr. Blackwell teaches. Predict how many students would bring in a photo taken at a theme park.

Let s represent the number of theme park photos.

$\frac{11}{28} = \frac{s}{560}$ Write an equivalent ratio.

$\frac{11}{28} = \frac{s}{560}$ (×20, ×20) Since $28 \times 20 = 560$, multiply 11 by 20 to find s.

$\frac{11}{28} = \frac{220}{560}$ $s = 220$

Of the 560 students, you can expect about 220 to bring a photo from a theme park.

Got it? **Do these problems to find out.**

A survey found that 6 out of every 10 students have a blog.

a. What is the probability that a student at the school has a blog?

b. Suppose there are about 250 students at the school. About how many have a blog?

a. ________

b. ________

Make Predictions Using Equations

You can also use the percent equation to make predictions.

Examples

3. A survey found that 85% of people use emoticons on their instant messengers. Predict how many of the 2,450 students at Washington Middle School use emoticons.

Words	What number of students is 85% of 2,450 students?
Variable	Let n represent the number of students.
Equation	$n = 0.85 \cdot 2{,}450$

$n = 0.85 \cdot 2{,}450$ Write the percent equation.

$n = 2{,}082.5$ Multiply.

About 2,083 of the students use emoticons.

4. The circle graph shows the results of a survey in which children ages 8 to 12 were asked whether they have a television in their bedroom. Predict how many out of 1,725 students would not have a television in their bedroom.

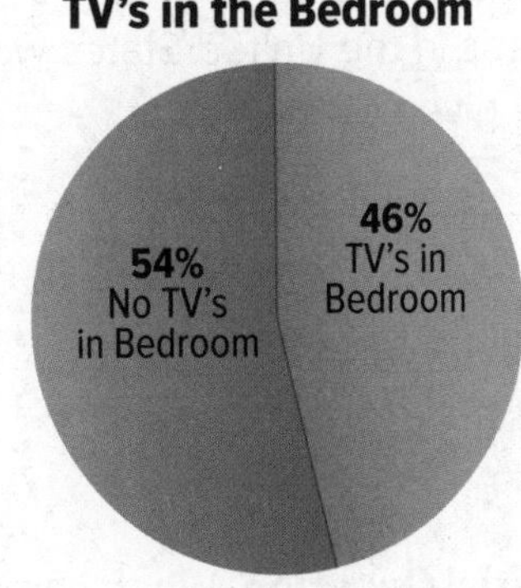

You can use the percent equation and the survey results to predict what part p of the 1,725 students have no TV in their bedroom.

part = percent · whole

$p = 0.54 \cdot 1{,}725$ Survey results: 54%

$p = 931.5$ Multiply.

About 932 students do not have a television in their bedroom.

What proportion could you use to solve Example 4? Write your answer below.

Got it? Do this problem to find out.

c. Refer to Example 4. Predict how many out of 1,370 students have a television in their bedroom.

c. ______________

Guided Practice

The table shows the results of a survey of Hamilton Middle School seventh graders. Use the table to find the following probabilities. (Examples 1 and 2)

Career Field	Students
Entertainment	17
Education	14
Medicine	11
Public service	6
Sports	2

1. the probability of choosing a career in public service

2. the probability of choosing a career in education

3. the probability of choosing a career in sports

4. Predict how many students out of 400 will enter the education field.

5. Predict how many students out of 500 will enter the medical field.

6. Use the circle graph that shows the results of a poll to which 60,000 teens responded. Predict how many of the approximately 28 million teens in the United States would buy a music CD if they were given $20. (Examples 3 and 4)

How Would You Spend a Gift of $20?

Category	Percent
Other	9%
Save it	33%
Go to movie	5%
Clothing/ jewelry	21%
Music CD	32%

7. **Building on the Essential Question** When can statistics be used to gain information about a population from a sample?

Rate Yourself!

How confident are you about making predictions? Check the box that applies.

For more help, go online to access a Personal Tutor.

Independent Practice

Go online for Step-by-Step Solutions

The table shows the results of a survey of 150 students. Use the table to find the probability of a student participating in each sport. (Example 1)

Show your work.

Sport	Students
Baseball/softball	36
Basketball	30
Football	45
Gymnastics	12
Tennis	18
Volleyball	9

1. football

2. tennis

3 gymnastics

4. volleyball

5 Three out of every 10 students ages 6–14 have a magazine subscription. Suppose there are 30 students in Annabelle's class. About how many will have a magazine subscription? (Example 2)

6. Use the graph that shows the percent of cat owners who train their cats in each category. (Examples 3 and 4)

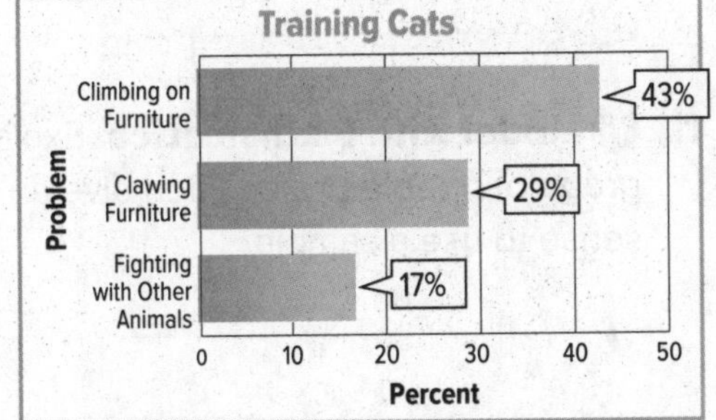

a. Out of 255 cat owners, predict how many owners trained their cat not to climb on furniture.

b. Out of 316 cat owners, predict how many cat owners trained their cat not to claw on furniture.

7. MP **Make a Prediction** The school librarian recorded the types of books students checked out on a typical day. Suppose there are 605 students enrolled at the school. Predict the number of students that prefer humor books. Compare this to the number of students at the school who prefer nonfiction.

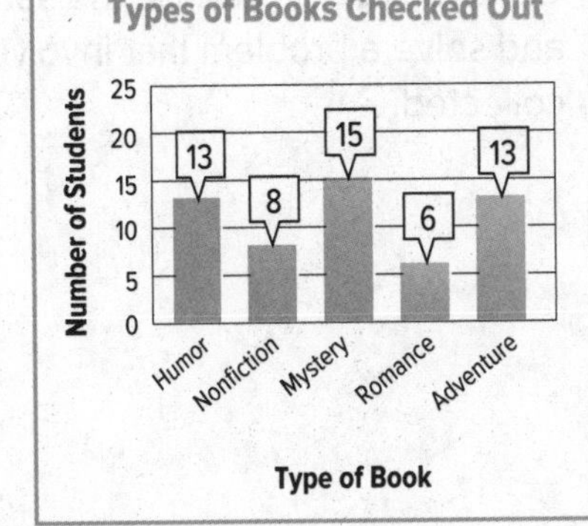

H.O.T. Problems Higher Order Thinking

8. **MP Find the Error** A survey of a seventh-grade class showed that 4 out of every 10 students are taking a trip during spring break. There are 150 students in the seventh grade. Caitlyn is trying to determine how many of the seventh-grade students can be expected to take a trip during spring break. Find her mistake and correct it.

9. **MP Persevere with Problems** One letter tile is drawn from the bag and replaced 300 times. Predict how many times a consonant will *not* be picked.

10. **MP Persevere with Problems** A survey found that 80% of teens enjoy going to the movies in their free time. Out of 5,200 teens, predict how many said that they do not enjoy going to the movies in their free time.

11. **MP Model with Mathematics** Explain how to use a sample to predict what a group of people prefer. Then give an example of a situation in which it makes sense to use a sample.

12. **MP Model with Mathematics** Design a survey to give to your classmates. Construct a bar graph in the space below to represent your data. Then write and solve a problem that involves a prediction based on the data you collected.

Name ______________________ My Homework ______________________

Extra Practice

Solve.

13. Luther won 12 of the last 20 video games he played. Find the probability of Luther winning the next game he plays. $\frac{3}{5}$, 0.6, or 60%

$$P(\text{winning}) = \frac{\text{number of games won}}{\text{number of games played}}$$
$$= \frac{12}{20} \text{ or } \frac{3}{5}$$

14. Refer to Exercise 13. Suppose Luther plays a total of 60 games with his friends over the next month. Predict how many of these games Luther will win. ________

15. Use the graph that shows the number of times teens volunteer.

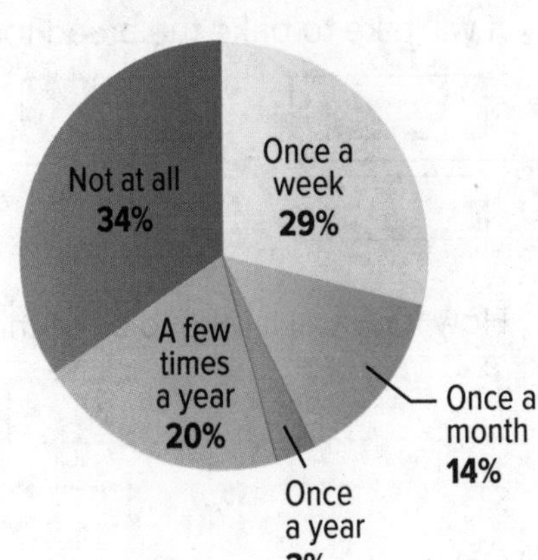

a. About 300,000 teens ages 12–14 live in Virginia. Predict the number of teens in this age group who volunteer a few times a year.

____

b. Tennessee has about 250,000 teens ages 12–14. Predict the number of teens in this age group who volunteer once a week.

____

c. About 240,000 teens ages 12–14 live in Missouri. Predict the number of teens in this age group who volunteer once a year.

____

16. MP **Make a Prediction** The probability of Jaden making a free throw is 15%. Predict the number of free throws that he can expect to make if he attempts 40 free throws.

____

Draw a line to match each situation with the appropriate equation or proportion.

17. 27 MP3s is what percent of 238 MP3s? — **a.** $n = 27 \cdot 2.38$

18. 238% of 27 is what number? — **b.** $\frac{27}{100} = \frac{p}{238}$

19. 27% of MP3 owners download music weekly. Predict how many MP3 owners out of 238 owners download music weekly. — **c.** $\frac{27}{238} = \frac{n}{100}$

CCSS Power Up! Common Core Test Practice

20. There were 515 students surveyed on how they spend time with their families. Which of the following estimates are accurate? Select all that apply.

- ☐ About 175 students spend time with their family eating dinner.
- ☐ About 72 students spend time with their family playing sports.
- ☐ About 50 students spend time with their family watching TV.
- ☐ About 38 students spend time with their family taking walks.

How Students Spend Time with Family	
Eating Dinner	34%
Watching TV	20%
Talking	14%
Playing Sports	14%
Talking Walks	4%
Other	14%

21. Yesterday a bakery baked 54 loaves of bread in 20 minutes. Today the bakery needs to bake 405 loaves of bread at the same rate. Select values to complete the model below to predict how long it will take to bake the bread today.

$$\frac{\square}{\square} = \frac{\square}{\square}$$

20
54
405
x

How long will it take to bake the bread today?

____

CCSS Common Core Spiral Review

22. A magazine rack contains 5 sports magazines, 7 news magazines, and 10 fashion magazines. After a magazine is chosen, it is *not* replaced. Find the probability of randomly choosing two fashion magazines. **7.SP.8a**

____

23. Each week, Ryan's mother has him randomly choose a chore that he must complete from the list shown. The first week, he chose washing the dishes. What is the probability that Ryan will choose washing the dishes two more weeks in a row? **7.SP.5**

Weekly Chores
Collecting the trash
Folding the laundry
Cleaning the house
Washing the dishes
Cutting the grass

____

24. In how many different orders can a person watch 3 different movies? Use a list to show the sample space. **7.SP.8b**

____

Lesson 2

Unbiased and Biased Samples

Real-World Link

Entertainment A T.V. programming manager wants to conduct a survey to determine which reality television show is the favorite of viewers in a certain viewing area. He is considering the three samples shown. Draw an X through the two samples that would not fairly represent all of the people in the viewing area.

Sample 1	Sample 2
100 people that are trying out for a reality show	100 students at your middle school

Sample 3
Every 100th person at a shopping mall

Explain why the two samples that you crossed out do *not* fairly represent all of the people in the viewing area? Explain.

____

____

____

____

Essential Question

HOW do you know which type of graph to use when displaying data?

Vocabulary

unbiased sample
simple random sample
systematic random sample
biased sample
convenience sample
voluntary response sample

Common Core State Standards

Content Standards
7.SP.1, 7.SP.2

MP Mathematical Practices
1, 3, 4, 5

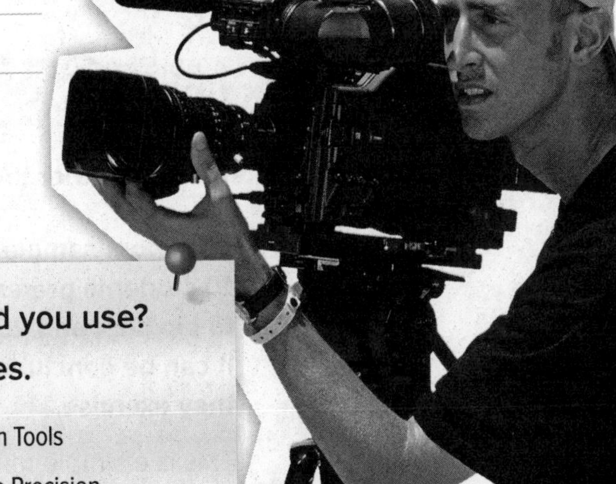

Which MP Mathematical Practices did you use?
Shade the circle(s) that applies.

① Persevere with Problems
② Reason Abstractly
③ Construct an Argument
④ Model with Mathematics
⑤ Use Math Tools
⑥ Attend to Precision
⑦ Make Use of Structure
⑧ Use Repeated Reasoning

Work Zone

Biased and Unbiased Samples

To get valid results, a sample must be chosen very carefully. An **unbiased sample** is selected so that it accurately represents the entire population. Two ways to pick an unbiased sample are listed below.

Unbiased Samples		
Type	**Description**	**Example**
Simple Random Sample	Each item or person in the population is as likely to be chosen as any other.	Each student's name is written on a piece of paper. The names are placed in a bowl, and names are picked without looking.
Systematic Random Sample	The items or people are selected according to a specific time or item interval.	Every 20th person is chosen from an alphabetical list of all students attending a school.

In a **biased sample**, one or more parts of the population are favored over others. Two ways to pick a biased sample are listed below.

Everyday Use
Bias is a tendency or prejudice

Math Use
Bias is an error introduced by selecting or encouraging a specific outcome

Biased Samples		
Type	**Description**	**Example**
Convenience Sample	A convenience sample consists of members of a population that are easily accessed.	To represent all the students attending a school, the principal surveys the students in one math class.
Voluntary Response Sample	A voluntary response sample involves only those who want to participate in the sampling.	Students at a school who wish to express their opinions complete an online survey.

Examples

Tutor

Determine whether the conclusion is valid. Justify your answer.

1. **A random sample of students at a middle school shows that 10 students prefer listening to rock, 15 students prefer listening to hip hop, and 25 students prefer no music while they exercise. It can be concluded that half the students prefer no music while they exercise.**

This is a simple random sample. So, the sample is unbiased and the conclusion is valid.

Determine whether each conclusion is valid. Justify your answer.

2. **Every tenth person who walks into a department store is surveyed to determine his or her music preference. Out of 150 customers, 70 stated that they prefer rock music. The manager concludes that about half of all customers prefer rock music.**

Since the population is every tenth customer of a department store, the sample is an unbiased, systematic random sample. The conclusion is valid.

3. **The customers of a music store are surveyed to determine their favorite leisure time activity. The results are shown in the graph. The store manager concludes that most people prefer to listen to music in their leisure time.**

Leisure Time Activities

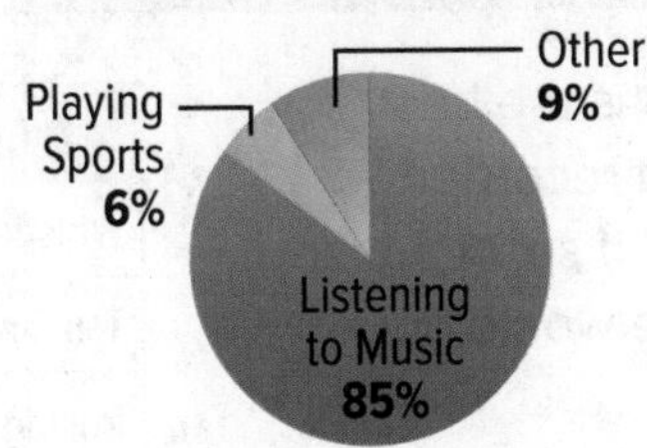

The customers of a music store probably like to listen to music in their leisure time. The sample is a biased, convenience sample since all of the people surveyed are in one specific location. The conclusion is not valid.

Got it? **Do this problem to find out.**

a. A radio station asks its listeners to indicate their preference for one of two candidates in an upcoming election. Seventy-two percent of the listeners who responded preferred candidate A, so the radio station announced that candidate A would win the election. Is the conclusion valid? Justify your answer.

Show your work.

a. ____________

Use Sampling to Predict

A valid sampling method uses unbiased samples. If a sampling method is valid, you can make generalizations about the population.

Example

4. A store sells 3 types of pants: jeans, capris, and cargos. The store workers survey 50 customers at random about their favorite type of pants. The survey responses are indicated at the right. If 450 pairs of pants are ordered, how many should be jeans?

Type	Number
Jeans	25
Capris	15
Cargos	10

First, determine whether the sample method is valid. The sample is a simple random sample since customers were randomly selected. Thus, the sample method is valid.

$\frac{25}{50}$ or 50% of the customers prefer jeans. So, find 50% of 450.

$0.5 \times 450 = 225$, so about 225 pairs of jeans should be ordered.

Guided Practice

1. Zach is trying to decide which of three golf courses is the best. He randomly surveyed people at a sports store and recorded the results in the table. Is the sample method valid? If so, suppose Zach surveyed 150 more people. How many people would be expected to vote for Rolling Meadows? (Example 4)

Course	Number
Whispering Trail	10
Tall Pines	8
Rolling Meadows	7

2. To find how much money the average American family spends to cool their home, 100 Alaskan families are surveyed at random. Of the families, 85 said that they spend less than $75 per month on cooling. The researcher concluded that the average American family spends less than $75 on cooling per month. Is the conclusion valid? Explain. (Examples 1–3)

3. **Building on the Essential Question** How is using a survey one way to determine experimental probability?

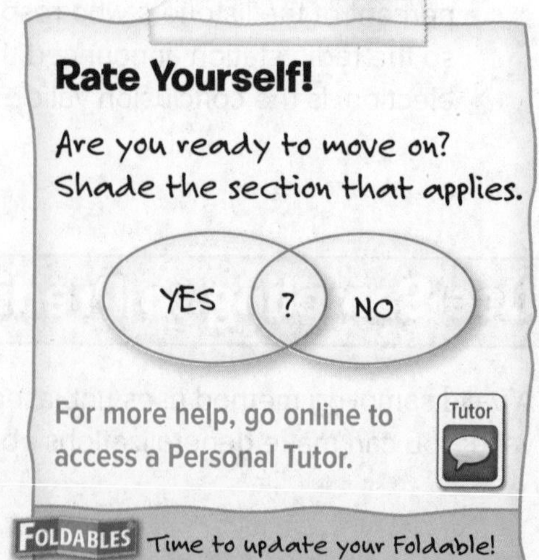

Independent Practice

Go online for Step-by-Step Solutions 

Determine whether each conclusion is valid. Justify your answer.

(Examples 1–3)

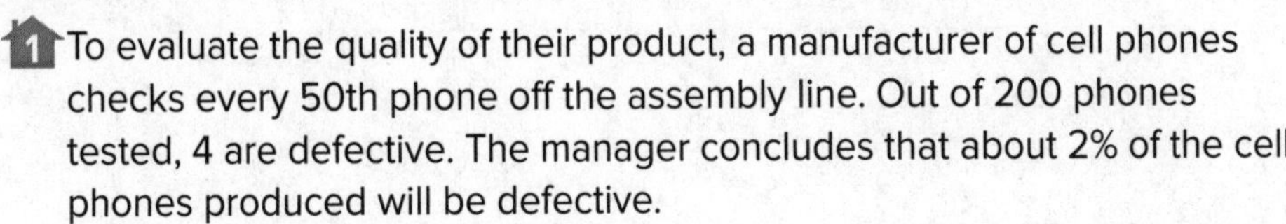

1 To evaluate the quality of their product, a manufacturer of cell phones checks every 50th phone off the assembly line. Out of 200 phones tested, 4 are defective. The manager concludes that about 2% of the cell phones produced will be defective.

2. To determine whether the students will attend an arts festival at the school, Oliver surveys his friends in the art club. All of Oliver's friends plan to attend. So, Oliver assumes that all the students at his school will also attend.

3 A random sample of people at a mall shows that 22 prefer to take a family trip by car, 18 prefer to travel by plane, and 4 prefer to travel by bus. Is the sample method valid? If so, how many people out of 500 would you expect to say they prefer to travel by plane? (Example 4)

Preferred Ways to Travel

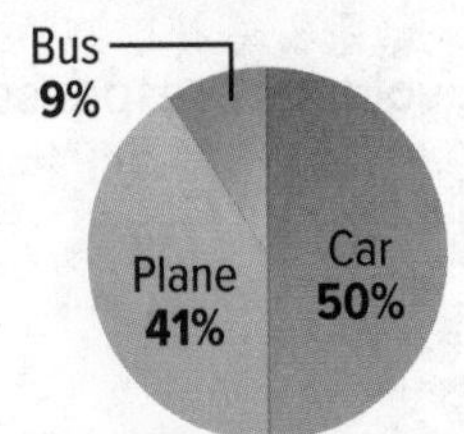

4. MP **Use Math Tools** Use the organizer to determine whether the conclusion is valid.

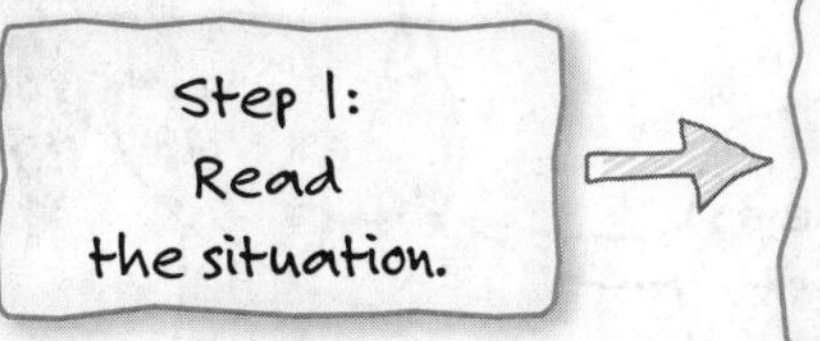

Marcus wants to predict the next student council president. He polls every fourth person from each grade level as they exit the cafeteria. In his poll, 65% chose Sophia. So, Marcus predicts Sophia will win the election.

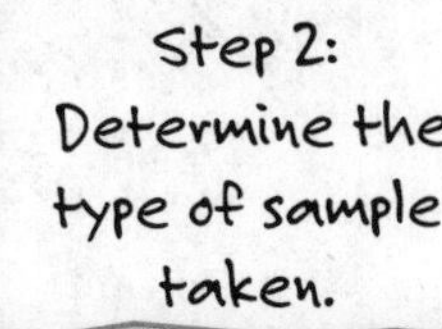

Step 3:
Determine if the
conclusion is valid.

H.O.T. Problems Higher Order Thinking

5. **MP Persevere with Problems** How could the wording of a question or the tone of voice of the interviewer affect a survey? Provide an example.

MP Justify Conculsions Determine whether each statement is *sometimes*, *always*, or *never* true. Explain your reasoning to a classmate.

6. A biased sample is valid.

7. A simple random sample is valid.

8. A voluntary response sample is valid.

9. **MP Find the Error** Marisol wants to determine how many students plan to attend the girls' varsity basketball game. Find her mistake and correct it.

10. **MP Model with Mathematics** Give an example of a data set from a random sample. Then make an inference about the population represented by the sample.

Name ______________________ My Homework ______________

Extra Practice

Determine whether each conclusion is valid. Justify your answer.

11. To determine what people in California think about a proposed law, 5,000 people from the state are randomly surveyed. Of the people surveyed, 58% are against the law. The legislature concludes that the law should not be passed.

This is an unbiased, simple random sample because randomly selected

Californians were surveyed. So, the conclusion is valid.

12. A magazine asks its readers to complete and return a questionnaire about popular television actors. The majority of those who replied liked one actor the most, so the magazine decides to write more articles about that actor.

____

____

13. The Student Council advisor asked every tenth student in the lunch line how they preferred to be contacted with school news. The results are shown in the table. Is this a random sample? If yes, suppose there are 684 students at the school. How many can be expected to prefer E-mail?

Method	Number
E-mail	16
Newsletter	12
Announcement	5
Telephone	3

____

____

MP Justify Conclusions **Each of the following surveys results in a biased sample. For each situation, explain why the survey is biased. Then explain how you would change the survey to obtain an unbiased sample.**

14. A store manager sends an E-mail survey to customers who have registered at the store's Web site.

____

____

____

____

15. A school district surveys the family of every tenth student to determine if they would vote in favor of the construction of a new school building.

____

____

____

____

CCSS Power Up! Common Core Test Practice

16. Maci surveyed all the members of her softball team about their favorite sport. Based on these results, Maci concludes that softball is the favorite sport among all her classmates. Explain why Maci's conclusion might *not* be valid. How could she change the survey to achieve a more valid conclusion?

Sport	Number of Members
Softball	12
Basketball	5
Soccer	3
Volleyball	8

17. Ms. Hernandez determined that 60% of the students in her classes brought an umbrella to school when the weather forecast predicted rain. She has a total of 150 students in her classes. Determine if each statement represents a valid or invalid conclusion.

a. On days when rain is forecasted, less than $\frac{2}{5}$ of her students bring an umbrella to school. ☐ Valid ☐ Invalid

b. On days when rain is forecasted, about 90 of her students bring an umbrella to school. ☐ Valid ☐ Invalid

c. On days when rain is forecasted, more than $\frac{1}{2}$ of her students bring an umbrella to school. ☐ Valid ☐ Invalid

Common Core Spiral Review

For Exercises 18 and 19, use the table that shows Alana's first six math test scores. 6.SP.3

Test	1	2	3	4	5	6
Score	88%	92%	70%	96%	84%	96%

18. Find the mean, median, and mode of Alana's test scores. Round to the nearest tenth if necessary.

mean: ________ median: ________ mode: ________

19. Determine which measure of center best represents Alana's performance. Justify your reasoning.

Inquiry Lab

Multiple Samples of Data

WHY is it important to analyze multiple samples of data before making predictions?

Content Standards
7.SP.2

Mathematical Practices
1, 3, 4, 5

A hostess at a restaurant randomly hands out crayons to young children. There are three different color crayons: green (G), red (R), and blue (B). The server gives out the green crayon 40% of the time, the red crayon 40% of the time, and the blue crayon 20% of the time.

Hands-On Activity 1

Tools

When you draw a conclusion about a population from a sample of data, you are making *inferences* about that population. Sometimes, making inferences about a population from only one sample is not as accurate as using multiple samples of data.

Use a spinner to simulate the situation above.

Step 1 Create a spinner with five equal sections. Label two sections G. Label another two sections R and label one section B.

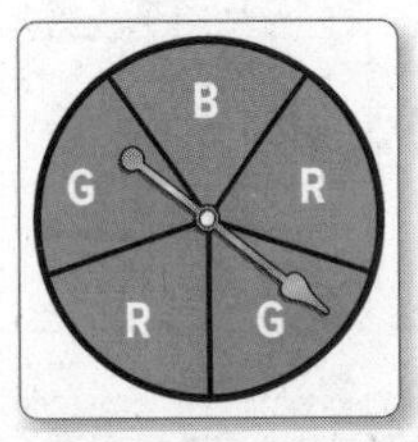

Step 2 Each spin of the spinner represents a young child receiving a crayon. Spin the spinner 20 times. Record the number of times each color of crayon was received in the column labeled Sample 1 in the table below. Repeat two more times. Record the results in the columns labeled Sample 2 and Sample 3 in the table.

Color	Sample 1 Frequency	Sample 2 Frequency	Sample 3 Frequency
Green			
Red			
Blue			

Compare the results of the 3 samples. Do you notice any differences?

The most commonly used keyboard is the QWERTY keyboard. However, there is another type of keyboard called the Dvorak keyboard that is based on letter frequency. Complete the Activity below about letter frequencies.

Hands-On Activity 2

The table at the right contains fifteen randomly selected words from the English language dictionary.

Sample 1		
airport	juggle	sewer
blueberry	lemon	standard
costume	mileage	thread
doorstop	percentage	vacuum
instrument	print	whale

Step 1 Find the frequency of each letter. Record the frequencies in the Sample 1 rows of the tables below.

Letter	a	b	c	d	e	f	g	h	i	j	k	l	m
Sample 1 Frequency													
Sample 2 Frequency													
Sample 3 Frequency													

Letter	n	o	p	q	r	s	t	u	v	w	x	y	z
Sample 1 Frequency													
Sample 2 Frequency													
Sample 3 Frequency													

Step 2 Randomly select another 15 words from a dictionary. Record the frequency of the letters in the rows labeled Sample 2 in the tables above.

Step 3 Repeat Step 2. Record the frequency of the letters in the rows labeled Sample 3.

Work with a partner to collect multiple samples based on the following situation.

Janet and Masao are making centerpieces for their school's fall dance. They randomly select a ribbon to use in each centerpiece. There are four different colors of ribbon to choose from: brown (B), green (G), orange (O), and yellow (Y).

1. MP **Model with Mathematics** Design a method to simulate how many times each ribbon will be selected. Describe your simulation.

Show your work.

2. Use the method you described in Exercise 1 to simulate the ribbon selection 20 times. Record the frequency of each color selection in the Sample 1 Frequency column of the table below.

Color	Sample 1 Frequency	Sample 2 Frequency	Sample 3 Frequency
Brown			
Green			
Orange			
Yellow			

3. Repeat the process described in Exercise 2 two more times. Record the frequencies of each color selection in the Sample 2 and Sample 3 columns.

4. Make an inference to determine which color was selected the most often in each sample.

5. The *relative frequency* of a color being selected is the ratio of the number of times the color was selected to the total number of selections. Find the relative frequency of an orange ribbon being selected for each sample.

Sample 1: _____ Sample 2: _____ Sample 3: _____

6. Masao predicts that 5 out of 10 centerpieces will have an orange ribbon. How far off is Masao's prediction? Explain.

Analyze and Reflect

Work with a partner to answer the following questions. Refer to Activity 2.

7. What is the relative frequency for the letter e for each sample? Round to the nearest hundredth.

   Sample 1: ________ Sample 2: ________ Sample 3: ________

8. What is the mean relative frequency of the letter e for the three samples? the median relative frequency? Round to the nearest tenth if necessary.

   mean relative frequency: ________ median relative frequency: ________

9. MP **Use Math Tools** Research on the Internet to find the actual relative frequency of the letter e for words in the English language. How do your sample results compare to the actual relative frequency?

10. MP **Reason Inductively** Write a few sentences describing the inferences you can make about the frequency of letters in the words in the English language using your three samples.

On Your Own

Create

11. MP **Justify Conclusions** Research on the Internet to find the relative frequency of other letters in words in the English language. Write how your sample results compare to the actual frequencies. Note any differences.

12. Inquiry WHY is it important to analyze multiple samples of data before making predictions?

Lesson 3

Misleading Graphs and Statistics

Real-World Link

Hockey The Stanley Cup is awarded annually to the champion team in the National Hockey League. The graph shows the total number of points scored in Stanley Cup playoff games by three players during their careers.

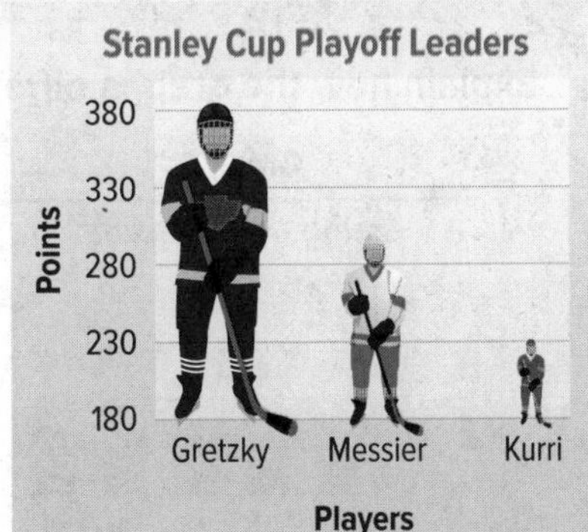

1. According to the size of the players, how many times more points does Messier appear to have than Kurri?

2. Do you think this is representative of the players' number of points? Explain.

3. What reason could someone have for intentionally creating a misleading Stanley Cup graph?

Essential Question

HOW do you know which type of graph to use when displaying data?

Common Core State Standards

Content Standards
Extension of 7.SP.1

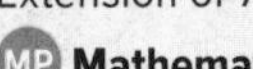

1, 3, 4

Which MP Mathematical Practices did you use? Shade the circle(s) that applies.

① Persevere with Problems
② Reason Abstractly
③ Construct an Argument
④ Model with Mathematics
⑤ Use Math Tools
⑥ Attend to Precision
⑦ Make Use of Structure
⑧ Use Repeated Reasoning

Work Zone

Identify a Misleading Graph

Graphs let readers analyze data easily, but are sometimes made to influence conclusions by misrepresenting the data.

Example

Changing Scales
To emphasize a change over time, reduce the scale interval on the vertical axis.

1. Explain how the graphs differ.

Graph A

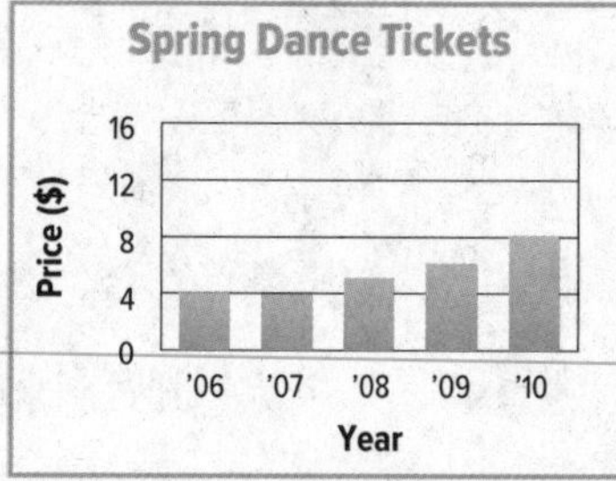

Graph B

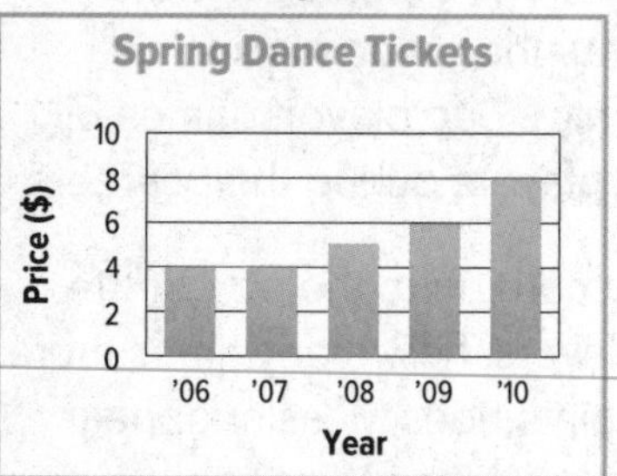

The graphs show the same data. However, the graphs differ in that Graph A uses an interval of 4, and Graph B uses an interval of 2.

Which graph appears to show a sharper increase in price?
Graph B makes it appear that the prices increased more rapidly even though the price increase is the same.

Which graph might the Student Council use to show that while ticket prices have risen, the increase is not significant? Why?
They might use Graph A. The scale used on the vertical axis of this graph makes the increase appear less significant.

Got it? **Do this problem to find out.**

a. The line graphs show monthly profits of a company from October to March. Which graph suggests that the business is extremely profitable? Is this a valid conclusion? Explain.

Show your work.

a. ______________

Graph A

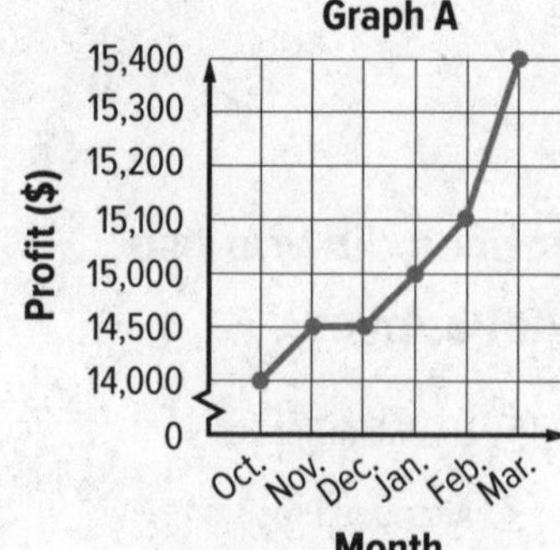

Graph B

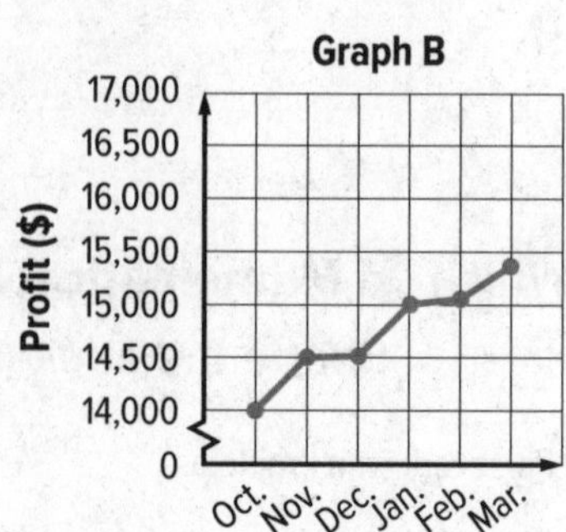

Misleading Statistics

Statistics can also be used to influence conclusions.

Example

2. An amusement park boasts that the average height of their roller coasters is 170 feet. Explain how this might be misleading.

Park Roller Coaster Heights	
Coaster	**Height (ft)**
Viper	109
Monster	135
Red Zip	115
Tornado	365
Riptide	126

Mean $\frac{109 + 135 + 115 + 365 + 126}{5} = \frac{850}{5}$

$= 170$

Median 109, 115, 126, 135, 365

Mode none

The average used by the park was the mean. This measure is much greater than most of the heights listed because of the coaster that is 365 feet. So, it is misleading to use this measure to attract visitors. A more appropriate measure to describe the data is the median, 126 feet, which is closer to the height of most of the coasters.

Mode
The mode is the number or numbers that appear most often in a set of data.

Got it? **Do this problem to find out.**

b. Find the mean, median, and mode of the sofa prices shown in the table. Which measurement might be misleading in describing the average cost of a sofa? Explain.

Sofa Prices	
Sofa Style	**Cost**
leather	\$1,700
reclining	\$1,400
DIY assembly	\$350
sectional	\$1,600
micro-fiber	\$1,400

b. ________________

Guided Practice

1. The graph suggests that Cy Young had three times as many wins as Jim Galvin. Is this a valid conclusion? Explain. (Example 1)

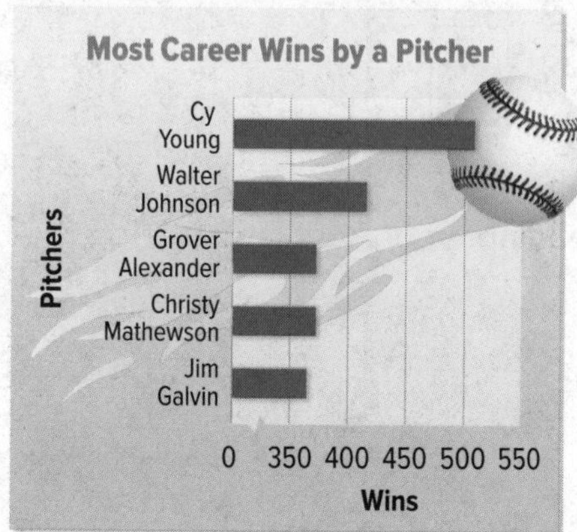

2. The graph at the right shows the results of a survey to determine students' favorite pets. Why is the graph misleading? (Example 1)

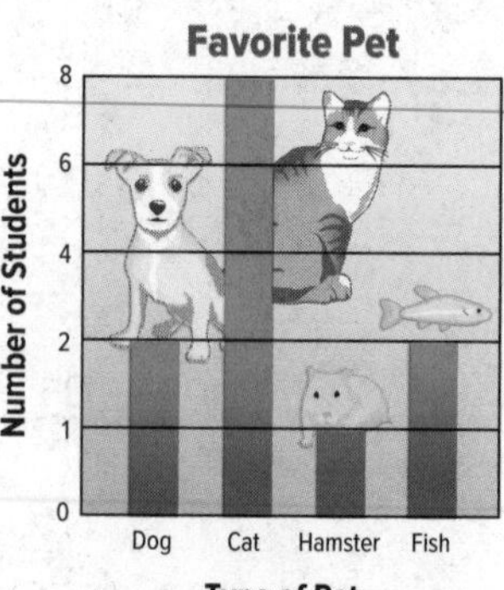

3. The table lists the five largest land vehicle tunnels in the United States. Write a convincing argument for which measure of center you would use to emphasize the average length of the tunnels. (Example 2)

U.S. Vehicle Tunnels	Length (ft)
Anton Anderson Memorial	13,300
E. Johnson Memorial	8,959
Eisenhower Memorial	8,941
Allegheny	6,072
Liberty Tubes	5,920

4. **Building on the Essential Question** Describe at least two ways in which the display of data can influence the conclusions reached.

Rate Yourself!

How well do you understand misleading graphs and statistics? Circle the image that applies.

Clear

Somewhat Clear

Not So Clear

For more help, go online to access a Personal Tutor.

Independent Practice

Go online for Step-by-Step Solutions

1 Which graph could be used to indicate a greater increase in monthly gas prices? Explain. (Example 1)

Graph A

Gas Prices

$3.50

$2.49

$2.30

3
2
1
0

Jan. Feb. Mar.

Graph B

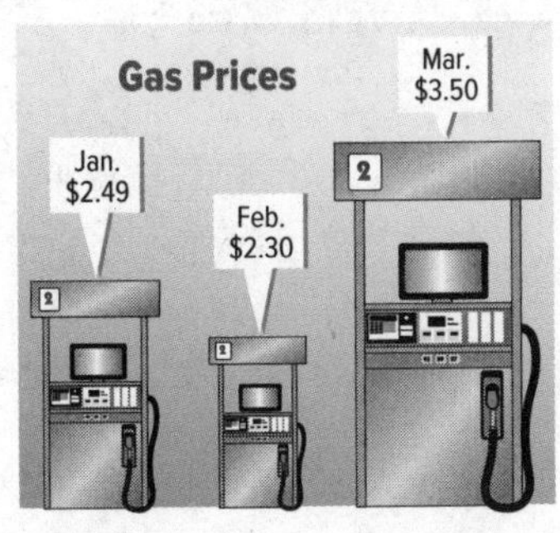

For Exercises 2 and 3, use the table. (Example 2)

2. Find the mean, median, and mode of the data. Which measure might be misleading in describing the average annual number of visitors who visit these sights? Explain.

Annual Sight-Seeing Visitors	
Sight	**Visitors**
Cape Cod	4,600,000
Grand Canyon	4,500,000
Lincoln Memorial	4,000,000
Castle Clinton	4,600,000
Smoky Mountains	10,200,000

3 Which measure would be best if you wanted a value close to the most number of visitors? Explain.

4. MP **Model with Mathematics** Refer to the graphic novel frame below. Which measure of center should the students use?

For Exercises 5 and 6, create a display that would support each argument. The monthly costs to rent an apartment for the last five years are $500, $525, $560, $585, and $605.

5. Rent has remained fairly stable.

6. Rent has increased dramatically.

H.O.T. Problems Higher Order Thinking

7. MP **Reason Inductively** How could the graph you created in Exercise 5 help influence someone's decision to rent the apartment?

8. MP **Persevere with Problems** Does adding values that are much greater or much less than the other values in a set of data affect the median of the set? Give an example to support your answer.

9. MP **Reason Inductively** The circle graph shows the results of a survey. In what way is this graph misleading? Explain.

Favorite Time of Year

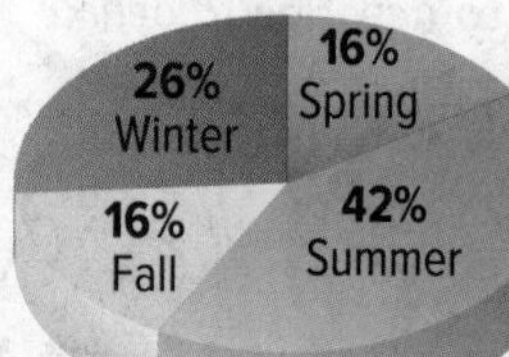

Extra Practice

10. To determine how often his students are tardy, Mr. Kessler considered the attendance record for his first period class. Why is this graph misleading?

Homework Help

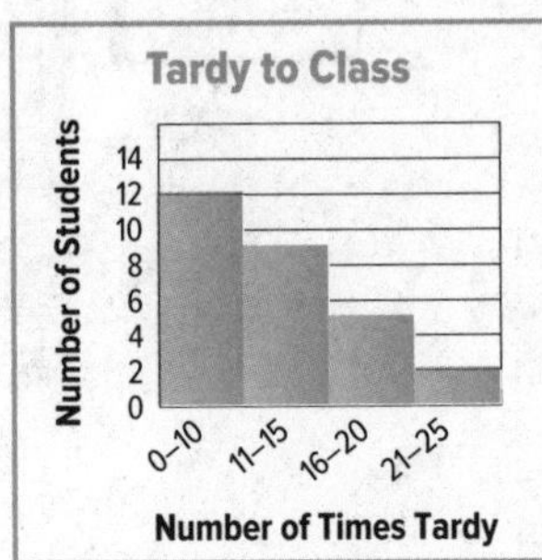

There are not equal intervals on the horizontal axis. So, the height of the bars is not representative of the sample.

11. The graph shows the height of a plant after 9 weeks of growth. Why is the graph misleading?

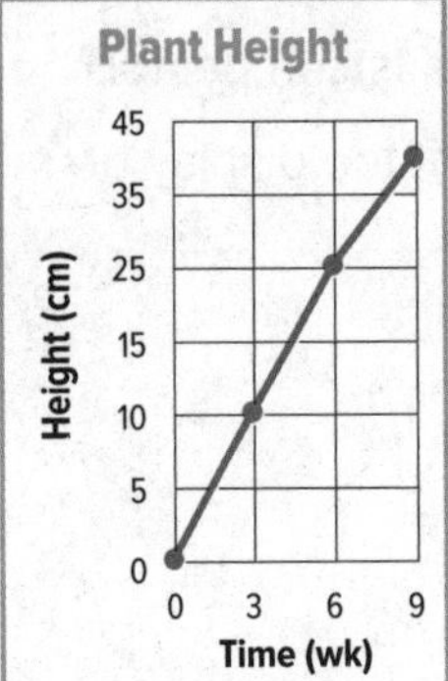

12. MP **Justify Conclusions** Each of the graphs below show the distance Romerio travels on his bike. Romerio wants to impress his friends with the distance he travels. Which graph should he show his friends? Explain.

Graph A

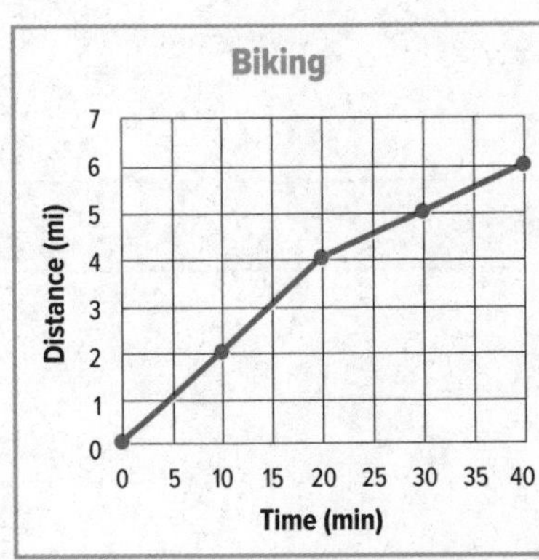

Graph B

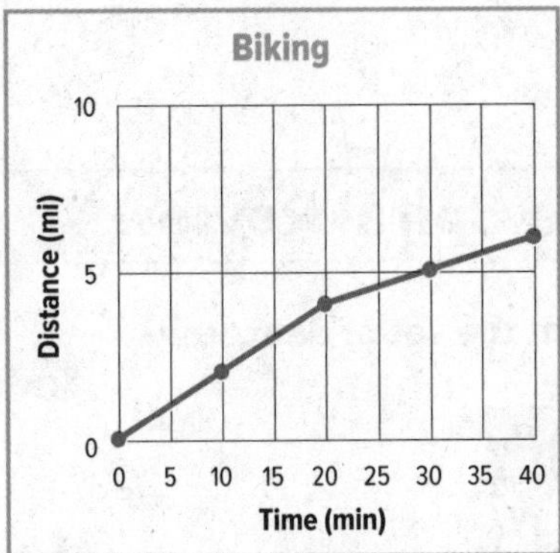

13. The scores Emily received on her math tests were 80, 90, 85, 100, 100, and 84. Why might it be misleading for Emily to say that most of the time she receives a score of 100?

Power Up! Common Core Test Practice

14. Phones For All uses the display shown at the right to compare the number of minutes that they offer per month versus their competitor.

a. How many more minutes per month does Phones For All offer than its competitor? ______

b. Why might the display be misleading?

15. The graph shows the average number of hours each week that certain students spend on extracurricular activities after school. Which of the following describe reasons why the graph may be misleading? Select all that apply.

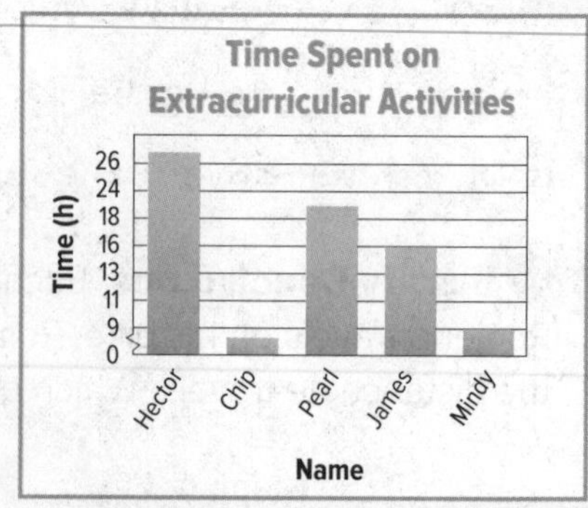

- ☐ The graph does not show the number of hours each student spent on extracurricular activities.
- ☐ The intervals on the vertical scale are inconsistent.
- ☐ The graph's title is misleading.

Common Core Spiral Review

Draw a histogram to represent the set of data. 6.SP.4

16.

Test Scores		
Percent	Tally	Frequency
50–59	|	1
60–69	||	2
70–79	||||	4
80–89	𝍸 𝍸 |	11
90–99	𝍸 |||	8

Problem-Solving Investigation
Use a Graph

Content Standards
7.SP.1

Mathematical Practices
1, 3, 4

Case #1 Fishy Waters

Tess recently purchased a saltwater aquarium. She needs to add 1 tablespoon of sea salt for every 5 gallons of water.

Sea Salt Requirements

Tablespoons of Sea Salt	1	2	3	4	5	6
Capacity of Tank (gallons)	5	10	15	20	25	30

How can she use a graph to show the number of tablespoons of salt required for a 50-gallon saltwater fish tank?

Understand *What are the facts?*

You know the number of gallons of the tank. You need to show the number of tablespoons of sea salt.

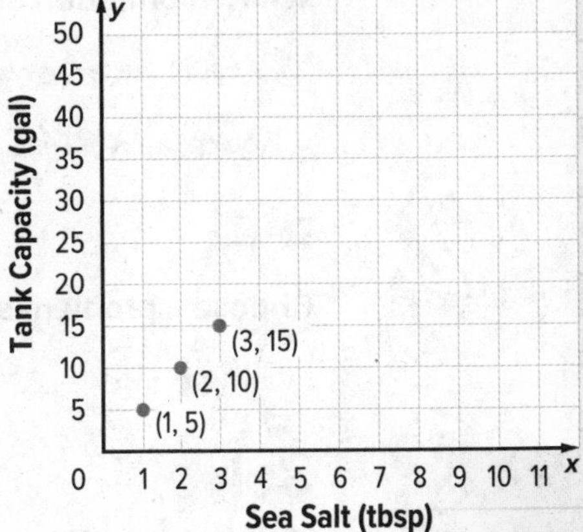

Plan *What is your strategy to solve this problem?*

Organize the rest of the data in a graph so you can easily see any trends.

Solve *How can you apply the strategy?*

Continue the graph until you align horizontally with 50 gallons. Graph a point. What value of sea salt corresponds with the point?

____

Check *Does the answer make sense?*

Find the unit rate of tablespoons of sea salt per gallon of water. Multiply the unit rate by the number of gallons to find the number of tablespoons of sea salt.

$$\frac{0.2 \text{ tbsp salt}}{1 \text{ gal water}} \times \frac{50 \text{ gal water}}{1} = \square \text{ tbsp salt} ✓$$

Analyze the Strategy

Tutor

Make a Prediction Suppose the tank holds 32 gallons. Predict how much sea salt is required.

____

Case #2 Calories

The table shows the average number of Calories burned while sleeping for various numbers of hours. Assume the trend continues.

Make a graph to determine the approximate number of Calories that are burned by sleeping for 10 hours.

Calories Burned While Sleeping	
Hours	Calories
6	386
7	450
8	514
9	579

Understand

Read the problem. What are you being asked to find?

I need to find ________________.

What information do you know?

There is an average of ________ Calories burned while sleeping for 6 hours and 514 Calories burned while sleeping for ________ hours.

Plan

Choose a problem-solving strategy.

I will use the ________________ strategy.

Solve

Use your problem-solving strategy to solve the problem.

Continue the graph until it is aligned vertically with 10 hours. Graph a point. Find what value of Calories corresponds with the point. So, about ________ Calories are burned while sleeping for 10 hours.

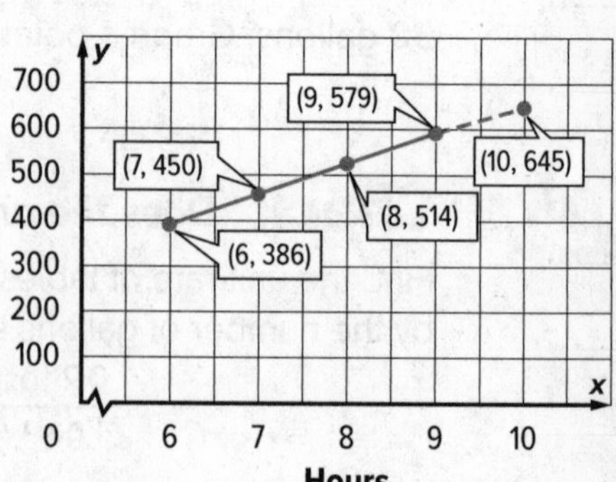

Check

Review the data in the table.

450 − 386 = 64; 514 − 450 = 64; 579 − 514 = 65. 645 − 579 = 66.

So, the answer seems reasonable.

Work with a small group to solve the following cases. Show your work on a separate piece of paper.

Case #3 Postage

The table shows the postage stamp rate from 1999 to 2009.

Make a graph of the data. Predict the year the postage rate will reach $0.52.

Postage Stamp Rates	
Year	**Cost ($)**
1999	0.33
2001	0.34
2002	0.37
2006	0.39
2007	0.41
2008	0.42
2009	0.44

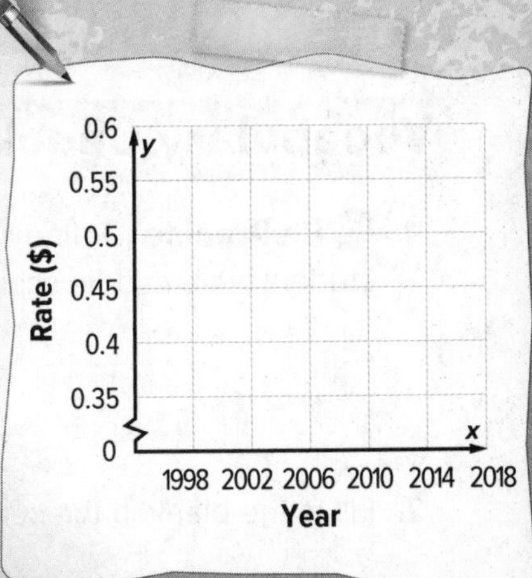

Case #4 Trains

The lengths of various train rides are 4, 1, 2, 3, 6, 2, 3, 2, 5, 8, and 4 hours.

Draw a box plot for the data set. What percent of the train rides are longer than 3 hours?

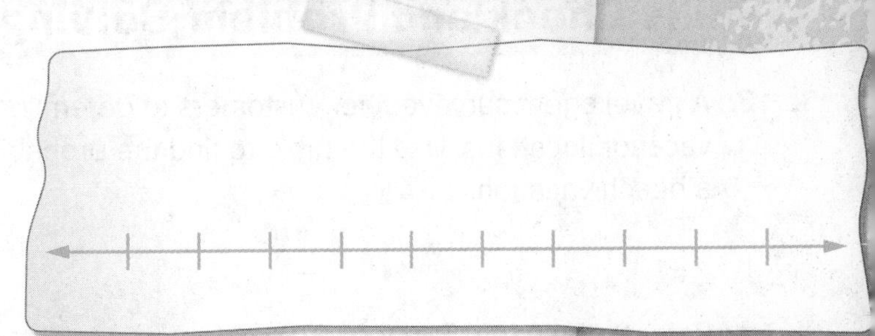

Case #5 Advertising

A local newspaper charges $14.50 for every three lines of a classified ad plus a 7% sales tax.

What is the cost of a 7-line ad? Round to the nearest hundredth.

Case #6 Anatomy

Each human hand has 27 bones. There are 6 more bones in the fingers than in the wrist. There are 3 fewer bones in the palm than in the wrist.

How many bones are in each part of the hand?

Mid-Chapter Check

Vocabulary Check

1. MP **Be Precise** Define *sample*. Give an example of a sample of the students in a middle school. (Lesson 1)

2. Fill in the blank in the sentence below with the correct terms. (Lesson 2)

   ______ and ______ are two types of unbiased samples.

Skills Check and Problem-Solving

3. A travel agent surveyed her customers to determine their favorite vacation locations. Use the table to find the probability of choosing a beach vacation. (Lesson 1)

Vacation Locations	
Location	**Customers**
amusement park	2
beach	11
campground	8
national park	4

4. Refer to the table. Suppose 120 customers are planning vacations. Predict how many will plan a national park vacation. (Lesson 1)

5. The number of points Emerson scored in 5 basketball games is 10, 8, 9, 8, and 30. Why might it be misleading for Emerson to say that she averages 13 points per game? (Lesson 3)

6. MP **Persevere with Problems** An online gaming site conducted a survey to determine the types of games people play online. The results are shown in the circle graph. If 1,500 people participated in the study, how many more would play card games than arcade games? (Lesson 1)

Games People Play Online

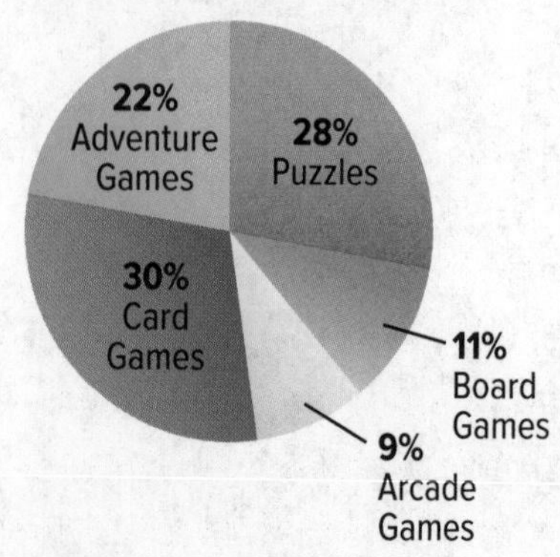

7. An owner of a restaurant wants to conduct a survey about possible menu changes. Give an example of a sampling method that would produce a valid sample. (Lesson 2)

Inquiry Lab

Collect Data

HOW can you use the measures of center and the range to compare two populations?

 **Content Standards**
7.SP.3, 7.SP.4

 **Mathematical Practices**
1, 3, 4

Studies show that teens need around 9 hours of sleep each night to stay healthy.

Hands-On Activity

Step 1 The results of a survey that asked 24 teens how many hours they slept last night are shown below. The teens were split into two populations, male and female.

Males	7	7	6	8	6	8	7	6	7	6	8	6
Females	8	8	7	6	8	7	6	6	7	8	9	7

Step 2 Graph the data for each population on a single line plot.

Number of Hours of Sleep

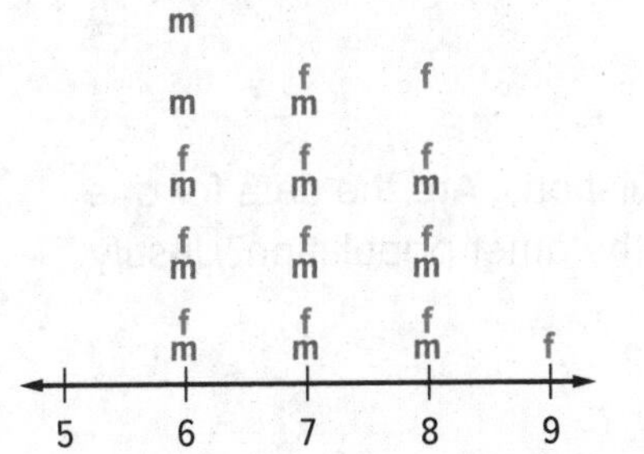

m male
f female

Step 3 Find the measures of center and range for each population.

	Mean	Median	Mode	Range
Males	$6.8\overline{3}$			
Females				

Are the data for males more or less varied than females?

____

Which measure most accurately represents the data of the whole class?

Explain. _______________

____

____

Investigate

Collaborate

1. Work with a partner to design your own survey that meets the following guidelines.
   - Create a survey question that involves two populations. For example, you might want to know about how many hours of sleep per night male students get in your school versus female students. Write your survey question below.

   ____

   - Survey a random sample that is representative of your school's population. Survey at least 25 people. Collect the data and record your results in a table on a separate piece of paper.
   - Create a display of your data. Be sure that the display shows the two populations.

Analyze and Reflect

Collaborate

Work with a partner to complete the exercises below based on the data you collected above.

2. Determine the measures of center (mean, median, and mode) and the range for each population's set of data.

   ____

3. MP **Reason Inductively** Compare the two populations. Are the data for one population more or less varied than the data for the other population? Justify your response.

   ____

4. Describe any other comparative inferences, or conclusions, you can make about differences in the two populations.

   ____

Create

On Your Own

5. Inquiry HOW can you use the measures of center and the range to compare two populations?

   ____

   ____

   ____

Lesson 4

Compare Populations

Real-World Link

Exercise Mr. Singh surveyed the students in his first period gym class to find out how many times they exercised this month. The box plot below shows the results.

How Many Times Have You Exercised This Month?

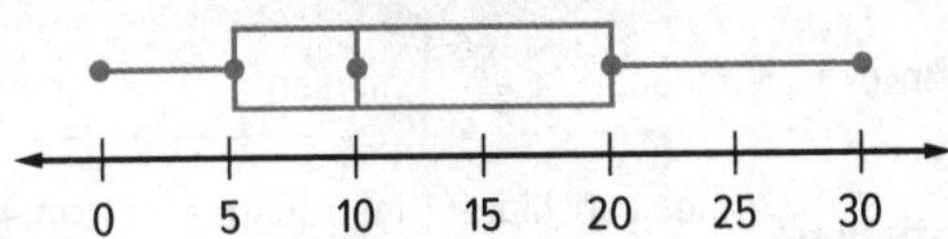

1. Find the following values.

Minimum: ☐ First Quartile: ☐

Maximum: ☐ Third Quartile: ☐

Range: ☐ Interquartile Range: ☐

2. What is the median? What does the median represent?

______

______

______

3. Write a conclusion that you can make from the box plot.

______

______

Essential Question

HOW do you know which type of graph to use when displaying data?

Vocabulary

double box plot
double dot plot

Common Core State Standards

Content Standards
7.SP.4

MP Mathematical Practices
1, 3, 4, 6

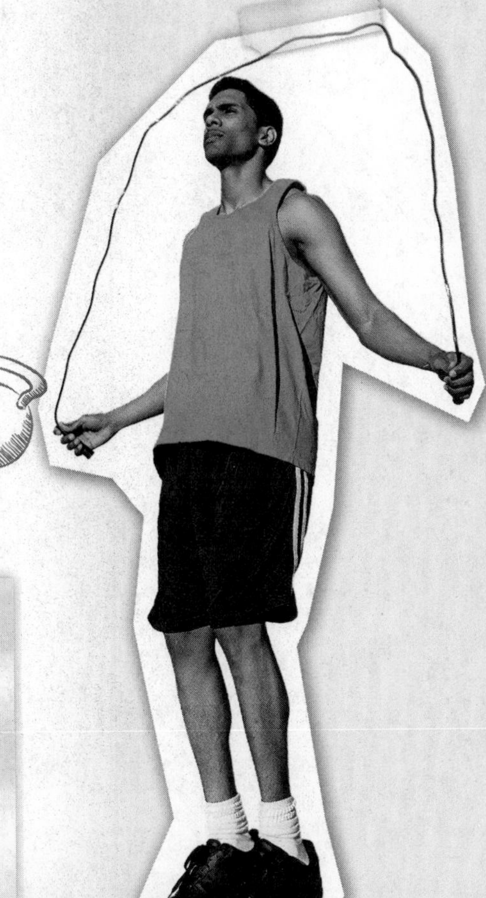

Which **MP Mathematical Practices** did you use? Shade the circle(s) that applies.

① Persevere with Problems
② Reason Abstractly
③ Construct an Argument
④ Model with Mathematics
⑤ Use Math Tools
⑥ Attend to Precision
⑦ Make Use of Structure
⑧ Use Repeated Reasoning

Work Zone

Box plots

A box plot is symmetric if the data are balanced at the center.

Symmetric

Not Symmetric

Compare Two Populations

A **double box plot** consists of two box plots graphed on the same number line. A **double dot plot** consists of two dot plots that are drawn on the same number line. You can draw inferences about two populations in a double box plot or double dot plot by comparing their centers and variations. The centers and variations to use are shown.

Most Appropriate Measures			
	Both sets of data are symmetric.	Neither set of data is symmetric.	Only one set of data is symmetric.
Measure of Center	mean	median	median
Measure of Variation	mean absolute deviation	interquartile range	interquartile range

Real World Example

1. Kacey surveyed a different group of students in her science and math classes. The double box plot shows the results for both classes. Compare their centers and variations. Write an inference you can draw about the two populations.

How Many Times Have You Posted A Blog This Month?

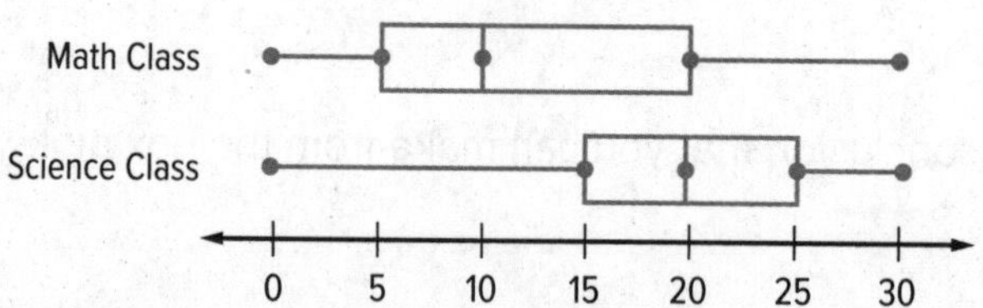

Neither box plot is symmetric. Use the median to compare the centers and the interquartile range to compare the variations.

	Math Class	Science Class
Median	10	20
Interquartile Range	20 − 5, or 15	25 − 15, or 10

Overall, the science students posted more blogs than the math students. The median for the science class is twice the median for the math class. There is a greater spread of data around the median for the math class than the science class.

Got it? Do this problem to find out.

a. The double box plot shows the costs of MP3 players at two different stores. Compare the centers and variations of the two populations. Write an inference you can draw about the two populations.

Cost of MP3 Players ($)

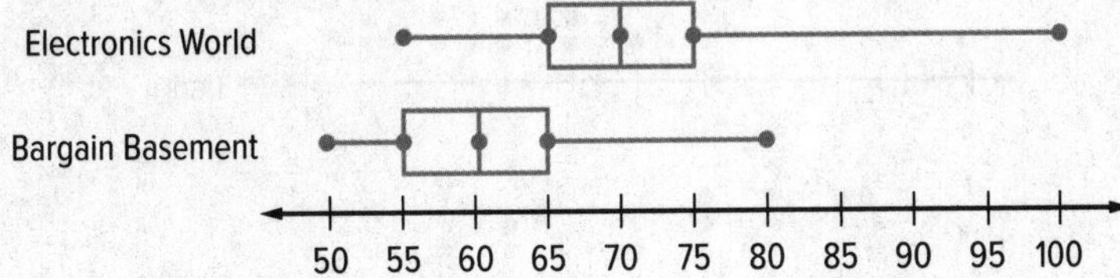

Show your work.

a. ____________

Example

2. The double dot plot below shows the daily high temperatures for two cities for thirteen days. Compare the centers and variations of the two populations. Write an inference you can draw about the two populations.

Daily High Temperatures (°F)

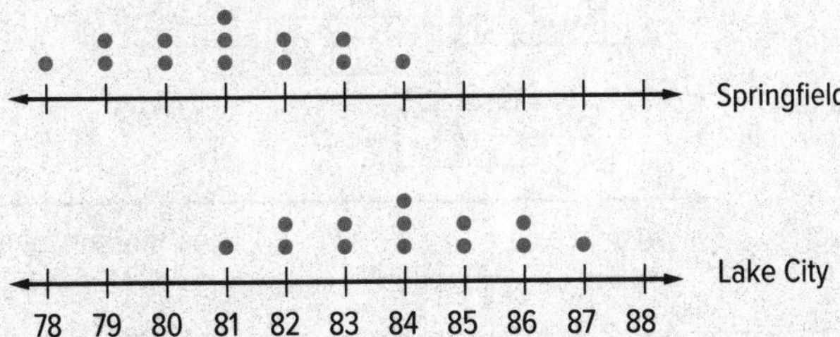

Both dot plots are symmetric. Use the mean to compare the centers and use the mean absolute deviation, rounded to the nearest tenth, to compare the variations.

	Springfield	Lake City
Mean	81	84
Mean Absolute Deviation	1.4	1.4

While both cities have the same variation, or spread of data about each of their means, Lake City has a greater mean temperature than Springfield.

Mean Absolute Deviation

To find the mean absolute deviation, find the absolute values of the differences between each value and the mean. Then find the average of those differences.

b. ____________

Got it? **Do this problem to find out.**

b. The double dot plot shows the number of new E-mails in each of Pedro's and Annika's inboxes for sixteen days. Compare the centers and variations of the two populations. Write an inference you can draw about the two populations.

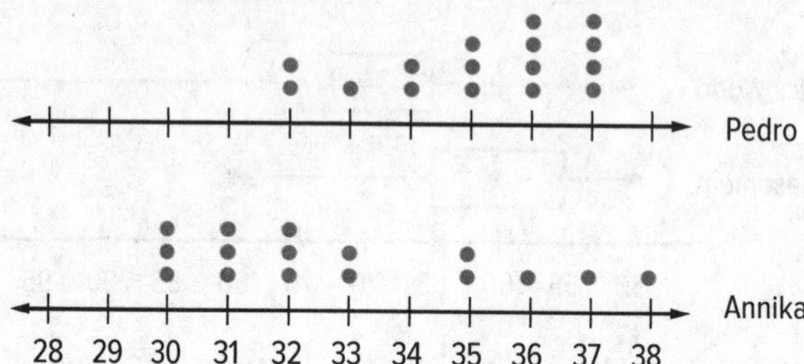

Examples

3. The double box plot shows the daily participants for two zip line companies for one month. Compare the centers and variations of the two populations. Which company has the greater number of daily participants?

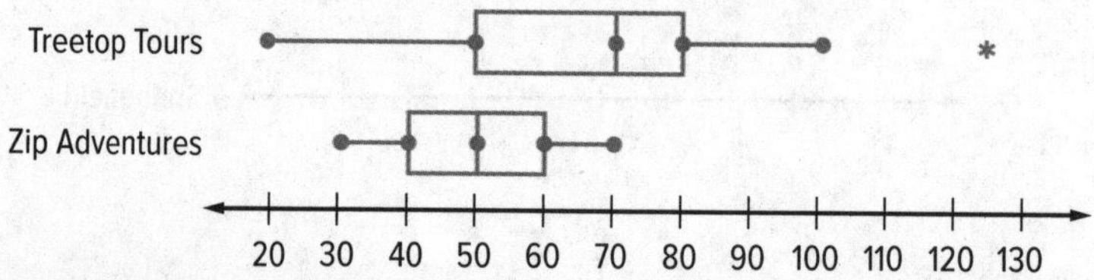

The distribution for Zip Adventures is symmetric, while the distribution for Treetop Tours is not symmetric. Use the median and the interquartile range to compare the populations.

	Treetop Tours	Zip Adventures
Median	70	50
Interquartile Range	30	20

Overall, Treetop Tours has a greater number of daily participants. However, Treetop Tours also has a greater variation, so it is more difficult to predict how many participants they may have each day. Zip Adventures has a greater consistency in their distribution.

STOP and Reflect

What can you tell about the set of data for Zip Adventures by looking at its box plot? Write your answer in the space below.

4. The double dot plot shows Jada's and Angel's number of hours worked in two weeks at their part-time jobs. Compare the centers and variations of the two populations. Who typically works the greater number of hours in a week?

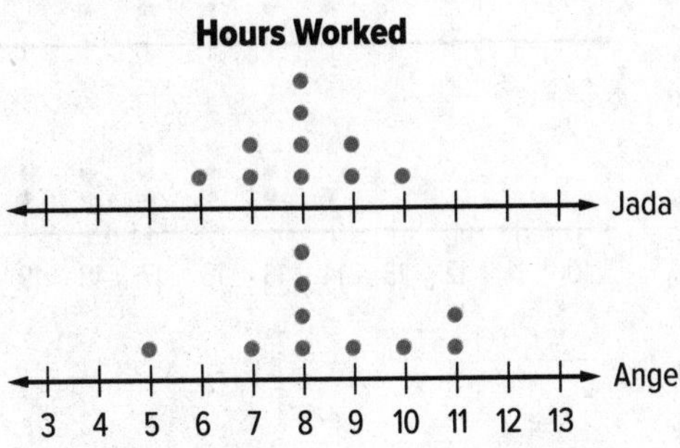

The distribution for Jada's number of hours is symmetric, while the distribution for Angel's number of hours is not symmetric. Use the median and interquartile range to compare the populations.

	Jada	Angel
Median	8	8
Interquartile Range	2	2

The median and interquartile range for both sets of data are the same. However, the interquartile range for Angel's number of hours worked is the difference of 10 and 8, while the interquartile range for Jada's number of hours is the difference of 9 and 7. So, Angel typically works more per week.

Got it? Do this problem to find out.

c. The double dot plot shows Kareem's and Martin's race times for a three-mile race. Compare the centers and variations of the two populations. Which runner is more likely to run a faster race?

c. ____________

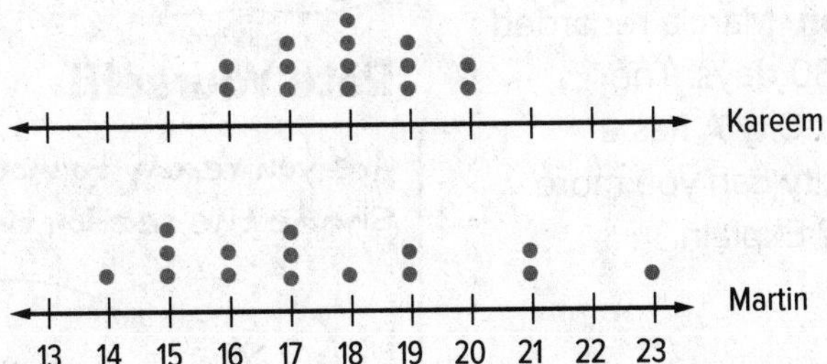

Guided Practice

1. The double dot plot at the right shows the quiz scores out of 20 points for two different class periods. Compare the centers and variations of the two populations. Round to the nearest tenth. Write an inference you can draw about the two populations. (Examples 1 and 2)

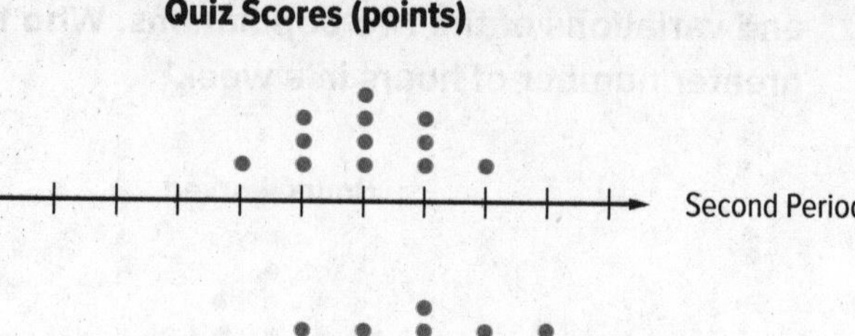

Fifth Period

10 11 12 13 14 15 16 17 18 19 20

2. The double box plot shows the speeds of cars recorded on two different roads in Hamilton County. Compare the centers and variations of the two populations. On which road are the speeds greater? (Examples 3 and 4)

Speed of Cars (mph)

Hayes Road

Jefferson Road

30 35 40 45 50 55 60 65 70 75 80

3. **Building on the Essential Question** Marcia recorded the daily temperatures for two cities for 30 days. The two populations have similar centers, but City A has a greater variation than City B. For which city can you more accurately predict the daily temperature? Explain.

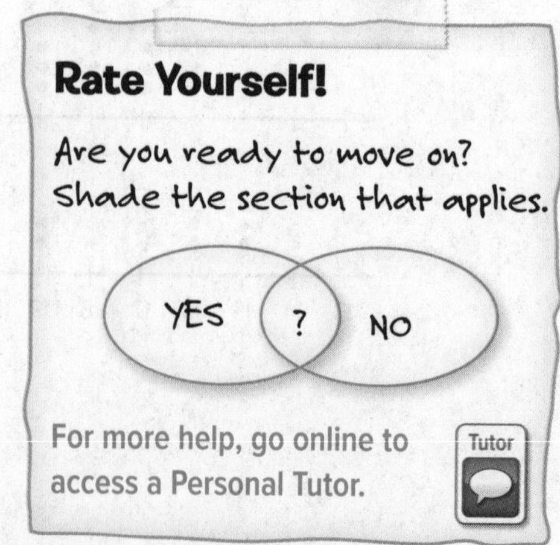

Name ______________________ My Homework ______________________

Independent Practice

Go online for Step-by-Step Solutions

1. Jordan randomly asked customers at two different restaurants how long they waited for a table before they were seated. The double box plot shows the results. Compare their centers and variations. Write an inference you can draw about the two populations. (Examples 1 and 2)

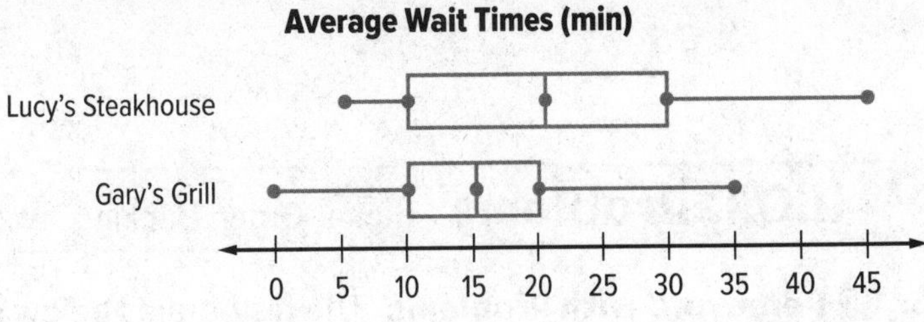

2. The double dot plot shows the times, in hours, for flights of two different airlines flying out of the same airport. Compare the centers and variations of the two populations. Which airline's flights had shorter flight times? (Examples 3 and 4)

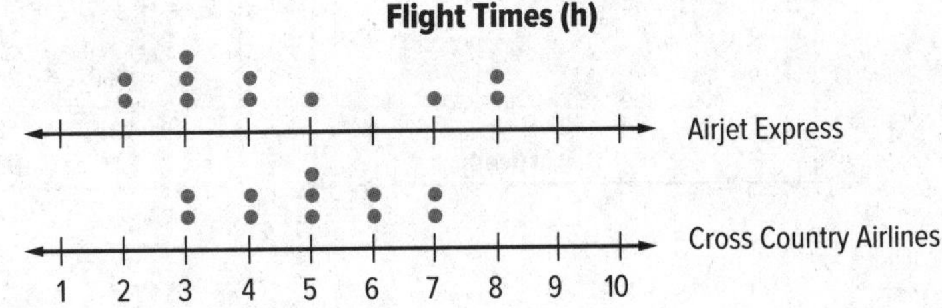

Copy and Solve **Write your answers for Exercise 3 on a separate piece of paper.**

3. **Multiple Representations** For a science project, Mackenzie is measuring the growth of two plants.

Weekly Plant Growth (cm)								
	Week 1	**Week 2**	**Week 3**	**Week 4**	**Week 5**	**Week 6**	**Week 7**	**Week 8**
Plant A	2	3	2	2.5	3.4	3	2.5	3
Plant B	3	2.5	3	3.4	3.2	3.8	3.5	2.5

a. **Numbers** Find the median and interquartile range for both plants.

b. **Graphs** Graph the data using a double box plot.

c. **Words** Write an inference you can draw about the two populations.

4. The median and interquartile range of a set of data is shown. Write a set of data consisting of seven values for the pair of measures.

Median: 6 Interquartile Range: 5

H.O.T. Problems Higher Order Thinking

5. **MP Persevere with Problems** The histograms below show the number of tall buildings for two cities. Explain why you cannot describe the specific location of the centers and spreads of the histograms.

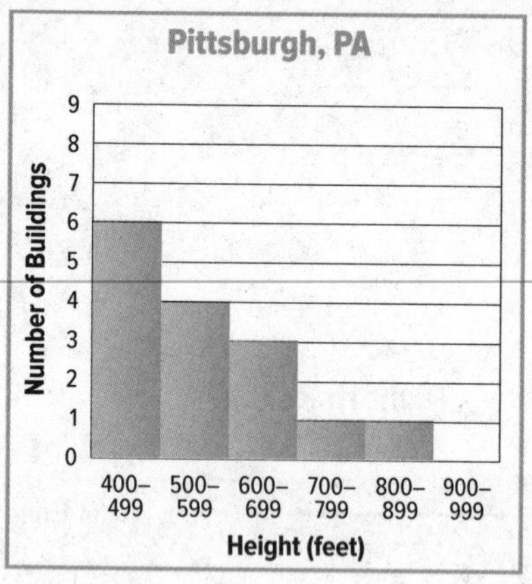

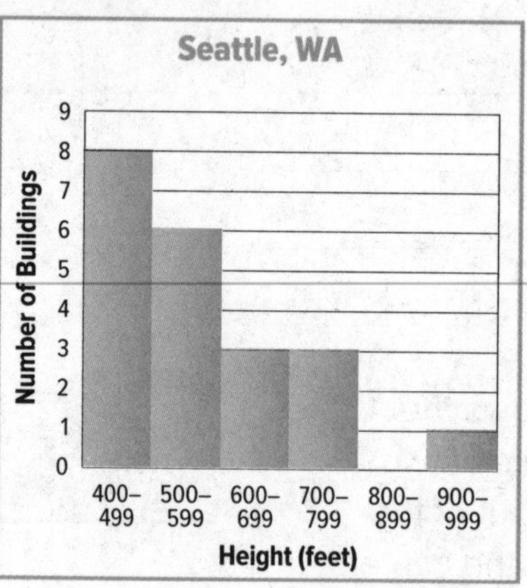

6. **MP Model with Mathematics** Refer to Exercise 1. What is a specific question you could ask about the two populations?

7. **MP Model with Mathematics** Two hockey teams, the Warriors and the Bulldogs, played 15 games each during a month. Both scored a minimum of 0 goals and a maximum of 8 goals. The Bulldogs generally scored fewer goals than the Warriors. Draw a double box plot that could represent the situation.

Name ______ My Homework ______

Extra Practice

8. The double dot plot shows the heights in inches for the girls and boys in Franklin's math class. Compare the centers and variations of the two populations. Round to the nearest tenth. Write an inference you can draw about the two populations.

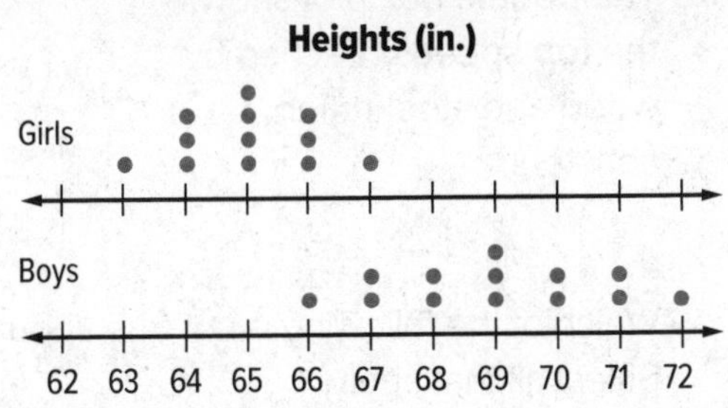

Homework Help

Both plots are symmetric. The girls' heights have a mean of 65 inches with a mean absolute deviation of about 0.8 inch. The boys' heights have a mean of 69 inches with a mean absolute deviation of about 1.4 inches. Overall, the girls' heights are lower than the boys' heights and are also more consistently grouped together.

9. The double box plot shows the number of points scored by the football team for two seasons. Compare the centers and variations of the two populations. During which season was the team's performance more consistent?

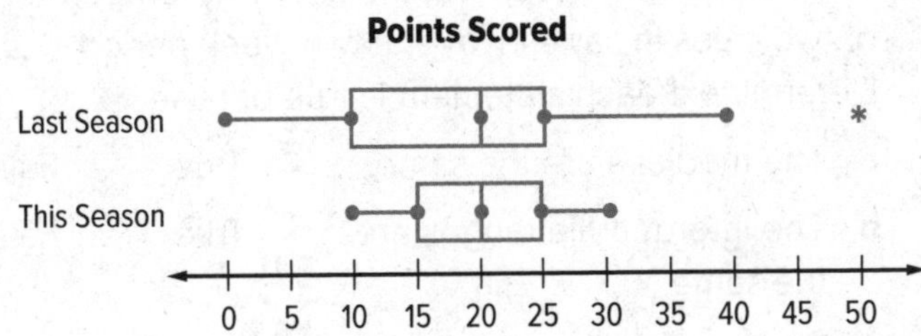

10. The double box plot shows the number of daily visitors to two different parks. Compare the centers and variations of the two populations. In general, which park has more daily visitors?

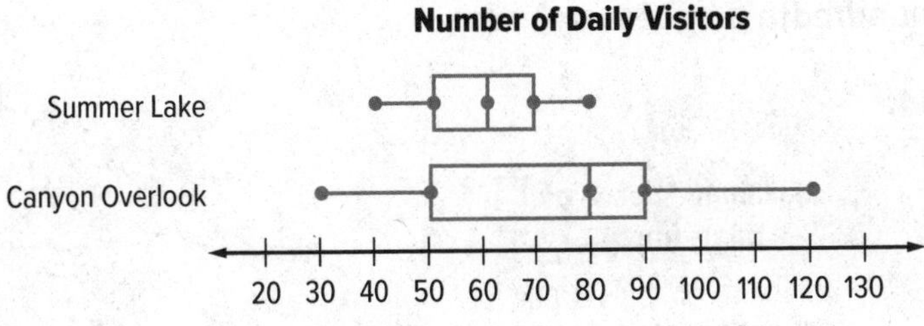

11. MP **Be Precise** The median and interquartile range of a set of data is shown. Write a set of data consisting of seven values for the pair of measures.

Median: 5 Interquartile Range: 5

CCSS Power Up! Common Core Test Practice

12. The double box plot shows the top speeds reached by wood and steel roller coasters.

Speed (miles per hour) of Roller Coasters

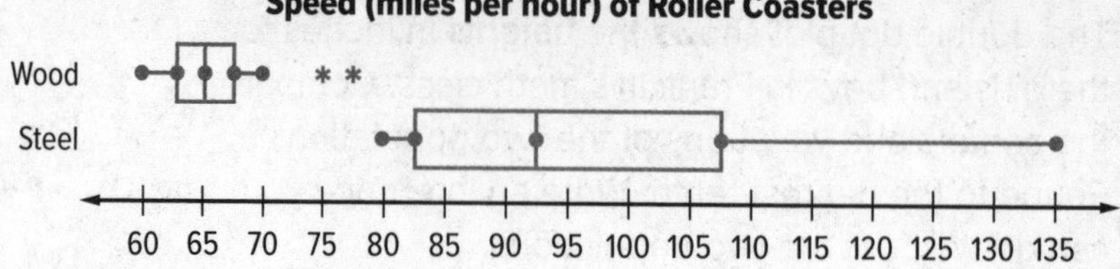

Which of the following is not true about the double box plot? Select all that apply.

- ☐ The data for steel coasters is symmetric.
- ☐ The data for wood coasters is symmetric.
- ☐ The top speed of the fastest steel coaster is 135 miles per hour.
- ☐ The top speed of the slowest wooden coaster is 60 miles per hour.

13. The double dot plot shows the daily low temperatures of two cities in January over a two week period. Determine if each statement is true or false

Daily Low Temperatures (°F)

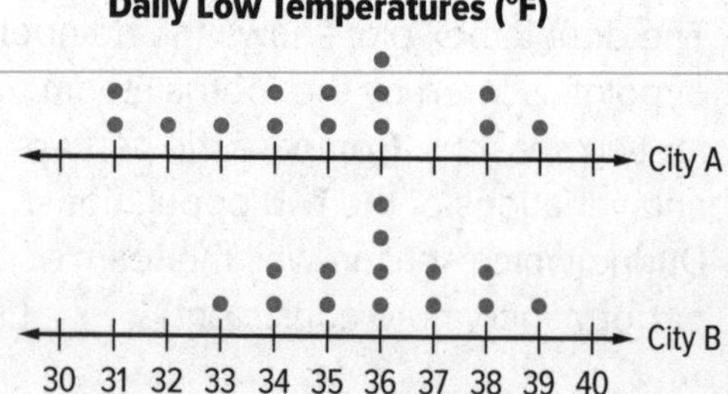

a. The medians are the same. ☐ True ☐ False

b. The interquartile ranges are the same. ☐ True ☐ False

c. The temperatures for City B are more consistent. ☐ True ☐ False

Common Core Spiral Review

Find the mean absolute deviation of each set of data. Round to the nearest hundredth if necessary. 6.SP.5c

14. ______

Maximum Speeds of Boats (mph)			
40	48	58	60
66	72	80	88

15. ______

Populations of Largest U.S. Cities (millions)			
1.3	3.8	1.5	8.4
0.9	1.4	2.3	1.3

16. Refer to the graph in Exercise 2. Describe the shape of the distribution of the data for Airjet Express. 6.SP.5d

17. Refer to the graph in Exercise 10. Describe the shape of the distribution of the data for Canyon Overlook. 6.SP.5d

Inquiry Lab

Visual Overlap of Data Distributions

WHAT does the ratio $\frac{\text{difference in means}}{\text{mean absolute deviation}}$ tell you about how much visual overlap there is between two distributions with similar variation?

Content Standards
7.SP.3

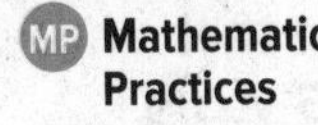

Mathematical Practices
1, 3

A survey was done. The tables below show the number of text messages sent and received daily for two different age groups.

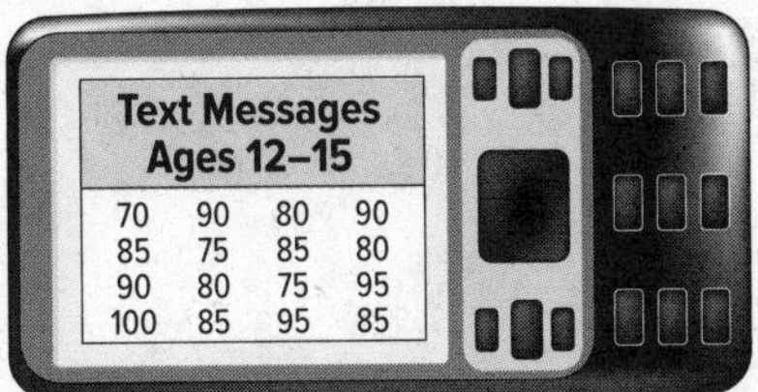

Text Messages Ages 12–15			
70	90	80	90
85	75	85	80
90	80	75	95
100	85	95	85

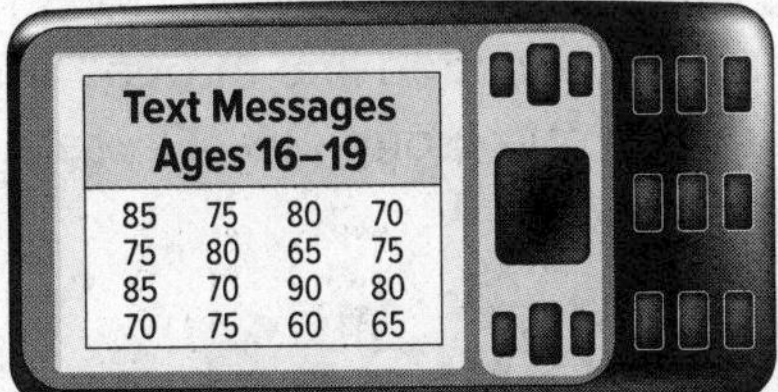

Text Messages Ages 16–19			
85	75	80	70
75	80	65	75
85	70	90	80
70	75	60	65

Hands-On Activity

You can compare two numerical data sets by comparing the shape of their distributions. The **visual overlap** of two distributions with similar variation is a visual demonstration that compares their centers to their variation, or spread.

Step 1 Use a double dot plot to display the data in each table.

Text Messages Sent and Received

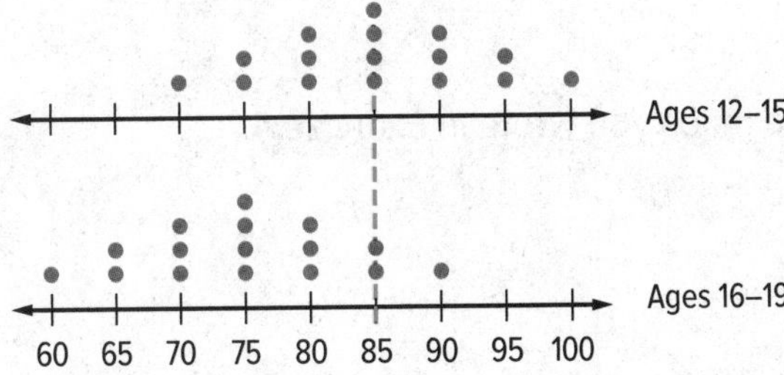

Step 2 Find the mean number of text messages for each age group.

Ages 12–15 mean = ☐ **Ages 16–19 mean =** ☐

Step 3 A red dotted line has been drawn through both dot plots that corresponds to the mean for the age group, 12–15 years. Draw a vertical dotted line through both dot plots that corresponds to the mean for the age group, 16–19 years. The dotted lines show the visual overlap between the centers.

Investigate

Work with a partner. The double dot plot compares the number of text messages sent and received by a third age group to the age group, 12–15 years.

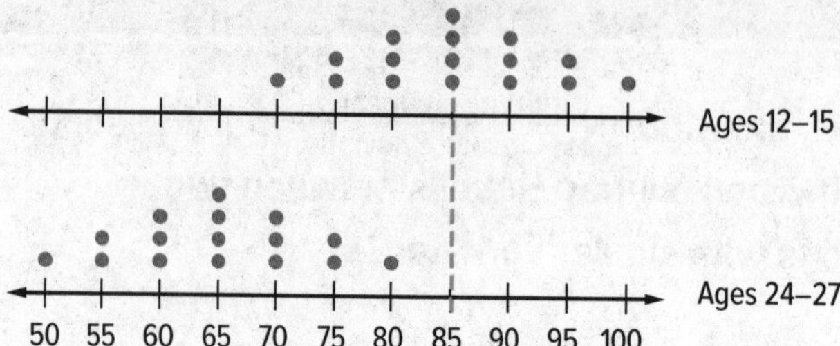

1. What is the mean number of texts for the age group, 24–27 years?

2. In the graph above, draw a vertical dotted line through both dot plots that corresponds to the mean for the age group, 24–27 years.

Analyze and Reflect

Work with a partner.

3. What is the difference between the means of the distributions for the Activity? for Exercise 1?

4. The mean absolute deviation of each distribution is 6.25 texts. For the Activity and Exercise 1, write the difference between the means and the mean absolute deviation as a ratio. Express the ratio as a decimal.

5. MP **Reason Inductively** Compare the ratios you wrote in Exercise 4.

Create

6. Inquiry What does the ratio $\frac{\text{difference in means}}{\text{mean absolute deviation}}$ tell you about how much visual overlap there is between two distributions with similar variation?

Lesson 5

Select an Appropriate Display

Real-World Link

There are many different types of graphs that are used to display all kinds of statistical data. List all of the types of graphs you can think of below.

The graphs below display the total number of pounds of plastic recycled each week during a ten-week period in different ways.

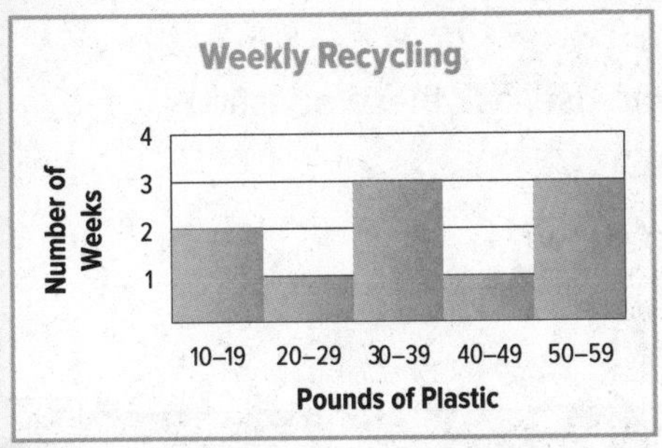

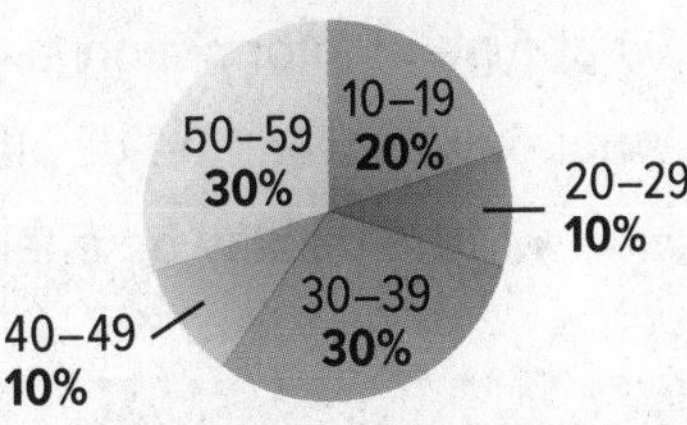

1. On the line below each graph, write the type of graph used.
2. Which display more easily shows the number of weeks the class collected between 30 and 39 pounds of plastic? ______
3. Which display more easily shows the percent of time that 40 to 49 pounds of plastic was recycled? ______

Which MP Mathematical Practices did you use? Shade the circle(s) that applies.

1. Persevere with Problems
2. Reason Abstractly
3. Construct an Argument
4. Model with Mathematics
5. Use Math Tools
6. Attend to Precision
7. Make Use of Structure
8. Use Repeated Reasoning

Essential Question

HOW do you know which type of graph to use when displaying data?

Common Core State Standards

Content Standards
Extension of 7.SP.1

MP Mathematical Practices
1, 3, 4

Key Concept

Select an Appropriate Display

Type of Display	Best Used to...
Bar Graph	show the number of items in specific categories
Box Plot	show measures of variation for a set of data; also useful for very large sets of data
Circle Graph	compare parts of the data to the whole
Double Bar Graph	compare two sets of categorical data
Histogram	show frequency of data divided into equal intervals
Line Graph	show change over a period of time
Line Plot	show frequency of data with a number line

Work Zone

Data Displays

Many situations have more than one appropriate display.

When deciding what type of display to use, ask these questions.

- What type of information is given?
- What do you want the display to show?
- How will the display be analyzed?

Example

1. Select an appropriate display to show the number of boys of different age ranges that participate in athletics.

Since the display will show an interval, a histogram like the one below would be an appropriate display to represent this data.

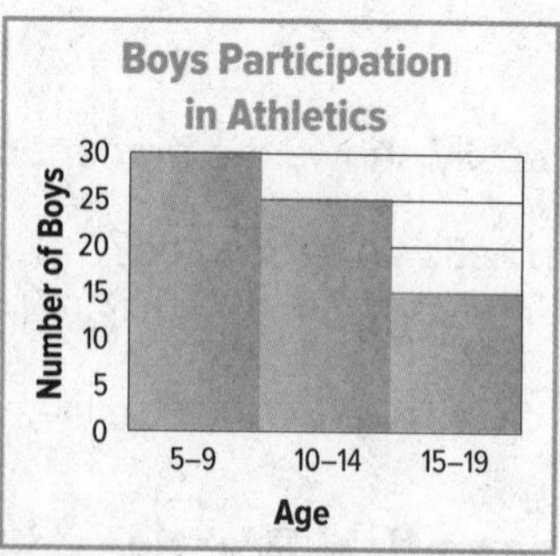

Got it? **Do this problem to find out.**

Show your work.

a. Select an appropriate display for the percent of students in each grade at a middle school.

a. ________

Example

2. Select an appropriate type of display to compare the percent of ethanol production by state. Justify your reasoning. Then construct the display. What can you conclude from your display?

Ethanol Production by State Per Year						
State	Iowa	Nebraska	Illinois	Minnesota	Indiana	Other
Gallons (millions)	3,534	1,665	1,135	1,102	1,074	5,098

You are asked to compare parts to a whole. A circle graph would be an appropriate display.

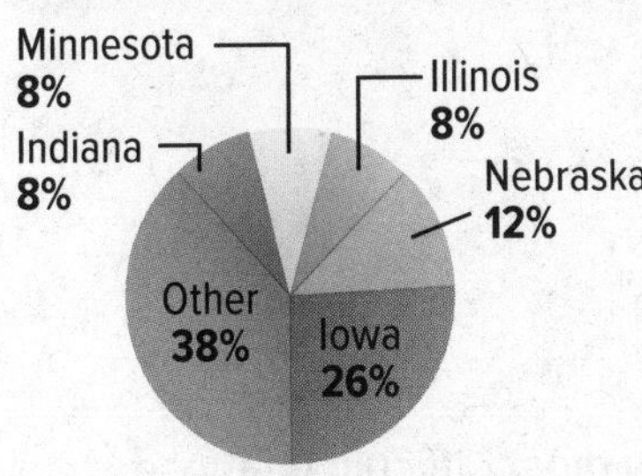

Indiana, Minnesota, and Illinois produce about the same amount of ethanol.

Got it? Do this problem to find out.

b. The table lists the ticket prices for school musicals during recent years. Select an appropriate display to predict the price of a ticket in 2013. Justify your reasoning. Then construct the display. What can you conclude from your display?

Ticket Prices	
Year	**Price ($)**
2009	5.00
2010	5.50
2011	6.50
2012	7.00

Guided Practice

Select an appropriate display for each situation. Justify your reasoning. (Example 1)

1. the number of people who have different kinds of pets

2. the percent of different ways electricity is generated

3. The prices of sandwiches at a restaurant are $4.50, $5.59, $3.99, $2.50, $4.99, $3.75, $2.99, $3.29, and $4.19. Select an appropriate display to determine how many sandwiches range from $3.00 to $3.99. Justify your reasoning. Then construct the display. What can you conclude from your display? (Example 2)

4. A survey asked teens which subject they felt was most difficult. Of those who responded, 25 said English, 39 said social studies, 17 said science, and 19 said other. Construct an appropriate display of the data. Justify your reasoning. Then name one thing you can conclude from the display. (Example 2)

5. **Building on the Essential Question** What are some of the factors to consider when selecting an appropriate display for a set of data?

Rate Yourself!

How confident are you about selecting an appropriate display? Shade the ring on the target.

I'm on target.

I need help.

For more help, go online to access a Personal Tutor.

Tutor

Independent Practice

Go online for Step-by-Step Solutions 

Select an appropriate display for each situation. Justify your reasoning.
(Example 1)

1. the median age of members in a community band

2. the number of students that favor chocolate or vanilla as a frosting

3. Select an appropriate display for the data. Justify your reasoning. Then construct the display. What can you conclude from your display? (Example 2)

Number of Push-ups			
45	35	42	37
44	40	36	42
45	40	42	39
44	43	36	39

4. MP **Model with Mathematics** Refer to the graphic novel frame below. What is the best type of display to use for this data? Explain.

5. Refer to the situations described below.

Situation A	Situation B
the number of customers ages 12–19 compared to all age groups	the number of customers ages 12, 13, 14, 15, and 16 who made a purchase

a. Which situation involves data that is best displayed in a bar graph? Explain your reasoning.

b. Refer to the situation you selected in part **a.** Could you display the data using another type of display? If so, which display? Explain.

H.O.T. Problems Higher Order Thinking

6. MP **Model with Mathematics** Give an example of a data set that would be best represented in a line graph.

7. MP **Reason Inductively** Determine if the following statement is *always*, *sometimes*, or *never* true. Justify your response.

A circle graph can be used to display data from a bar graph.

8. MP **Persevere with Problems** Determine if the following statement is *true* or *false*. Explain your reasoning.

A line plot can be used to display data from a histogram.

9. MP **Reason Inductively** Compare and contrast bar graphs and histograms. Explain when it is appropriate to use a histogram rather than a bar graph.

Extra Practice

MP **Justify Conclusions** **Select an appropriate display for each situation. Justify your reasoning.**

10. the resale value of a person's car over time

line graph; A line graph compares change over time.

11. the percent of people that drink 0, 1, 2, 3, or more than 3 glasses of water a day

____

12. the number of different colored cars at a car dealership

____

13. The circle graph shows the approximate percent of the total volume of each Great Lake.

a. Display the data using another type of display.

Volume of the Great Lakes

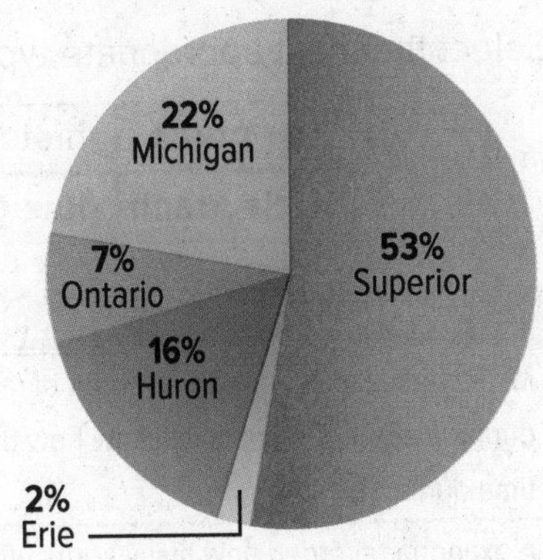

b. Write a convincing argument telling which display is more appropriate.

____

____

____

Copy and Solve **Select an appropriate display for each situation. Then justify your reasoning and construct the display on a separate sheet of paper. What can you conclude from your display?**

14.

Favorite Movies	
Type of Movie	**Number of People**
Comedy	48
Action	17
Drama	5
Horror	2

15.

Age Group	Number of Texts per Day
11–15	25
16–20	23
21–25	17
26–30	10

CCSS Power Up! Common Core Test Practice

16. The number of home runs hit by each player of a professional baseball team is shown in the table.

Home Runs					
10	15	5	10	12	5
12	12	4	5	10	7

Determine if each statement is true or false.

a. A line plot would be most appropriate for showing the frequency of the data on a number line. ☐ True ☐ False

b. A histogram would be most appropriate for showing the frequency of the data in equal intervals. ☐ True ☐ False

c. A circle graph would be most appropriate for showing how the number of home runs changes over time. ☐ True ☐ False

17. Select the most appropriate type of display for each situation.

bar graph	histogram	line plot
circle graph	line graph	

Situation	Type of Graph
Mr. Reynolds measured the amount of rain that had fallen every 15 minutes during a storm. He wants to show how the amount of rain fallen changed over time during the storm.	
Alexandria recorded how many hours of her free time she spent playing sports, watching TV, talking with her friends, or playing video games. She wants to compare the percentages of her free time that she spends on each activity.	
Rachel collected data on how many of her classmates ride the bus, get a ride, or walk to school. She wants to compare how many students are in each category.	

CCSS Common Core Spiral Review

Use the graph to answer Exercises 18–20. The graph shows the number of male and the number of female students that chose certain occupations to research. 6.SP.5

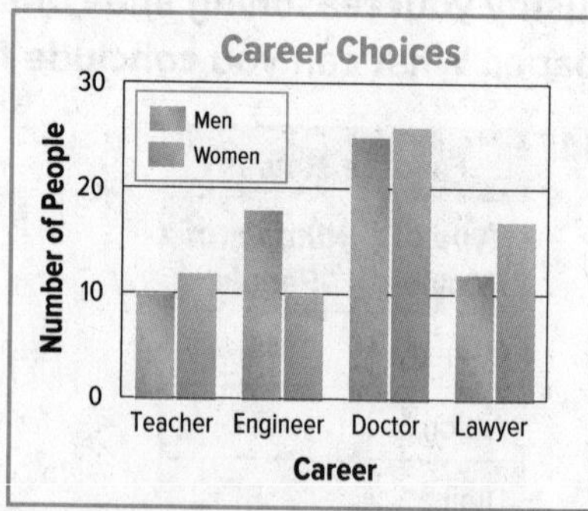

18. About how many people are represented in the graph? ______

19. About how many men and how many women are represented in the graph? ______

20. How many more women chose to research law? ______

21ST CENTURY CAREER in Market Research

Market Research Analyst

Do you think that gathering and analyzing information about people's opinions, tastes, likes, and dislikes sounds interesting? If so, then you should consider a career in market research. Market research analysts help companies understand what types of products and services consumers want. They design Internet, telephone, or mail response surveys and then analyze the data, identify trends, and present their conclusions and recommendations. Market research analysts must be analytical, creative problem-solvers, have strong backgrounds in mathematics, and have good written and verbal communication skills.

Is This the Career for You?

Are you interested in a career as a market research analyst? Take some of the following courses in high school.

- Algebra
- Calculus
- Computer Science
- English
- Statistics

Find out how math relates to a career in Market Research.

MP Keeping Your Eye on the Target Market!

Use the results of the survey in the table below to solve each problem.

1. At Hastings Middle School, 560 of the students use social networking sites. Predict how many of them use the sites to make plans with friends. ______

2. Suppose 17.9 million teens use online social networks. Predict how many will be using the sites to make new friends.

______

3. According to the survey, what percent of a teen's networking site friends are people they regularly see? ______

4. Landon randomly selects a friend from his social networking site. What is the probability that it is someone he never sees in person? Write as a percent. ______

5. Paris wants to leave a message on 8 of her friends' social networking sites. In how many ways can she leave a message on her friends' sites? ______

Survey Results: Teens and Social Networking	
Reason to Use Social Networks	**Percent of Respondents**
Stay in touch with friends	91%
Make plans with friends	72%
Make new friends	49%
Friends on Social Networking Sites	**Average Number**
People who are regularly seen	43
People who are occasionally seen	23
People who are never seen in person	33
Total	99

MP Career Project

It's time to update your career portfolio! Use the Internet or another source to research a career as a market research analyst. Write a paragraph that summarizes your findings.

______

______

______

______

______

______

What skills would you need to improve to succeed in this career?

- ______
- ______
- ______
- ______
- ______

Chapter Review

Check

Vocabulary Check

Complete the crossword puzzle using the vocabulary list at the beginning of the chapter.

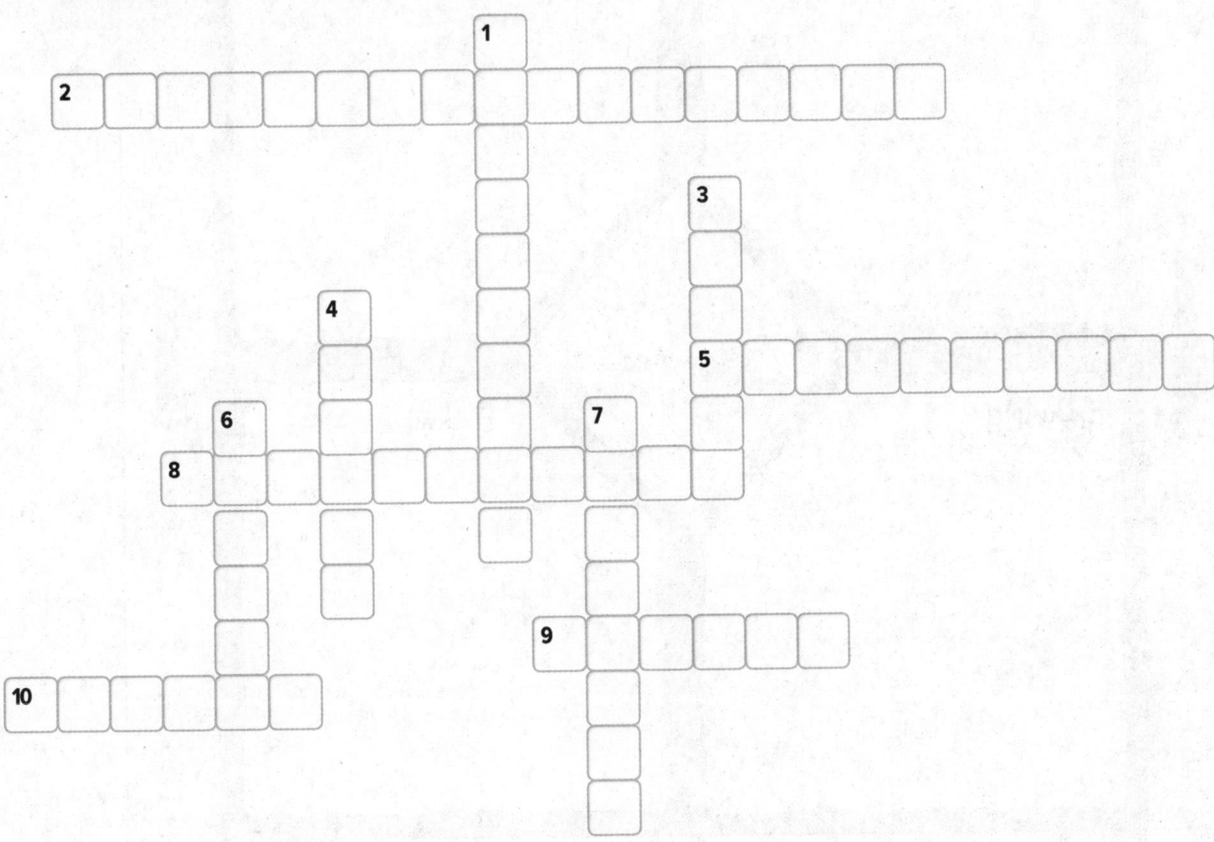

Across

2. sample involving only those who want to participate (two words)

5. the group being studied

8. sample in which members of a population are easily accessed

9. part of a group

10. sample in which one or more parts of the population are favored over other parts

Down

1. random sample in which items are selected according to a specific time or interval

3. random sample in which each item is as likely to be chosen as any other item

4. a method of collecting information

6. two box plots on the same number line

7. sample that represents the entire population

Key Concept Check

Use Your FOLDABLES

Use your Foldable to help review the chapter.

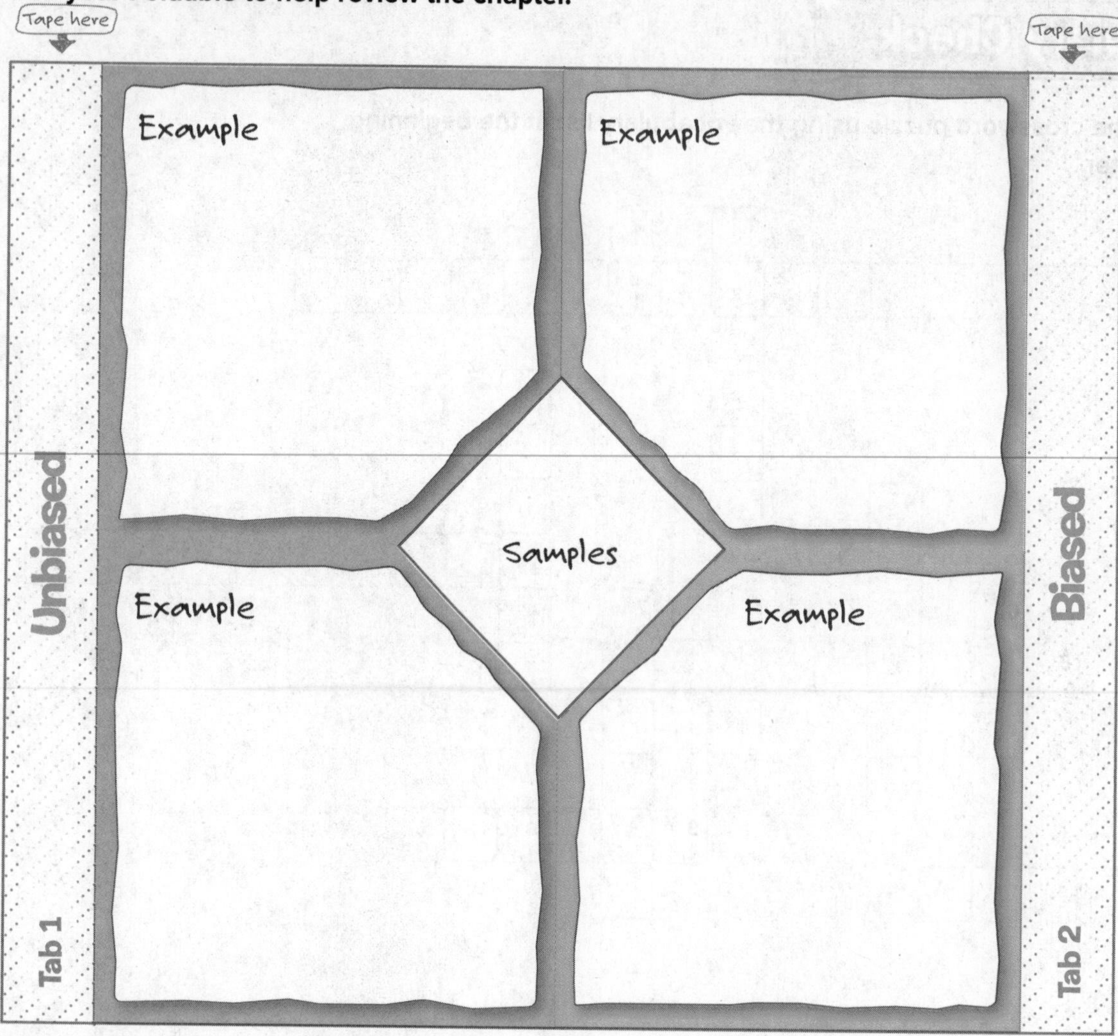

Got it?

Match each phrase with the correct term.

1. a method of collecting information
2. the group being studied
3. when one or more parts of the population is favored
4. a sample that involves only those who want to participate

a. voluntary response sample
b. biased sample
c. survey
d. population
e. convenience sample

Class Evaluation

Mr. Fuentes is analyzing his student's grades over the last three years. He has had approximately 65 students each year. To simplify his analysis, he has decided to use a random sample of data from only ten students from 2012 and 2013. He had no records of the grades from 2014, so he asked students to bring in transcripts. He used the first few transcripts he received for his 2014 data.

	Student Grades									
2012	?	58	86	78	82	79	84	83	82	72
2013	83	85	85	85	87	87	88	90	91	91
2014	79	83	84	88	88	90	93	93	94	95

Write your answers on another piece of paper. Show all of your work to receive full credit.

Part A
Is the above information likely to be a legitimate representation of all the students of each respective year? Is the 2012 data biased or unbiased? Explain your answers.

Part B
In the year 2012, one of the grades is missing. If the mean was 80.1, what is the missing grade?

Part C
Considering the years 2013 and 2014, in which year were the grades the most consistent? Which of the two years had the better scores? What type of display would best show the data? Justify your responses.

Part D
Mr. Fuentes wants to submit the data from one year for a Teacher of the Year award. Choose a year. Could the data be seen as misleading? Explain.

Reflect

Answering the Essential Question

Use what you learned about statistics to complete the graphic organizer.

Bar Graph

Line Graph

Essential Question

HOW do you know which type of graph to use when displaying data?

Double Dot Plot

Double Box Plot

Answer the Essential Question. HOW do you know which type of graph to use when displaying data?

UNIT PROJECT

Watch

Math Genes A Punnett Square is a graphical way to predict the genetic traits of offspring. In this project you will:

- **Collaborate** with your classmates as you research genetics and the Punnet Square.
- **Share** the results of your research in a creative way.
- **Reflect** on why learning mathematics is important.

Complete the activities below and discover the fun you can have with genetics.

Collaborate

Go Online Work with your group to research and complete each activity. You will use your results in the Share section on the following page.

1. Use the Internet to research Punnett Squares and their role in genetics. Write a paragraph describing your findings.

2. Create sample genes for pet traits. Then create a Punnett Square using those traits. Describe what each outcome represents. Include a graph with your explanation.

3. Refer to Exercise 2. How many different genetic outcomes are possible according to your Punnett Square? What is the probability of each outcome occurring?

4. Create three word problems that involve using probability and the Punnett Squares to help answer the questions.

5. Collect two or more genetic-related information samples about students in your class. For example, you can collect data on attached/unattached earlobes. Analyze the data and make a prediction about the genetics of the entire school. Draw an appropriate graph of your results.

With your group, decide on a way to share what you have learned about genetics and Punnett Squares. Some suggestions are listed below, but you can also think of other creative ways to your present your information. Remember to show how you used mathematics to complete each of the activities in this project.

- Create a digital presentation of the facts you learned about genetics.
- Act as a genetic scientist. Write a journal entry that explains your current research on predicting traits passed down from generations.

with Health

Health Literacy Select a health condition or disease and research how genetics may play a part in the disease. Write 1-2 paragraphs explaining how genetics may influence someone's risk of getting the disease and steps that can be taken to reduce the risk factors.

Check out the note on the right to connect this project with other subjects.

6. **Answer the Essential Question** Why is learning mathematics important?

   a. How did what you learned about probability help you to understand why learning mathematics is important?

   b. How did what you learned about statistics help you to understand why learning mathematics is important?

Glossary/Glosario

Go online for the eGlossary.

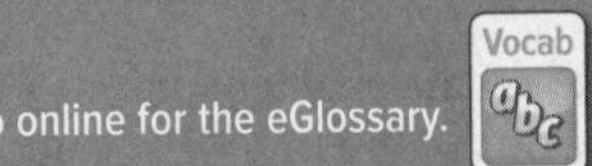

The eGlossary contains words and definitions in the following 13 languages:

Arabic
Bengali
Brazilian Portuguese
Cantonese
English
Haitian Creole
Hmong
Korean
Russian
Spanish
Tagalog
Urdu
Vietnamese

English

Aa

absolute value The distance the number is from zero on a number line.

acute angle An angle with a measure greater than 0° and less than 90°.

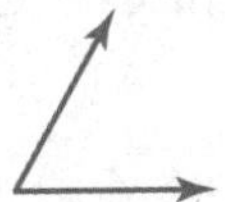

acute triangle A triangle having three acute angles.

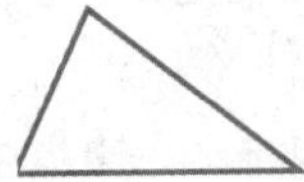

Addition Property of Equality If you add the same number to each side of an equation, the two sides remain equal.

Addition Property of Inequality If you add the same number to each side of an inequality, the inequality remains true.

Additive Identity Property The sum of any number and zero is the number.

additive inverse Two integers that are opposites. The sum of an integer and its additive inverse is zero.

adjacent angles Angles that have the same vertex, share a common side, and do not overlap.

algebra A branch of mathematics that involves expressions with variables.

algebraic expression A combination of variables, numbers, and at least one operation.

Español

valor absoluto Distancia a la que se encuentra un número de cero en la recta numérica.

ángulo agudo Ángulo que mide más de 0° y menos de 90°.

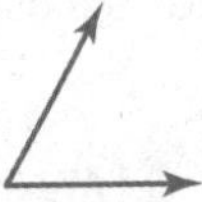

triángulo acutángulo Triángulo con tres ángulos agudos.

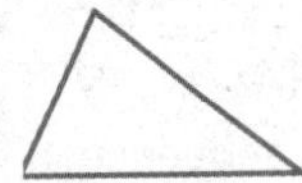

propiedad de adición de la igualdad Si sumas el mismo número a ambos lados de una ecuación, los dos lados permanecen iguales.

propiedad de desigualdad en la suma Si se suma el mismo número a cada lado de una desigualdad, la desigualdad sigue siendo verdadera.

propiedad de identidad de la suma La suma de cualquier número y cero es el mismo número.

inverso aditivo Dos enteros opuestos.

ángulos adyacentes Ángulos que comparten el mismo vértice y un común lado, pero no se sobreponen.

álgebra Rama de las matemáticas que trata de las expresiones con variables.

expresión algebraica Combinación de variables, números y por lo menos una operación.

alternate exterior angles Angles that are on opposite sides of the transversal and outside the parallel lines.

ángulos alternos externos Ángulos en lados opuestos de la trasversal y afuera de las rectas paralelas.

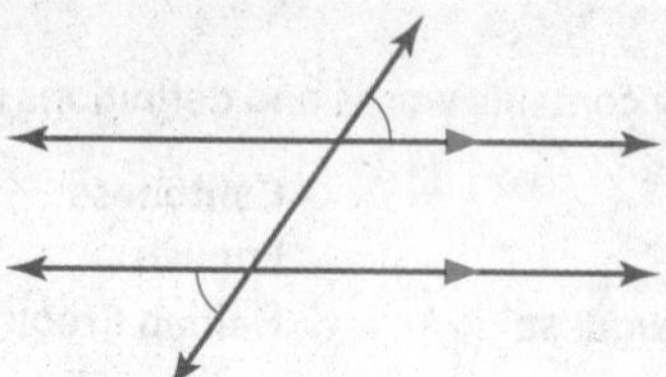

alternate interior angles Angles that are on opposite sides of the transversal and inside the parallel lines.

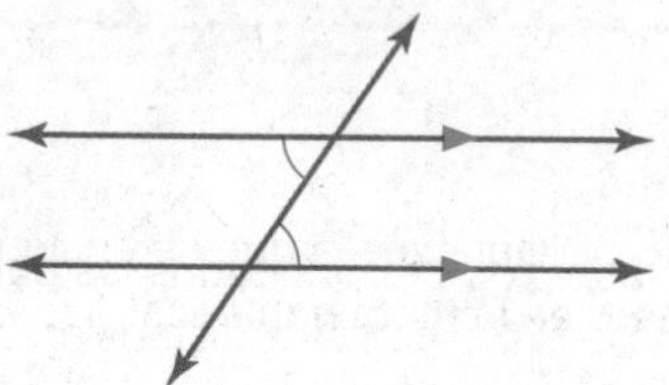

ángulos alternos internos Ángulos en lados opuestos de la trasversal y dentro de las rectas paralelas.

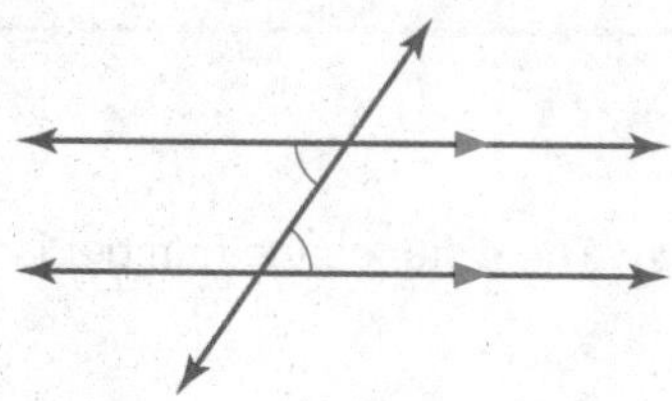

angle Two rays with a common endpoint form an angle. The rays and vertex are used to name the angle.

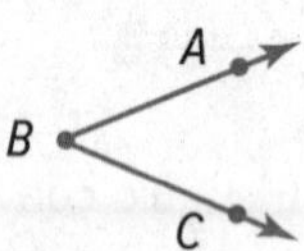

$\angle ABC$, $\angle CBA$, or $\angle B$

ángulo Dos rayos con un extremo común forman un ángulo. Los rayos y el vértice se usan para nombrar el ángulo.

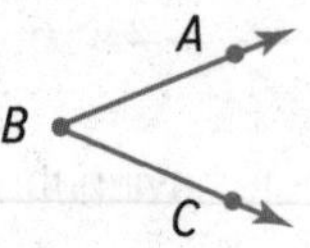

$\angle ABC$, $\angle CBA$ o $\angle B$

arithmetic sequence A sequence in which the difference between any two consecutive terms is the same.

sucesión aritmética Sucesión en la cual la diferencia entre dos términos consecutivos es constante.

Associative Property The way in which numbers are grouped does not change their sum or product.

propiedad asociativa La forma en que se agrupan números al sumarlos o multiplicarlos no altera su suma o producto.

Bb

bar notation In repeating decimals, the line or bar placed over the digits that repeat. For example, $2.\overline{63}$ indicates that the digits 63 repeat.

notación de barra Línea o barra que se coloca sobre los dígitos que se repiten en decimales periódicos. Por ejemplo, $2.\overline{63}$ indica que los dígitos 63 se repiten.

base In a power, the number used as a factor. In 10^3, the base is 10. That is, $10^3 = 10 \times 10 \times 10$.

base En una potencia, el número usado como factor. En 10^3, la base es 10. Es decir, $10^3 = 10 \times 10 \times 10$.

base One of the two parallel congruent faces of a prism.

base Una de las dos caras paralelas congruentes de un prisma.

biased sample A sample drawn in such a way that one or more parts of the population are favored over others.

muestra sesgada Muestra en que se favorece una o más partes de una población.

box plot A method of visually displaying a distribution of data values by using the median, quartiles, and extremes of the data set. A box shows the middle 50% of the data.

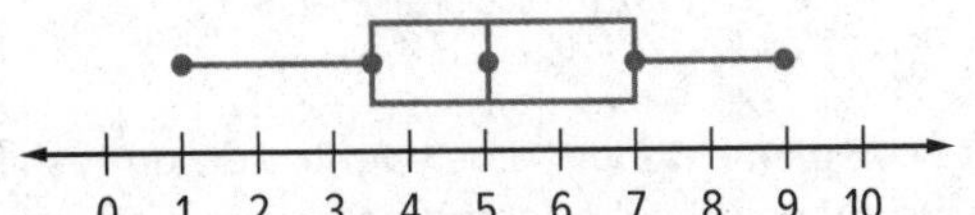

diagrama de caja Un método de mostrar visualmente una distribución de valores usando la mediana, cuartiles y extremos del conjunto de datos. Una caja muestra el 50% del medio de los datos.

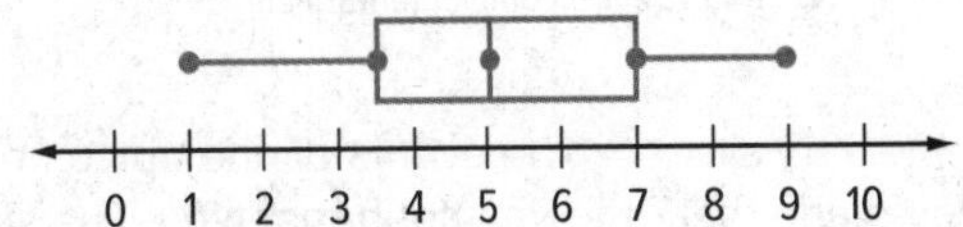

Cc

center The point from which all points on circle are the same distance.

centro El punto desde el cual todos los puntos en una circunferencia están a la misma distancia.

circle The set of all points in a plane that are the same distance from a given point called the center.

círculo Conjunto de todos los puntos de un plano que están a la misma distancia de un punto dado denominado "centro".

circle graph A graph that shows data as parts of a whole. In a circle graph, the percents add up to 100.

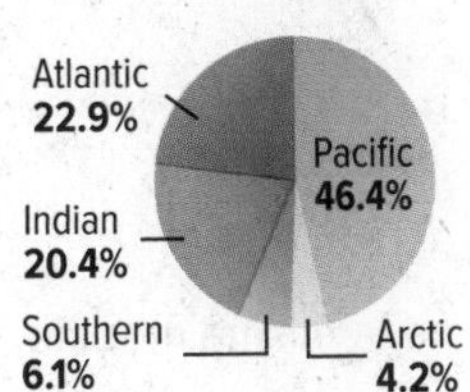

gráfica circular Gráfica que muestra los datos como partes de un todo. En una gráfica circular los porcentajes suman 100.

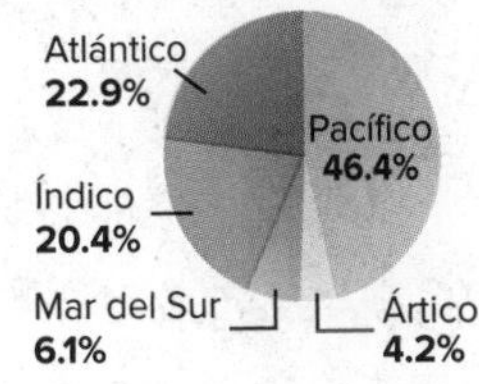

circumference The distance around a circle.

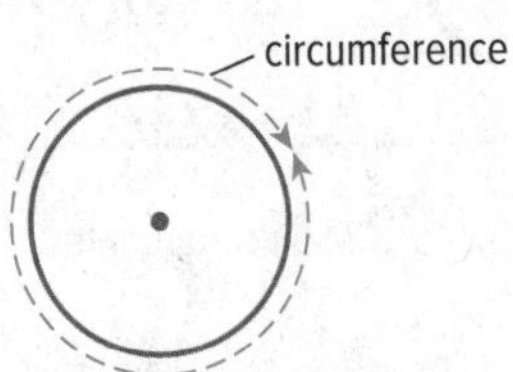

circunferencia Distancia en torno a un círculo.

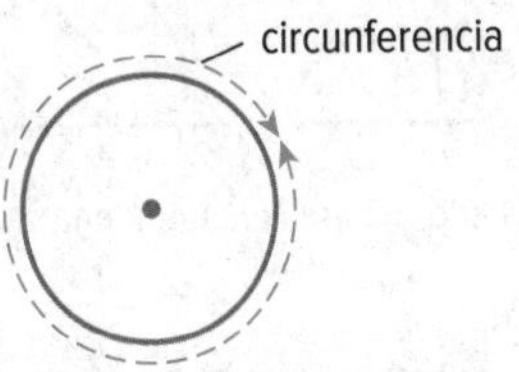

coefficient The numerical factor of a term that contains a variable.

coeficiente El factor numérico de un término que contiene una variable.

common denominator A common multiple of the denominators of two or more fractions. 24 is a common denominator for $\frac{1}{3}$, $\frac{5}{8}$, and $\frac{3}{4}$ because 24 is the LCM of 3, 8, and 4.

común denominador El múltiplo común de los denominadores de dos o más fracciones. 24 es un denominador común para $\frac{1}{3}$, $\frac{5}{8}$ y $\frac{3}{4}$ porque 24 es el mcm de 3, 8 y 4.

Commutative Property The order in which two numbers are added or multiplied does not change their sum or product.

propiedad conmutativa El orden en que se suman o multiplican dos números no altera el resultado.

complementary angles Two angles are complementary if the sum of their measures is 90°.

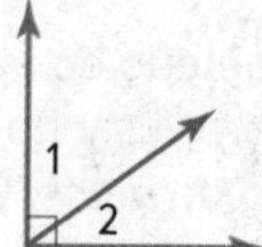

$\angle 1$ and $\angle 2$ are complementary angles.

ángulos complementarios Dos ángulos son complementarios si la suma de sus medidas es 90°.

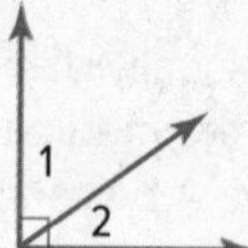

$\angle 1$ y $\angle 2$ son complementarios.

complementary events The events of one outcome happening and that outcome not happening. The sum of the probabilities of an event and its complement is 1 or 100%. In symbols, $P(\text{A}) + P(\textit{not}\ \text{A}) = 1$.

eventos complementarios Los eventos de un resultado que ocurre y ese resultado que no ocurre. La suma de las probabilidades de un evento y su complemento es 1 ó 100. En símbolos $P(\text{A}) + P(\textit{no}\ \text{A}) = 1$.

complex fraction A fraction $\frac{A}{B}$ where A or B are fractions and B does not equal zero.

fracción compleja Una fracción $\frac{A}{B}$ en la cual A o B son fracciones y B no es igual a cero.

composite figure A figure that is made up of two or more three-dimensional figures.

figura compuesta Figura formada por dos o más figuras tridimensionales.

compound event An event consisting of two or more simple events.

evento compuesto Un evento que consiste en dos o más eventos simples.

cone A three-dimensional figure with one circular base connected by a curved surface to a single vertex.

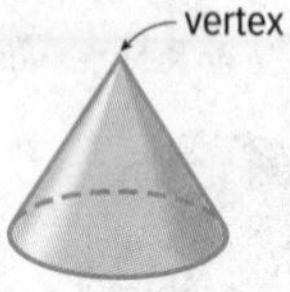

cono Una figura tridimensional con una base circular conectada por una superficie curva para un solo vértice.

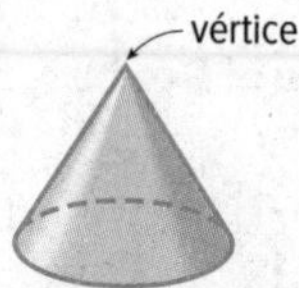

congruent Having the same measure.

congruente Que tiene la misma medida.

congruent angles Angles that have the same measure.

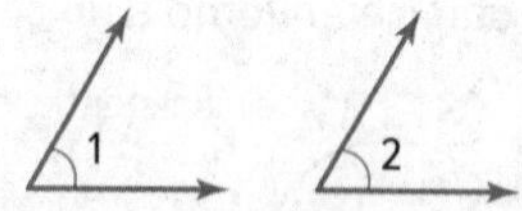

$\angle 1$ and $\angle 2$ are congruent angles.

ángulos congruentes Ángulos que tienen la misma medida.

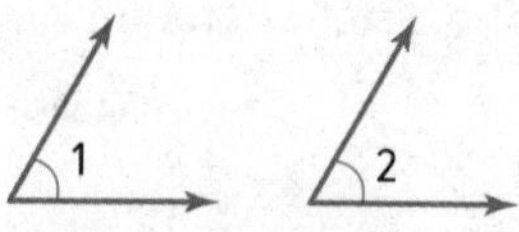

$\angle 1$ y $\angle 2$ son congruentes.

congruent figures Figures that have the same size and same shape and corresponding sides and angles with equal measure.

figuras congruentes Figuras que tienen el mismo tamaño y la misma forma y los lados y los ángulos correspondientes tienen igual medida.

congruent segments Sides with the same length.

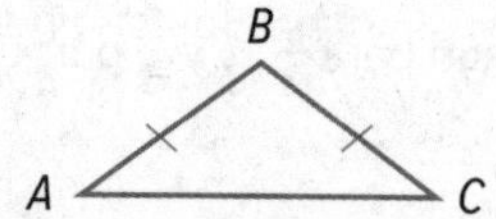

Side $\overline{AB}$ is congruent to side $\overline{BC}$.

segmentos congruentes Lados con la misma longitud.

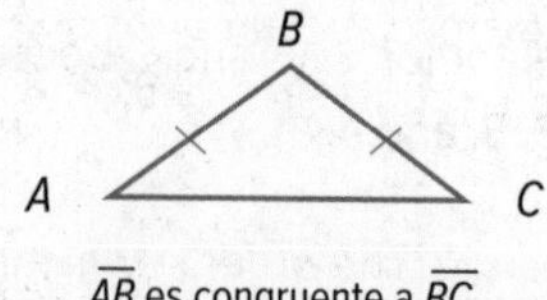

$\overline{AB}$ es congruente a $\overline{BC}$.

constant A term that does not contain a variable.

constant of proportionality A constant ratio or unit rate of two variable quantities. It is also called the constant of variation.

constant of variation The constant ratio in a direct variation. It is also called the constant of proportionality.

constant rate of change The rate of change in a linear relationship.

continuous data Data that take on any real number value. It can be determined by considering what numbers are reasonable as part of the domain.

convenience sample A sample which consists of members of a population that are easily accessed.

coordinate plane A plane in which a horizontal number line and a vertical number line intersect at their zero points. Also called a coordinate grid.

coplanar Lines or points that lie in the same plane.

corresponding angles Angles in the same position on parallel lines in relation to a transversal.

corresponding sides The sides of similar figures that are in the same relative position.

counterexample A specific case which proves a statement false.

cross product The product of the numerator of one ratio and the denominator of the other ratio. The cross products of any proportion are equal.

cross section The cross section of a solid and a plane.

constante Término que no contiene ninguna variable.

constante de proporcionalidad Una razón constante o tasa por unidad de dos cantidades variables. También se llama constante de variación.

constante de variación Una razón constante o tasa por unidad de dos cantidades variables. También se llama constante de proporcionalidad.

razón constante de cambio Tasa de cambio en una relación lineal.

datos continuos Datos que asumen cualquier valor numérico real. Se pueden determinar al considerar qué números son razonables como parte del dominio.

muestra de conveniencia Muestra que incluye miembros de una población fácilmente accesibles.

plano de coordenadas Plano en el cual se han trazado dos rectas numéricas, una horizontal y una vertical, que se intersecan en sus puntos cero. También conocido como sistema de coordenadas.

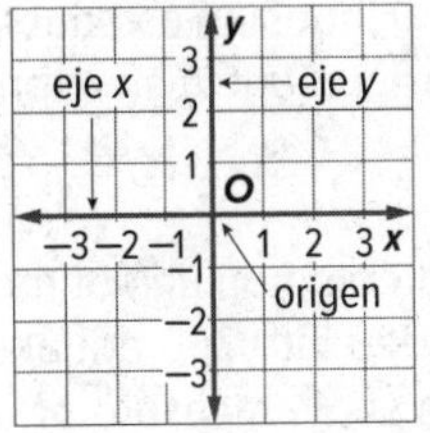

coplanar Líneas o puntos situados en el mismo plano.

ángulos correspondientes Ángulos que están en la misma posición sobre rectas paralelas en relación con la transversal.

lados correspondientes Lados de figuras semejantes que estan en la misma posición.

contraejemplo Caso específico que demuestra la falsedad de un enunciado.

producto cruzado Producto del numerador de una razón por el denominador de la otra razón. Los productos cruzados de cualquier proporción son iguales.

sección transversal Intersección de un sólido con un plano.

cube root One of three equal factors of a number. If $a^3 = b$, then a is the cube root of b. The cube root of 125 is 5 since $5^3 = 125$.

raíz cúbica Uno de tres factores iguales de un número. Si $a^3 = b$, entonces a es la raíz cúbica de b. La raíz cúbica de 125 es 5, dado que $5^3 = 125$.

cubed The product in which a number is a factor three times. Two cubed is 8 because $2 \times 2 \times 2 = 8$.

al cubo El producto de un número por sí mismo, tres veces. Dos al cubo es 8 porque $2 \times 2 \times 2 = 8$.

cylinder A three-dimensional figure with two parallel congruent circular bases connected by a curved surface.

cilindro Una figura tridimensional con dos paralelas congruentes circulares bases conectados por una superficie curva.

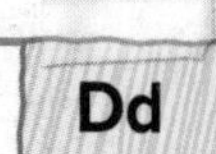

Dd

decagon A polygon having ten sides.

decágono Un polígono con diez lados.

defining a variable Choosing a variable and a quantity for the variable to represent in an expression or equation.

definir una variable El eligir una variable y una cantidad que esté representada por la variable en una expresión o en una ecuacion.

degrees The most common unit of measure for angles. If a circle were divided into 360 equal-sized parts, each part would have an angle measure of 1 degree.

grados La unidad más común para medir ángulos. Si un círculo se divide en 360 partes iguales, cada parte tiene una medida angular de 1 grado.

dependent events Two or more events in which the outcome of one event affects the outcome of the other event(s).

eventos dependientes Dos o más eventos en que el resultado de un evento afecta el resultado de otro u otros eventos.

dependent variable The variable in a relation with a value that depends on the value of the independent variable.

variable dependiente La variable en una relación cuyo valor depende del valor de la variable independiente.

derived unit A unit that is derived from a measurement system base unit, such as length, mass, or time.

unidad derivada Unidad que se deriva de una unidad básica de un sistema de medidas, como la longitud, la masa o el tiempo.

diagonal A line segment that connects two nonconsecutive vertices.

diagonal Segmento de recta que une dos vértices no consecutivos de un polígono.

diameter The distance across a circle through its center.

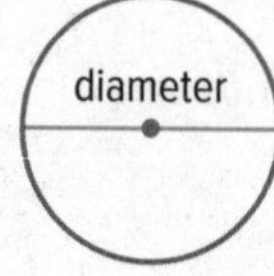

diámetro Segmento que pasa por el centro de un círculo y lo divide en dos partes iguales.

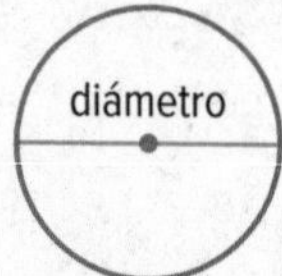

dimensional analysis The process of including units of measurement when you compute.

análisis dimensional Proceso que incluye las unidades de medida al hacer cálculos.

direct variation The relationship between two variable quantities that have a constant ratio.

variación directa Relación entre las cantidades de dos variables que tienen una tasa constante.

discount The amount by which the regular price of an item is reduced.

descuento Cantidad que se le rebaja al precio regular de un artículo.

discrete data When solutions of a function are only integer values. It can be determined by considering what numbers are reasonable as part of the domain.

datos discretos Cuando las soluciones de una función son solo valores enteros. Se pueden determinar considerando qué números son razonables como parte del dominio.

disjoint events Events that cannot happen at the same time.

eventos disjuntos Eventos que no pueden ocurrir al mismo tiempo.

Distributive Property To multiply a sum by a number, multiply each addend of the sum by the number outside the parentheses. For any numbers a, b, and c, $a(b + c) = ab + ac$ and $a(b - c) = ab - ac$.

Example: $2(5 + 3) = (2 \times 5) + (2 \times 3)$ and $2(5 - 3) = (2 \times 5) - (2 \times 3)$

propiedad distributiva Para multiplicar una suma por un número, multiplíquese cada sumando de la suma por el número que está fuera del paréntesis. Sean cuales fuere los números a, b, y c, $a(b + c) = ab + ac$ y $a(b - c) = ab - ac$.

Ejemplo: $2(5 + 3) = (2 \cdot 5) + (2 \cdot 3)$ y $2(5 - 3) = (2 \cdot 5) - (2 \cdot 3)$

Division Property of Equality If you divide each side of an equation by the same nonzero number, the two sides remain equal.

propiedad de igualdad de la división Si divides ambos lados de una ecuación entre el mismo número no nulo, los lados permanecen iguales.

Division Property of Inequality When you divide each side of an inequality by a negative number, the inequality symbol must be reversed for the inequality to remain true.

propiedad de desigualdad en la división Cuando se divide cada lado de una desigualdad entre un número negativo, el símbolo de desigualdad debe invertirse para que la desigualdad siga siendo verdadera.

domain The set of input values for a function.

dominio El conjunto de valores de entrada de una función.

double box plot Two box plots graphed on the same number line.

doble diagrama de caja Dos diagramas de caja sobre la misma recta numérica.

double dot plot A method of visually displaying a distribution of two sets of data values where each value is shown as a dot above a number line.

doble diagrama de puntos Un método de mostrar visualmente una distribución de dos conjuntos de valores donde cada valor se muestra como un punto arriba de una recta numérica.

Ee

edge The line segment where two faces of a polyhedron intersect.

borde El segmento de línea donde se cruzan dos caras de un poliedro.

enlargement An image larger than the original.

ampliación Imagen más grande que la original.

equation A mathematical sentence that contains an equals sign, =, stating that two quantities are equal.

ecuación Enunciado matemático que contiene el signo de igualdad = indicando que dos cantidades son iguales.

equiangular In a polygon, all of the angles are congruent.

equiangular En un polígono, todos los ángulos son congruentes.

equilateral In a polygon, all of the sides are congruent.

equilátero En un polígono, todos los lados son congruentes.

equilateral triangle A triangle having three congruent sides.

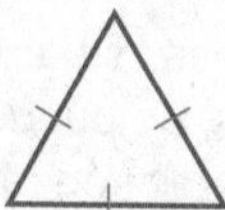

triángulo equilátero Triángulo con tres lados congruentes.

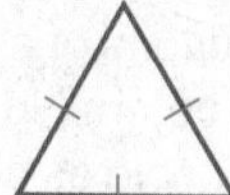

equivalent equations Two or more equations with the same solution.

ecuaciones equivalentes Dos o más ecuaciones con la misma solución.

equivalent expressions Expressions that have the same value.

expresiones equivalentes Expresiones que tienen el mismo valor.

equivalent ratios Two ratios that have the same value.

razones equivalentes Dos razones que tienen el mismo valor.

evaluate To find the value of an expression.

evaluar Calcular el valor de una expresión.

experimental probability An estimated probability based on the relative frequency of positive outcomes occurring during an experiment. It is based on what *actually* occurred during such an experiment.

probabilidad experimental Probabilidad estimada que se basa en la frecuencia relativa de los resultados positivos que ocurren durante un experimento. Se basa en lo que *en realidad* ocurre durante dicho experimento.

exponent In a power, the number that tells how many times the base is used as a factor. In 5^3, the exponent is 3. That is, $5^3 = 5 \times 5 \times 5$.

exponente En una potencia, el número que indica las veces que la base se usa como factor. En 5^3, el exponente es 3. Es decir, $5^3 = 5 \times 5 \times 5$.

exponential form Numbers written with exponents.

forma exponencial Números escritos usando exponentes.

Ff

face A flat surface of a polyhedron.

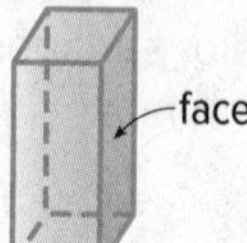

cara Una superficie plana de un poliedro.

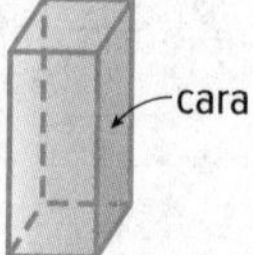

factor To write a number as a product of its factors.

factorizar Escribir un número como el producto de sus factores.

factored form An expression expressed as the product of its factors.

forma factorizada Una expresión expresada como el producto de sus factores.

factors Two or more numbers that are multiplied together to form a product.

factores Dos o más números que se multiplican entre sí para formar un producto.

fair game A game where each player has an equally likely chance of winning.

juego justo Juego donde cada jugador tiene igual posibilidad de ganar.

first quartile For a data set with median *M*, the first quartile is the median of the data values less than *M*.

primer cuartil Para un conjunto de datos con la mediana *M*, el primer cuartil es la mediana de los valores menores que *M*.

formula An equation that shows the relationship among certain quantities.

fórmula Ecuación que muestra la relación entre ciertas cantidades.

function A relationship which assigns exactly one output value for each input value.

función Relación que asigna exactamente un valor de salida a cada valor de entrada.

function rule The operation performed on the input of a function.

regla de función Operación que se efectúa en el valor de entrada.

function table A table used to organize the input numbers, output numbers, and the function rule.

tabla de funciones Tabla que organiza las entradas, la regla y las salidas de una función.

Fundamental Counting Principle Uses multiplication of the number of ways each event in an experiment can occur to find the number of possible outcomes in a sample space.

Principio Fundamental de Contar Este principio usa la multiplicación del número de veces que puede ocurrir cada evento en un experimento para calcular el número de posibles resultados en un espacio muestral.

gram A unit of mass in the metric system equivalent to 0.001 kilogram. The amount of matter an object can hold.

gramo Unidad de masa en el sistema métrico que equivale a 0.001 de kilogramo. La cantidad de materia que puede contener un objeto.

graph The process of placing a point on a number line or on a coordinate plane at its proper location.

graficar Proceso de dibujar o trazar un punto en una recta numérica o en un plano de coordenadas en su ubicación correcta.

gratuity Also known as a tip. It is a small amount of money in return for a service.

gratificación También conocida como propina. Es una cantidad pequeña de dinero en retribución por un servicio.

heptagon A polygon having seven sides.

heptágono Polígono con siete lados.

hexagon A polygon having six sides.

hexágono Polígono con seis lados.

histogram A type of bar graph used to display numerical data that have been organized into equal intervals.

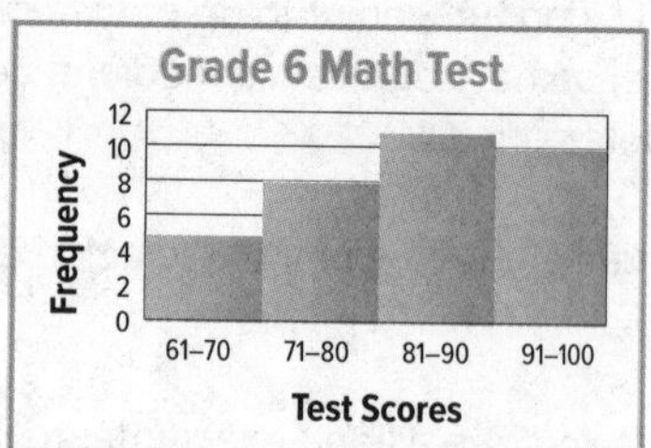

histograma Tipo de gráfica de barras que se usa para exhibir datos que se han organizado en intervalos iguales.

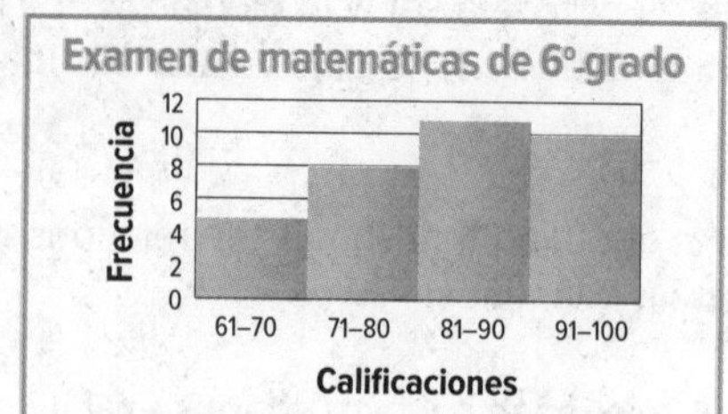

Ii

Identity Property of Zero The sum of an addend and zero is the addend. Example: $5 + 0 = 5$

propiedad de identidad del cero La suma de un sumando y cero es igual al sumando. Ejemplo: $5 + 0 = 5$

independent events Two or more events in which the outcome of one event does not affect the outcome of the other event(s).

eventos independientes Dos o más eventos en los cuales el resultado de uno de ellos no afecta el resultado de los otros eventos.

independent variable The variable in a function with a value that is subject to choice.

variable independiente Variable en una función cuyo valor está sujeto a elección.

indirect measurement Finding a measurement using similar figures to find the length, width, or height of objects that are too difficult to measure directly.

medición indirecta Hallar una medición usando figuras semejantes para calcular el largo, ancho o altura de objetos que son difíciles de medir directamente.

inequality An open sentence that uses $<$, $>$, $\neq$, $\leq$, or $\geq$ to compare two quantities.

desigualdad Enunciado abierto que usa $<$, $>$, $\neq$, $\leq$ o $\geq$ para comparar dos cantidades.

integer Any number from the set $\{..., -4, -3, -2, -1, 0, 1, 2, 3, 4, ...\}$, where ... means continues without end.

entero Cualquier número del conjunto $\{..., -4, -3, -2, -1, 0, 1, 2, 3, 4, ...\}$, donde ... significa que continúa sin fin.

interquartile range A measure of variation in a set of numerical data. It is the distance between first and third quartiles of the data set.

rango intercuartil Una medida de la variación en un conjunto de datos numéricos. Es la distancia entre el primer y el tercer cuartiles del conjunto de datos.

inverse variation A relationship where the product of x and y is a constant k. As x increases in value, y decreases in value, or as y decreases in value, x increases in value.

variación inversa Relación en la cual el producto de x y y es una constante k. A medida que aumenta el valor de x, disminuye el valor de y o a medida que disminuye el valor de y, aumenta el valor de x.

irrational number A number that cannot be expressed as the ratio of two integers.

número irracional Número que no se puede expresar como el razón de dos enteros.

isosceles triangle A triangle having at least two congruent sides.

triángulo isósceles Triángulo que tiene por lo menos dos lados congruentes.

Kk

kilogram The base unit of mass in the metric system. One kilogram equals 1,000 grams.

kilogramo Unidad básica de masa del sistema métrico. Un kilogramo equivale a 1,000 gramos.

Ll

lateral face In a polyhedron, a face that is not a base.

cara lateral En un poliedro, las caras que no forman las bases.

lateral surface area The sum of the areas of all of the lateral faces of a solid.

área de superficie lateral Suma de las áreas de todas las caras de un sólido.

least common denominator (LCD) The least common multiple of the denominators of two or more fractions. You can use the LCD to compare fractions.

mínimo común denominador (mcd) El menor de los múltiplos de los denominadores de dos o más fracciones. Puedes usar el mínimo común denominador para comparar fracciones.

like fractions Fractions that have the same denominators.

fracciones semejantes Fracciones que tienen los mismos denominadores.

like terms Terms that contain the same variables raised to the same power. Example: $5x$ and $6x$ are like terms.

términos semejante Términos que contienen las mismas variables elevadas a la misma potencia. Ejemplo: $5x$ y $6x$ son *términos semejante*.

line graph A type of statistical graph using lines to show how values change over a period of time.

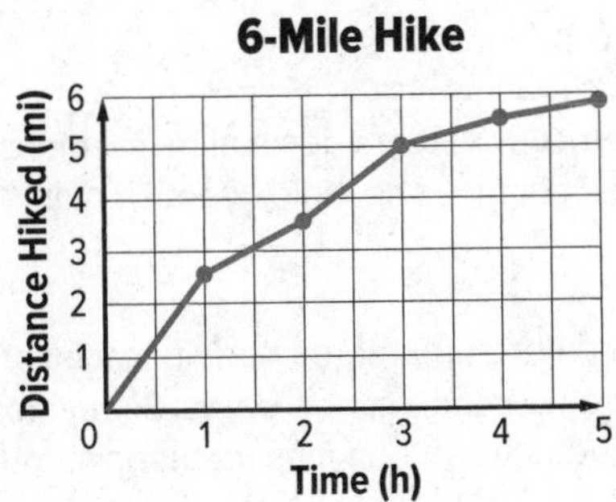

gráfica lineal Tipo de gráfica estadística que usa segmentos de recta para mostrar cómo cambian los valores durante un período de tiempo.

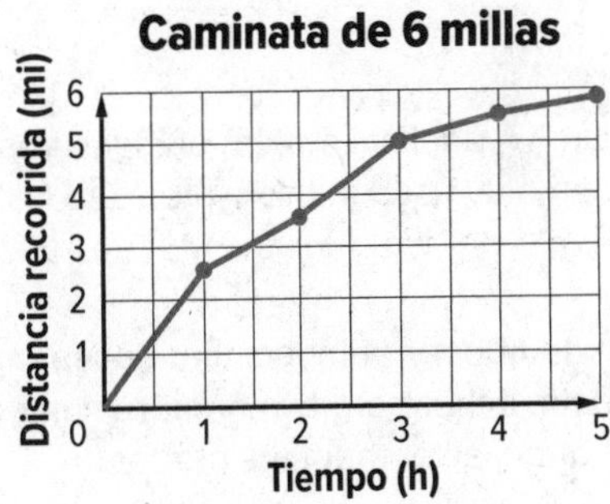

linear expression An algebraic expression in which the variable is raised to the first power, and variables are not multiplied nor divided.

expresión lineal Expresión algebraica en la cual la variable se eleva a la primera potencia.

linear function A function for which the graph is a straight line.

función lineal Función cuya gráfica es una recta.

linear relationship A relationship for which the graph is a straight line.

relación lineal Una relación para la cual la gráfica es una línea recta.

liter The base unit of capacity in the metric system. The amount of dry or liquid material an object can hold.

litro Unidad básica de capacidad del sistema métrico. La cantidad de materia líquida o sólida que puede contener un objeto.

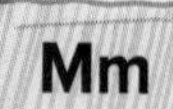

markdown An amount by which the regular price of an item is reduced.

rebaja Una cantidad por la cual el precio regular de un artículo se reduce.

markup The amount the price of an item is increased above the price the store paid for the item.

margen de utilidad Cantidad de aumento en el precio de un artículo por encima del precio que paga la tienda por dicho artículo.

mean The sum of the data divided by the number of items in the data set.

media La suma de los datos dividida entre el número total de artículos en el conjunto de datos.

mean absolute deviation A measure of variation in a set of numerical data, computed by adding the distances between each data value and the mean, then dividing by the number of data values.

desviación media absoluta Una medida de variación en un conjunto de datos numéricos que se calcula sumando las distancias entre el valor de cada dato y la media, y luego dividiendo entre el número de valores.

measures of center Numbers that are used to describe the center of a set of data. These measures include the mean, median, and mode.

medidas del centro Números que se usan para describir el centro de un conjunto de datos. Estas medidas incluyen la media, la mediana y la moda.

measures of variation A measure used to describe the distribution of data.

medidas de variación Medida usada para describir la distribución de los datos.

median A measure of center in a set of numerical data. The median of a list of values is the value appearing at the center of a sorted version of the list—or the mean of the two central values, if the list contains an even number of values.

mediana Una medida del centro en un conjunto de dados númericos. La mediana de una lista de valores es el valor que aparace en el centro de una versíon ordenada de la lista, o la media de dos valores centrales si la lista contiene un número par de valores.

meter The base unit of length in the metric system.

metro Unidad fundamental de longitud del sistema métrico.

metric system A decimal system of measures. The prefixes commonly used in this system are kilo-, centi-, and milli-.

sistema métrico Sistema decimal de medidas. Los prefijos más comunes son kilo-, centi- y mili-.

mode The number or numbers that appear most often in a set of data. If there are two or more numbers that occur most often, all of them are modes.

moda El número o números que aparece con más frecuencia en un conjunto de datos. Si hay dos o más números que ocurren con más frecuencia, todosellos son modas.

monomial A number, variable, or product of a number and one or more variables.

monomio Número, variable o producto de un número y una o más variables.

Multiplication Property of Equality If you multiply each side of an equation by the same nonzero number, the two sides remain equal.

propiedad de multiplicación de la igualdad Si multiplicas ambos lados de una ecuación por el mismo número no nulo, lo lados permanecen iguales.

Multiplication Property of Inequality When you multiply each side of an inequality by a negative number, the inequality symbol must be reversed for the inequality to remain true.

propiedad de desigualdad en la multiplicación Cuando se multiplica cada lado de una desigualdad por un número negativo, el símbolo de desigualdad debe invertirse para que la desigualdad siga siendo verdadera.

Multiplicative Identity Property The product of any number and one is the number.

propiedad de identidad de la multiplicación El producto de cualquier número y uno es el mismo número.

Multiplicative Property of Zero The product of any number and zero is zero.

propiedad del cero en la multiplicación El producto de cualquier número y cero es cero.

multiplicative inverse Two numbers with a product of 1. For example, the multiplicative inverse of $\frac{2}{3}$ is $\frac{3}{2}$.

inverso multiplicativo Dos números cuyo producto es 1. Por ejemplo, el inverso multiplicativo de $\frac{2}{3}$ es $\frac{3}{2}$.

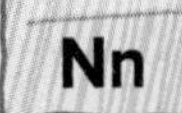

Nn

negative exponent Any nonzero number to the negative *n* power. It is the multiplicative inverse of its *n*th power.

exponente negativo Cualquier número que no sea cero a la potencia negative de *n*. Es el inverso multiplicativo de su *enésimo* potencia.

negative integer An integer that is less than zero. Negative integers are written with a − sign.

entero negativo Número menor que cero. Se escriben con el signo −.

net A two-dimensional figure that can be used to build a three-dimensional figure.

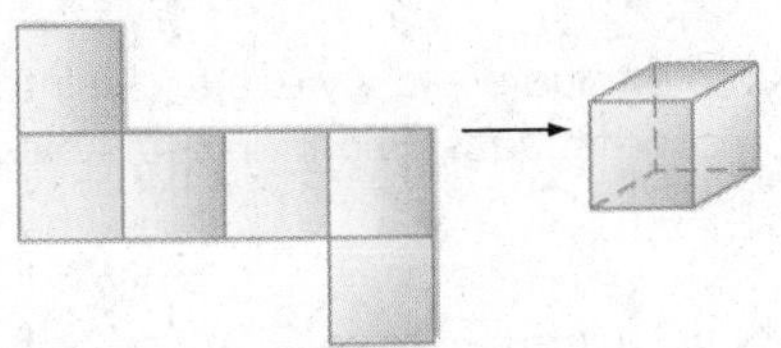

red Figura bidimensional que sirve para hacer una figura tridimensional.

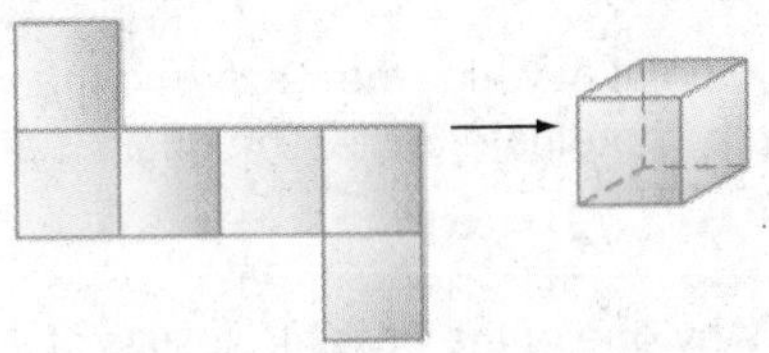

nonagon A polygon having nine sides.

enágono Polígono que tiene nueve lados.

nonlinear function A function for which the graph is *not* a straight line.

nonlinear function Función cuya gráfica *no* es una línea recta.

nonproportional The relationship between two ratios with a rate or ratio that is not constant.

no proporcional Relación entre dos razones cuya tasa o razón no es constante.

numerical expression A combination of numbers and operations.

expresión numérica Combinación de números y operaciones.

Oo

obtuse angle Any angle that measures greater than 90° but less than 180°.

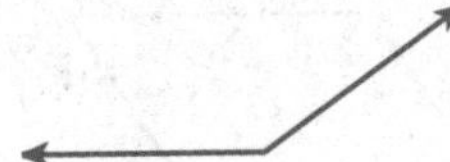

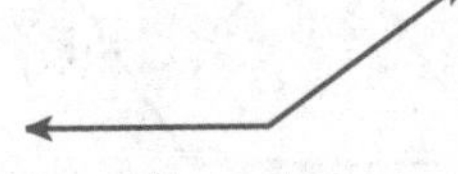

ángulo obtuso Cualquier ángulo que mide más de 90° pero menos de 180°.

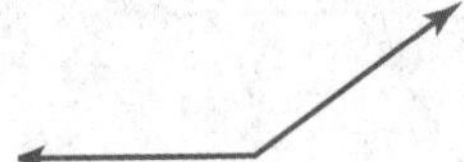

obtuse triangle A triangle having one obtuse angle.

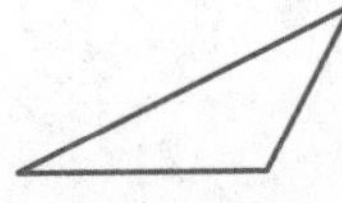

triángulo obtusángulo Triángulo que tiene un ángulo obtuso.

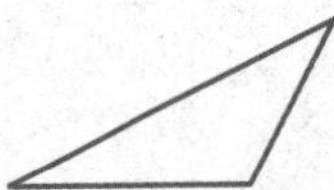

octagon A polygon having eight sides.

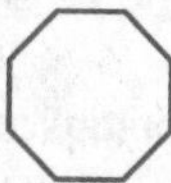

octágono Polígono que tiene ocho lados.

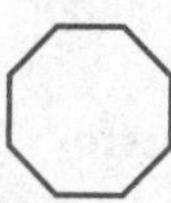

opposites Two integers are opposites if they are represented on the number line by points that are the same distance from zero, but on opposite sides of zero. The sum of two opposites is zero.

opuestos Dos enteros son opuestos si, en la recta numérica, están representados por puntos que equidistan de cero, pero en direcciones opuestas. La suma de dos opuestos es cero.

order of operations The rules to follow when more than one operation is used in a numerical expression.

1. Evaluate the expressions inside grouping symbols.
2. Evaluate all powers.
3. Multiply and divide in order from left to right.
4. Add and subtract in order from left to right.

orden de las operaciones Reglas a seguir cuando se usa más de una operación en una expresión numérica.

1. Primero, evalúa las expresiones dentro de los símbolos de agrupación.
2. Evalúa todas las potencias.
3. Multiplica y divide en orden de izquierda a derecha.
4. Suma y resta en orden de izquierda a derecha.

ordered pair A pair of numbers used to locate a point in the coordinate plane. An ordered pair is written in the form (*x*-coordinate, *y*-coordinate).

par ordenado Par de números que se utiliza para ubicar un punto en un plano de coordenadas. Se escribe de la siguiente forma: (coordenada *x*, coordenada *y*).

origin The point at which the *x*-axis and the *y*-axis intersect in a coordinate plane. The origin is at (0, 0).

origen Punto en que el eje *x* y el eje *y* se intersecan en un plano de coordenadas. El origen está ubicado en (0, 0).

outcome Any one of the possible results of an action. For example, 4 is an outcome when a number cube is rolled.

resultado Cualquiera de los resultados posibles de una acción. Por ejemplo, 4 puede ser un resultado al lanzar un cubo numerado.

outlier A data value that is either much *greater* or much *less* than the median.

valor atípico Valor de los datos que es mucho *mayor* o mucho *menor* que la mediana.

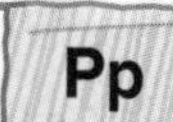

Pp

parallel lines Lines in a plane that never intersect.

rectas paralelas Rectas en un plano que nunca se intersecan.

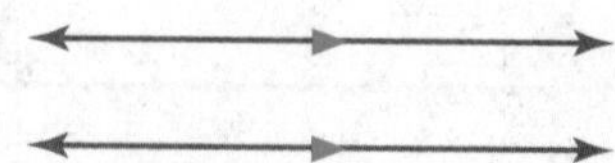

parallelogram A quadrilateral with opposite sides parallel and opposite sides congruent.

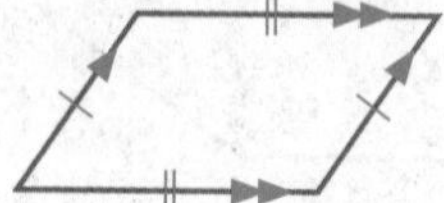

paralelogramo Cuadrilátero cuyos lados opuestos son paralelos y congruentes.

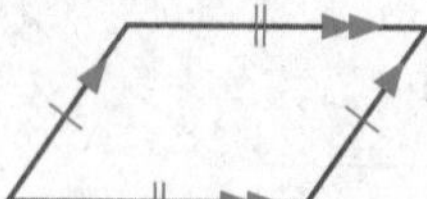

pentagon A polygon having five sides.

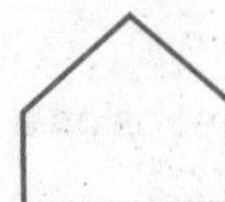

pentágono Polígono que tiene cinco lados.

percent equation An equation that describes the relationship between the part, whole, and percent.

$$\text{part} = \text{percent} \cdot \text{whole}$$

ecuación porcentual Ecuación que describe la relación entre la parte, el todo y el por ciento.

$$\text{parte} = \text{por ciento} \cdot \text{todo}$$

percent error A ratio that compares the inaccuracy of an estimate (amount of error) to the actual amount.

porcentaje de error Una razón que compara la inexactitud de una estimación (cantidad del error) con la cantidad real.

percent of change A ratio that compares the change in a quantity to the original amount.

$$\text{percent of change} = \frac{\text{amount of change}}{\text{original amount}}$$

porcentaje de cambio Razón que compara el cambio en una cantidad a la cantidad original.

$$\text{porcentaje de cambio} = \frac{\text{cantidad del cambio}}{\text{cantidad original}}$$

percent of decrease A negative percent of change.

porcentaje de disminución Porcentaje de cambio negativo.

percent of increase A positive percent of change.

porcentaje de aumento Porcentaje de cambio positivo.

percent proportion One ratio or fraction that compares part of a quantity to the whole quantity. The other ratio is the equivalent percent written as a fraction with a denominator of 100.

$$\frac{\text{part}}{\text{whole}} = \frac{\text{percent}}{100}$$

proporción porcentual Razón o fracción que compara parte de una cantidad a toda la cantidad. La otra razón es el porcentaje equivalente escrito como fracción con 100 de denominador.

$$\frac{\text{parte}}{\text{todo}} = \frac{\text{porcentaje}}{100}$$

perfect squares Numbers with square roots that are whole numbers. 25 is a perfect square because the square root of 25 is 5.

cuadrados perfectos Números cuya raíz cuadrada es un número entero. 25 es un cuadrado perfecto porque la raíz cuadrada de 25 es 5.

permutation An arrangement, or listing, of objects in which order is important.

permutación Arreglo o lista de objetos en la cual el orden es importante.

perpendicular lines Lines that meet or cross each other to form right angles.

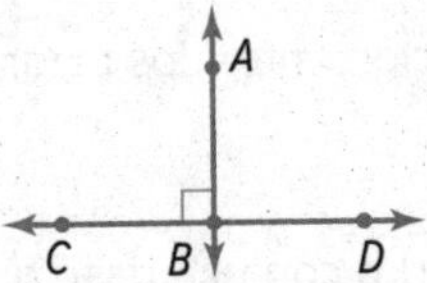

rectas perpendiculares Rectas que al encontrarse o cruzarse forman ángulos rectos.

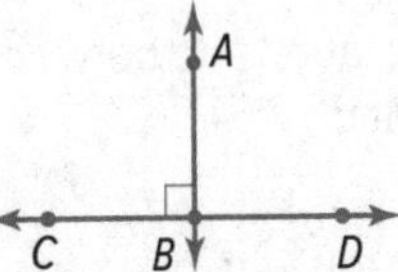

pi The ratio of the circumference of a circle to its diameter. The Greek letter π represents this number. The value of pi is 3.1415926. . . . Approximations for pi are 3.14 and $\frac{22}{7}$.

pi Relación entre la circunferencia de un círculo y su diámetro. La letra griega π representa este número. El valor de pi es 3.1415926. . . . Las aproximaciones de pi son 3.14 y $\frac{22}{7}$.

plane A two-dimensional flat surface that extends in all directions.

plano Superficie bidimensional que se extiende en todas direcciones.

polygon A simple closed figure formed by three or more straight line segments.

polígono Figura cerrada simple formada por tres o más segmentos de recta.

polyhedron A three-dimensional figure with faces that are polygons.

poliedro Una figura tridimensional con caras que son polígonos.

population The entire group of items or individuals from which the samples under consideration are taken.

población El grupo total de individuos o de artículos del cual se toman las muestras bajo estudio.

positive integer An integer that is greater than zero. They are written with or without a + sign.

entero positivo Entero que es mayor que cero; se escribe con o sin el signo +.

powers Numbers expressed using exponents. The power 3^2 is read *three to the second power,* or *three squared.*

potencias Números que se expresan usando exponentes. La potencia 3^2 se lee *tres a la segunda potencia o tres al cuadrado.*

precision The ability of a measurement to be consistently reproduced.

precisión Capacidad que tiene una medición de poder reproducirse consistentemente.

principal The amount of money deposited or borrowed.

capital Cantidad de dinero que se deposita o se toma prestada.

prism A polyhedron with two parallel congruent faces called bases.

prisma Un poliedro con dos caras congruentes paralelas llamadas bases.

probability The chance that some event will happen. It is the ratio of the number of favorable outcomes to the number of possible outcomes.

probabilidad La posibilidad de que suceda un evento. Es la razón del número de resultados favorables al número de resultados posibles.

probability model A model used to assign probabilities to outcomes of a chance process by examining the nature of the process.

modelo de probabilidad Un modelo usado para asignar probabilidades a resultados de un proceso aleatorio examinando la naturaleza del proceso.

properties Statements that are true for any number or variable.

propiedades Enunciados que son verdaderos para cualquier número o variable.

proportion An equation stating that two ratios or rates are equivalent.

proporción Ecuación que indica que dos razones o tasas son equivalentes.

proportional The relationship between two ratios with a constant rate or ratio.

proporcional Relación entre dos razones con una tasa o razón constante.

pyramid A polyhedron with one base that is a polygon and three or more triangular faces that meet at a common vertex.

pirámide Un poliedro con una base que es un polígono y tres o más caras triangulares que se encuentran en un vértice común.

quadrant One of the four regions into which the two perpendicular number lines of the coordinate plane separate the plane.

y-axis
Quadrant II
Quadrant I
O
x-axis
Quadrant III
Quadrant IV

cuadrante Una de las cuatro regiones en que dos rectas numéricas perpendiculares dividen el plano de coordenadas.

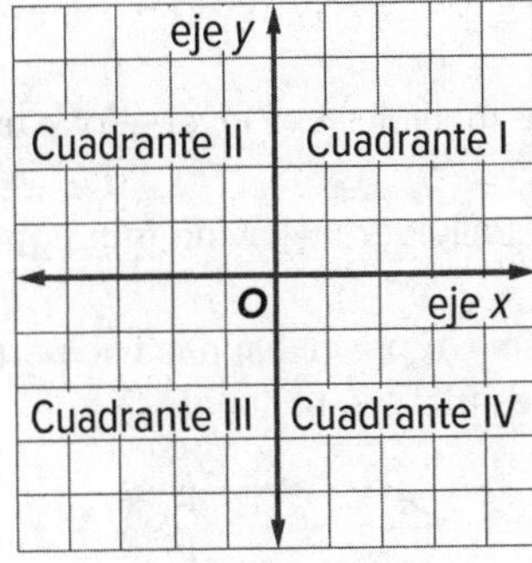

quadrilateral A closed figure having four sides and four angles.

cuadrilátero Figura cerrada que tiene cuatro lados y cuatro ángulos.

quartile A value that divides the data set into four equal parts.

cuartil Valor que divide el conjunto de datos en cuatro partes iguales.

radical sign The symbol used to indicate a nonnegative square root, $\sqrt{\ }$.

signo radical Símbolo que se usa para indicar una raíz cuadrada no negativa, $\sqrt{\ }$.

radius The distance from the center of a circle to any point on the circle.

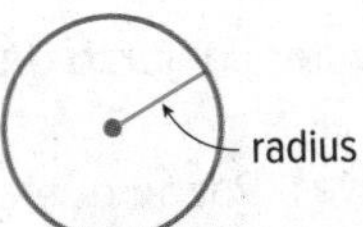

radio Distancia desde el centro de un círculo hasta cualquiera de sus puntos.

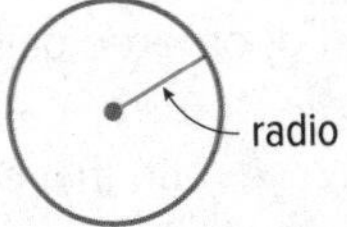

random Outcomes occur at random if each outcome occurs by chance. For example, rolling a number on a number cube occurs at random.

azar Los resultados ocurren aleatoriamente si cada resultado ocurre por casualidad. Por ejemplo, sacar un número en un cubo numerado ocurre al azar.

range The set of output values for a function.

rango Conjunto de valores de salida para una función.

range The difference between the greatest and least data value.

rango La diferencia entre el número mayor y el menor en un conjunto de datos.

rate A ratio that compares two quantities with different kinds of units.

tasa Razón que compara dos cantidades que tienen distintas unidades de medida.

rate of change A rate that describes how one quantity changes in relation to another. A rate of change is usually expressed as a unit rate.

tasa de cambio Tasa que describe cómo cambia una cantidad con respecto a otra. Por lo general, se expresa como tasa unitaria.

rational numbers The set of numbers that can be written in the form $\frac{a}{b}$, where a and b are integers and $b \neq 0$.
Examples: $1 = \frac{1}{1}, \frac{2}{9}, -2.3 = -2\frac{3}{10}$

real numbers A set made up of rational and irrational numbers.

reciprocal The multiplicative inverse of a number.

rectangle A parallelogram having four right angles.

rectangular prism A prism that has two parallel congruent bases that are rectangles.

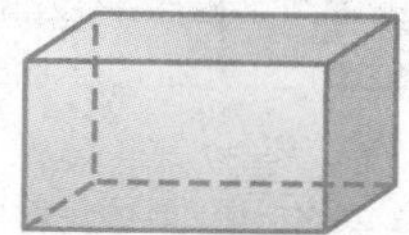

reduction An image smaller than the original.

regular polygon A polygon that has all sides congruent and all angles congruent.

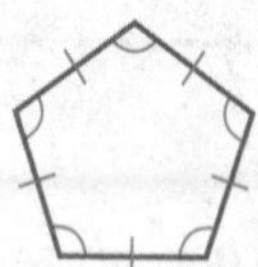

regular pyramid A pyramid whose base is a regular polygon and in which the segment from the vertex to the center of the base is the altitude.

relation Any set of ordered pairs.

relative frequency A ratio that compares the frequency of each category to the total.

repeating decimal The decimal form of a rational number.

rhombus A parallelogram having four congruent sides.

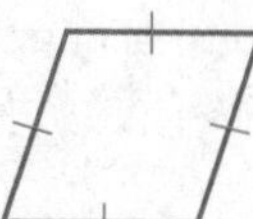

right angle An angle that measures exactly 90°.

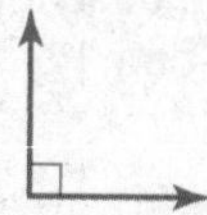

números racionales Conjunto de números que puede escribirse en la forma $\frac{a}{b}$ donde a y b son números enteros y $b \neq 0$.
Ejemplos: $1 = \frac{1}{1}, \frac{2}{9}, -2.3 = -2\frac{3}{10}$

números reales Conjunto de números racionales e irracionales.

recíproco El inverso multiplicativo de un número.

rectángulo Paralelogramo con cuatro ángulos rectos.

prisma rectangular Un prisma con dos bases paralelas congruentes que son rectángulos.

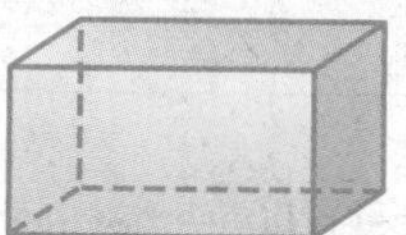

reducción Imagen más pequeña que la original.

polígono regular Polígono con todos los lados y todos los ángulos congruentes.

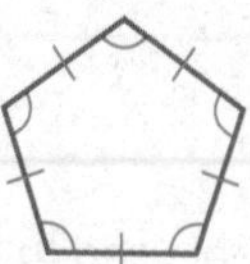

pirámide regular Pirámide cuya base es un polígono regular y en la cual el segmento desde el vértice hasta el centro de la base es la altura.

relación Cualquier conjunto de pares ordenados.

frecuencia relativa Razón que compara la frecuencia de cada categoría al total.

decimal periódico La forma decimal de un número racional.

rombo Paralelogramo que tiene cuatro lados congruentes.

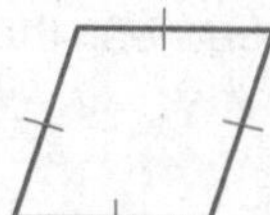

ángulo recto Ángulo que mide exactamente 90°.

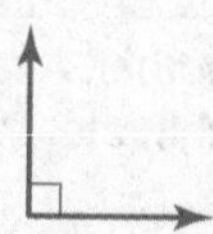

right triangle A triangle having one right angle.

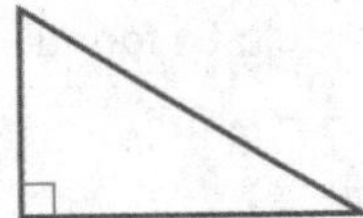

triángulo rectángulo Triángulo que tiene un ángulo recto.

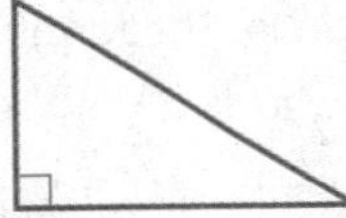

Ss

sales tax An additional amount of money charged on items that people buy.

impuesto sobre las ventas Cantidad de dinero adicional que se cobra por los artículos que se compran.

sample A randomly selected group chosen for the purpose of collecting data.

muestra Grupo escogido al azar o aleatoriamente que se usa con el propósito de recoger datos.

sample space The set of all possible outcomes of a probability experiment.

espacio muestral Conjunto de todos los resultados posibles de un experimento probabilístico.

scale The scale that gives the ratio that compares the measurements of a drawing or model to the measurements of the real object.

escala Razón que compara las medidas de un dibujo o modelo a las medidas del objeto real.

scale drawing A drawing that is used to represent objects that are too large or too small to be drawn at actual size.

dibujo a escala Dibujo que se usa para representar objetos que son demasiado grandes o demasiado pequeños como para dibujarlos de tamaño natural.

scale factor A scale written as a ratio without units in simplest form.

factor de escala Escala escrita como una razón sin unidades en forma simplificada.

scale model A model used to represent objects that are too large or too small to be built at actual size.

modelo a escala Réplica de un objeto real, el cual es demasiado grande o demasiado pequeño como para construirlo de tamaño natural.

scalene triangle A triangle having no congruent sides.

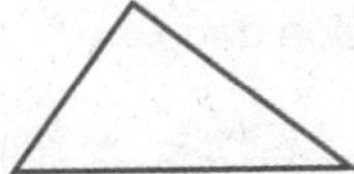

triángulo escaleno Triángulo sin lados congruentes.

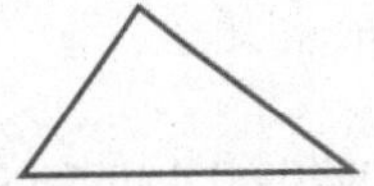

scatter plot In a scatter plot, two sets of related data are plotted as ordered pairs on the same graph.

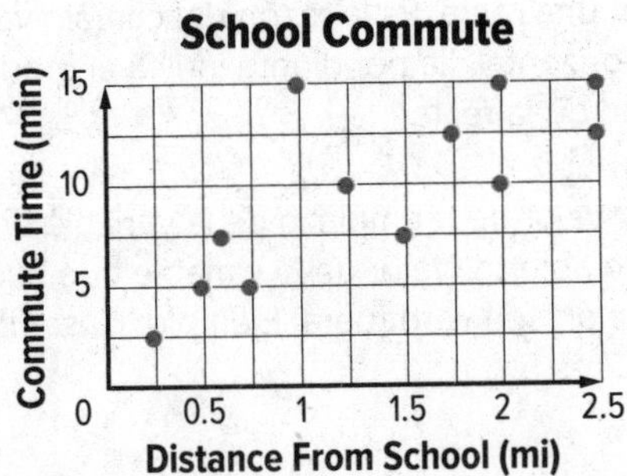

diagrama de dispersión Diagrama en que dos conjuntos de datos relacionados aparecen graficados como pares ordenados en la misma gráfica.

Tiempo para llegar a la escuela

Tiempo (min)

0 5 10 15

0.5 1 1.5 2 2.5

Distancia a la escuela (mi)

selling price The amount the customer pays for an item.

precio de venta Cantidad de dinero que paga un consumidor por un artículo.

semicircle Half of a circle. The formula for the area of a semicircle is $A = \frac{1}{2}\pi r^2$.

semicírculo Medio círculo La fórmula para el área de un semicírculo es $A = \frac{1}{2}\pi r^2$.

sequence An ordered list of numbers, such as 0, 1, 2, 3 or 2, 4, 6, 8.

sucesión Lista ordenada de números, como 0, 1, 2, 3 ó 2, 4, 6, 8.

similar figures Figures that have the same shape but not necessarily the same size.

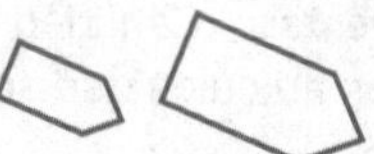

figuras semejantes Figuras que tienen la misma forma, pero no necesariamente el mismo tamaño.

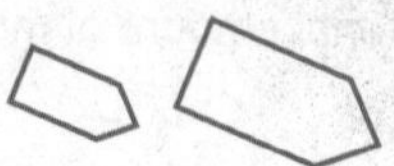

similar solids Solids with the same shape. Their corresponding linear measures are proportional.

sólidos semejantes Sólidos con la misma forma. Sus medidas lineales correspondientes son proporcionales.

simple event One outcome or a collection of outcomes.

eventos simples Un resultado o una colección de resultados.

simple interest The amount paid or earned for the use of money. The formula for simple interest is $I = prt$.

interés simple Cantidad que se paga o que se gana por el uso del dinero. La fórmula para calcular el interés simple es $I = prt$.

simple random sample An unbiased sample where each item or person in the population is as likely to be chosen as any other.

muestra aleatoria simple Muestra de una población que tiene la misma probabilidad de escogerse que cualquier otra.

simplest form An expression is in simplest form when it is replaced by an equivalent expression having no like terms or parentheses.

expresión mínima Expresión en su forma más simple cuando es reemplazada por una expresión equivalente que no tiene términos similares ni paréntesis.

simplify Write an expression in simplest form.

simplificar Escribir una expresión en su forma más simple.

simulation An experiment that is designed to model the action in a given situation.

simulación Un experimento diseñado para modelar la acción en una situación dada.

skew lines Lines that do not intersect and are not coplanar.

rectas alabeadas Rectas que no se intersecan y que no son coplanares.

slant height The height of each lateral face.

altura oblicua Altura de cada cara lateral.

slope The rate of change between any two points on a line. It is the ratio of vertical change to horizontal change. The slope tells how steep the line is.

pendiente Razón de cambio entre cualquier par de puntos en una recta. Es la razón del cambio vertical al cambio horizontal. La pendiente indica el grado de inclinación de la recta.

solution A replacement value for the variable in an open sentence. A value for the variable that makes an equation true. Example: The *solution* of $12 = x + 7$ is 5.

solución Valor de reemplazo de la variable en un enunciado abierto. Valor de la variable que hace que una ecuación sea verdadera. Ejemplo: La *solución* de $12 = x + 7$ es 5.

square The product of a number and itself. 36 is the square of 6.

cuadrado Producto de un número por sí mismo. 36 es el cuadrado de 6.

square A parallelogram having four right angles and four congruent sides.

cuadrado Paralelogramo con cuatro ángulos rectos y cuatro lados congruentes.

square root The factors multiplied to form perfect squares.

al cuadrado Factores multiplicados para formar cuadrados perfectos.

squared The product of a number and itself. 36 is the square of 6.

raíz cuadrada El producto de un número por sí mismo. 36 es el cuadrado de 6.

standard form Numbers written without exponents.

forma estándar Números escritos sin exponentes.

statistics The study of collecting, organizing, and interpreting data.

estadística Estudio que consiste en recopilar, organizar e interpretar datos.

straight angle An angle that measures exactly 180°.

ángulo llano Ángulo que mide exactamente 180°.

Subtraction Property of Equality If you subtract the same number from each side of an equation, the two sides remain equal.

propiedad de sustracción de la igualdad Si restas el mismo número de ambos lados de una ecuación, los dos lados permanecen iguales.

Subtraction Property of Inequality If you subtract the same number from each side of an inequality, the inequality remains true.

propiedad de desigualdad en la resta Si se resta el mismo número a cada lado de una desigualdad, la desigualdad sigue siendo verdadera.

supplementary angles Two angles are supplementary if the sum of their measures is 180°.

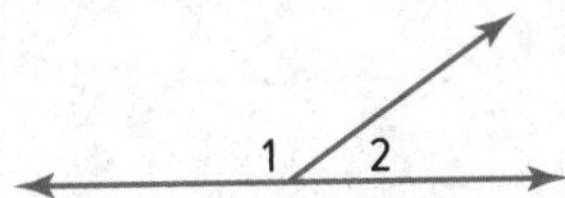

$\angle 1$ and $\angle 2$ are supplementary angles.

ángulos suplementarios Dos ángulos son suplementarios si la suma de sus medidas es 180°.

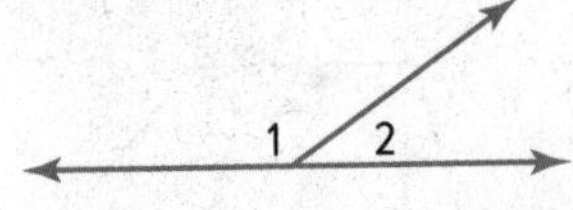

$\angle 1$ y $\angle 2$ son suplementarios.

surface area The sum of the areas of all the surfaces (faces) of a three-dimensional figure.

área de superficie La suma de las áreas de todas las superficies (caras) de una figura tridimensional.

survey A question or set of questions designed to collect data about a specific group of people, or population.

encuesta Pregunta o conjunto de preguntas diseñadas para recoger datos sobre un grupo específico de personas o población.

systematic random sample A sample where the items or people are selected according to a specific time or item interval.

muestra aleatoria sistemática Muestra en que los elementos o personas se eligen según un intervalo de tiempo o elemento específico.

Tt

term Each number in a sequence.

término Cada número en una sucesión.

term A number, a variable, or a product or quotient of numbers and variables.

término Número, variable, producto o cociente de números y de variables.

terminating decimal A repeating decimal which has a repeating digit of 0.

decimal finito Un decimal periódico que tiene un dígito que se repite que es 0.

theoretical probability The ratio of the number of ways an event can occur to the number of possible outcomes. It is based on what *should* happen when conducting a probability experiment.

probabilidad teórica Razón del número de maneras en que puede ocurrir un evento al número de resultados posibles. Se basa en lo que *debería* pasar cuando se conduce un experimento probabilístico.

three-dimensional figure A figure with length, width, and height.

figura tridimensional Figura que tiene largo, ancho y alto.

third quartile For a data set with median M, the third quartile is the median of the data values greater than M.

tercer cuartil Para un conjunto de datos con la mediana M, el tercer cuartil es la mediana de los valores mayores que M.

tip Also known as a gratuity, it is a small amount of money in return for a service.

propina También conocida como gratificación; es una cantidad pequeña de dinero en recompensa por un servicio.

transversal The third line formed when two parallel lines are intersected.

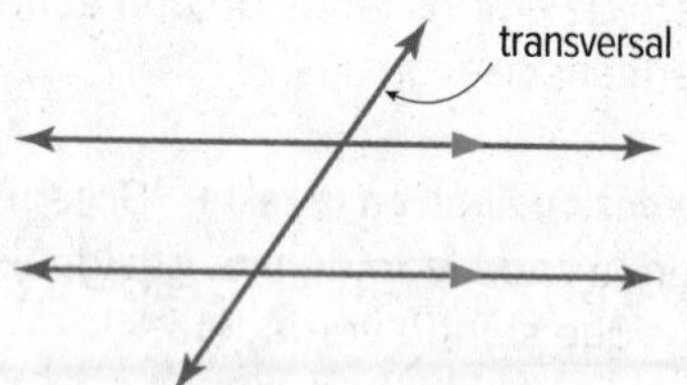

transversal Tercera recta que se forma cuando se intersecan dos rectas paralelas.

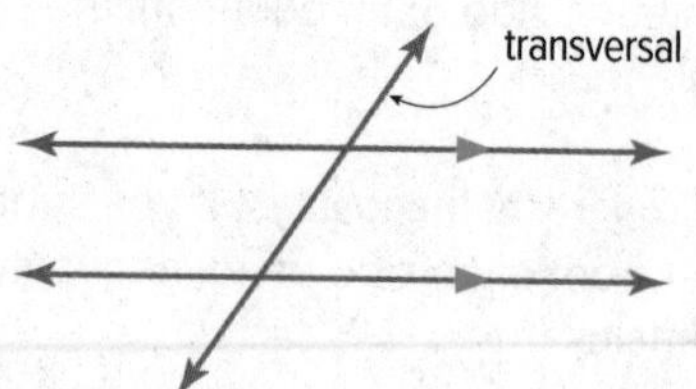

trapezoid A quadrilateral with one pair of parallel sides.

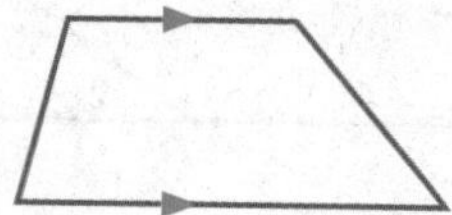

trapecio Cuadrilátero con un único par de lados paralelos.

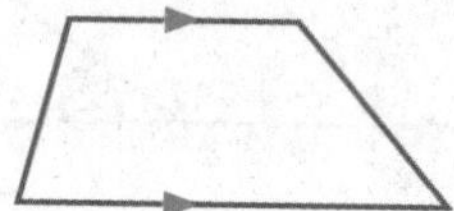

tree diagram A diagram used to show the sample space.

diagrama de árbol Diagrama que se usa para mostrar el espacio muestral.

triangle A figure with three sides and three angles.

triángulo Figura con tres lados y tres ángulos.

triangular prism A prism that has two parallel congruent bases that are triangles.

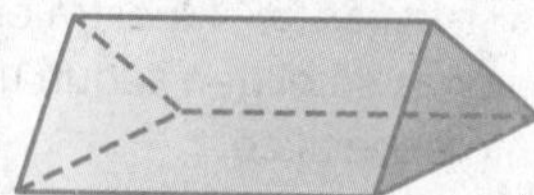

prisma triangular Un prisma que tiene dos bases congruentes paralelas que triángulos.

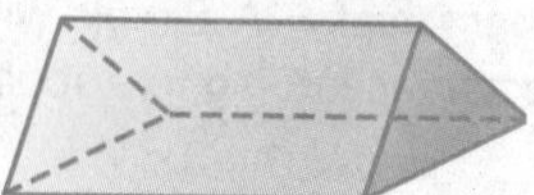

two-step equation An equation having two different operations.

ecuación de dos pasos Ecuación que contiene dos operaciones distintas.

two-step inequality An inequality than contains two operations.

desigualdad de dos pasos Desigualdad que contiene dos operaciones.

Uu

unbiased sample A sample representative of the entire population.

muestra no sesgada Muestra que se selecciona de modo que se representativa de la población entera.

unfair game A game where there is not a chance of each player being equally likely to win.

juego injusto Juego donde cada jugador no tiene la misma posibilidad de ganar.

uniform probability model A probability model which assigns equal probability to all outcomes.

modelo de probabilidad uniforme Un modelo de probabilidad que asigna igual probabilidad a todos los resultados.

unit rate A rate that is simplified so that it has a denominator of 1 unit.

tasa unitaria Tasa simplificada para que tenga un denominador igual a 1.

unit ratio A unit rate where the denominator is one unit.

razón unitaria Tasa unitaria en que el denominador es la unidad.

unlike fractions Fractions with different denominators.

fracciones con distinto denominador Fracciones cuyos denominadores son diferentes.

Vv

variable A symbol, usually a letter, used to represent a number in mathematical expressions or sentences.

variable Símbolo, por lo general una letra, que se usa para representar un número en expresiones o enunciados matemáticos.

vertex A vertex of an angle is the common endpoint of the rays forming the angle.

vertex

vértice El vértice de un ángulo es el extremo común de los rayos que lo forman.

vértice

vertex The point where three or more faces of a polyhedron intersect.

vértice El punto donde tres o más caras de un poliedro se cruzan.

vertex The point at the tip of a cone.

vértice El punto en la punta de un cono.

vertical angles Opposite angles formed by the intersection of two lines. Vertical angles are congruent.

1 2

$\angle 1$ and $\angle 2$ are vertical angles.

ángulos opuestos por el vértice Ángulos opuestos formados por la intersección de dos rectas. Los ángulos opuestos por el vértice son congruentes.

1 2

$\angle 1$ y $\angle 2$ son ángulos opuestos por el vértice.

visual overlap A visual demonstration that compares the centers of two distributions with their variation, or spread.

superposición visual Una demostración visual que compara los centros de dos distribuciones con su variación, o magnitud.

volume The number of cubic units needed to fill the space occupied by a solid.

volumen Número de unidades cúbicas que se requieren para llenar el espacio que ocupa un sólido.

voluntary response sample A sample which involves only those who want to participate in the sampling.

muestra de respuesta voluntaria Muestra que involucra sólo aquellos que quieren participar en el muestreo.

Xx

x-axis The horizontal number line in a coordinate plane.

eje x La recta numérica horizontal en el plano de coordenadas.

x-coordinate The first number of an ordered pair. It corresponds to a number on the *x*-axis.

coordenada x El primer número de un par ordenado. Corresponde a un número en el eje *x*.

Yy

y-axis The vertical number line in a coordinate plane.

eje y La recta numérica vertical en el plano de coordenadas.

y-coordinate The second number of an ordered pair. It corresponds to a number on the *y*-axis.

coordenada y El segundo número de un par ordenado. Corresponde a un número en el eje *y*.

Zz

zero pair The result when one positive counter is paired with one negative counter. The value of a zero pair is 0.

par nulo Resultado de hacer coordinar una ficha positiva con una negativa. El valor de un par nulo es 0.

Chapter 5 Expressions

Page 348 Chapter 5 Are You Ready?

1. 16 **3.** 16 **5.** −50 **7.** −25

Pages 353–354 Lesson 5-1 Independent Practice

1. 34 **3** 3 **5.** 3 **7.** 2 **9.** −1 **11** 50 + 0.17*m*; \$75.50 **13.** 9.1 **15.** 37.85 **17.** Sample answer: The fee to rent a bicycle is \$10 plus \$5 for each hour. The expression $5x + 10$ represents the total cost for renting a bicycle for *x* hours. **19.** Sample answer: $2n + 4$; $2(n + 2)$

Pages 355–356 Lesson 5-1 Extra Practice

21. 4 **23.** −12 **25.** 5 **27.** \$8.75 **29a.** True **29b.** False **29c.** True **31.** Let *p* = the number of hours Paida worked; $p + 8$ **33.** Let *n* = Nathan's age; $n - 3$

Pages 361–362 Lesson 5-2 Independent Practice

1. 7 is added to the previous term; 28, 35, 42 **3** 8 is added to the previous term; 58, 66, 74 **5.** 0.8 is added to the previous term; 5.6, 6.4, 7.2 **7** 3*n*; 36 in.

9a.

x	1	2	3	4	5
y	3	6	9	12	15

9b. 3*n*

9c.

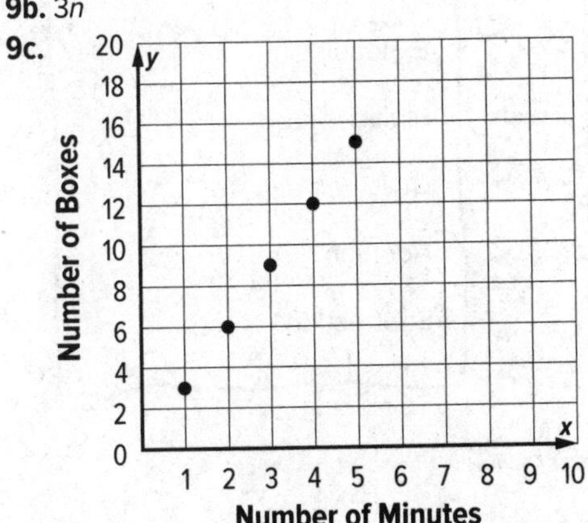

Sample answer: The number of boxes increases by 3 each minute. The points appear to fall in a straight line passing through the origin. **9d.** 135 boxes **11.** + 1, + 2, + 3, + 4, ...; 16, 22, 29 **13.** 81; Sample answer: The multiples of 6 from 41 to 523 can be represented by the sequence 42, 48, 54, ... 522. The expression $6n + 36$ represents this sequence. When $n = 81$, the value of the expression is 522. So, the 81st term of the sequence is 522. There are 81 multiples of 6 between 41 and 523.

Pages 363–364 Lesson 5-2 Extra Practice

15. 10 is added to the previous term; 46, 56, 66 **17.** 1.5 is added to the previous term; 10.5, 12.0, 13.5 **19.** 4 is added to the previous term; 20.6, 24.6, 28.6 **21.** 25 is added to the previous term; 120, 145, 170 **23a.** Each figure is 8 less than the previous figure. **23b.** 40, 32 **25.** 33, 30, 27 **27a.** True **27b.** True **27c.** False **29.** 1 **31.** 64 **33.** 5 **35.** \$1.50

Pages 371–372 Lesson 5-3 Independent Practice

1. Commutative (+) **3** Associative (+) **5.** false; Sample answer: $(24 \div 4) \div 2 \neq 24 \div (4 \div 2)$

7. $= (15 + 12) + 8a$ Associative (+)
$= 27 + 8a$ Simplify.

9 $= 3x \cdot (x \cdot 7)$ Commutative (×)
$= (3x \cdot x) \cdot 7$ Associative (×)
$= 3x^2 \cdot 7$ Simplify.
$= 3 \cdot 7 \cdot x^2$ Commutative (×)
$= (3 \cdot 7) \cdot x^2$ Associative (×)
$= 21x^2$ Simplify.

11. $[7 + (47 + 3)][5 \cdot (2 \cdot 3)]$, Associative (+); $(7 + 50)[5 \cdot (2 \cdot 3)]$, Simplify; $57[5 \cdot (2 \cdot 3)]$, Simplify; $57[(5 \cdot 2) \cdot 3]$, Associative (×); $57 \cdot 10 \cdot 3$, Simplify; $(57 \cdot 10) \cdot 3$, Associative (×); $570 \cdot 3$, Simplify; 1,710 **13.** Blake incorrectly multiplied both the 5 and *m* by 4. He should have used the Associative Property to group the 5 and 4 together, simplify, and then multiply by *m*. $4 \cdot (5 \cdot m) = 20m$ **15a.** no; Sample answer: $2 - 3 = -1$ and −1 is not a whole number **15b.** no; Sample answer: $1 + 1 = 2$ and 2 is not a member of the set.

Pages 373–374 Lesson 5-3 Extra Practice

17. Commutative (×) **19.** Associative (+) **21.** 48 s; Sample answer: $12.4 + 12.6 = 25$ and $11.8 + 11.2 = 23$, $25 + 23 = 48$

23. $= (18 + 5) + 6m$ Associative (+)
$= 23 + 6m$ Simplify.

25. $= 10 \cdot 7 \cdot y$ Commutative (×)
$= (10 \cdot 7) \cdot y$ Associative (×)
$= 70y$ Simplify.

27. $2(2.29) + 2(2.21) + 2.50$; $2(2.29) + 2.50 + 2(2.21)$; $2.50 + 2(2.21 + 2.29)$ **29.** 36 **31.** 226 **33.** 74 **35.** \$1.25(3) + \$0.45(2); \$4.65

Pages 379–380 Lesson 5-4 Independent Practice

1. 33 **3** −30 **5.** 4 **7.** $-12x + 24$ **9.** $30 - 6q$ **11.** $-15 + 3b$ **13** \$27.40; 4(\$7.00 − \$0.15) = $4 \cdot 7 - 4 \cdot 0.15$ **15.** 315;

$$9(30 + 5) = 9(30) + 9(5)$$
$$= 270 + 45$$

17. 672;
$$(100 + 12)6 = 100(6) + 12(6) \\ = 600 + 72$$
19. 488;
$$4(120 + 2) = 4(120) + 4(2) \\ = 480 + 8$$
21. Sample answer: $6(2a + 3b - c)$ **23.** $2a + ay + 2b + by$ **25.** No; $3 + (4 \cdot 5) = 23$ but $(3 + 4) \cdot (3 + 5) = 56$

Pages 381–382 Lesson 5-4 Extra Practice

27. 24 **29.** $-8a - 8b$ **31.** $-2p - 14$ **33.** $n(4.75 + 2.50) + 30$; $7.25n + 30$ **35.** $-12ab - 30ac$ **37.** $6y + 12z$ **39.** $-72p + 48n$ **41.** 3 × ($18.95 + $14.95 + $9.95); $131.55; Sample answer: The ticket prices can be added first and then the sum can be multiplied by 3. This requires fewer steps and easier computations than multiplying each price by 3 and then adding the resulting products. **43.** −46 **45.** −47

Page 385 Problem-Solving Investigation Make a Table

Case 3. 26 containers **Case 5.** $2n + 2$; 18 toothpicks

Pages 391–392 Lesson 5-5 Independent Practice

1. terms: 2, $3a$, $9a$; like terms: $3a$, $9a$; coefficients: 3, 9; constant: 2 **3.** terms: 9, $-z$, 3, $-2z$; like terms: 9 and 3, $-z$ and $-2z$; coefficients: −1, −2; constants: 9, 3 **5.** $11c$ **7.** $1.03t$; $74.16 **9.** $2x + 30$ **11. a.** $7 + 5x + 4y + 2z$ **b.** $43 **13.** $16a + 8b + 4$ **15.** Sample answer: $3x + x - 7$; coefficients: 3, 1; constant: −7 **17.** $18x - 3$; $18x - 3 = 18(2) - 3 = 33$ and $8x - 2x + 12x - 3 = 8(2) - 2(2) + 12(2) - 3 = 33$

Pages 393–394 Lesson 5-5 Extra Practice

19. terms: 4, $5y$, $-6y$, y; like terms: $5y$, $-6y$, y; coefficients: 5, −6, 1; constant: 4 **21.** terms: $-3d$, 8, $-d$, −2; like terms: $-3d$ and $-d$, 8 and −2; coefficients: −3, −1; constants: 8, −2 **23.** $2 + 4d$ **25.** $m - 2$ **27.** $7m - 20$ **29.** $20x + 9$ **31.** $38g + 36h - 38$ **33.** $5a + 9b$ **35.** t = hours Tricia volunteered; $t + 9$ **37.** 10 **39.** 13

Pages 399–400 Lesson 5-6 Independent Practice

1. $11x + 11$ **3.** $4x - 16$ **5.** $4x + 14$ **7.** $(22x + 10)$ mm; 230 mm **9.** $-x + 2$ **11.** $8.7x - 1.6$ **13.** Sample answer: $(10x + 2)$ and $(-15x + 2)$ **15.** $2x + 1$; The expression $2x + 1$ will always be odd when x is an integer because when an integer is doubled, the result is always even. Adding one to the result will give an odd number.

Pages 401–402 Lesson 5-6 Extra Practice

17. $-4x + 16$ **19.** $-2x - 2$ **21.** $-6x + 5$ **23.** $(24x + 9)$ yd; 177 yd **25a.** False **25b.** False **25c.** True **27.** 35 **29.** 85

Pages 407–408 Lesson 5-7 Independent Practice

1. $5x + 2$ **3.** $2x + 2$ **5.** $8x - 12$ **7.** $5x - 2$; 248 customers **9.** $x + 0.51$ **11.** Sample answer: The additive inverse of $(2x + 1)$ is $(-2x - 1)$.
$$(5x + 3) - (2x + 1) = (5x + 3) + (-2x - 1) \\ = 5x + 3 + (-2x) + (-1) \\ = 5x + (-2x) + 3 + (-1) \\ = 3x + 2$$
13. $-x + 5$ **15.** Sample answer: The rule is to add the inverse when subtracting integers, and is applied to each term in the linear expression that is subtracted.

Pages 409–410 Lesson 5-7 Extra Practice

17. $-3x + 6$ **19.** $-16x + 2$ **21.** $-4x - 7$ **23.** $0.8x + 0.6$ **25.** $-x + 2$ **27.** $12 + 1.50t - (10 + 1.25t) = 2 + 0.25t$ **29.** $(12x - 4)$ ft; 32 ft **31.** $-\frac{1}{4}$ **33.** $\frac{1}{8}$ **35.** $\frac{2}{3}$

Pages 419–420 Lesson 5-8 Independent Practice

1. 24 **3.** $36k$ **5.** cannot be factored **7.** 4 units by $(x - 2)$ units **9.** $(x + 2)$ dollars **11.** $5(x + 4)$ units^2 **13.** $4(5x + 19)$ units^2 **15.** Sample answer: $20m$ and $12mn$ **17.** $6(4x - y)$

Pages 421–422 Lesson 5-8 Extra Practice

19. $6rs$ **21.** $20x$ **23.** $25xy$ **25.** $6(3x + 1)$ **27.** $5(2x - 7)$ **29.** $10(3x - 4)$ **31.** $(2x + 5)$ in. **33.** $\frac{2}{3}(x + 9)$ **35.** $\frac{5}{6}(x - 36)$ **37.** $\frac{3}{8}(x + 48)$ **39.** $16ab$, $12a$; $28a$, $20a$ **41.** $3a + 30$

43.

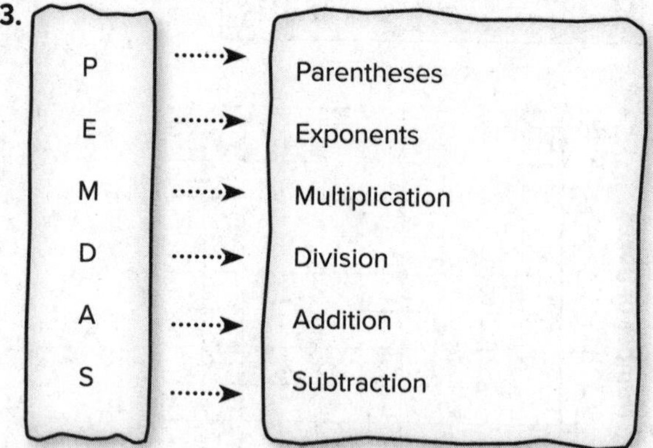

Page 425 Chapter Review Vocabulary Check

Across

3. simplest form **7.** sequence **11.** counterexample **13.** define

Down

1. equivalent **5.** variable **9.** term

Page 426 Chapter Review Key Concept Check

1. 1 + 3 **3.** $2x - 4$ **5.** $3(x + 7)$

Chapter 6 Equations and Inequalities

Page 432 Chapter 6 Are You Ready?

1. $p + 3$ **3.** $g + 10$ **5.** 17 **7.** 1 **9.** 35

Pages 441–442 Lesson 6-1 Independent Practice

1. 7 **3** 17 **5.** -1

7

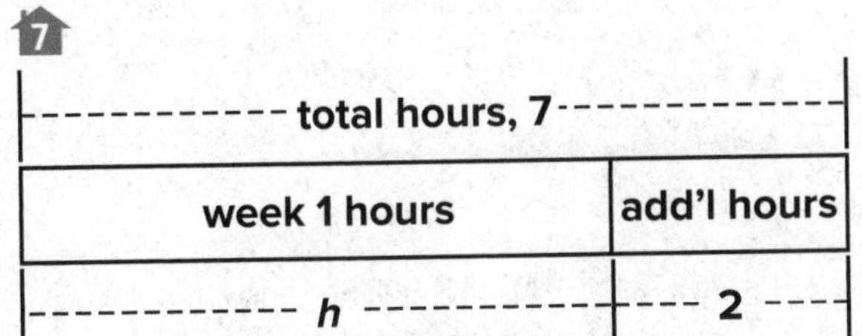

$7 = h + 2$; 5 h **9a.** $s - 65 = 5$; 70 mph

9b.

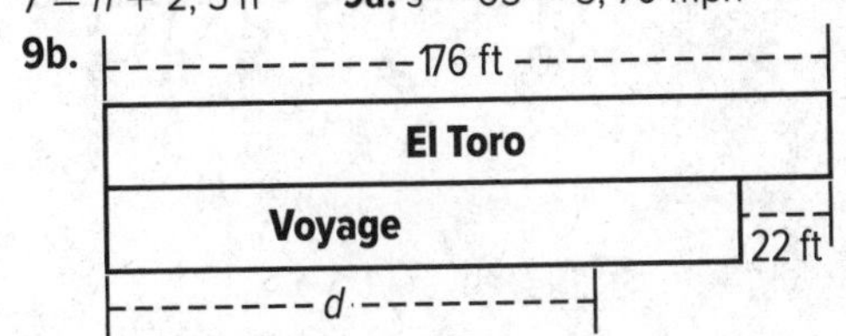

$d + 22 = 176$; 154 ft **9c.** The solution of each equation is 197; Colossos is 197 feet tall. **11.** $115 + 115 + 65 + x = 360$; 65 **13.** She should have subtracted 5 from each side; -13 **15.** $x + 2 = 8$; The solution for the other equations is -6.

Pages 443–444 Lesson 6-1 Extra Practice

17. -9 **19.** -12 **21.** 7

23.

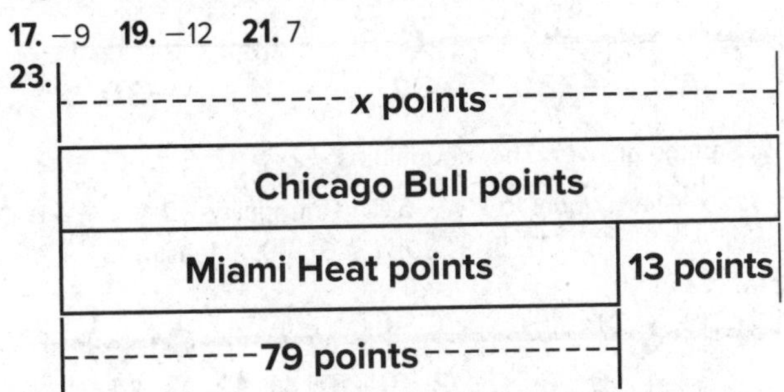

$x - 13 = 79$; 92 points **25.** $-1 + (-3) + s + 2 = 0$; $+2$ **27.** -1.25 **29.** 12.3 **31.** 5.8 **33.** $-\frac{1}{12}$ **35a.** True **35b.** False **35c.** True **37.** -20 **39.** 60 **41.** -36 **43.** $3h = -3$; $h = -1$

Pages 451–452 Lesson 6-2 Independent Practice

1. 7 **3.** 8 **5.** 80 **7** -5 **9.** -90 **11** $205 = \frac{d}{3}$; 615 mi

13a.

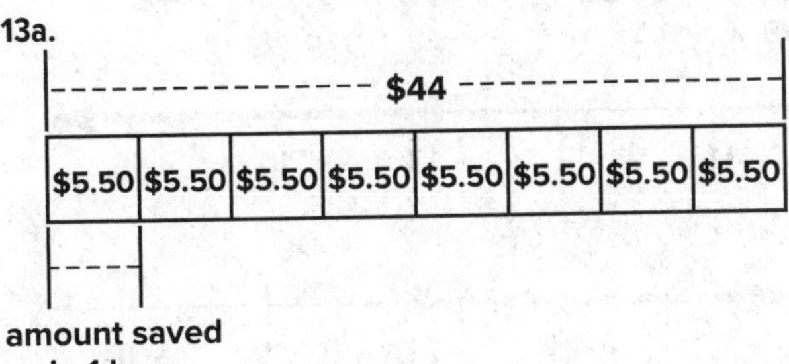

13b. $5.5x = 44$ **13c.** Sample answer: Divide each side by 5.5. Then simplify. $x = 8$ **15.** True; Sample answer: Multiply each side of the equation by $\frac{1}{5}$ instead of dividing each side by 5. **17.** Sample answer: Multiply both sides by x, then divide both sides of the equation by 6; -5.

Pages 453–454 Lesson 6-2 Extra Practice

19. 4 **21.** 70 **23.** -120 **25.** $50 = 25t$; 2 s **27.** 8 in. **29.** $3\frac{1}{3}$ **31.** $1\frac{1}{100}$ **33.** $\frac{13}{4}$ **35.** 4 **37.** 2.1 **39.** $\frac{21}{4}$ or $5\frac{1}{4}$

Pages 461–462 Lesson 6-3 Independent Practice

1. 5 **3** 3 **5.** $\frac{20}{3}$ or $6\frac{2}{3}$ **7** $\frac{3}{4}p = 46.50$; \$62 **9.** Emily's homeroom class; Sample answer: Write and solve the equations $0.75e = 15$ and $\frac{2}{3}s = 12$; $e = 20$ and $s = 18$; Since $20 > 18$, Emily's homeroom class has more students. **11.** 20; Sample answer: Solve $8 = \frac{m}{4}$ to find that $m = 32$. So, replace m with 32 to find $32 - 12 = 20$. **13.** Sample answer: Multiply each side by 2. Then divide each side by $(b_1 + b_2)$. So, $\frac{2A}{b_1 + b_2} = h$.

Pages 463–464 Lesson 6-3 Extra Practice

15. 7 **17.** -3.8 **19.** $-\frac{125}{12}$ or $-10\frac{5}{12}$

21.

$\frac{7}{15}$ of elevation, 140 ft

20	20	20	20	20	20	20	20	20	20	20	20	20	20	20

total elevation, x ft

$140 = \frac{7}{15}x$; 300 ft

23. a train that travels 100 miles in $\frac{2}{3}$ hour; a train that travels 90 miles in $\frac{3}{5}$ hour **25.** 22 **27.** 2 **29.** $3 \times \$0.25 + 5 \times \$0.50$; \$3.25

Pages 473–474 Lesson 6-4 Independent Practice

1. 3 **3** 7 **5.** -3

7.

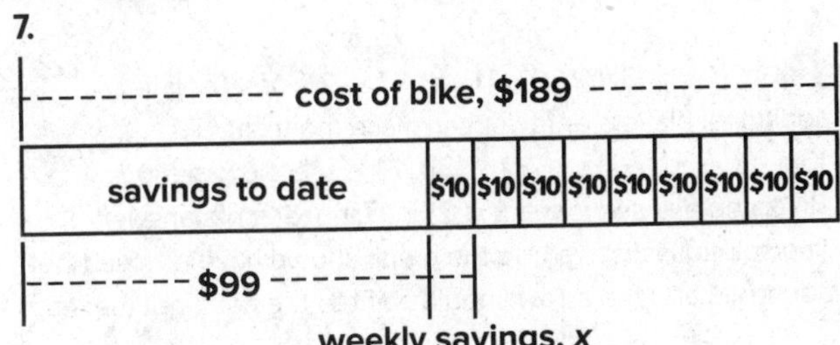

$189 = 10x + 99$; 9 weeks

9. 2.25 **11** **a.** -9°C **b.** 92.2°F **13.** No, none of the Fahrenheit temperatures convert to the same temperature in Celsius. Only -40°F $= -40$°C. **15.** Sample answer: Cameron found the area of a trapezoid to be 52 square inches. One base was 12 inches long and the other was 14 inches long. What is the height h of the trapezoid?; 4 in.

Pages 475–476 Lesson 6-4 Extra Practice

17. −4 **19.** 4 **21.** 36

23a.

perimeter, 48 cm			
width	width	16	16

23b. $48 = 32 + 2w$; 8 cm **23c.** Sample answer: Using either method, you would subtract first and then divide

25. $c = 30 + 0.05m$; 395 mi **27.** $6 \cdot 10 + 6 \cdot n$ or $60 + 6n$

29. $5(x + 7)$ **31.** $10(t + 3)$

Pages 485–486 Lesson 6-5 Independent Practice

1. 6 **3** −14 **5.** −3.2 **7** $3(\ell + 5) = 60$; 15 in.

9a. $12(m - 2.57) = 0.36$ **9b.** Sample answer: I first divided each side by 12 and then added 2.57 to each side; $2.60.

11. Sample answer: Marisol should have divided by six before subtracting three; $6(x + 3) = 21$, $x + 3 = 3.5$, $x = 3.5 - 3$, $x = 0.5$ **13.** $3(x - 8) = 12$; Sample answer: Valeria bought a new collar for each of her three dogs. She paid $8 for each necklace. Suppose she had $12 left. How much money did Valeria have initially to spend on each dog collar?; $12 per collar

Pages 487–488 Lesson 6-5 Extra Practice

15. 16 **17.** −2 **19.** 78 **21.** $5\frac{3}{4}$ or 5.75

23. $1.20\left(n + 2\frac{1}{2}\right) = 4.50$; 1.25 or $1\frac{1}{4}$ pounds **25.** Divide both sides by *p*; Add *q* to both sides. **27.** −4 **29.** 4 **31.** −3

33. 2 **35.** −3, −2, −1, 0

Page 491 Problem-Solving Investigation Work Backward

Case 3. 1,250 ft **Case 5.** 6:25 A.M.

Pages 501–502 Lesson 6-6 Independent Practice

1. $h \leq -8$ **3** $5 < n$ **5.** $x > -1$

7. $m \geq -6$;

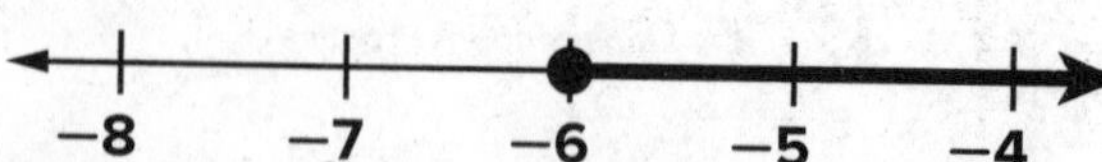

9 $n + 4 > 13$; $n > 9$ **11.** $p + 17 \leq 26$; $p \leq 9$; Nine additional players or fewer can make the team.

13a. $42 + x \geq 74$; $x \geq 32$ **13b.** $74 + y \geq 110$; $y \geq 36$

15. Sample answer: $x + 3 < 25$ **17.** no; Sample answer: The solution is $x \geq -1$, so the graph should have a closed dot above −1 and the arrow should point to the right, not the left.

Pages 503–504 Lesson 6-6 Extra Practice

19. $m \leq 4.3$

21. $-5 < a$

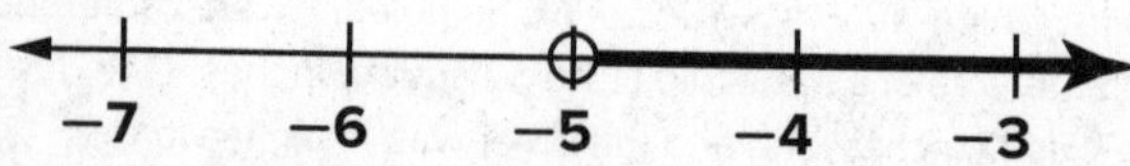

23. $n - 8 < 10$; $n < 18$ **25.** $68 + c \leq 125$; $c \leq 57$; The salesman has 57 cars or less left to sell.

27. $2\frac{2}{3} > x$ or $x < 2\frac{2}{3}$

2 $2\frac{1}{3}$ $2\frac{2}{3}$ 3

29. $m \geq 11\frac{1}{5}$

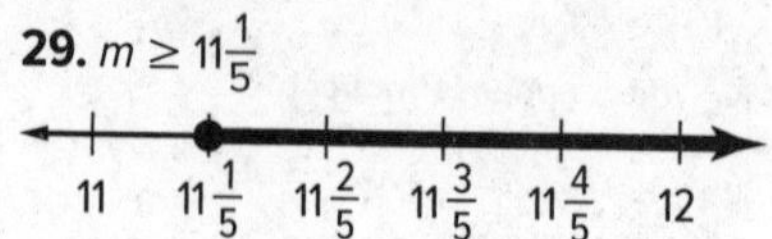

31. $n \geq -4\frac{3}{16}$

$-4\frac{3}{16}$ $-3\frac{6}{16}$

33. $x + 4 \leq 7$; $-7 \geq x - 10$ **35.** −4; See answer 39 for graph. **37.** −2; See answer 39 for graph. **39.** −6;

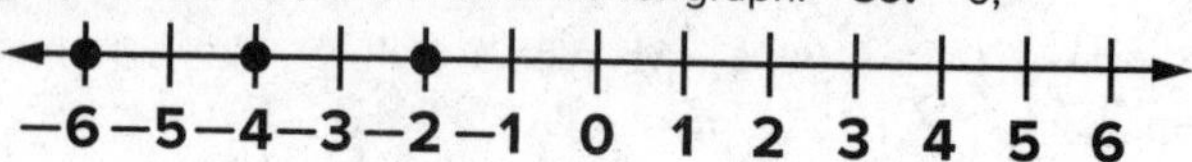

Pages 509–510 Lesson 6-7 Independent Practice

1. $y < 3$ **3** $180 \leq m$ **5.** $m \geq 56$ **7.** $n \leq 4.5$

9. $w \leq -45$

11 $4 < t$

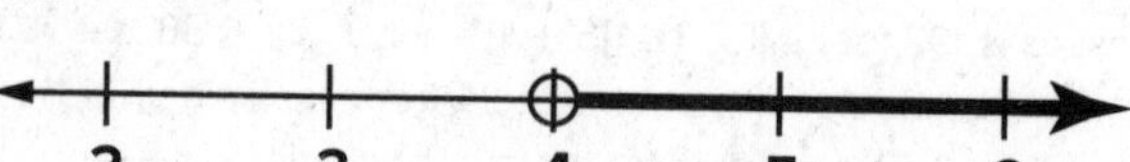

13. $x \leq -32$

−34 −33 −32 −31 −30

15. Sample answer: The inequalities $-2x > 12$, $\frac{x}{2} < -3$, and $-7 > x - 1$ are equal to $x < -6$. The inequality $-2 < x + 4$ is equal to $x > -6$. **17.** $5n \leq 30$; $n \leq 6$ **19.** at least a 15

21a.

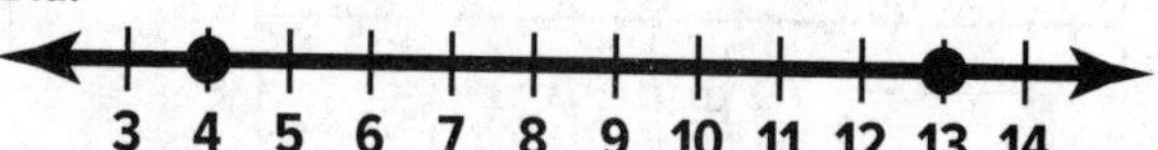

21b. yes; It represents the solutions that satisfy both inequalities. **21c.** $4 \leq b \leq 13$

21d.

3 4 5 6 7 8 9 10 11 12 13 14

Pages 511–512 Lesson 6-7 Extra Practice

23. $y > -5$ **25.** $p \geq -6$ **27.** $-40 < y$ or $y > -40$

29. $w > 13$

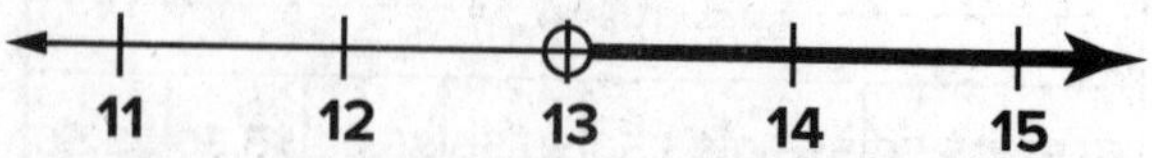

31. $-20 \geq t$ or $t \leq -20$

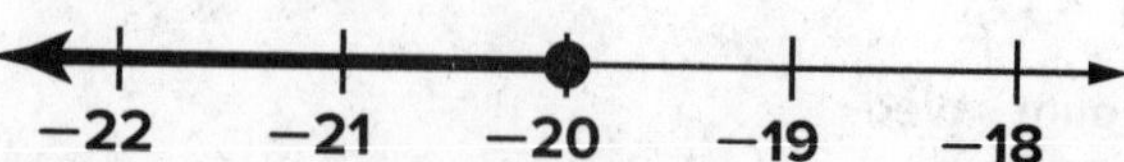

33. $4n \geq -12$; $n \geq -3$ **35a.** False **35b.** True **35c.** False

37. 2 **39.** −2.5 **41.** 12

Pages 517–518 Lesson 6-8 Independent Practice

1. $x \geq 1$; 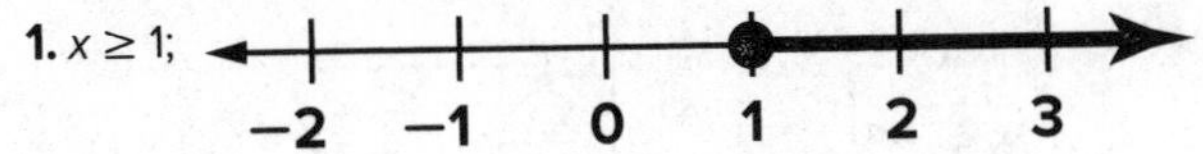

3 $x > 12$

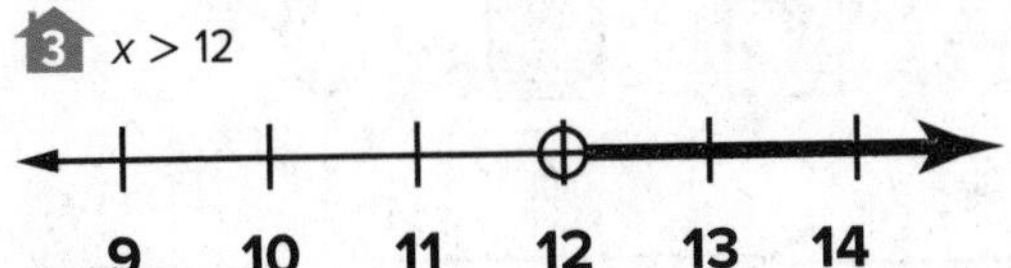

5 $30 + 7x \geq 205$; $x \geq 25$ hours; He will have to work at least 25 hours **7.** $\frac{x}{-5} + 1 \leq 7$; $x \geq -30$ **9.** $-2x - 6 > -18$; $x < 6$ **11.** Sample answer: $-2x + 5 > -7$ **13.** Sample answer: $\frac{x}{2} + 5 \geq 30$ **15.** $\frac{73 + x}{6} \geq 15$; $x \geq 17$; at least 17 points **17.** Sample answer: At the electronics store, CDs were marked down $2.80 from their regular price. Orlando has $45 to spend on 4 CDs. How much can he spend on each CD?; $14.05

Pages 519–520 Lesson 6-8 Extra Practice

19. $x \leq -8$

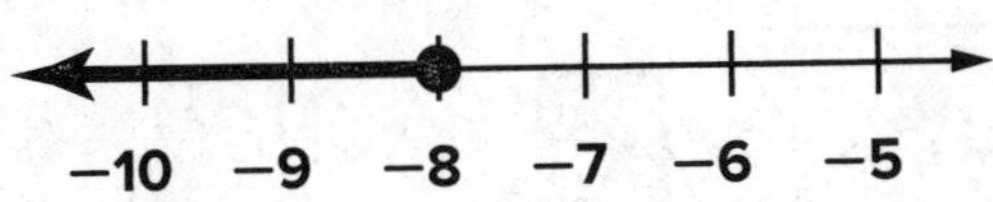

21. $x \geq 42$

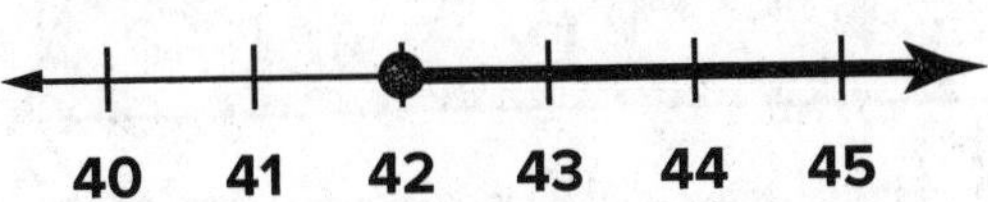

23. $75 + 5s \geq 125$; $s \geq 10$; Audrey needs to make at least 10 sales for her pay to be $125. **25.** Add 5 to both sides; Divide both sides by −2; Reverse the inequality symbol.

27. $n > -3$

29. $t < 2$

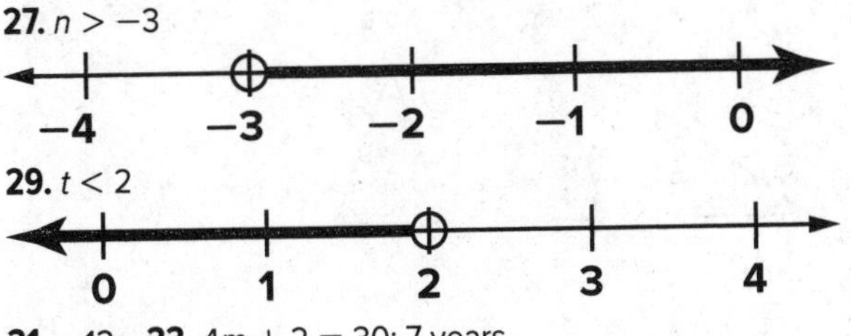

31. −12 **33.** $4m + 2 = 30$; 7 years

Page 524 Chapter Review Key Concept Check

1. solution **3.** equivalent equations

Chapter 7 Geometric Figures

Page 534 Chapter 7 Are You Ready?

1. 40° **3.** 90° **5.** 6.72 yd^2

Pages 539–540 Lesson 7-1 Independent Practice

1. ∠ABC, ∠CBA, ∠B, ∠4; acute **3** ∠MNP, ∠PNM, ∠N, ∠1; obtuse **5** neither **7.** adjacent **9.** vertical **11.** 11 **15.** True; Sample answer:

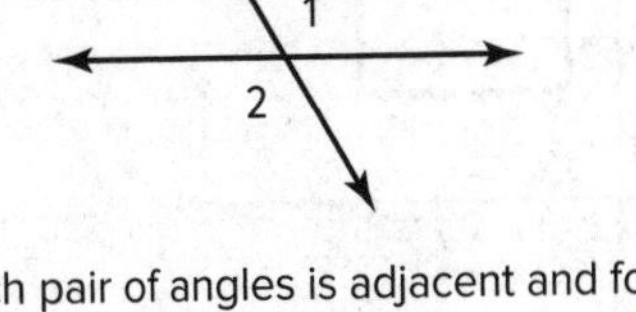

17. Sample answer: Each pair of angles is adjacent and forms a straight angle. So $(2x + 8) + (5x - 10) = 180$ and $(3x + 42) + (x + 34) = 180$. When you solve both equations, $x = 26$. The angle measures are 60°, 120°, 120°, and 60°.

Pages 541–542 Lesson 7-1 Extra Practice

19. ∠HKI, ∠IKH, ∠K, ∠8; obtuse **21a.** Sample answer: ∠1 and ∠3; Since ∠1 and ∠3 are opposite angles formed by the intersection of two lines, they are vertical angles. **21b.** Sample answer: ∠1 and ∠2; Since ∠1 and ∠2 share a common vertex, a common side, and do not overlap, they are adjacent angles. **23.** 9 **25a.** vertical **25b.** adjacent **25c.** adjacent **25d.** vertical **27.** 134° **29.** 38° **31.** rectangle

Pages 547–548 Lesson 7-2 Independent Practice

1. neither **3** supplementary **5.** 20 **7** 23 **9.** Sample answer: ∠CGK, ∠KGJ **11a.** adjacent; adjacent; vertical **11b.** $m\angle 1 + m\angle 2 = 180°$; $m\angle 2 + m\angle 3 = 180°$ **11c.** $m\angle 1 = 180° - m\angle 2$; $m\angle 3 = 180° - m\angle 2$; Sample answer: $m\angle 1$ and $m\angle 3$ are equal. **11d.** Sample answer: Vertical angles are congruent. **13a.** $m\angle E = 39°$; $m\angle F = 51°$ **13b.** $m\angle B = 60°$; $m\angle C = 120°$ **15.** Sample answer: Right angles have a measure of 90°, so two right angles will always have a sum of 180°. This is the definition of supplementary angles.

Pages 549–550 Lesson 7-2 Extra Practice

17. complementary **19.** 15 **21.** never; Sample answer: Since an obtuse angle is greater than 90°, the sum of two obtuse angles must be greater than, not equal to, 180°.

23a.

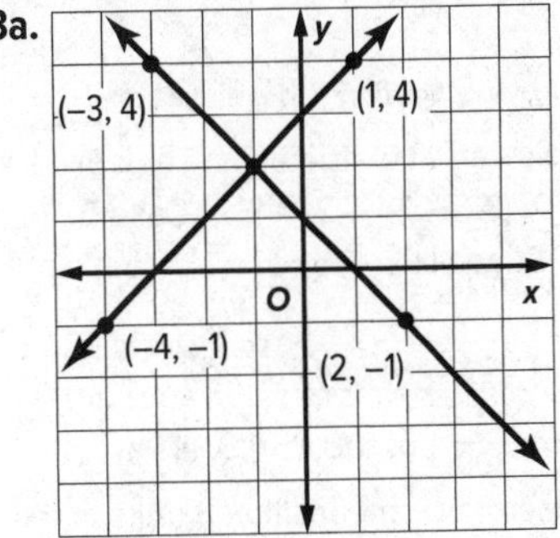

23b. The lines appear to be perpendicular. **23c.** line *a*: 1; line *b*: −1 **25.** x°; 90°; 180°; 45°; 45°

27. 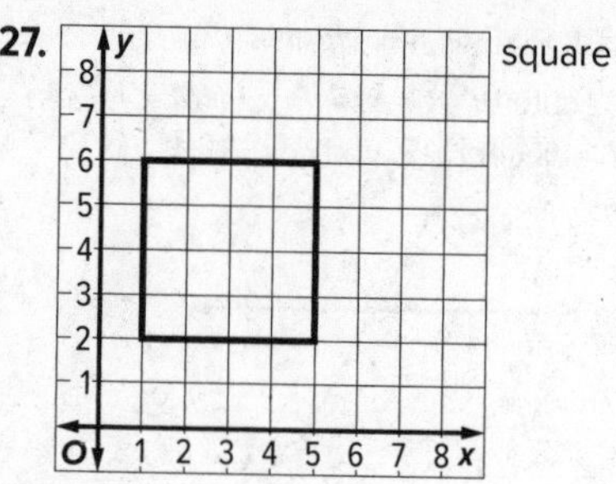

square

Pages 559–560 Lesson 7-3 Independent Practice

1 acute equilateral; Sample answer:

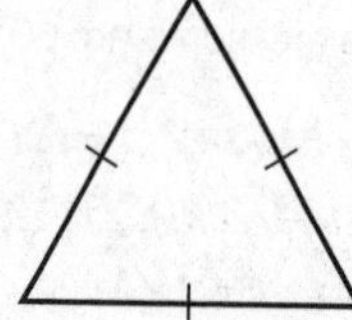

3 acute equilateral **5.** obtuse isosceles **7.** 118 **9.** acute isosceles **11.** $125 + a = 180$, so $a = 55$; $a + b + 60 = 180$, so $b = 65$; $60 + d = 90$, so $d = 30$; $c + d + 90 = 180$, so $c = 60$ **13a.** never; Sample answer: The sum of the interior angles of a triangle is 180°. Two right angles have a sum of 180°. This means the third angle would equal 0°, which is not possible. **13b.** never; Sample answer: The sum of the interior angles of a triangle is 180°. The measure of an obtuse angle is greater than 90°. So, a triangle cannot have more than one obtuse angle.

Pages 561–562 Lesson 7-3 Extra Practice

15. acute isosceles **17.** right scalene **19.** obtuse isosceles; Sample answer:

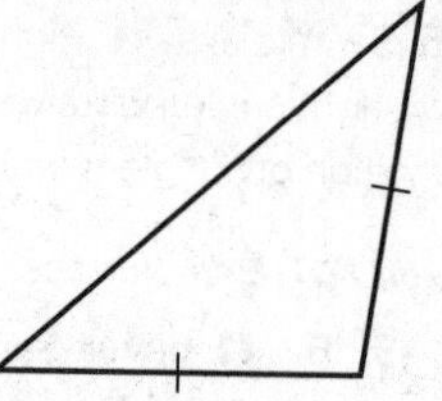

21. 90 **23.** 53° **25.** 30 **27a.** False **27b.** True **27c.** False **29.** 25 in^2 **31.** 35 cm^2 **33.** 9 yd^2

Page 569 Problem-Solving Investigation Make a Model

Case 3. 15 tables; 2 people can sit on the ends. Then divide the remaining people by 2. $(32 - 2) \div 2 = 15$. **Case 5.** Faith: Spanish; Sarah: German; Guadalupe: French

Pages 579–580 Lesson 7-4 Independent Practice

1 102.6 mi **3** 12 cm; $\frac{1}{300}$ **5.** 108 ft^2 **9.** always; Sample answer: A scale factor of $\frac{3}{1}$ means that 3 units of the drawing is equal to 1 unit of the object, so the scale drawing or model will be larger than the actual object.

Pages 581–582 Lesson 7-4 Extra Practice

11. 30 km **13.** 102.5 km **15.** $109\frac{3}{8}$ ft **17.** 3,420 ft^2 **19.** 40 ft by 60 ft; Sample answer: Set up and solve proportions to find the actual length and width: $\frac{1 \text{ in.}}{20 \text{ ft}} = \frac{2 \text{ in.}}{w \text{ ft}}$ and $\frac{1 \text{ in.}}{20 \text{ ft}} = \frac{3 \text{ in.}}{\ell \text{ ft}}$ **21.** 10 **23.** 22

Pages 589–590 Lesson 7-5 Independent Practice

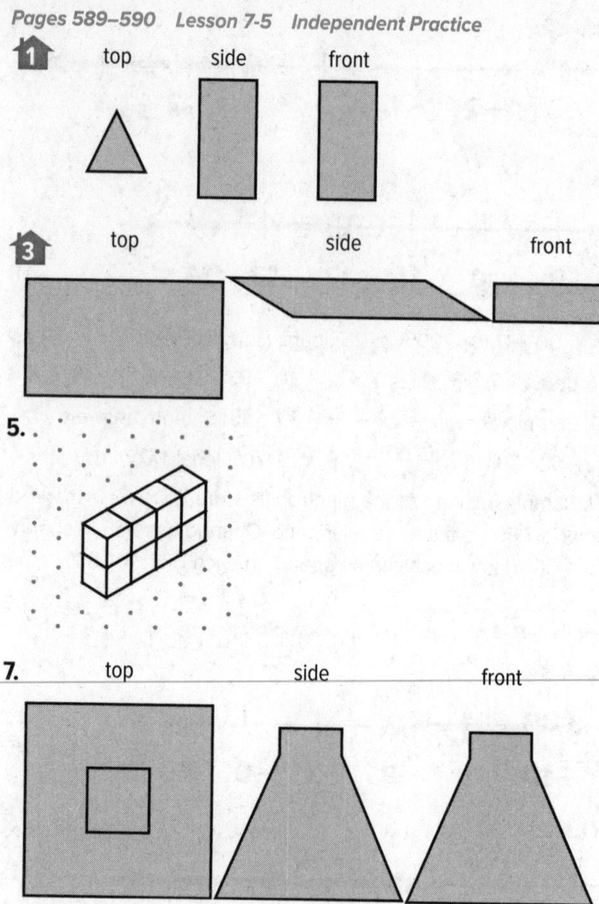

9. triangle; It is the only two-dimensional figure. **11a.** never **11b.** never **11c.** sometimes

Pages 591–592 Lesson 7-5 Extra Practice

13. top side front

15.

17.

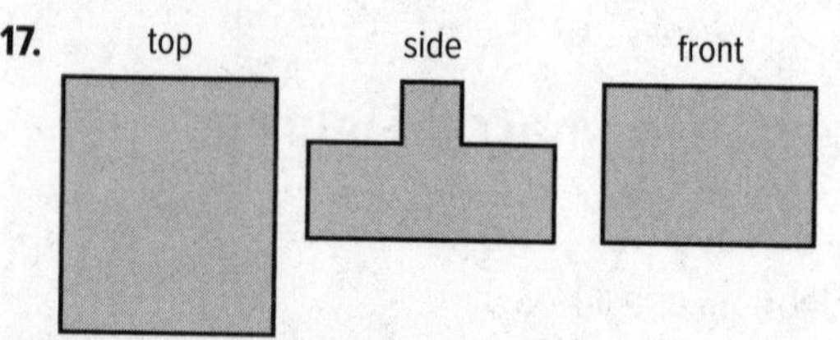

19.  ;

21. line; $\overleftrightarrow{WX}$ or $\overleftrightarrow{XW}$ **23.** line segment; $\overline{EF}$ or $\overline{FE}$ **25.** parallel

Pages 597–598 Lesson 7-6 Independent Practice

1 **figure name:** triangular pyramid
bases: *ACD*
faces: *ACD, ABD, ABC, DBC*
edges: $\overline{AB}, \overline{BC}, \overline{CD}, \overline{AD}, \overline{AC}, \overline{BD}$
vertices: *A, B, C, D*
skew lines: Sample answer: $\overline{BD}$ and $\overline{AC}$

3 rectangle **5.** triangle **7.** False; two planes intersect at a line, which is an infinite number of points. **11.** sometimes; A rectangular prism has 2 bases and 4 faces, but a triangular prism has 2 bases and 3 faces. **13.** sometimes; A triangular pyramid has a triangle for its base.

Pages 599–600 Lesson 7-6 Extra Practice

15. figure name: rectangular prism
bases: *ABCD, EFGH, ABFE, DCGH, ADHE, BCGF*
faces: *ABCD, EFGH, ABFE, DCGH, ADHE, BCGF*
edges: $\overline{AB}, \overline{BC}, \overline{CD}, \overline{AD}, \overline{EF}, \overline{FG}, \overline{GH}, \overline{EH}, \overline{AE}, \overline{BF}, \overline{CG}, \overline{DH}$
vertices: *A, B, C, D, E, F, G, H*
skew lines: Sample answer: $\overline{DH}$ and $\overline{GF}$

17. curve **19.** Because there are two parallel, congruent triangular bases, it is a triangular prism. **21a.** Figure 2 **21b.** Figure 1 **21c.** Figure 3 **23.** hexagon **25.** 90°

Page 603 Chapter Review Vocabulary Check

Across
11. equilateral **15.** complementary
Down
1. adjacent **3.** supplementary **5.** triangle **7.** vertical **9.** acute **13.** right

Page 604 Chapter Review Key Concept Check

1. vertex **3.** 90° **5.** scale drawing

Chapter 8 Measure Figures

Page 610 Chapter 8 Are You Ready?

1. 42 sq m **3.** 76.5 sq mm

Pages 617–618 Lesson 8-1 Independent Practice

1. 2.5 mm **3.** 34 cm **5** 3.14 × 13 = 40.8 cm **7** 19 people **9a.** 30 mm **9b.** 31.4 mm **9c.** 31.4159 mm **9d.** Sample answer: The more decimal places of the estimate of π, the more precise the circumference. **11.** 18 in. **13.** 257 cm **15.** Greater than; Sample answer: Since the radius is 4 feet, the diameter is 8 feet. Since π is a little more than 3, the circumference will be a little more than 3 times 8, or 24 feet. **17.** The circumference would double. For example, with a diameter of 4 feet, the circumference is about 12.6 feet. With a diameter of 8 feet, the circumference is about 25.1 feet.

Pages 619–620 Lesson 8-1 Extra Practice

19. 3.5 in. **21.** 72 ft **23.** $\frac{22}{7} \times 21 = 66$ ft **25.** $\frac{22}{7} \times 42 =$ 132 mm **27.** 37.7 cm **29.** Each is π, or about 3.14, units longer than the previous circle. **31a.** False **31b.** True **31c.** False **33.** 12.74 m^2 **35.** 36,976 ft^2

Pages 627–628 Lesson 8-2 Independent Practice

1. 3.14 × 6 × 6 = 113.0 cm^2 **3** 3.14 × 5.5 × 5.5 = 95.0 ft^2 **5.** 3.14 × 6.3 × 6.3 = 124.6 mm^2 **7.** 254.3 ft^2 **9** 226.1 in^2 **11.** 163.3 yd^2 **13.** The large pizza; the medium pizza's area is 78.5 square inches and costs $0.102 per square inch. The large pizza's area is 153.86 square inches and costs $0.097 per square inch. **15.** When the radius of a circle is doubled, the circumference doubles and the area is 4 times as large. In the formula for area of a circle, the radius is squared, so when the radius of a circle is doubled, the area is 2^2 or 4 times as large. **17.** 5.9 in^2 **19.** Sample answer: To find the area of the quarter circle, multiply the area of the entire circle by 14; $A = \frac{1}{4}\pi r^2$; 19.6 in^2

Pages 629–630 Lesson 8-2 Extra Practice

21. 3.14 × 6.3 × 6.3 = 124.6 cm^2 **23.** 3.14 × 5.4 × 5.4 = 91.6 yd^2 **25.** 3.14 × 9.3 × 9.3 = 271.6 mm^2 **27.** 144.7 ft^2 **29.** 64.3 in^2 **31.** circle; $\frac{1}{2} \cdot 100 \cdot 100 < 3 \cdot 50 \cdot 50$ **33.** 154 in^2; Sample answer: Using $\frac{22}{7}$ makes the computation easier since the radius is 7. The 7s cancel out in the multiplication. **35.** 210 in^2 **37.** 39.5 cm^2

Pages 635–636 Lesson 8-3 Independent Practice

1. 64 cm^2 **3.** 220.5 cm^2 **5** 38.6 ft^2 **7** 119.5 ft^2 **9.** 77 cm^2 **11.** 44.6 ft^2; 30.3 ft **13.** 110.8 ft^2

Pages 637–638 Lesson 8-3 Extra Practice

15. 87.5 m^2 **17.** 180 cm^2 **19.** 9 cm^2 **21.** 240 ft^2 **23a.** 36 **23b.** 14.14 **23c.** 92.56 **25.** 3.7 cm^2 **27.** 4.7 m

Pages 643–644 Lesson 8-4 Independent Practice

1 192 m^3 **3** 108 m^3 **5a.** 96 ft^3; 128 ft^3; 168 ft^3; 160 ft^3; 120 ft^3 **5b.** The height must allow the water to be deep enough for someone to get wet and the length and width must allow a person to fit. So the first and last sets of dimensions would not work. **7a.** Sample answer: There is a direct relationship between the volume and the length. Since the length is doubled, the volume is also doubled. **7b.** The volume is eight times greater. **7c.** Neither; Sample answer: doubling the height will result in a volume of 4 • 4 • 10 or 160 in^3; doubling the width will result in a volume of 4 • 8 • 5 or 160 in^3.

Pages 645–646 Lesson 8-4 Extra Practice

11. 236.3 cm^3 **13.** 20.4 mm^3 **15.** $306.52 = 19.4h$; 15.8 m **17.** $166\frac{1}{4}$ yd^3 **19.** 2 in. by 1.5 in. by 0.5 in.; 3 in. by 1 in. by 0.5 in. **21.** 25.8 m **23.** 29.2 cm

Page 649 Problem-Solving Investigation Solve a Simpler Problem

Case 3. 80 chairs **Case 5.** 7,763,270.6 mi^2; Sample answer: The area of Asia is about 17,251,712.4 mi^2 and the area of North America is about 9,488,441.8 mi^2. 17,251,712.4 − 9,488,441.8 = 7,763,270.6

Pages 657–658 Lesson 8-5 Independent Practice

1. 80 ft^3 **3.** 42 ft^3 **5.** 14 in. **7.** 10 in^3 **9.** The volume is eight times greater; Sample answer: Since each dimension is two times greater, the volume is 2 × 2 × 2 or eight times greater. **11.** Sample answer: first set: area of the base, 40 ft^2; height of the pyramid, 12 ft; second set: area of the base, 30 ft^2; height of the pyramid, 16 ft **13.** The volumes are the same.

Pages 659–660 Lesson 8-5 Extra Practice

15. 60 in^3 **17.** 195 yd^3 **19.** 11 ft **21.** 22 in. **23.** 1,234.2 m^3 **25.** 24 in.; Sample answer: Replace V with 1,560 and B with 13 × 15 in the formula $V = \frac{1}{3}Bh$. Then solve for h. **27.** 1.5 ft^2 **29.** 28.75 ft^2

Pages 669–670 Lesson 8-6 Independent Practice

1. 314 cm^2 **3.** 207 in^2 **5.** 180 in^2 **7.** $S.A. = 6x^2$
9. False; Sample answer: A 9 × 7 × 13 rectangular prism has a surface area of 2(9 × 13) + 2(9 × 7) + 2(13 × 7) or 542 square units. Doubling the length, the surface area is 2(18 × 13) + 2(18 × 7) + 2(13 × 7) or 902 square units. 2 × 542 ≠ 902
11. 1,926 cm^2

Pages 671–672 Lesson 8-6 Extra Practice

13. 833.1 mm^2 **15.** 96 ft^2 **17.** Yes; there are 2,520 ft^2 of fencing. Since 8 gallons of paint will cover 350 • 8 or 2,800 ft^2 and 2,800 ft^2 > 2,520 ft^2, 8 gallons is enough paint.
19. 64.5 in^2 **21a.** 12.5 **21b.** 50 **21c.** 71 **21d.** 196
23. triangle; triangle; triangle **25.** rectangle; circle; oval

Pages 681–682 Lesson 8-7 Independent Practice

1. 95 in^2 **3.** 328 in^2 **5.** 0.52 ft^2 **7.** 78 in^2
9. 6.5 cm
11.

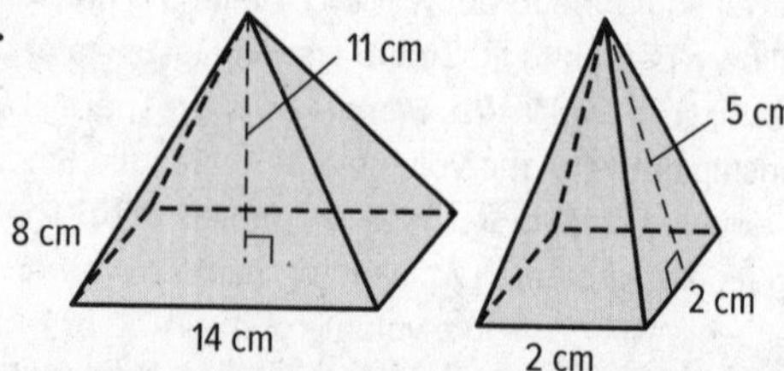

Sample answer: Both a square pyramid and a rectangular pyramid have isosceles triangles as their lateral faces. All the lateral faces are congruent on a square pyramid but, on a rectangular pyramid, the opposite pairs of lateral faces are congruent.

Pages 683–684 Lesson 8-7 Extra Practice

13. 197.1 m^2 **15.** 765 cm^2 **17.** 26.1 ft^2
19.

Area of the base

Slant height

$S.A. = B + \frac{1}{2}P\ell$

Perimeter of the base

21a. True **21b.** True **21c.** False **23.** 456 ft^2 **25.** 10 ft

Pages 693–694 Lesson 8-8 Independent Practice

1. 2.3 m^3 **3.** 2,600 ft^2 **5.** 0.5 ft^3 **7.** 10.4 m^2
9. 100 in^3 **13.** less than; Sample answer: The combined surface area of the two prisms is 180 in^2. Since they share a common surface, the area of that surface is not included in the total surface area.

Pages 695–696 Lesson 8-8 Extra Practice

15. 100 in^3 **17.** 280.2 cm^2 **19a.** 1.68 ft^3 **19b.** 19.28 ft^2
21.

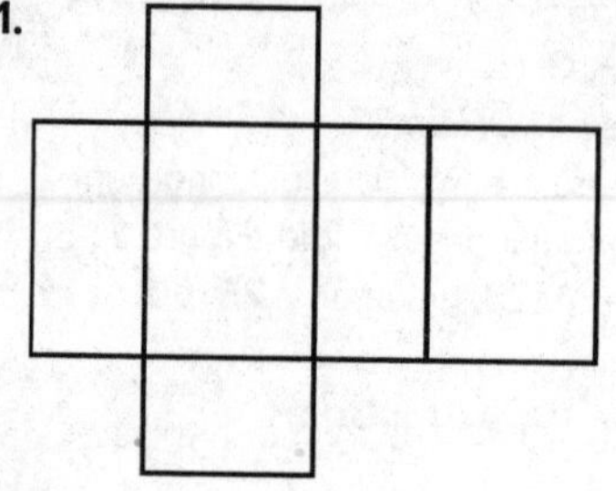

23.

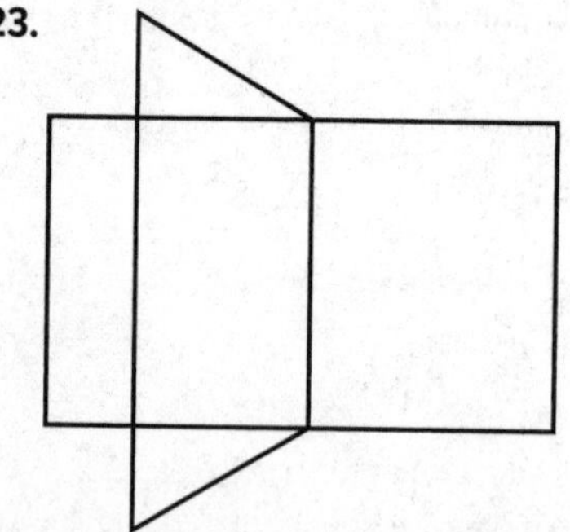

Page 699 Chapter Review Vocabulary Check

1. diameter **3.** circle **5.** circumference **7.** semicircle
9. volume **11.** lateral

Page 700 Chapter Review Key Concept Check

1. twice **3.** height

Chapter 9 Probability

Page 710 Chapter 9 Are You Ready?

1. $\frac{1}{3}$ **3.** $\frac{2}{3}$ **5.** 30 **7.** 24

Pages 715–716 Lesson 9-1 Independent Practice

1. $\frac{1}{4}$, 25%, or 0.25 **3** $\frac{1}{1}$, 100%, or 1 **5** $\frac{1}{5}$, 0.2, or 20%; Sample answer: Since 80% arrive on time, that means that 20% do not arrive on time. **7.** Picking a black jelly bean is impossible since the probability of picking a black jelly bean is 0%. **9a.** $\frac{1}{8}$, 0.125, 12.5%; $\frac{1}{2}$, 0.5, 50% **9b.** $\frac{3}{4}$, 0.75, 75%
11. 70%, $\frac{1}{3}$; Sample answer: 70% and $\frac{1}{3}$ are probabilities that are not complementary because $0.7 + 0.\overline{3} \neq 1$. The other sets of probability are complementary.

Pages 717–718 Lesson 9-1 Extra Practice

13. $\frac{1}{5}$, 20%, or 0.2 **15.** $\frac{7}{10}$, 70%, or 0.7 **17.** $\frac{1}{2}$, 50%, or 0.5
19. $\frac{3}{5}$, 60%, or 0.6 **21.** The complement of selecting a girl is selecting a boy. The probability of the complement is $\frac{37}{100}$, 0.37, or 37%. **23.** $\frac{124}{125}$, 99.2%, or 0.992; It is very likely that card 13 will *not* be chosen. **25.** *P*(orange) $= \frac{1}{5}$; *P*(green) $= \frac{2}{5}$ **27.** >
29. $\frac{3}{25}$, $\frac{1}{5}$; Bryan misses more foul shots than Dwayne.

Pages 725–726 Lesson 9-2 Independent Practice

1 **a.** $\frac{1}{5}$; The experimental probability is close to the theoretical probability of $\frac{1}{6}$. **b.** $\frac{9}{10}$; The experimental probability is close to the theoretical probability of $\frac{5}{6}$.
3a. 162 people **3b.** about 134 people **5** **a.** $\frac{1}{3}$
b. $\frac{6}{25}$, $\frac{13}{50}$
c.

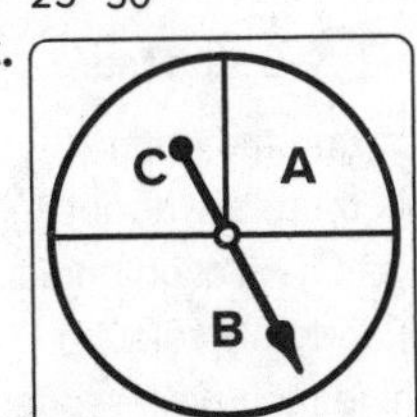

Sample answer: Section B should be one half of the spinner and sections A and C should each be one fourth of the spinner.
7. Yes; Sample answer: $\frac{5 \text{ sharpened}}{10 \text{ unsharpened}} = \frac{20 \text{ sharpened}}{x \text{ unsharpened}}$. So, $x = 40$.

Pages 727–728 Lesson 9-2 Extra Practice

9. $\frac{9}{20}$; The experimental probability of $\frac{9}{20}$ is close to the theoretical probability of $\frac{1}{2}$. **11.** 50 customers
13. experimental; 40; less **15.** *P*(not red) **17.** vanilla sundae, vanilla cone, chocolate sundae, chocolate cone, strawberry sundae, strawberry cone; equally likely

Pages 737–738 Lesson 9-3 Independent Practice

1. H1, H2, H3, H4, H5, T1, T2, T3, T4, T5
3 purple 10, purple 18, purple 21, purple 24, green 10, green 18, green 21, green 24, black 10, black 18, black 21, black 24, silver 10, silver 18, silver 21, silver 24

5. $\frac{1}{36}$;

1, 1	1, 2	1, 3	1, 4	1, 5	1, 6
2, 1	2, 2	2, 3	2, 4	2, 5	2, 6
3, 1	3, 2	3, 3	3, 4	3, 5	3, 6
4, 1	4, 2	4, 3	4, 4	4, 5	4, 6
5, 1	5, 2	5, 3	5, 4	5, 5	5, 6
6, 1	6, 2	6, 3	6, 4	6, 5	6, 6

7 *P*(Player 1) $= \frac{6}{8}$ or $\frac{3}{4}$; *P*(Player 2) $= \frac{2}{8}$ or $\frac{1}{4}$; RRB, RYB, RRY, RYY, BRB, BYB, BYY, BRY **9.** The first outcome in the I bracket should be IC.

Pages 739–740 Lesson 9-3 Extra Practice

11.

Appetizer	Entree	Dessert	Sample Space
S	S	C	SSC
		A	SSA
	C	C	SCC
		A	SCA
Sa	S	C	SaSC
		A	SaSA
	C	C	SaCC
		A	SaCA

13a. 16 combinations **13b.** $\frac{1}{16}$
13c.

Shoes	Socks	Sample Space
black	green	black, green
	yellow	black, yellow
	black	black, black
	white	black, white
yellow	green	yellow, green
	yellow	yellow, yellow
	black	yellow, black
	white	yellow, white

8 combinations

15. (Ava, Brooke); (Greg, Brooke); (Antoine, Mario) **17.** $\frac{1}{8}$
19. $\frac{1}{2}$ **21.** $\frac{1}{3}$; There are 2 numbers out of 6 on a number cube that are greater than 4. $\frac{2}{6} = \frac{1}{3}$

Pages 745–746 Lesson 9-4 Independent Practice

1. Sample answer: Spin a spinner with 4 equal-size sections 50 times. **3.** Sample answer: Spin a spinner divided into 3 equal sections and roll a number cube. Repeat the simulation until all types of cookies are obtained. **5.** Sample answer: Use 3 red marbles to represent winning and 7 blue marbles to represent losing. Draw 1 marble 4 times, replacing the marble each time. **7.** Sample answer: a survey of 100 people voting on whether or not to enact a tax increase, where each person is equally likely to vote yes or no. Toss a coin 100 times. **9.** Sample answer: sometimes; The spinner must have equal-sized sections.

Pages 747–748 Lesson 9-4 Extra Practice

11. Sample answer: Use a spinner with 5 equal sections to represent the 5 discounts. Spin 4 times to represent 4 customers receiving cards. **13.** Sample answer: Toss a coin. Heads represents one color and tails represents the other color. Repeat until both colors are selected. **15.** Sample answer: Spin a spinner with 4 equal sections. Each section represents one of the magazines. Repeat the simulation until all possible magazines are selected. **17.** Spin a spinner with equal size spaces labeled A, B, C, D, E, and F. Let spinning A represent winning a prize and let spinning other letters represent not winning a prize; Roll a number cube. Let rolling a 1 represent winning a prize and let rolling a 2, 3, 4, 5, or 6 represent not winning a prize.

Page 755 Problem-Solving Investigation Act It Out

Case 3. 31 **Case 5.** unfair; Sample answer: There are 20 out of 36 outcomes that are multiples of 3 and only 15 that are multiples of 4. Jason has a greater chance of winning.

Pages 761–762 Lesson 9-5 Independent Practice

1. 12 **3.** 84 **5.** 6 possible routes; $\frac{1}{6}$ or about 17% **7.** $\frac{1}{50}$; very unlikely **9.** No; the number of selections is 32 • 11 or 352, which is less than 365. **11.** 10 groups, 8 activities have 80 outcomes; the other two have 72 outcomes. **13.** 6^x

Pages 763–764 Lesson 9-5 Extra Practice

15. 8 **17.** 27 **19.** 16 **21.** $\frac{4}{48}$ or $\frac{1}{12}$; Sample answer: There are 3 • 4 • 4 or 48 different possible outcomes of a phone plan. There are 1 • 4 • 1 or 4 different possible outcomes of a phone plan that includes Brand B and has a headset. **23.** $108 = 9 \times c \times 2$; 6 colors **25.** $\frac{1}{2}$ **27.** Sample answer: Assign each number of a number cube to a toy. Roll the number cube. Repeat until all numbers are rolled.

Pages 769–770 Lesson 9-6 Independent Practice

1. 24 **3.** 840 **5.** 40,320 **7.** 120 ways **9.** 6 **11.** Sample answer: The number of ways you can order 3 books on a shelf is 3 • 2 • 1 or 6. **13a.** 15 **13b.** 120 **13c.** 10 **13d.** 28

Pages 771–772 Lesson 9-6 Extra Practice

15. 60 **17.** 120 **19.** $\frac{1}{90}$ **21.** $\frac{1}{120}$ **23a.** False **23b.** True **23c.** True **25.** $\frac{29}{30}$ **27.** WB, WG, RB, RG, GB, GG

Pages 779–780 Lesson 9-7 Independent Practice

1. $\frac{1}{24}$ **3.** $\frac{1}{8}$ **5.** $\frac{1}{144}$ **7.** $\frac{7}{95}$ **9.** $\frac{1}{19}$ **11.** $\frac{3}{8}$; dependent event; after the first piece of paper is chosen, there is one less from which to choose. **13.** Sample answer: Spinning the spinner twice represents two independent events. The probability of getting an even number is $\frac{2}{5}$ each time; $\frac{2}{5} \cdot \frac{2}{5}$ or $\frac{4}{25}$. **15a.** 0.0004 or 0.04% **15b.** about 400 packages

Pages 781–782 Lesson 9-7 Extra Practice

17. $\frac{5}{14}$ **19.** $\frac{92}{287}$ **21.** $\frac{3}{20}$ **23.** $\frac{7}{60}$ **25.** $\frac{3}{55}$ **27.** $\frac{6}{55}$ **29a.** True **29b.** True **29c.** False **31.** 18 **33.** 51 **35.** $9 = 0.15x$; 60

Page 785 Chapter Review Vocabulary Check

1. sample space **3.** theoretical

Page 786 Chapter Review Key Concept Check

1. experimental probability **3.** compound event

Chapter 10 Statistics

Page 792 Chapter 10 Are You Ready?

1. Rihanna **3.** 75

Pages 797–798 Lesson 10-1 Independent Practice

1. $\frac{3}{10}$, 0.3, or 30% **3.** $\frac{2}{25}$, 0.08, or 8% **5.** 9 students **7.** About 143 students prefer humor books, and the number of students that prefer nonfiction is 88. So, there are about 55 more students who prefer humor books to nonfiction books. **9.** about 100 times **11.** Sample answer: Randomly select a part of the group to get a sample. Determine their preferences and use the results to determine the percent of the total group. It makes sense to use a sample when surveying the population of a city.

Pages 799–800 Lesson 10-1 Extra Practice

13. $\frac{3}{5}$, 0.6, or 60% **15a.** about 60,000 **15b.** about 72,500 **15c.** about 7,200 **17.** $\frac{27}{238} = \frac{n}{100}$ **19.** $\frac{27}{100} = \frac{p}{238}$ **21.** $\frac{54}{20} = \frac{405}{x}$; 150 minutes or 2.5 hours **23.** $\frac{1}{25}$

Pages 805–806 Lesson 10-2 Independent Practice

1 The conclusion is valid. This is an unbiased systematic random sample. **3** This is a simple random sample. So, the sample is valid; about 205 people. **5.** Sample answer: Questions should be asked in a neutral manner. For example, the question "You really don't like Brand X, do you?" might not get the same answer as the question "Do you prefer Brand X or Brand Y?" **7.** Sometimes; Sample answer: The sample needs to represent the entire population to be valid. **9.** Sample answer: The sample will be biased because it is a convenience sample. Marisol will be asking only basketball fans.

Pages 807–808 Lesson 10-2 Extra Practice

11. This is an unbiased, simple random sample because randomly selected Californians were surveyed. So, the conclusion is valid. **13.** This is an unbiased, systematic random sample. So, the conclusion is valid; 304 students. **15.** The survey results in a convenience sample; Sample answer: The school district should survey every tenth family living within the school district's boundaries. **17a.** Invalid **17b.** Valid **17c.** Valid **19.** median; Sample answer: She scored better than the mean on four of the tests. She scored lower than the mode on four of the tests.

Pages 817–818 Lesson 10-3 Independent Practice

1 Graph B; Sample answer: The ratio of the area of the gas pumps in the graph on the right are not proportional to the cost of gas. **3** The median or the mode because they are much closer in value to most of the data.

5.

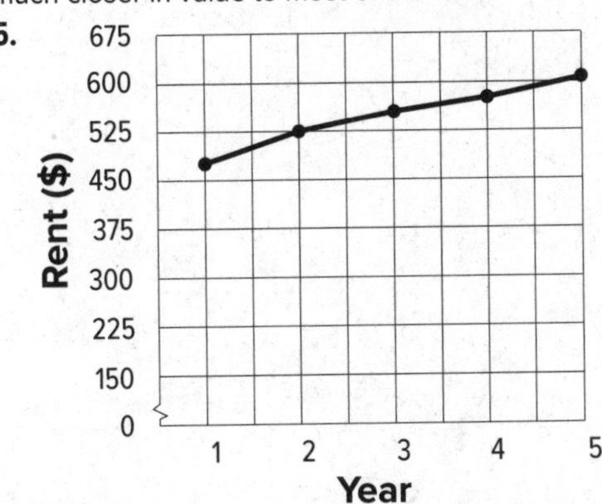

7. Sample answer: Since the graph makes it seem as if rent has been stable, a person may choose to become a tenant. **9.** Sample answer: The graph makes it appear that the Fall section is greater than the Spring section. This is because the perspective of the graph makes it appear greater, when, in fact, they are equal in size.

Pages 819–820 Lesson 10-3 Extra Practice

11. Sample answer: The scale of the graph is not divided into equal intervals, so differences in heights appear less than they actually are. **13.** Sample answer: The mode is 100, but she only received 100 two times out of 6 tests. **15.** The intervals on the vertical scale are inconsistent.

Page 823 Problem-Solving Investigation Use a Graph

Case 3. Sample answer: 2017

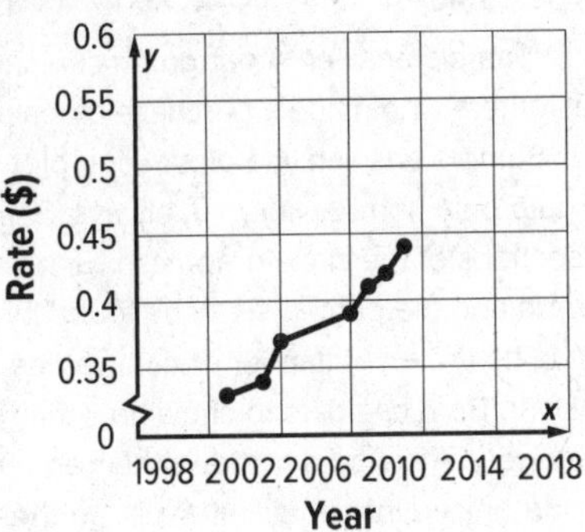

Case 5. $36.20

Pages 833–834 Lesson 10-4 Independent Practice

1 Sample answer: The times at Lucy's Steakhouse have a median of 20 minutes with an interquartile range of 20 minutes. The times at Gary's Grill have a median of 15 minutes with an interquartile range of 10 minutes. In general, a customer will wait longer at Lucy's Steakhouse. **3a.** Plant A: 2.75, 0.75; Plant B: 3.1; 0.7

3b.

Plant Growth

Plant A

Plant B

2 3 4

3c. Sample answer: Both populations have similar interquartile ranges. The median for Plant B is greater. So, Plant B generally showed more growth. **5.** The data shown in the histograms are only shown in intervals. Specific values are not shown.

Pages 835–836 Lesson 10-4 Extra Practice

9. this season; Sample answer: Both seasons' scores have a median of 20 points, but last season's scores have an interquartile range of 15 points while this season's interquartile range is 10 points. So, the football team's performance was more consistent this season. **11.** Sample answer: 2, 4, 4, 5, 8, 9, 10 **13a.** False **13b.** False **13c.** True **15.** 1.74 million **17.** Sample answer: The shape of the distribution is not symmetric since the lengths of each box and whisker are not the same. There are no outliers.

Pages 843–844 Lesson 10-5 Independent Practice

1 box plot; shows the median

3. A box plot is an appropriate graph because there is a large set of data and it will show the measures of variation of the data set. This graph has a median of 41.

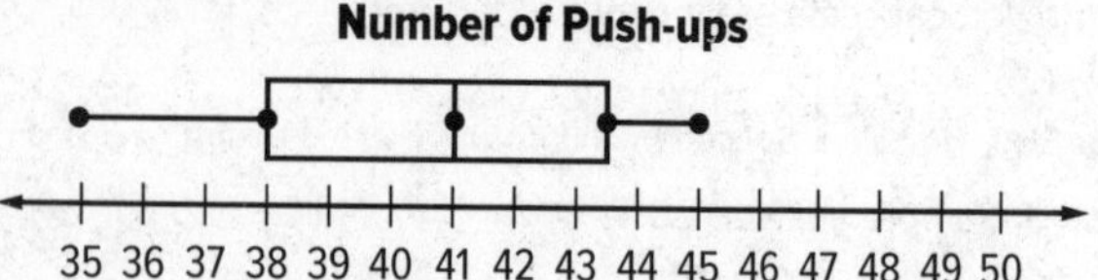

5a. Situation B; Sample answer: A bar graph can show the number of customers who made a purchase by each individual age. **5b.** Yes; Sample answer: line plot; A line plot shows the frequency of data on a number line. **7.** always; Sample answer: The sections of the circle graph can be taken from the bars of the graph and the percents can be found by dividing each bar's value by the total number of data values.

9. Sample answer: Both use bars to show how many things are in each category. A histogram shows the frequency of data that has been organized into equal intervals, so there is no space between the bars. It would be appropriate to use a histogram instead of a bar graph when the data can be organized into equal intervals.

Pages 845–846 Lesson 10-5 Extra Practice

11. circle graph; compares parts to a whole

13a.

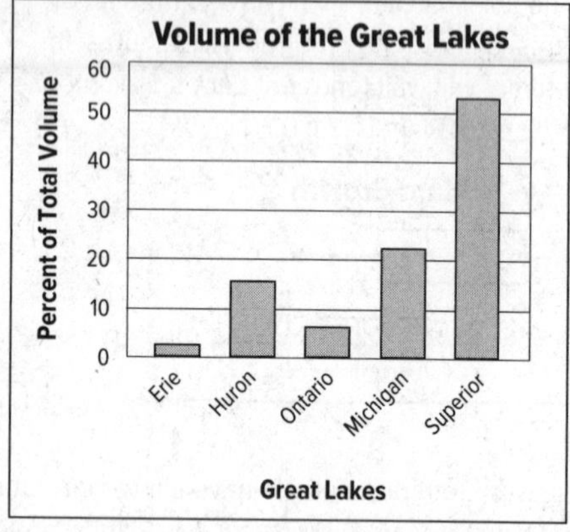

13b. Sample answer: The circle graph is most appropriate because it shows how each lake compares to the whole.

15.

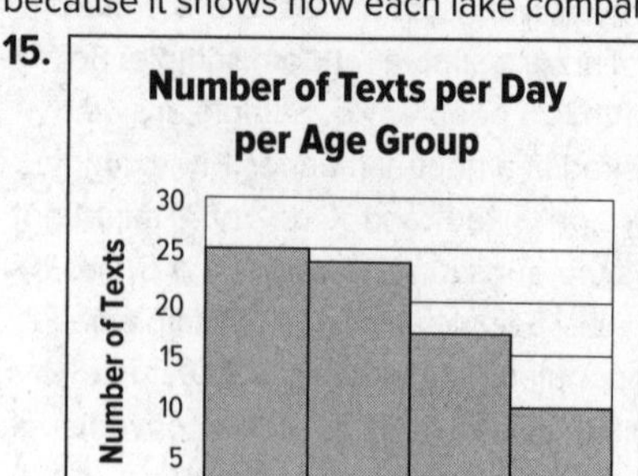

A histogram is an appropriate graph because the data is given in intervals. The graph shows people ages 26–30 text the least amount. **17.** line graph; circle graph; bar graph

19. 65 men; 65 women

Page 849 Chapter Review Vocabulary Check

Across

5. population **9.** sample

Down

1. systematic **3.** simple **7.** unbiased

Page 850 Chapter Review Key Concept Check

1. survey **3.** biased sample

Index

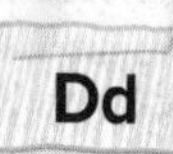

Dd

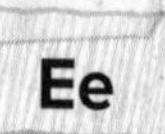

Ee

Ff

Gg

Hh

Ii

Kk

Ll

Mm

Nn

Oo

Pp

Qq

Rr

Ss

Yy

Zz

Name ____________________

Work Mats

=

Work Mats

Name ____________________

Name ______________________________

Name ______________________________

Work Mats

0 1 2 3 4 5 6 7 8 9

−11 −10 −9 −8 −7 −6 −5 −4 −3 −2 −1 0 1 2 3 4 5 6 7 8 9 10 11

What Are Foldables and How Do I Create Them?

Foldables are three-dimensional graphic organizers that help you create study guides for each chapter in your book.

Step 1 Go to the back of your book to find the Foldable for the chapter you are currently studying. Follow the cutting and assembly instructions at the top of the page.

Step 2 Go to the Key Concept Check at the end of the chapter you are currently studying. Match up the tabs and attach your Foldable to this page. Dotted tabs show where to place your Foldable. Striped tabs indicate where to tape the Foldable.

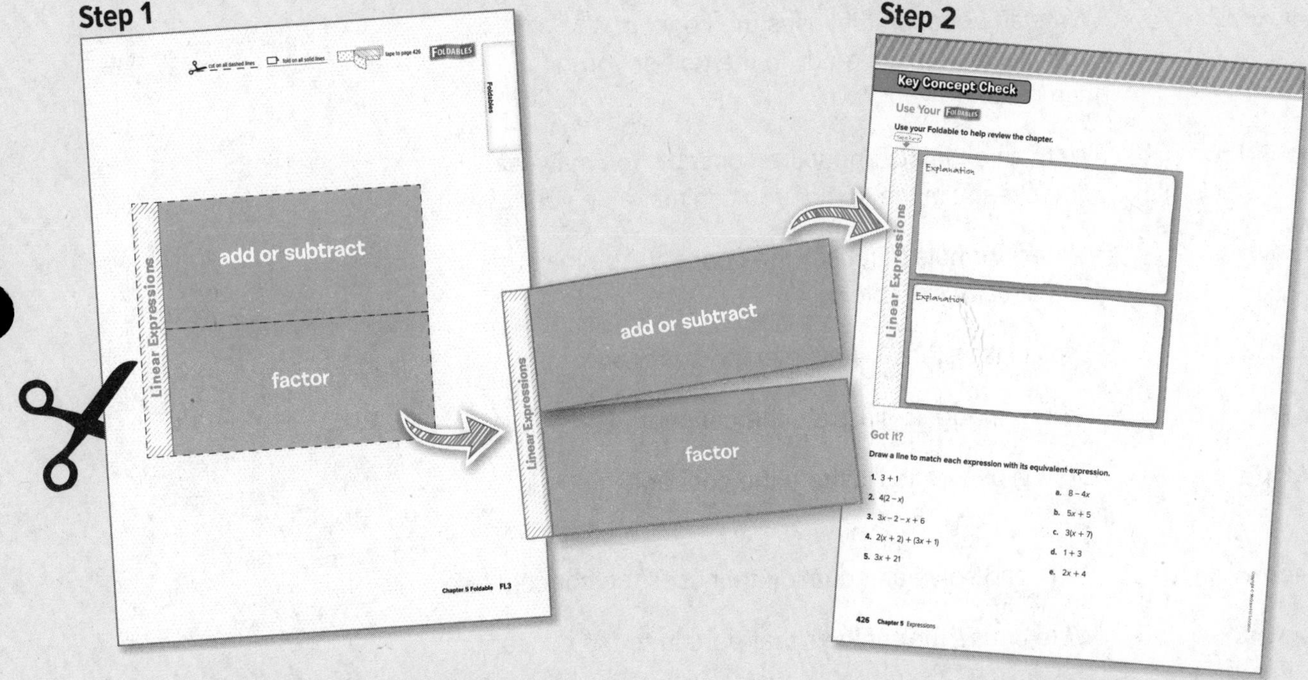

How Will I Know When to Use My Foldable?

When it's time to work on your Foldable, you will see a Foldables logo at the bottom of the **Rate Yourself!** box on the Guided Practice pages. This lets you know that it is time to update it with concepts from that lesson. Once you've completed your Foldable, use it to study for the chapter test.

How Do I Complete My Foldable?

No two Foldables in your book will look alike. However, some will ask you to fill in similar information. Below are some of the instructions you'll see as you complete your Foldable. **HAVE FUN** learning math using Foldables!

Instructions and what they mean

Instruction	Meaning
Best Used to...	Complete the sentence explaining when the concept should be used.
Definition	Write a definition in your own words.
Description	Describe the concept using words.
Equation	Write an equation that uses the concept. You may use one already in the text or you can make up your own.
Example	Write an example about the concept. You may use one already in the text or you can make up your own.
Formulas	Write a formula that uses the concept. You may use one already in the text.
How do I ...?	Explain the steps involved in the concept.
Models	Draw a model to illustrate the concept.
Picture	Draw a picture to illustrate the concept.
Solve Algebraically	Write and solve an equation that uses the concept.
Symbols	Write or use the symbols that pertain to the concept.
Write About It	Write a definition or description in your own words.
Words	Write the words that pertain to the concept.

Meet Foldables Author Dinah Zike

Dinah Zike is known for designing hands-on manipulatives that are used nationally and internationally by teachers and parents. Dinah is an explosion of energy and ideas. Her excitement and joy for learning inspires everyone she touches.

cut on all dashed lines fold on all solid lines tape to page 426 FOLDABLES

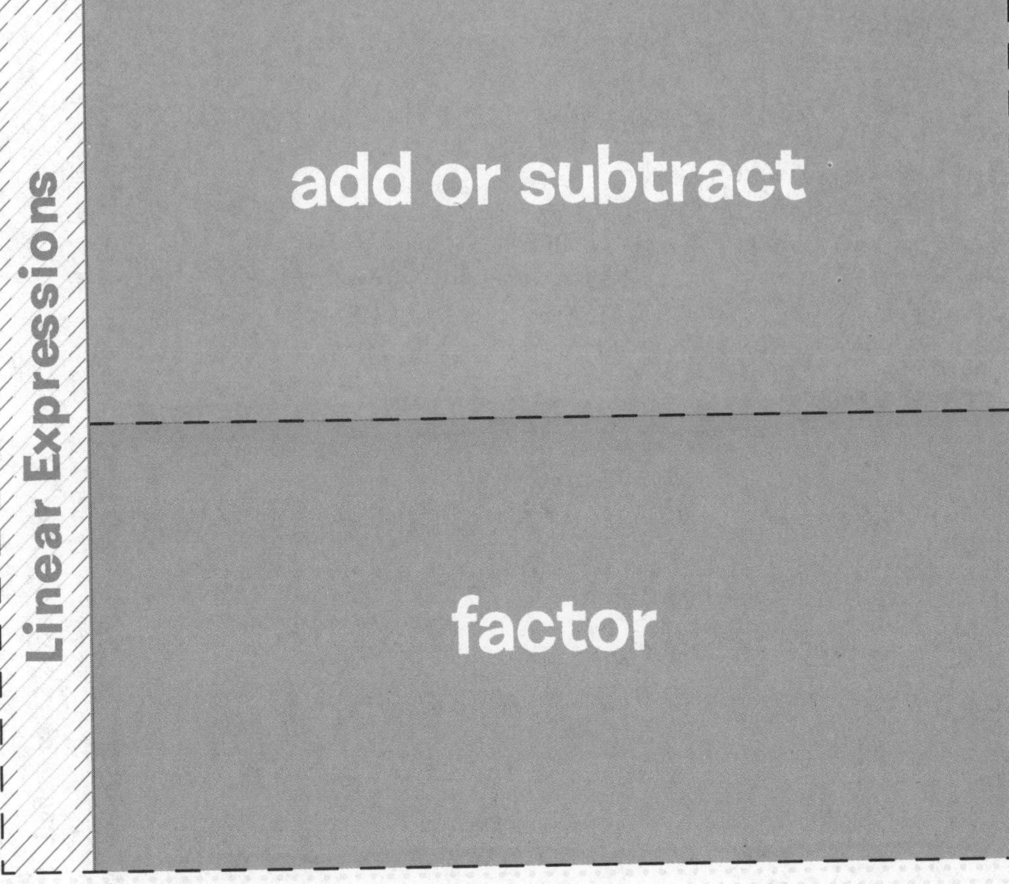

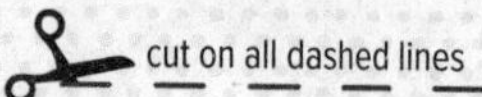

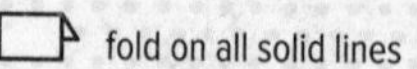

tape to page 426

Examples

Examples

page 426

cut on all dashed lines fold on all solid lines tape to page 524 FOLDABLES®

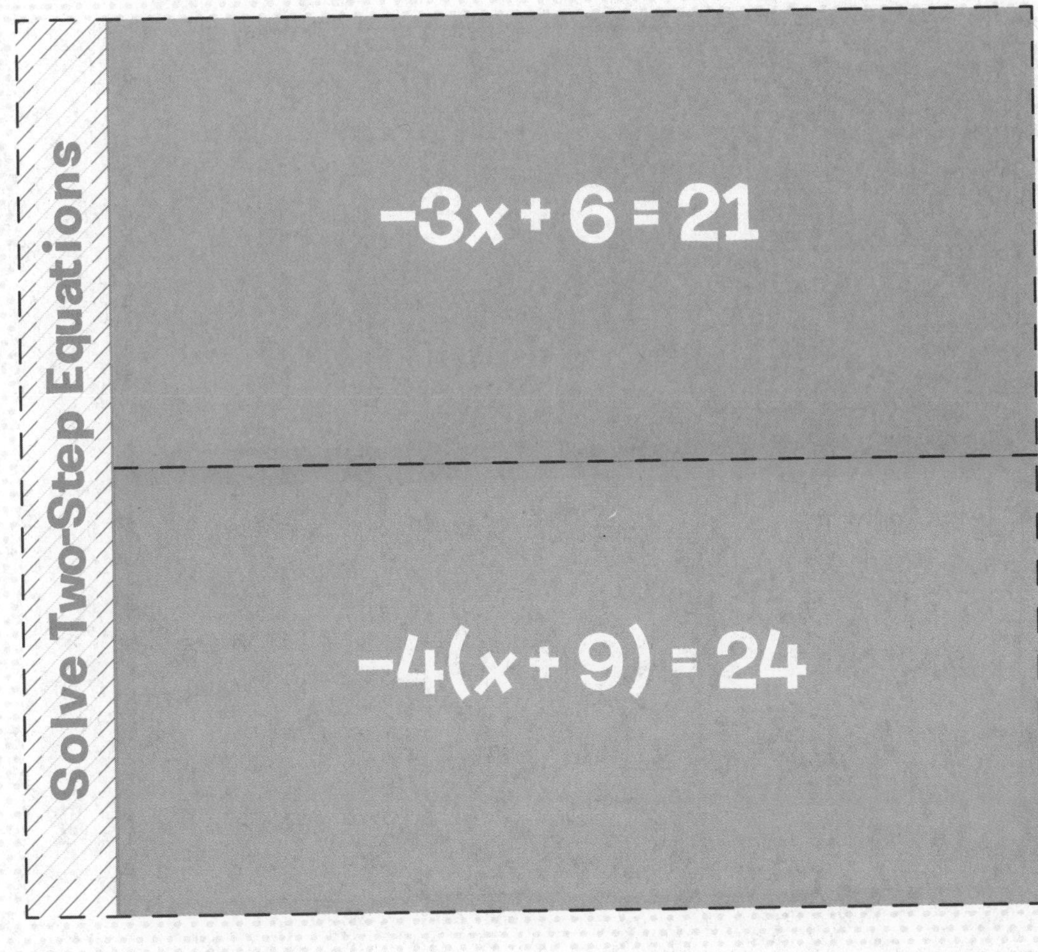

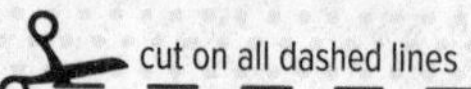

fold on all solid lines

tape to page 524

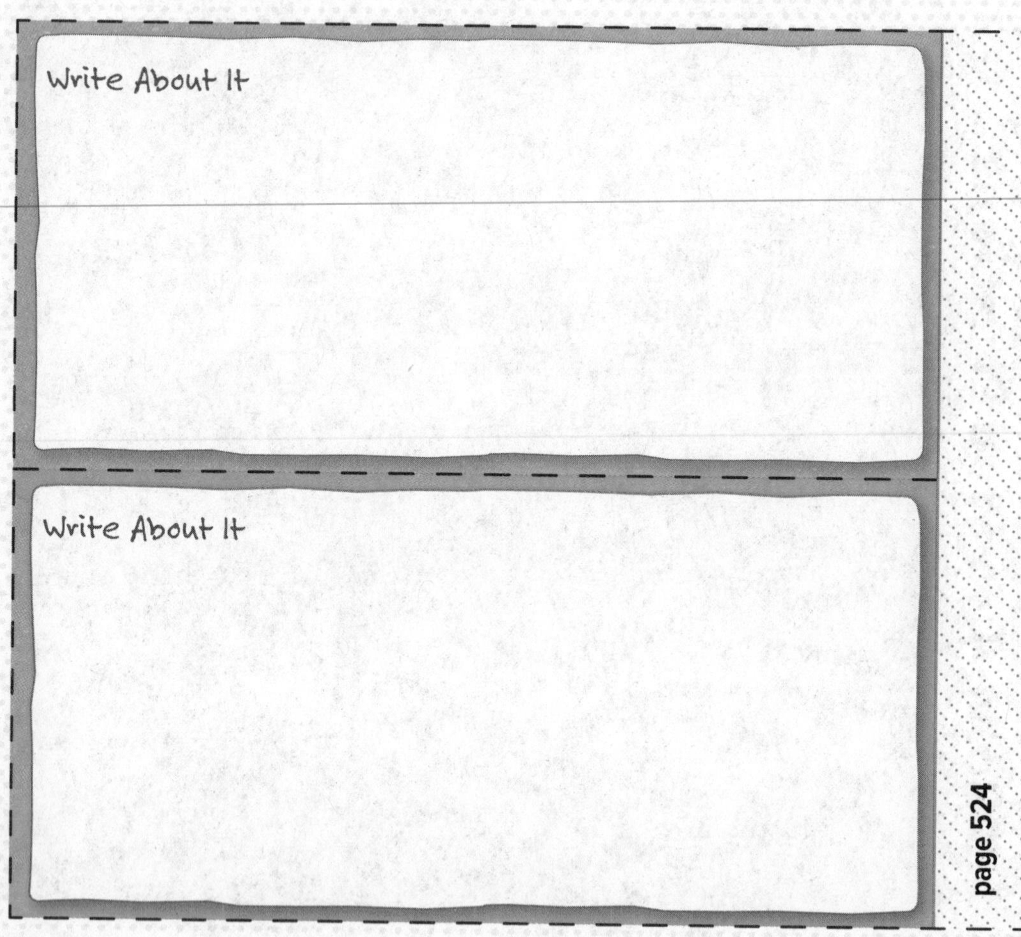

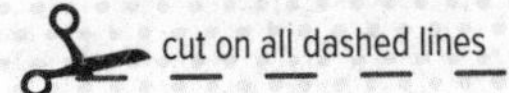

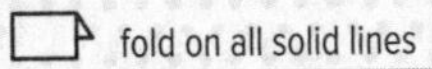

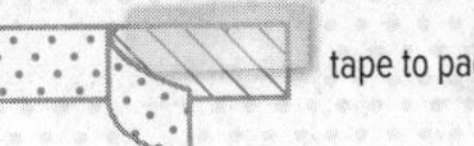

Angles

acute	scalene
obtuse	isosceles
right	equilateral

Triangles

cut on all dashed lines — fold on all solid lines — tape to page 604

FOLDABLES®

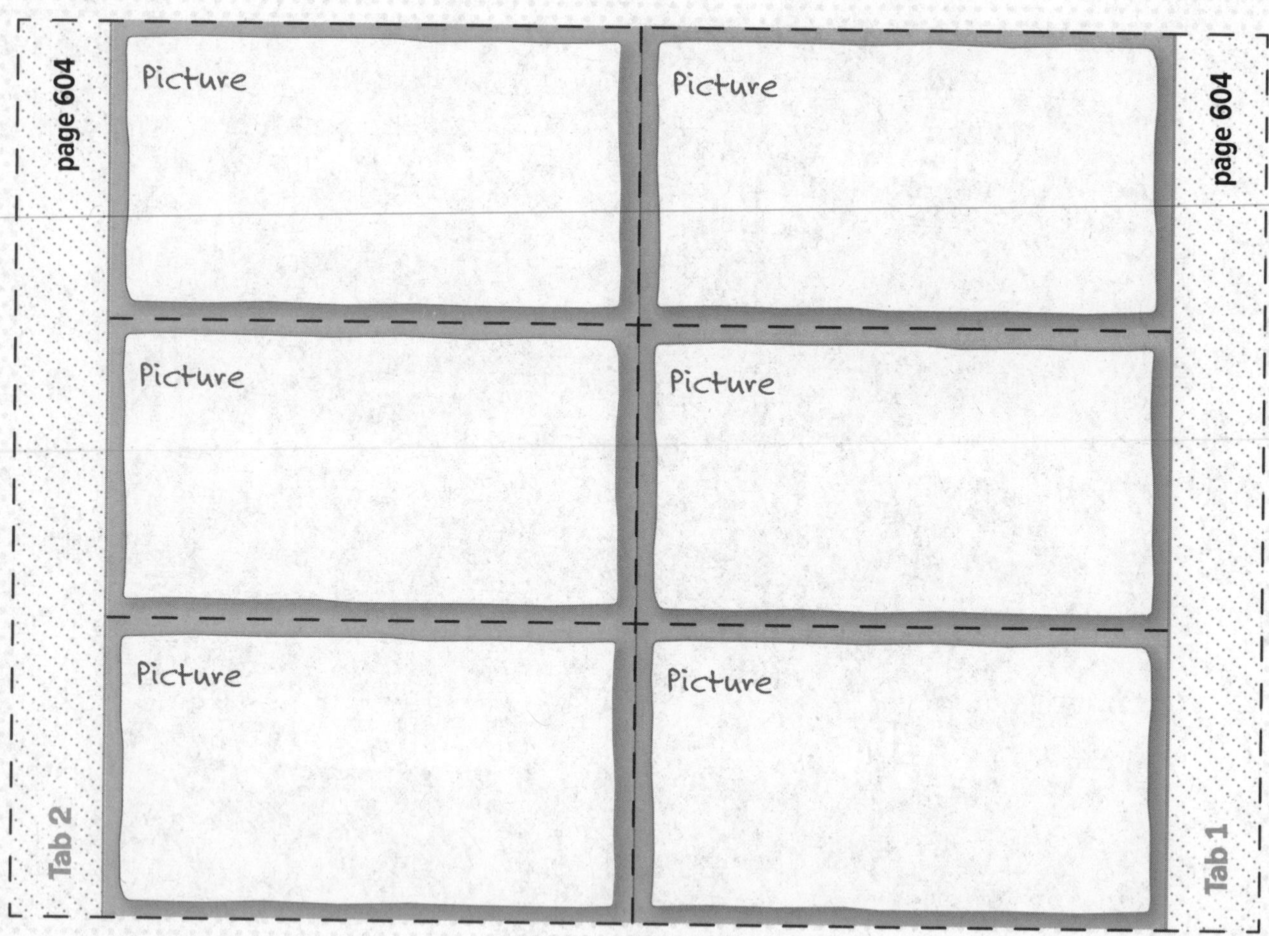

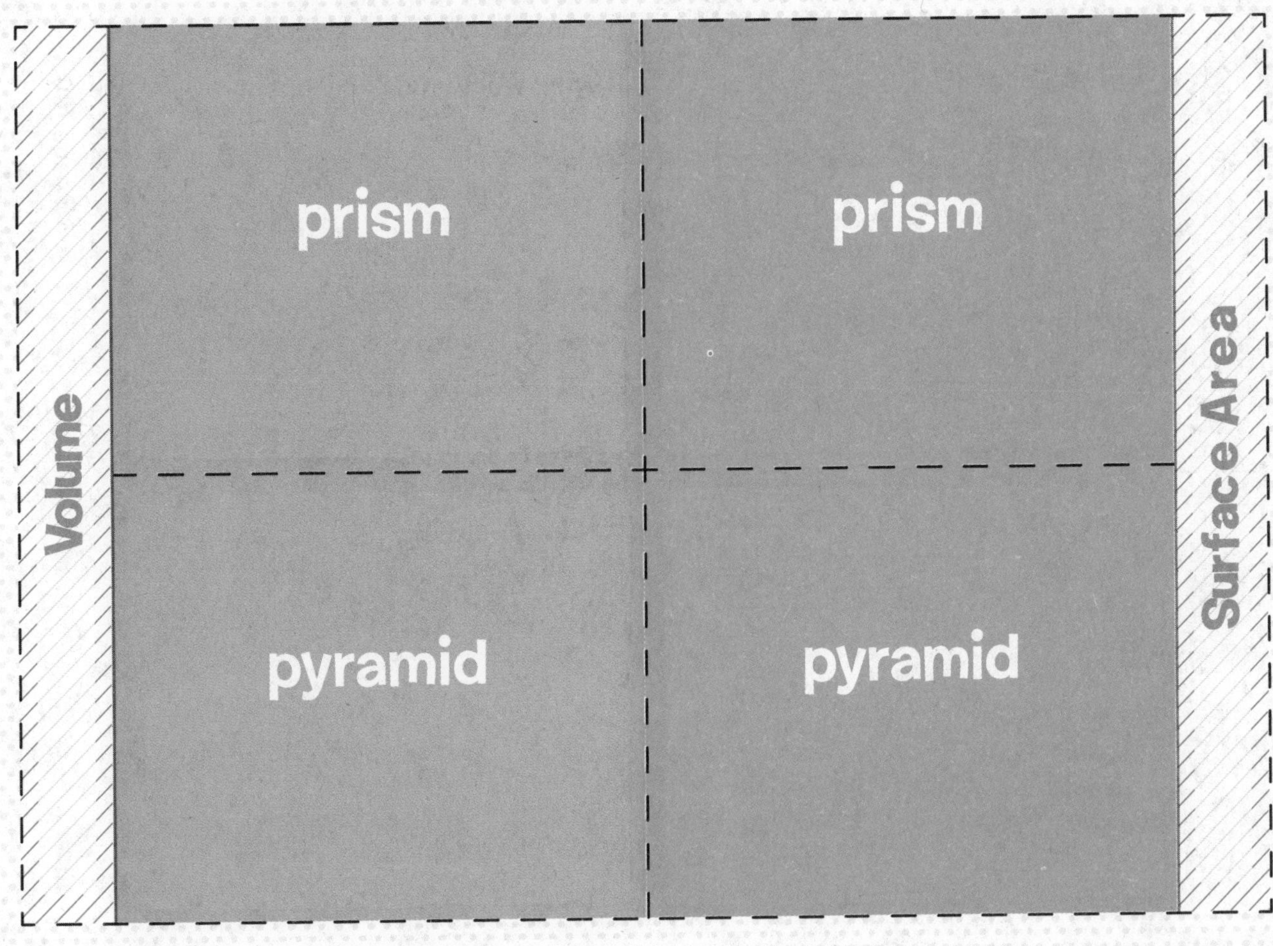
Volume
prism
prism
Surface Area
pyramid
pyramid

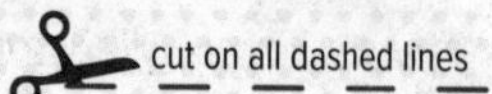
cut on all dashed lines

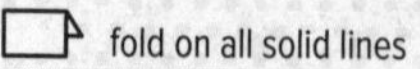
fold on all solid lines

tape to page 700

page 700

Write About It

Write About It

page 700

Write About It

Write About It

Tab 2

Tab 1

cut on all dashed lines fold on all solid lines tape to page 786 FOLDABLES

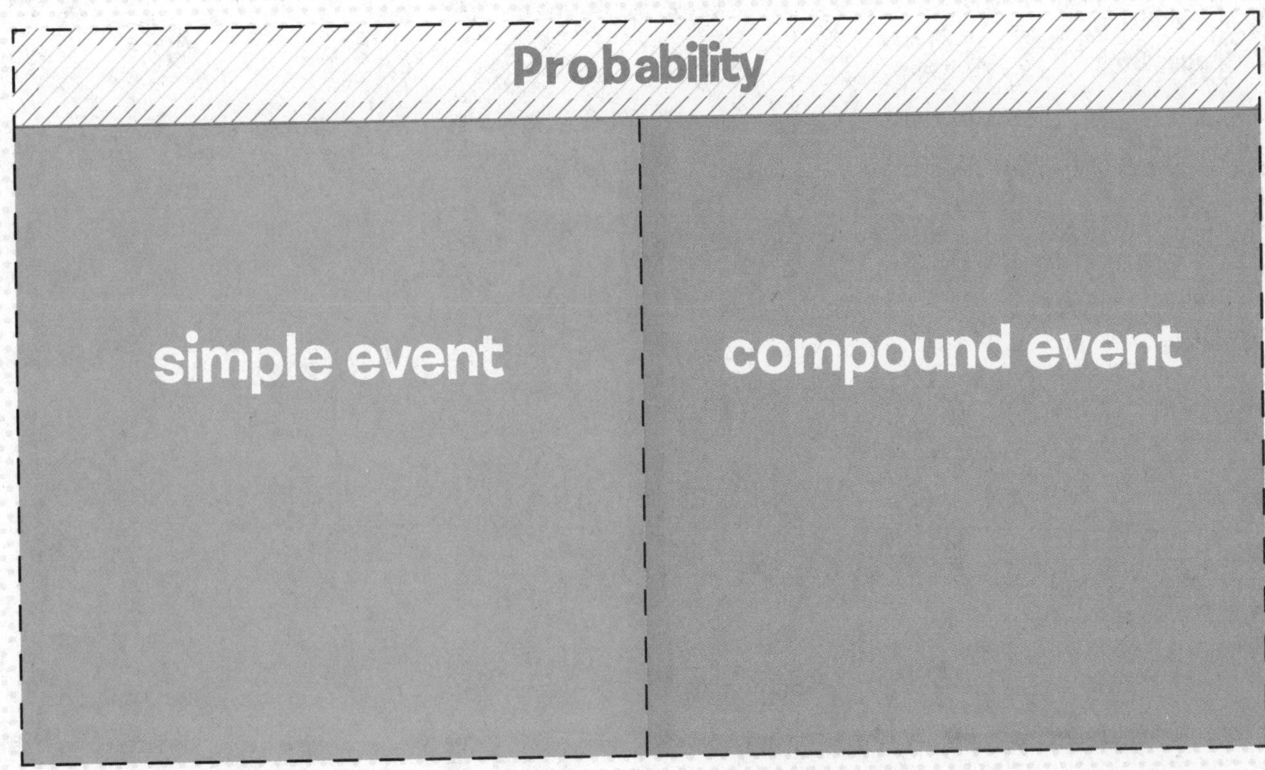

cut on all dashed lines   fold on all solid lines   tape to page 786

FOLDABLES®

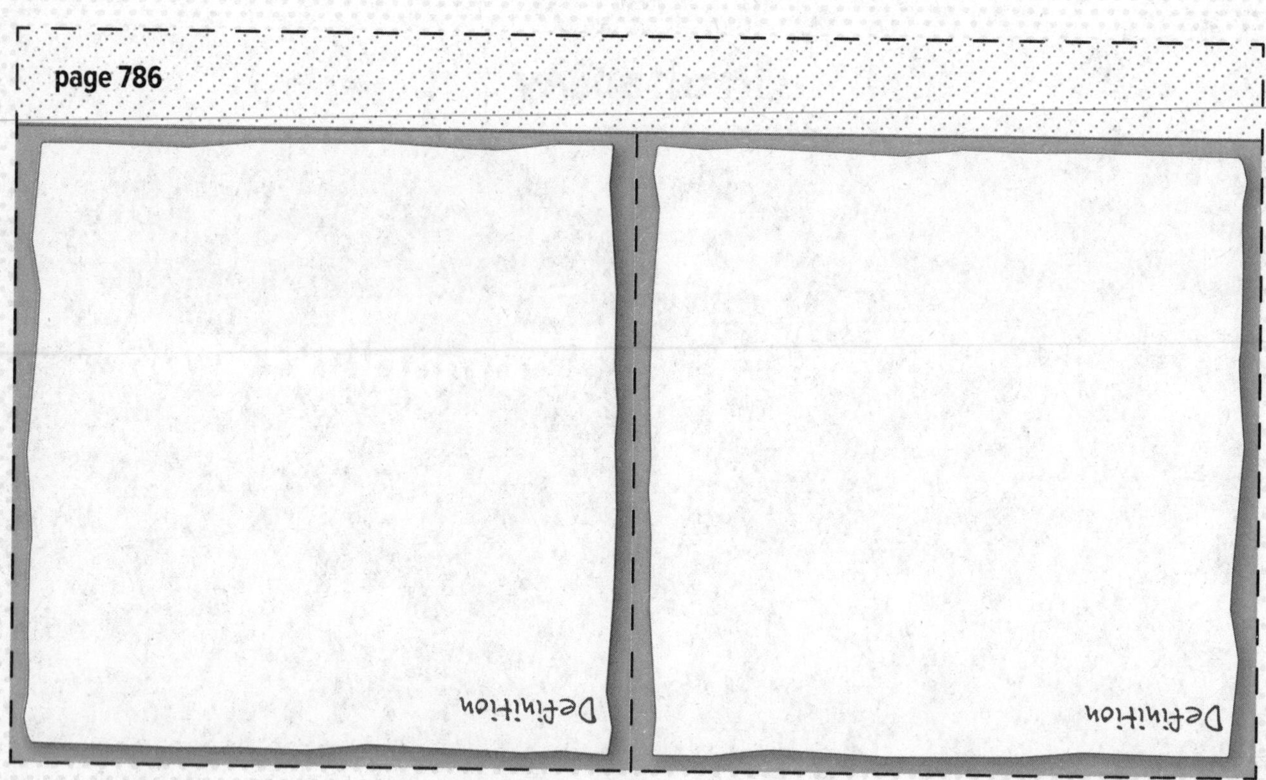

fold on all solid lines

tape to page 850

FOLDABLES®

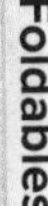

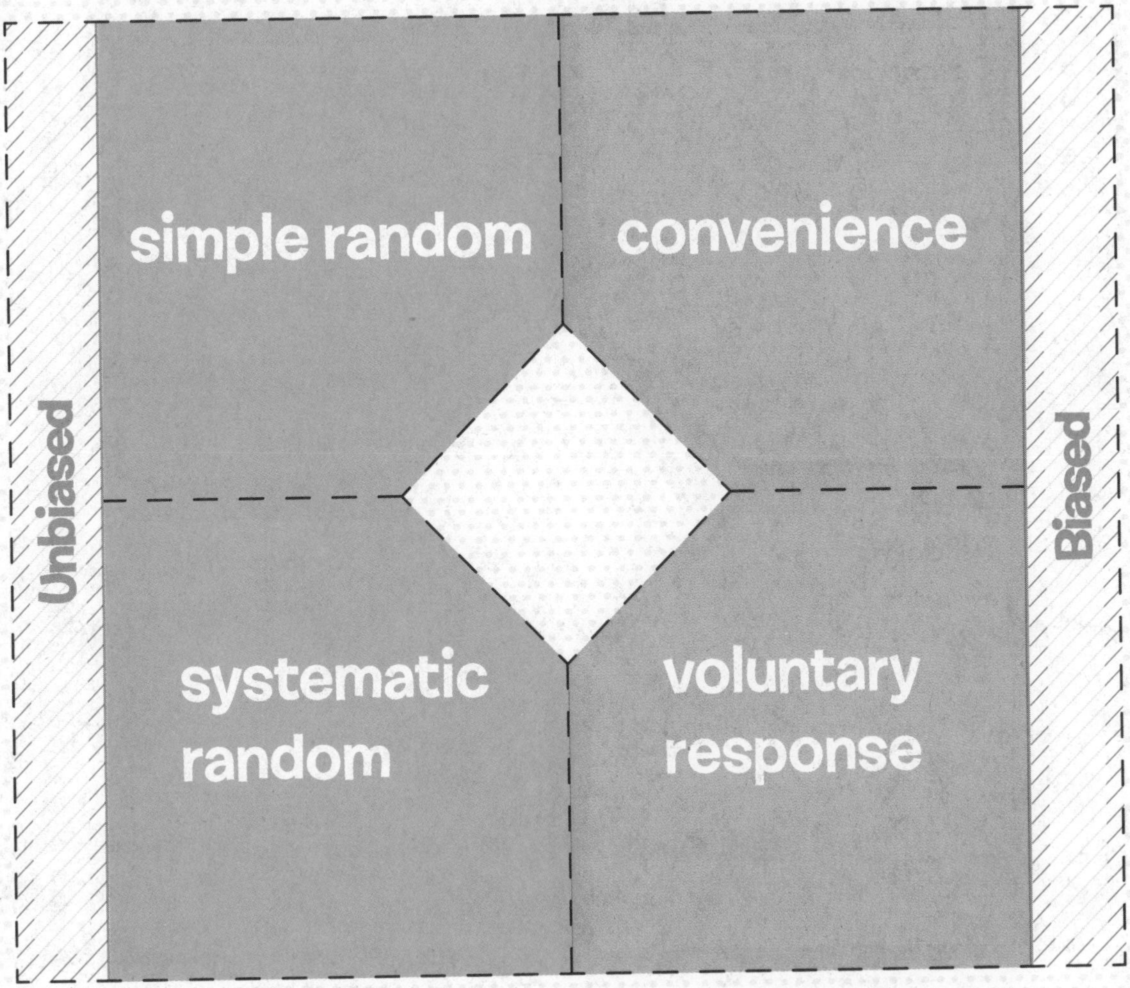

Unbiased

simple random

systematic random

convenience

voluntary response

Biased

cut on all dashed lines

fold on all solid lines

tape to page 850

FOLDABLES®

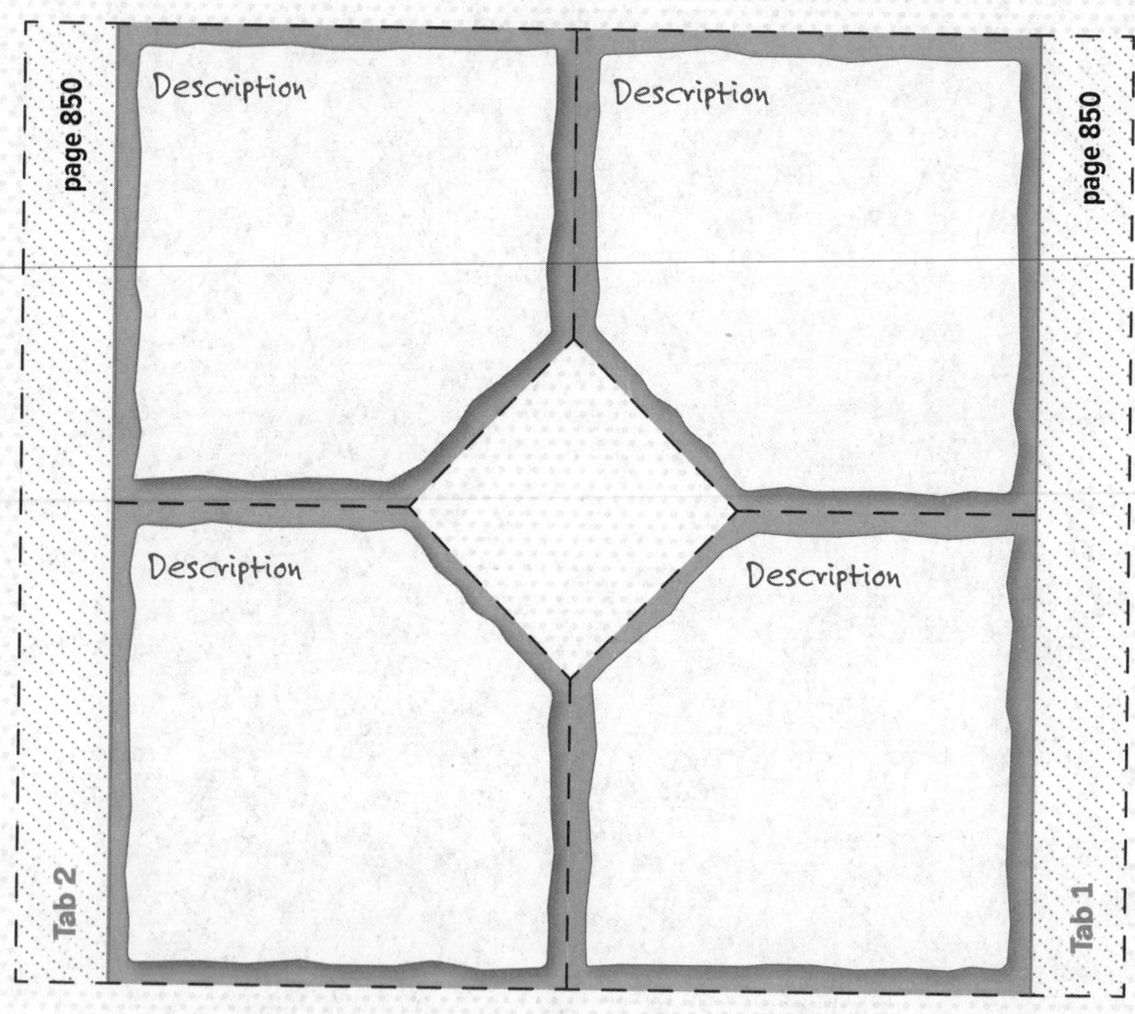